Structure Theory for Canonical Classes of Finite Groups

Wenbin Guo

Structure Theory for Canonical Classes of Finite Groups

 Springer

Wenbin Guo
School of Mathematical Sciences
University of Science and Technology of China
Hefei
China

ISBN 978-3-662-51622-5   ISBN 978-3-662-45747-4 (eBook)
DOI 10.1007/978-3-662-45747-4

Mathematics subject Classification: 20D10, 20D15, 20D20, 20D25, 20D30, 20D35, 20D40, 20E28

Springer Heidelberg New York Dordrecht London
© Springer-Verlag Berlin Heidelberg 2015
Softcover reprint of the hardcover 1st edition 2015

Printed on acid-free paper

Springer is part of Springer Science+Business Media (www.springer.com)

Preface

The central challenge of any mathematical theory is to provide reasonable classifications and constructive descriptions of the investigating objects that are most useful in diverse applications. At the same time, we realize that the purpose is to support new methods of investigation, which, in the end, constitute ideological riches of the given theory. For example, the development of the theory of finite non-simple groups in the past 50 years has clearly shown this tendency. Although the theory of finite groups has never been lacking in general methods, ideas, or unsolved problems, the large body of results has inevitably brought us to the point of needing to develop new methods to systemize the material.

One example of such systemizing is the idea of Gaschüts that the use of some given classes of groups, called saturated formations, is convenient for investigating the inner structure of finite groups. The theory of formation initially found wide application in the research of finite groups and infinite groups, as reflected in a series of classical monographs. The first is the famous book of Huppert [248]. The book, which described research in the structure of finite groups, attracted the attention of many specialists in algebra. The investigation involving saturated and partially saturated (ω-saturated or soluble ω-saturated) formations became one of the predominate directions of development within the contemporary theory of group classes, which indubitably testifies to its actual importance. Following the publication of monographs by L. A. Shemetkov [359], L. A. Shemetkov and A. N.Skiba [366], and Doerk and Hawkes [89], the direction of theoretical development was shaped by pithy and harmonious theory. The subsequent monographs by W. Guo [135] and A. Ballester-Bolinches and L. M. Ezquerro [34] reflect the modern status of the theory of formations. At the same time, it is necessary to mention that the swiftly augmented flow of articles about the theory of finite groups, particularly in the past 20 years, has brought to the forefront the necessity of further analysis and development of new methods of research. Additional developments have included the formation properties of the classes of all soluble (π-soluble) groups, all supersoluble (π-supersoluble) groups, all nilpotent (π-nilpotent) groups, all quasinilpotent groups, and other classes of finite groups with broad application in modern finite group theory research.

The first aim of this book is to introduce the subsequent development and application of the theory of classes of finite groups. Note also that many important results on the theory of classes of groups obtained during recent years are not well reflected in the monographs formerly cited, for example, the theory of subgroup functors, the theory of X-permutable subgroups, the theory of quasi-F-groups, the theory of F-cohypercentres for Fitting classes, and the theory of the algebra of formations, which is related to the research of semigroups and the lattices of formations. To eliminate such information deficiency is the second aim of this book. The third aim is to systemize and unify the results obtained during recent years and related research in various classes of non-simple groups.

The largest term $Z_\infty(G)$ of the upper central series of a finite group G is called the hypercenter of G. In fact, $Z_\infty(G)$ is the largest normal subgroup of G such that every chief factor H/K of G below $Z_\infty(G)$ is central; that is, $C_G(H/K) = G$. This elementary observation allows us to define the formation analog of the hypercenter. The $\mathcal{F}$-hypercenter $Z_{\mathcal{F}}(G)$ of G is the largest normal subgroup of G such that every chief factor H/K of G below $Z_{\mathfrak{F}}(G)$ is $\mathfrak{F}$-central in G, that is, $(H/K) \rtimes (G/C_G(H/K)) \in \mathcal{F}$.

The $\mathfrak{F}$-hypercenter and $\mathfrak{F}$-hypercentral subgroups (that are, the normal subgroups contained in $Z_{\mathfrak{F}}(G)$) have great influence on the structure of groups, and so have been investigated by a large number of researchers. Nevertheless, there are still some open problems concerning the $\mathfrak{F}$-hypercenter. The main goal of Chap. 1 is a further study of the $\mathfrak{F}$-hypercenter and the generalized $\mathfrak{F}$-hypercenter of a group. As a result, in Sect. 5 of Chap. 1, we provide solutions to some of these problems, in particular, the solution to Baer–Shemetkov's problem of the description of the subgroup $Z_{\mathfrak{F}}(G)$ as the intersection of all $\mathcal{F}$-maximal subgroups of G, and the solution to Agrawal's problem regarding the intersection of all maximal supersoluble subgroups.

In the first section of Chap. 1, we collect some base results on saturated and solubly saturated formations that are used in our proofs.

In Sect. 2 of Chap. 1, we study the $\mathfrak{F}$-hypercentral subgroups in detail. The theorems we prove in this section are used in many other sections of this book and develop some known results of many authors (e.g., Gaschütz [112]; Huppert [89 IV, 6.15]; Shemetkov [357]; Selkin [344]; Ballester-Balinches [24]; Skiba [384] and others).

In Sect. 3 of Chap. 1, we consider applications of the theory of generalized quasinilpotent groups. Recall that a group G is said to be quasinilpotent if for every chief factor H/K and every $x \in G$, x induces an inner automorphism on H/K (see [250, p. 124]). We note that for every central chief factor H/K, an element of G induces trivial automorphism on H/K; thus, we can say that a group G is quasinilpotent if for every noncentral chief factor H/K and every $x \in G$, x induces an inner automorphism on H/K. In a general case, we say that G is a quasi-$\mathfrak{F}$-group if for every $\mathfrak{F}$-eccentric chief factor H/K of G, every automorphism of H/K induced by an element of G is an inner automorphism. The theory of quasinilpotent groups is well represented in the book [250]. The theory of the quasi-$\mathfrak{F}$-groups, which covers the theory of quasinilpotent groups, was developed by W. Guo and A. N. Skiba [188,

189, 192]. In this section of Chap. 1, we give the complete theory of quasi-$\mathfrak{F}$-groups and consider some related applications.

In Sect. 4 of Chap. 1, we explain further applications of the results of the theoretical work described in Sect. 2. In particular, we study the groups G with factorization $G = AB$ such that $A \cap B \leq Z_{\mathfrak{F}}(A) \cap Z_{\mathfrak{F}}(B)$. Such an approach to studying groups allows us to generalize some results of Baer [23], Friesen [108], Wielandt [447], Kegel [264], Doerk [87] concerning factorizations of groups.

If for subgroups A and B of a group G we have $AB = BA$, then A, B are said to be permutable or A is said to be permutable with B. In this case, AB is a subgroup of G. Hence, the permutability of subgroups is one of the important relations among subgroups. The research of permutability in finite and infinite groups is still of broad interest. In fact, such a direction for group theory may go back to the classic works [231] and [324]. Hall [231] proved that group G is soluble if and only if it has at least one Sylow system, that is, a complete set of pairwise permutable Sylow subgroups. Ore [324] introduced to the mathematical practice the so-called quasinormal subgroups, that are, the subgroups which permute with all subgroups of the overall group. Increasing interest in that subject was characteristic of the 1960s and 70s (e.g., Huppert [246, 247]; Kegel [263]; Thompson [403]; Deskins [85], Čunihin [79]; Stonehewer [396]; Ito and Szé [252]; Maier and Schmid [307] and others). The results obtained during that period were well-written in the books [32, 79, 89, 248, 276, 343, 450].

Some new ideas about the research of permutable subgroups are described in Chaps. 2–4 of this book.

We often encounter a situation in which $AB \neq BA$ for subgroups A and B of a group G, but there exists an element $x \in X \subseteq G$ such that $AB^x = B^x A$. In this case, we say that A is X-permutable with B. The well-known examples are any two Hall subgroups of a soluble group and any two normal embedded subgroups of a group, which are G-permutable (see [89], Chap. 1). The concept of X-permutability of subgroups is extraordinarily useful when researching the problems of classifying groups, and the terminology of X-permutability provides a beautiful way to describe group classes. The second chapter in that monograph is devoted to the detailed analysis of the properties of X-permutable subgroups and their application to solve a series of open problems in group theory. The first of such problems concerns the generalization of the well-known Schur–Zassenhaus theorem.

The Schur–Zassenhaus theorem asserts that *If G has a normal Hall π-subgroup A, then G is an $E_{\pi'}$-group (that is, G has a Hall π'-group). Moreover, if either A or G/A is soluble, then G is a $C_{\pi'}$-subgroup (that is, any two Hall π'-subgroups of G are conjugate).*

Recall that a group G is said to be π-separable if G has a chief series

$$1 = G_0 \leq H_1 \leq \ldots \leq H_{t-1} \leq H_t = G, \qquad (*)$$

such that each index $|H_i : H_{i-1}|$ is either a π-number or a π'-number.

The most interesting generalization of the Schur–Zassenhaus theorem is the following classical result of P. Hall and S. A. Čunihin: Any π-separable group is a

D_π-group (that is, G is a C_π-group and any π-subgroup of G is contained in some Hall π-subgroup of G).

It is well known that the above Schur–Zassenhaus theorem and the Hall–Čunihin theorem are truly fundamental results of group theory. In connection with these important results, the following two problems have naturally arisen: (I) *Does the conclusion of the Schur–Zassenhaus theorem hold if the Hall subgroup A of G is not normal? In other words, can we weaken the condition of normality for the Hall subgroup A of G so that the conclusion of the Schur–Zassenhaus theorem is still true?* (II) *Can we replace the condition of normality for the members of series (*) by some weaker condition?* For example, can we replace the condition of normality by permutability of the members of series (*) with some systems of subgroups of G?

In Sect. 2 of Chap. 2, we provide a positive answer to these two problems based on a careful analysis of the properties of X-permutability. The results obtained are nontrivial generalizations of the Schur–Zassenhaus theorem and the Hall–Čunihin theorem.

Another open problem, which is solved in Chap. 2, is to describe the group in which every subgroup can be written as an intersection of the subgroups of prime power indexes. Note that the structure of groups in which every subgroup can be written as an intersection of pairwise coprime prime power indexes is known (see [450], Chap. 6). The problem is completely solved in Sect. 3 of Chap. 2. The resolution of this problem was found in the close connection with Johnson's problem [256] about the description of the group in which every meet-irreducible subgroup has a prime-power index. Partially this problem was solved in [450]. The complete resolution of Johnson's problem is also introduced in this section. In the remaining sections of Chap. 2, we discuss some new characteristics for many classes of finite groups.

For two subgroups A and B of a group G, if $G = AB$, then B is said to be a supplement of A in G. More often we consider the supplements with some restrictions on $A \cap B$. For example, we often encounter the situation in which $A \cap B = 1$. In this case, B is called a complement of A in G. If B is also normal in G, then B is called a normal complement of A in G.

In [324], Ore considered the subgroup H of the group G with the following (Ore's) condition: G has normal subgroups N and M such that $HM = G$ and $H \cap M \leq N \leq H$. It is clear that if a subgroup H of G is either normal in G or has a normal complement in G, then H satisfies Ore's condition. Note also that if H is either complemented in G or is normal in G, then H satisfies the following (Ballester-Bolinches–Guo–Wang's) condition: G has a subgroup M and a normal subgroup N such that $HM = G$ and $H \cap M \leq N \leq H$ [45, 435].

In the past 20 years, a large number of research publications have involved finding and applying other generalized complemented and generalized normally complemented subgroups. The papers of A. N. Skiba [383], W. Guo [141] and B. Li [277] have made the greatest impact on this direction of research.

Note that, at present, a large number of interesting and profound ideas and theorems in this direction that have accumulated require systemization and unification. The aim of Chap. 3 is to solve this non-simple problem. The main tool for solving this problem is the method of subgroup functors.

If a group G has a chain of subgroups

$$H = H_0 < H_1 < \ldots < H_n = G,$$

where H_{i-1} is a maximal subgroup of H_i, $i = 1, 2, \ldots, n$, then this chain is called a maximal (H, G)-chain, and the number n is said to be the length of the chain. In this case, H is said to be an n-maximal subgroup of G. If $K < H \leq G$ and K is a maximal subgroup of H, then (K, H) is said to be a maximal pair in G.

In Sect. 1 of Chap. 4, we construct the original theory of Σ-embedded subgroups. The basic tool of this theory is the concept of covering and avoiding subgroup pairs. Suppose that $A \leq G$ and $K \leq H \leq G$. Then we say that A covers the pair (K, H) if $AH = AK$; and A avoids (K, H) if $A \cap H = A \cap K$. Let A be a subgroup of G and $\Sigma = \{G_0 \leq G_1 \leq \ldots \leq G_n\}$ be some subgroup series of G. Then we say that A is Σ-embedded in G if A either covers or avoids every maximal pair (K, H) such that $G_{i-1} \leq K < H \leq G_i$, for some i. We show that the description of many important classes of groups may be based on the concept of the Σ-embedded subgroup.

In Sect. 2 of Chap. 4, we investigate the classes of groups with distinct restrictions on maximal, 2-maximal and 3-maximal subgroups. In particular, we provide complete descriptions of the groups in which any maximal subgroup, 2-maximal subgroups and 3-maximal subgroups are pair-permutable. In Sect. 3 of Chap. 4, we describe the group in which every maximal chain of G of length 3 contains a proper subnormal entry and there exists at least one maximal chain of G of length 2 that contains no proper subnormal entry. In Sect. 4, we establish the theory of $\hat{\theta}$-pairs for maximal subgroups of a finite group and introduce some new characterizations of the structure of finite groups.

The application of formations and Fitting classes to the investigation of group theory is inconceivable without the benefit of detailed research of formations and Fitting classes, themselves. This circumstance leads to the necessity of developing methods for building and investigating formations and Fitting classes. In particular, this leads to the investigation of algebraic systems connected with operations on the set of formations and Fitting classes. Important roles are played by the operations of the product of the classes (which lead to the semigroups of all formations and Fitting classes), and the operations of generation and intersection of classes of groups (which lead to research of the lattices of all formations and Fitting classes). Many difficult problems have arisen in this direction of research. We provide solutions to some of these problems in Chap. 5.

A formation $\mathfrak{F}$ is said to be ω-solubly saturated or ω-composition if for every $p \in \omega$, $\mathfrak{F}$ contains every group G with $G/\Phi(O_p(G)) \in \mathfrak{F}$. If there is a group G such that $\mathfrak{F}$ coincides with the intersection of all ω-composition formations containing G, then $\mathfrak{F}$ is called a one-generated ω-composition formation. Since one-generated ω-composition formations are compact elements of the algebraic lattice of all ω-composition formations, it is important to study problems associated with one-generated ω-composition formations. This circumstance has predetermined a wide range of interest in studying formations of this kind.

In Sects. 2 and 3 of Chap. 5, on the basis of the results in Chap. 1 and some technical results of Sect. 1 of Chap. 5, we provide solutions to the following two problems concerning factorizations of one-generated ω-composition formations: (I) Letting $\mathfrak{F} = \mathfrak{M}\mathfrak{H}$ be a one-generated ω-composition formation, we consider whether it is true that $\mathfrak{M}$ is an ω-composition formation if $\mathfrak{H} \neq \mathfrak{F}$ (see Problem 18 in [389]); (II) Describe non-cancellated factorizations of one-generated ω-composition formations ([389], Problem 21).

Some special cases of these two problems have been discussed in the papers of many authors (see, in particular, Skiba [377, 380, 382, 388]; Rizhik and Skiba [331]; Vishnevskaya [426]; Jahad [257]; Guo and Shum [170]; Guo and Skiba [186]; Ballester-Bolinches et al. [28, 30, 31]; Ballester-Bolinches and Perez-Ramos [43]; Vorob'ev [429]; Elovikov [95, 96]).

In Chap. 5, we give complete solutions to these two problems.

Considerable space in Chap. 5 is devoted to methods for constructing and analyzing the lattices of formations. In Sect. 4 of Chap. 5, based on a detailed study of the compact elements of the lattice of the formations, we provide a description of the formation in which the lattice of the subformations is a Boolean lattice. This provides solutions for two of the problems proposed in the book of A. N. Skiba [381, p. 192].

In Sect. 5 of Chap. 5, we provide a negative answer to the problem posed by O. V. Melnikov, regarding whether any graduated formation is a Baer-local formation.

Sections 6 and 7 of Chap. 5 are devoted to the subsequent development of the research methods of groups by means of the theory of formations and Fitting classes. In particular, we describe the formations and Fitting class as determined by the properties of Hall subgroups, and find new characterizations for diverse classes of finite subgroups (covering subgroups, injectors, etc.).

Sections 8 and 9 of Chap. 5 are devoted to introduce the theory of $\mathfrak{F}$-cohypercentre for Fitting classes and ω-local Fitting classes.

In Sect. 10 of Chap. 5, we provide a negative answer to the Doerk-Howkes-Shemetkov problem [363], which considered whether it was proved that if the formation $\mathfrak{F}$ is hereditary or soluble or saturated, then the class $(G^{\mathfrak{F}}|G$ is a group) is not necessarily closed under subdirect products in general.

In the last section of each chapter, we add some additional information and present some open problems.

I would like to acknowledge my indebtedness to Professors K. P. Shum, Victor D. Mazurov, Danila O. Revin, Aexander N. Skiba, Evgeny P. Vdovin, Nikolai T. Vorob'ev, and L. A. Shemetkov who have assisted me with this work. In particular, I am very grateful that Professor A. N. Skiba provided many constructive suggestions and help for this book. I also thank my Ph.D students who have read parts of the manuscript and made some helpful suggestions, as well as The National Natural Science Foundation of China, Wu Wen-Tsun Key Laboratory of Mathematics of

Chinese Academy of Sciences, and University of Science and Technology of China on my research work have been supported by. Finally I am pleased to express my thanks to Springer for converting this project into a reality and for their help while writing this book.

Wenbin Guo

University of Science and Technology of China
September 2014

Contents

Chapter 1
The $\mathfrak{F}$-Hypercenter and Its Generalizations

1.1 Basic Concepts and Results

Classes of Groups A collection $\mathfrak{X}$ of groups is called a *class* of groups if every isomorphic image of every group in $\mathfrak{X}$ belongs to $\mathfrak{X}$. The empty-set $\varnothing$ is a class of groups by definition. If $\mathcal{X}$ is any collection of groups, then we write $(\mathcal{X})$ to denote the intersection of all classes of groups containing $\mathcal{X}$. In particular, (1) is the class of all identity groups.

If $G \in \mathfrak{X}$, then we say that G is an $\mathfrak{X}$-group. We say, following Mal'cev [308], that a class $\mathfrak{X}$ of groups is *(normally) hereditary* if it contains every (normal) subgroup of every group in $\mathcal{X}$.

Let π be a set of some primes and π' the complement of π in the set $\mathbb{P}$ of all primes.

In this book, we use the following notation for the classes of groups:

$\mathfrak{A}$ is the class of all abelian groups.

$\mathfrak{A}(n)$ is the class of all abelian groups of exponent dividing n.

$\mathfrak{G}$ is the class of all groups.

$\mathfrak{S}$ is the class of all soluble groups.

$\mathfrak{N}$ is the class of all nilpotent groups.

$\mathfrak{U}$ is the class of all supersoluble groups.

$\mathfrak{G}_\pi$ is the class of all π-groups ($\mathfrak{G}_\varnothing = (1)$).

$\mathfrak{S}_\pi$ is the class of all soluble π-groups ($\mathfrak{S}_\varnothing = (1)$).

$\mathfrak{N}_\pi$ is the class of all nilpotent π-groups.

$\mathfrak{G}_p$ is the class of all p-groups (p is a prime).

$\mathfrak{G}_{cp}$ is the class of all groups in which every p-chief factor, that is, a chief factor of p power order (where p is a prime), is central.

$\mathfrak{N}^r$ is the class of all soluble groups G with Fitting length (or nilpotent length) $l(G) \le r$ (see p. 519 in [89]).

Let $\mathfrak{X}$ be a nonempty class of groups containing (1). Then for any group G we write $G^{\mathfrak{X}}$ to denote the intersection of all normal subgroups N of G with $G/N \in \mathfrak{X}$ and say that $G^{\mathfrak{X}}$ is the *$\mathfrak{X}$-residual* of G; we write $G_{\mathfrak{X}}$ to denote the product of all normal subgroup N of G with $N \in \mathfrak{X}$ and say that $G_{\mathfrak{X}}$ is the *$\mathfrak{X}$-radical* of G. For a class $\mathfrak{F}$ of groups, we let $\pi(\mathfrak{F}) = \bigcup_{G \in \mathfrak{F}} \pi(G)$.

© Springer-Verlag Berlin Heidelberg 2015
W. Guo, *Structure Theory for Canonical Classes of Finite Groups*,
DOI 10.1007/978-3-662-45747-4_1

For any two classes $\mathfrak{M}$ and $\mathfrak{H}$ of groups, the symbol $\mathfrak{M}\mathfrak{H}$ denotes the class of all groups G with $G^{\mathfrak{H}} \in \mathfrak{M}$. We use the symbol $\mathfrak{M} \diamond \mathfrak{H}$ to denote the class of all groups G with $G/G_{\mathfrak{M}} \in \mathfrak{H}$.

A *formation* is a class $\mathfrak{F}$ of groups with the following properties:

(i) Every homomorphic image of an $\mathfrak{F}$-group is an $\mathfrak{F}$-group.
(ii) If G/M and G/N are $\mathfrak{F}$-groups, then $G/(M \cap N)$ belongs to $\mathfrak{F}$.

The empty-set $\emptyset$ is a formation by definition.

A *Fitting class* is a class $\mathfrak{F}$ of groups with the following properties:

(i) Every normal subgroup of an $\mathfrak{F}$-group is an $\mathfrak{F}$-group.
(ii) If M and N are $\mathfrak{F}$-groups and M and N are normal in G, then MN belongs to $\mathfrak{F}$.

The empty-set $\emptyset$ is a Fitting class by definition.

A Fitting formation is a class $\mathfrak{F}$ which is both a formation and a Fitting class.

From the definitions of the formations and the Fitting classes, we see that for every nonempty formation (Fitting class) $\mathfrak{X}$ we have $G/G^{\mathfrak{X}} \in \mathfrak{X}$ ($G_{\mathfrak{X}} \in \mathfrak{X}$, respectively). Moreover, for any two formations $\mathfrak{M}$ and $\mathfrak{H}$ (for any two Fitting classes $\mathfrak{M}$ and $\mathfrak{H}$), the class $\mathfrak{M}\mathfrak{H}$ is a formation (the class $\mathfrak{M} \diamond \mathfrak{H}$ is a Fitting class) (see p. 338 and 566 in [89]).

The study of the products of formations and Fitting classes are based on the following obvious facts.

Lemma 1.1 *Let N be a normal subgroup of a group G.*

(1) If $\mathfrak{F}$ is a nonempty formation, then $(G/N)^{\mathfrak{F}} = G^{\mathfrak{F}}N/N$.
(2) If $\mathfrak{F}$ is a nonempty Fitting class, then $N_{\mathfrak{F}} = G_{\mathfrak{F}} \cap N$.

Saturated Formations A formation $\mathfrak{F}$ is said to be *saturated* or *local* if $G \in \mathfrak{F}$ whenever $G/\Phi(G) \in \mathfrak{F}$.

For any formation function

$$f : \mathbb{P} \to \{\text{formations of groups}\}, \tag{1.1}$$

the symbol $LF(f)$ denotes the collection of all groups G such that either $G = 1$ or $G \neq 1$ and $G/C_G(H/K) \in f(p)$ for every chief factor H/K of G and every $p \in \pi(H/K)$. Clearly, the class $LF(f)$ is a nonempty formation. If $\mathfrak{F} = LF(f)$ for some formation function f, then f is called a *local definition* or a *local satellite* (L.A. Shemetkov) of the formation $\mathfrak{F}$.

Let $\mathfrak{X}$ be a class of groups and p be a prime. We write $\mathfrak{X}(p)$ to denote the intersection of all formations containing the set $\{G/O_{p'p}(G)|G \in \mathfrak{X}\}$ if $p \in \pi(\mathfrak{X})$; otherwise, we put $\mathfrak{X}(p) = \emptyset$.

Theorem 1.2 (Gaschutz, Lubeseder, and Schmid [89, Chap. IV, Theorem 4.6]).

1) For any function f of the form (1.1), the class $LF(f)$ is a saturated formation.
2) For any nonempty saturated formation $\mathfrak{F}$, $\mathfrak{F} = LH(h)$, where $h(p) = \mathfrak{F}(p)$ for all primes p.

3) *For any nonempty saturated formation $\mathfrak{F}$, there is a unique formation function F of the form (1.1) such that $\mathfrak{F} = LF(F)$ and $F(p) = \mathfrak{G}_p F(p) = \mathfrak{G}_p \mathfrak{F}(p) \subseteq \mathfrak{F}$ for all primes p.*

The formation function F in this theorem is called the *canonical* local satellite of $\mathcal{F}$. It is well known that

$$O_{p'p}(G) = \cap\{C_G(H/K) \mid H/K \text{ is a chief factor of } G \text{ and } p \in \pi(H/K)\}.$$

Therefore, $G \in \mathfrak{F} = LF(f)$ if and only if either $G = 1$ or $G \neq 1$ and $G/O_{p'p}(G) \in f(p)$ for all $p \in \pi(G)$.

(1.3) Description of the canonical local satellite of some known saturated formations. Let $\mathfrak{F} = LF(F)$ be a saturated formation, where F is the canonical local satellite of $\mathcal{F}$. For almost all known concrete saturated formations $\mathcal{F}$, the description of their canonical local satellite F is known (see, for example, [135, Chap. III] and [89, Chap. IV]). In particular,

1) If $\mathfrak{F}$ is the class of all (soluble) π-groups, then $F(p) = \mathcal{F}$ for all primes $p \in \pi$ and $F(p) = \varnothing$ for all primes $p \notin \pi$;
2) If $\mathfrak{F}$ is the class of all nilpotent π-groups, then $F(p) = \mathfrak{G}_p$ for all primes $p \in \pi$ and $F(p) = \varnothing$ for all primes $p \notin \pi$;
3) If $\mathfrak{F}$ is the class of all π-nilpotent groups, then $F(p) = \mathfrak{G}_p$ for all primes $p \in \pi$ and $F(p) = \mathfrak{F}$ for all primes $p \notin \pi$;
4) If $\mathfrak{F}$ is the class of all π-supersoluble groups, then $F(p) = \mathfrak{G}_p\mathfrak{A}(p-1)$ for all primes $p \in \pi$ and $F(p) = \mathfrak{F}$ for all primes $p \notin \pi$;
5) If $\mathfrak{F}$ is the class of all π-soluble groups, then $F(p) = \mathfrak{F}$ for all primes p;
6) If $\mathfrak{F}$ is the class of all groups G with $G' \leq F(G)$, then $F(p) = \mathfrak{G}_p\mathfrak{A}$ for all primes p;
7) If $\mathfrak{F}$ is the class of all p-decomposable groups, then $F(p) = \mathfrak{G}_p$ and $F(q) = \mathfrak{G}_{p'}$ for all primes $q \neq p$;
8) If $\mathfrak{F} = \mathfrak{N}^r$, then $F(p) = \mathfrak{G}_p\mathfrak{N}^{r-1}$ for all primes p.

Solubly Saturated Formation A formation $\mathcal{F}$ is said to be *solubly saturated, composition,* or a *Baer-local* formation if it contains each group G with $G/\Phi(N) \in \mathfrak{F}$ for some soluble normal subgroup N of G. It is easy to see that any saturated formation is solubly saturated.

A group G is called *primitive* if it has a maximal subgroup M such that $M_G = 1$, where M_G is the largest normal subgroup of G contained in M.

Proposition 1.4 (Baer, [89, A, 15.2]). *Let G be a primitive group and M a maximal subgroup of G such that $M_G = 1$. Then exactly one of the following statements holds:*

(1)G has a unique minimal normal subgroup N, $C_G(N) = N = F(G)$, and $G = N \rtimes M$.
(2)G has a unique minimal normal subgroup N, N is non-abelian, and $G = NM$.
(3)G has exactly two minimal normal subgroups N and N^*, $C_G(N) = N^*$, $C_G(N^*) = N$, and $G = N \rtimes M = N^* \rtimes M$. Also $N \simeq N^* \simeq NN^* \cap M$. Moreover, if $V < G$ and $VN = VN^* = G$, then $V \cap N = V \cap N^* = 1$.

We use $R(G)$ to denote the $\mathfrak{S}$-radical $G_{\mathfrak{S}}$ of G.

Proposition 1.5 (L.A. Shemetkov). *The intersection of the centralizers of all non-abelian chief factors of a group G coincides with its $\mathfrak{S}$-radical $R(G)$.*

Proof Let I be the intersection of the centralizers of all non-abelian chief factors of G. Then for any G-chief factor H/K of I we have $H/K \subseteq C_G(H/K)$. Hence, $I \le R(G)$. Now let H/K be any non-abelian chief factor of G and $C = C_G(H/K)$. Let M be a maximal subgroup of G such that $K \le M$ and $HM = G$. Then the chief factors H/K and HM_G/M_G of G are G-isomorphic. Hence, $M_G \le C$ and HM_G/M_G is the unique minimal normal subgroup of G/M_G by Proposition 1.4. Suppose that $R(G) \not\le M_G$. Then for some chief factors T/L of G we have $L \le M_G, T \le R(G)$, and $TM = G$. Hence, T/L and TM_G/M_G are G-isomorphic. This implies that T/L is non-abelian, which contradicts $T \le R(G)$. Thus, $R(G) \le M_G \le C$. It follows that $R(G) \le I$. Therefore, $R(G) = I$.

Recall that $C^p(G)$ denotes the intersection of the centralizers of all abelian p-chief factors of a group G ($C^p(G) = G$ if G has no such chief factors) [89, p. 371].

For any collection of groups $\mathfrak{X}$ we write $\mathrm{Com}(\mathfrak{X})$ to denote the class of all groups L such that L is isomorphic to some abelian composition factors of some group in $\mathfrak{X}$. If $\mathfrak{X}$ is the set of one group G, then we write $\mathrm{Com}(G)$ instead of $\mathrm{Com}(\mathfrak{X})$.

For a formation function of the form

$$f : \mathbb{P} \cup \{0\} \to \{\text{formations of groups}\} \tag{1.2}$$

we put, following [360],

$$CLF(f) = \{G \text{ is a group} \mid G/R(G) \in f(0)$$

and

$$G/C^p(G) \in f(p) \text{ for any prime } p \in \pi(\mathrm{Com}(G))\}.$$

If $\mathfrak{F} = CLF(f)$ for some formation function f, then we say that f is a *composition satellite* of the formation $\mathfrak{F}$.

From Proposition 1.4, [359, Chap. 1, Theorem 3.2] and Baer's Theorem [89, IV, 4.17], the following result follows.

Theorem 1.6

(1) *For any function f of the form (1.2), the class $CLF(f)$ is a solubly saturated formation.*
(2) *For any nonempty solubly saturated formation $\mathfrak{F}$, there is a unique function F of the form (1.1) such that $\mathfrak{F} = CLF(F)$, $F(p) = \mathfrak{S}_p F(p) \subseteq \mathfrak{F}$ for all primes p, and $F(0) = \mathfrak{F}$.*
(3) *If $\mathfrak{F} = LF(H)$ is a saturated formation, where H is the canonical local satellite of $\mathfrak{F}$, then $\mathfrak{F} = CLF(F)$, where $F(p) = H(p)$ for all primes p.*

The function F in Theorem 1.6 (2) is called the *canonical* composition satellite of $\mathcal{F}$.

Proposition 1.7 (see [135, 3.1.9] and [32, II, 3.1.40]). *Let $\mathfrak{F}$ be (solubly) saturated formation and F is the canonical local (the canonical composition, respectively) satellite of $\mathfrak{F}$.*

(a) *$\mathfrak{F}$ is (normally) hereditary if and only if $F(p)$ is (normally) hereditary for all primes p.*

(b) *$\mathfrak{F}$ is a Fitting formation if and only if $F(p)$ is a Fitting formation for all primes p.*

$\mathfrak{X}$-Critical Groups Let $\mathfrak{X}$ be a class of groups. A group G is called a *minimal non-$\mathfrak{X}$-group* (or *$\mathfrak{X}$-critical group*) if G is not in $\mathfrak{X}$ but all proper subgroups of G are in $\mathfrak{X}$. A maximal subgroup M of a group G is called *$\mathfrak{X}$-normal* in G provided $G/M_G \in \mathfrak{X}$. Otherwise, it is called *$\mathfrak{X}$-abnormal* in G.

Proposition 1.8 (Semenchuk, [135, 3.4.2]). *Let $\mathfrak{F}$ be a saturated formation and G be a group whose $\mathcal{F}$-residual $G^{\mathfrak{F}}$ is soluble. Suppose that every maximal subgroup of G not containing $G^{\mathfrak{F}}$ belongs to $\mathfrak{F}$. Write $P = G^{\mathfrak{F}}$. Then:*

(a) *P is a p-group for some prime p and P is of exponent p or of exponent 4 (if P is a non-abelian 2-group).*

(b) *$P/\Phi(P)$ is a $\mathfrak{F}$-eccentric (see the following page) chief factor of G.*

(c) *$\Phi(P) = P \cap \Phi(G) \leq Z(P)$.*

(d) *If P is abelian, then $\Phi(P) = 1$.*

(e) *$\Phi(G) \leq Z_{\mathfrak{F}}(G)$.*

(f) *If P is non-abelian, then $\Phi(P) = Z(P) = P'$.*

(g) *Any two $\mathfrak{F}$-abnormal maximal subgroup of G are conjugate in G.*

An $\mathfrak{N}$-critical group is called a *Schmidt group*.

Proposition 1.9 (Schmidt, see [135, 3.4.11] and [359, 26.1]). *Let G be a Schmidt group. Then the following statements hold.*

(1) *G is a soluble, $|G| = p^a q^b$ for two different primes p and q (where $a, b \in \mathbb{N}$), and $G = P \rtimes Q$, where $P = G^{\mathfrak{N}}$ is a Sylow p-subgroup of G and $Q = \langle a \rangle$ a cyclic Sylow q-subgroup of G.*

(2) *$a^q \in Z(G)$.*

(3) *$\Phi(G) = Z_{\infty}(G)$.*

(4) *$\Phi(P) \subseteq Z(P)$ and $c(P) \leq 2$, where $c(P)$ is the nilpotent class of P.*

(5) *G has precisely two classes of maximal subgroups: $< a^q > \times P$ and $\{P' \times Q^x | x \in G\}$.*

(6) *$G' = P$ and $P/\Phi(P)$ is a non-central chief factor of G.*

(7) *If P is not abelian, then the center, the commutators subgroup, and the Frattini subgroup of P coincide, and their exponent is p.*

(8) *If P is an abelian, then $\Phi(P) = 1$.*

(9) *If $|P/\Phi(P)| = p^{\alpha}$, then p^{α} is congruent to 1 modulo q.*

Proposition 1.10 (Ito [248, IV, 5.4]). *Every minimal non-p-nilpotent group is a p-closed Schmidt subgroup.*

Let $p_1 > p_2 > \ldots > p_r$ be the set of all primes dividing $|G|$, P_i a Sylow p_i-subgroup of G. Then we say that G is dispersive in the sense of Ore or simply G is dispersive if all subgroups P_1, $P_1 P_2$, $\ldots$, $P_1 P_2 \ldots P_{r-1}$ are normal in G.

In general, let φ be any linear ordering on the set of all primes $\mathbb{P}$. A group G of order $p_1^{\alpha_1} p_2^{\alpha_2} \cdots p_r^{\alpha_r}$, where $p_1 \varphi p_2 \varphi \cdots \varphi p_r$, is called φ-dispersive if G_{p_1}, $G_{p_1} G_{p_2}, \cdots, G_{p_1} G_{p_2} G_{p_r}$ are normal in G.

Proposition 1.11 (Huppert, see [135, 3.11.8]). *Let G be a $\mathfrak{U}$-critical group.*

(1) G is soluble and $\pi(G)| \leq 3$.
(2) If G is not a Schmidt group, then G is dispersive.
(3) G has a unique nonidentity normal Sylow subgroup.

We say that G is a *primary* group provided G is of prime power order. We say that G is a *p-primary* group provided $|G|$ is a power of p.

Proposition 1.12 (Doerk, see [135, 3.11.9]). *Let G be a $\mathfrak{U}$-critical group and $P = G^{\mathfrak{U}}$.*

(1) P is a Sylow p-subgroup of G for some prime p.
(2) If S is a complement of P in G, then $S/S \cap \Phi(G)$ is either a cyclic primary group or a minimal non-abelian group.

$\mathfrak{X}$-Central Chief Factors   Let $K \leq H \leq G$, where K and H are normal in G. Then we write $Aut_G(H/K)$ to denote the automorphism group induced by G on H/K, that is, $Aut_G(H/K) = G/C_G(H/K)$.

Let $\mathfrak{X}$ be a class of groups. A chief factor H/K of a group G is said to be $\mathfrak{X}$-*central* in G provided $(H/K) \rtimes Aut_G(H/K) \in \mathfrak{X}$. Otherwise, it is said to be $\mathfrak{X}$-*eccentric*.

Every $\mathfrak{N}$-central factor H/K of G is central, that is, $C_G(H/K) = G$; every $\mathfrak{U}$-central factor H/K of G is cyclic; for every $\mathfrak{S}$-central factor H/K of G the group $Aut_G(H/K) = G/C_G(H/K)$ is soluble.

The importance of this concept is connected with the following fact.

Proposition 1.13 (Barnes and Kegel, see [89, IV, 1.5]). *If $\mathfrak{F}$ is a formation and $G \in \mathfrak{F}$, then every chief factor of G is $\mathfrak{F}$-central.*

The following lemma can be proved by direct calculations.

Lemma 1.14 *Let H/K and E/T be chief factors of a group G. If H/K and E/T are G-isomorphic, then*

$$(H/K) \rtimes Aut_G(H/K) \simeq (E/T) \rtimes Aut_G(E/T).$$

Proposition 1.15 *Let $\mathfrak{F}$ be a (solubly) saturated formation and F the canonical local (the canonical composition, respectively) satellite of $\mathcal{F}$. Let H/K be a chief factor of G and p be a prime dividing $|H/K|$.*

(1) If $\mathfrak{F}$ is saturated, then H/K is $\mathfrak{F}$-central in G if and only if $G/C_G(H/K) \in F(p)$.

(2) If $\mathfrak{F}$ is solubly saturated, then H/K is $\mathfrak{F}$-central in G if and only if $G/C_G(H/K) \in F(p)$ in the case where H/K is abelian, and $G/C_G(H/K) \in \mathfrak{F}$ in the case where H/K is non-abelian.

Proof Without loss we can assume that $K = 1$. If H is abelian, then $C_{H \rtimes Aut_G(H)}(H) = H$, so the assertions follow from Theorem 1.6(2)(3).

Now suppose that H is non-abelian. Then $C_G(H) \cap H = 1$. Hence, in view of Theorem 1.14 and the G-isomorphism $C_G(H)H/C_G(H) \simeq H$, we may assume that $C_G(H) = 1$. Therefore, we only need to show that if $G \simeq G/C_G(H) \in \mathfrak{F}$, then $H \rtimes G \in \mathfrak{F}$. But this directly follows from Theorem 1.13.

1.2 Generalized $\mathfrak{X}$-Hypercentral Subgroups

The Subgroup $\Phi^*(G)$ We put $\Phi^*(G) = \Phi(R(G))$. Clearly $\Phi^*(G)$ is a characteristic subgroup of G.

We say that a chief factor H/K of G is *Frattini* (*solubly-Frattini*) if $H/K \le \Phi(G/K)$ (if $H/K \le \Phi^*(G/K)$, respectively).

Theorem 2.1 (Guo and Skiba [200]). *Let N and K be subgroups of a group G, where N is normal and K is subnormal in G. Then:*

(a) $\Phi^*(K) \le \Phi^*(G)$.
(b) $\Phi^*(G)N/N \le \Phi^*(G/N)$.
(c) If $N \le \Phi^*(G)$, then $\Phi^*(G/N) = \Phi^*(G)/N$.
(d) If $G = G_1 \times \ldots \times G_t$, then $\Phi^*(G) = \Phi^*(G_1) \times \ldots \times \Phi^*(G_t)$.
(e) Let H/K be a (solubly) Frattini chief factor of G. If $HN \ne KN$, then HN/KN is a (solubly) Frattini chief factor of G/KN.

Proof

(a) Let $K = K_t \trianglelefteq \ldots \trianglelefteq K_0 = G$. We prove the assertion by induction on t. Since $R(K_1)$ is characteristic in K_1, it is normal in G. Hence,

$$\Phi^*(K_1) = \Phi(R(K_1)) \le \Phi(R(G)) = \Phi^*(G)$$

by [89, A, 9.2(d)]. On the other hand, $\Phi^*(K) \le \Phi^*(K_1)$ by induction, so $\Phi^*(K) \le \Phi^*(G)$.

(b) Let $f : R(G)/R(G) \cap N \to R(G)N/N$ be the canonical isomorphism from $R(G)/R(G) \cap N$ onto $R(G)N/N$. Then $f(\Phi(R(G)/R(G) \cap N)) = \Phi(R(G)N/N)$ and

$$f(\Phi(R(G))(R(G) \cap N)/(R(G) \cap N))) = \Phi(R(G))N/N.$$

But by [89, A, 9.2 (e)] we have

$$\Phi(R(G))(R(G) \cap N)/(R(G) \cap N) \le \Phi(R(G)/R(G) \cap N)).$$

Then, since $R(G)N/N \leq R(G/N)$, we have

$$\Phi^*(G)N/N = \Phi(R(G))N/N \leq \Phi(R(G)N/N) \leq \Phi(R(G/N)) = \Phi^*(G/N)$$

by (a).

(c) Since $N \leq \Phi^*(G) = \Phi(R(G))$, N is nilpotent, and so $R(G)/N = R(G/N)$. Hence,

$$\Phi^*(G)/N = \Phi(R(G))/N = \Phi(R(G/N)) = \Phi^*(G/N).$$

(d) This follows from in [89, A, 9.4] and the fact that $R(G) = R(G_1) \times \ldots \times R(G_t)$.

(e) Suppose that H/K is solubly Frattini. By (b),

$$(KN/K)(H/K)/(KN/K) = (NH/K)/(KN/K) \leq \Phi^*((G/K)/(KN/K)).$$

Hence, from the G-isomorphism $(G/K)/(KN/K) \simeq G/KN$ we get that $NH/KN \leq \Phi^*(G/KN)$. Finally, in view of the G-isomorphism $HN/KN \simeq H/K(H \cap N)$, HN/KN is a chief factor of G. Hence, HN/KN is a solubly Frattini chief factor of G/KN. The second statement of (e) may be proved similarly.

Let θ be some group-theoretical property, that is, if a subgroup H of a group G has the property θ in G, then $\varphi(H)$ has this property in any isomorphic image $\varphi(G)$ of G. We say that a chief factor H/K of a group G is a θ-*chief factor* of G if H/K has the property θ in G/K.

Definition 2.2 Let $\mathfrak{X}$ be a class of groups. A normal subgroup N of a group G is said to be θ-*hypercentral* in G if either $N = 1$ or $N \neq 1$ and there exists a chief series

$$1 = N_0 < N_1 < \ldots < N_t = N \tag{1.3}$$

of G below N such that every θ-chief factor N_i/N_{i-1} of this series is $\mathfrak{X}$-central in G.

In particular, we say that N is

(a) $\mathfrak{X}$-*hypercentral* in G provided every factor of series (1.3) is $\mathfrak{X}$-central in G.

(b) $\mathfrak{X}\Phi$-*hypercentral* in G provided every non-Frattini factor of series (1.3) is $\mathfrak{X}$-central in G.

(c) $\mathfrak{X}\Phi^*$-*hypercentral* in G provided every non-solubly-Frattini factor of series (1.3) is $\mathfrak{X}$-central in G.

(d) $\pi\mathfrak{X}$-*hypercentral* in G provided every factor of series (1.3) of order divisible by at least one prime in π is $\mathfrak{X}$-central in G.

The θ-*hypercenter* $Z_\theta(G)$ of the group G is the product of all normal θ-hypercentral subgroups of G. In this book, the following special cases of this concept play important role:

(I) The $\mathfrak{X}$-*hypercenter* $Z_{\mathfrak{X}}(G)$ of G is the product of all normal $\mathfrak{X}$-hypercentral subgroups of G.

(II) The $\mathfrak{X}\Phi$-*hypercenter* $Z_{\mathfrak{X}\Phi}(G)$ of G is the product of all normal $\mathfrak{X}\Phi$-hypercentral subgroups of G.

(III) The $\mathfrak{X}\Phi^*$-*hypercenter* $Z_{\mathfrak{X}\Phi^*}(G)$ of G is the product of all normal $\mathfrak{X}\Phi^*$-hypercentral subgroups of G.

(IV) The $\pi\mathfrak{X}$-*hypercenter* $Z_{\pi\mathfrak{X}}(G)$ of G is the product of all normal $\pi\mathfrak{X}$-hypercentral subgroups of G.

It is clear that the subgroups $Z_{\mathfrak{X}}(G)$, $Z_{\pi\mathfrak{X}}(G)$, $Z_{\mathfrak{X}\Phi}(G)$, and $Z_{\mathfrak{X}\Phi^*}(G)$ are all characteristic subgroups of G. It is also clear that subgroups $Z_{\mathfrak{X}}(G) \leq Z_{\pi\mathfrak{X}}(G)$, $Z_{\mathfrak{X}}(G) = Z_{\mathbb{P}\mathfrak{X}}(G)$, and $Z_{\mathfrak{X}}(G) \leq Z_{\mathfrak{X}\Phi^*}(G) \leq Z_{\mathfrak{X}\Phi}(G)$.

Example 2.3 Let V be a simple $\mathbb{F}_3 A_4$-module which is faithful for the alternating group A_4. Then we may consider V as an $\mathbb{F}_3 SL_2(3)$-module with $C_{SL_2(3)}(V) = Z$, where Z is the unique minimal normal subgroup of $SL_2(3)$. Let $E = V \rtimes SL_2(3)$. Then $Z_{2\mathfrak{N}}(E) = Z_{2\mathfrak{U}}(E) = VZ$ and $Z_{\mathfrak{N}}(E) = Z_{3\mathfrak{N}}(E) = Z_{3\mathfrak{U}}(E) = Z_{\mathfrak{N}}(E) = Z$.

Let $A = A_3(E)$ be the 3-Frattini module of E ([89, p. 853]), and let G be a non-splitting extension of A by E. By Corollary 1 in [118], $VZ = O_{3'3}(E) = C_E(A/Rad(A))$. Hence, for some normal subgroup N of G we have $A/N \leq \Phi(G/N)$ and $G/C_G(A/N) \simeq A_4$. Thus, $|A/N| > 3$ and so

$$Z_{\mathfrak{U}\Phi}(G/N) = Z_{\Phi}(G/N) = (A/N)(D/N),$$

where D/N is the unique minimal normal subgroup of G/N with order 2. On the other hand, $Z_{\mathfrak{U}}(G/N) = Z_{\infty}(G/N) = D/N$.

Example 2.4 Let $r < p < q$ be primes and suppose that r divides $p - 1$. Let S be a non-abelian group of order pq, C_p and C_q groups of order p and q, respectively. Let $A = R \rtimes C_p$, where R is a simple $\mathbb{F}_r C_p$-module which is faithful for C_p, and $B = P \rtimes C_q$, where P is a simple $\mathbb{F}_p C_q$-module which is faithful for C_q. Let $H = A \times B \times S$, and let V be a projective envelope of a trivial $\mathbb{F}_p H$-module. Let $G = V \rtimes H$. Let $C = C_H(V)$, and let C_0 be the intersection of the centralizers in H of all G-chief factors of V. Then $\Phi(G) = Rad(V)$ by Lemma 3.14 in [89, Chap. B], and $C = O_{p'}(H)$ by Theorem 14.6 in [249, Chap. VII]. Suppose that all G-chief factors of V are cyclic. Then H/C_0 is an abelian group of exponent dividing $p - 1$. Since q does not divide $p - 1$, $C_q \leq C_0$. Hence, $C_q \leq R$. This contradiction shows that G has a Frattini chief factor K/L such that $|K/L| > p$ and for every G-chief factors M/N between K and V we have $|M/N| = p$. Hence,

$$Z_{\mathfrak{U}\Phi}(G/L) = (V/L)(ZL/L) \neq Z_{\Phi}(G/L) = V/L,$$

where Z is the unique normal subgroup of H with order p.

Lemma 2.5 *Let H be a normal subgroup of a group G. Let $\mathcal{K}_1$ and $\mathcal{K}_2$ be G-chief series of H. Then there exists a one-to-one correspondence between the chief factors of $\mathcal{K}_1$ and those of $\mathcal{K}_2$ such that corresponding factors are G-isomorphic and such that the Frattini (in G) chief factors of $\mathcal{K}_1$ corresponds to the Frattini (in G) chief factors of $\mathcal{K}_2$.*

Proof The assertion is a strengthened form of Theorem 9.13 in [89, Chap. A] with the same proof.

We use $Z_n(A)$ to denote is the n-th member of the upper central series of A.

If $K < H \leq E \leq G$, where H/K is a chief factor of a group G, then we say that H/K is a *G-chief factor* of E.

Theorem 2.6 (Guo, Skiba [200, 198, 194]). *Let $\mathfrak{X}$ be a nonempty class of groups, A and B normal subgroups of G. Let $\theta \in \{\mathfrak{X}\Phi, \mathfrak{X}\Phi^*, \pi\mathfrak{X}\}$.*

(a) *If A is θ-hypercentral in G, then AB/B is θ-hypercentral in G/B.*
(b) *A and B are θ-hypercentral in G, then AB is θ-hypercentral in G.*
(c) *$Z_\theta(G)$ is θ-hypercentral in G.*
(d) *$Z_\theta(G)A/A \leq Z_\theta(G/A)$.*
(e) *$Z_\theta(H)A/A \leq Z_\theta(HA/A)$ for any subgroup H of G.*
(f) *If A is θ-hypercentral in G, then*

$$Z_\theta(G)/A = Z_\theta(G/A).$$

(h) *$Z_\theta(G/Z_\theta(G)) = 1$.*
(i) *If $\mathcal{F}$ is a formation and $G \in \mathfrak{F}$, then $Z_\mathfrak{F}(G) = G$.*
(j) *Every non-Frattini G-chief factor of $Z_{\mathfrak{X}\Phi}(G)$ is $\mathfrak{X}$-central.*
(k) *Every G-chief factor of $Z_{\pi\mathfrak{X}}(G)$ of order divisible by at least one a prime in π is $\mathfrak{X}$-central.*
(l) *If $\mathfrak{X}$ is a solubly saturated formation and $G/Z_{\mathfrak{X}\Phi^*}(G) \in \mathfrak{X}$, then $G \in \mathfrak{X}$.*
(m) *If $\mathfrak{X}$ is a saturated formation and $G/Z_{\mathfrak{X}\Phi}(G) \in \mathfrak{X}$, then $G \in \mathfrak{X}$.*
(n) *$Z_n(Z_\infty(A))N/N) \leq Z_n(Z_\infty(AN/N))$.*

Proof We shall prove this theorem for the case, where $\theta = \mathfrak{X}\Phi^*$. Other cases may be proved similarly.

(a) Let $1 = A_0 < A_1 < \ldots < A_t = A$ (*) be a chief series of G below A such that every non-solubly-Frattini factor A_i/A_{i-1} of Series (*) is $\mathfrak{X}$-central in G. Let H/K be a chief factor of G such that for some i we have $K = BA_{i-1}$ and $H = BA_i$. Then H/K is G-isomorphic to $A_i/A_{i-1}(A_i \cap B) = A_i/A_{i-1}$. Suppose that H/K is not solubly Frattini. Then, by Lemma 2.1(e), A_i/A_{i-1} is also not solubly Frattini. Hence, A_i/A_{i-1} is $\mathcal{X}$-central in G, so H/K $\mathcal{X}$-central in G by Lemma 1.14. Therefore, every non-solubly-Frattini factor of the series $1 \leq BA_0/B \leq BA_1/B \leq \ldots \leq BA_t/B = BA/B$ is $\mathcal{X}$-central in G/B.
(b) This follows from (a).
(c) This follows from (b).
(d) This follows from (a) and (b).
(e) Let $f : H/H \cap A \to HA/A$ be the canonical isomorphism from $H/H \cap A$ onto HA/A. Then $f(Z_{\mathfrak{X}\Phi^*}(H/H \cap A)) = Z_{\mathfrak{X}\Phi^*}(HA/A)$ and

$$f(Z_{\mathfrak{X}\Phi^*}(H)(H \cap A)/(H \cap A)) = Z_{\mathfrak{X}\Phi^*}(H)A/A.$$

By (d) we have

$$Z_{\mathfrak{X}\Phi^*}(H)(H \cap A)/(H \cap A) \leq Z_{\mathfrak{X}\Phi^*}(H/H \cap A).$$

Hence, $Z_{\mathfrak{X}\Phi^*}(H)A/A \leq Z_{\mathfrak{X}\Phi^*}(HA/A)$.

(f) By (d), $Z_{\mathfrak{X}\Phi^*}(G)/A \leq Z_{\mathfrak{X}\Phi^*}(G/A)$. On the other hand, for any $\mathfrak{X}\Phi^*$-hypercentral subgroup V/A of G/A the subgroup V is $\mathfrak{X}\Phi^*$-hypercentral in G since A is $\mathfrak{X}\Phi^*$-hypercentral in G. Hence, $Z_{\mathfrak{X}\Phi^*}(G/A) \leq Z_{\mathfrak{X}\Phi^*}(G)/A$, so we have (e).

(h) This follows from (c) and (f).

(i) This follows from Proposition 1.13

(j) , (k) This follows from (c) and Lemma 2.5.

(l) By (c), there is a minimal normal subgroup of G such that $N \leq Z_{\mathfrak{X}\Phi^*}(G)$ and either $N \leq \Phi^*(G)$ or N is $\mathfrak{X}$-central in G. If N is $\mathfrak{X}$-central in G, then by (f),

$$Z_{\mathfrak{X}\Phi^*}(G)/N = Z_{\mathfrak{X}\Phi^*}(G/N).$$

Hence, $G/N \in \mathfrak{X}$ by induction, which implies that $G \in \mathfrak{X}$.

(m) See the proof of (l).

(n) See the proof of (d).

In the case when $\mathfrak{X}$ is a saturated formation we may obtain some additional information on the subgroup $Z_{\pi\mathfrak{F}}(G)$.

Theorem 2.7 (Guo, Skiba [198, 200]). *Let $\mathfrak{F}$ be a saturated formation. Let $\pi \subseteq \pi(\mathfrak{F})$ and $A \leq G$.*

(a) If $\mathfrak{F}$ is (normally) hereditary and H is a (normal) subgroup of G, then $Z_{\pi\mathfrak{F}}(A) \cap H \leq Z_{\pi\mathfrak{F}}(H \cap A)$.

(b) If $\mathfrak{G}_{\pi'}\mathfrak{F} = \mathfrak{F}$ and $G/Z_{\pi\mathfrak{F}}(G) \in \mathfrak{F}$, then $G \in \mathfrak{F}$. In particular, if $G/Z_{\mathfrak{F}}(G) \in \mathfrak{F}$, then $G \in \mathfrak{F}$.

(c) Suppose that $\mathfrak{F}$ is (normally) hereditary and let H be a (normal) subgroup of G. If $\mathfrak{G}_{\pi'}\mathfrak{F} = \mathfrak{F}$ and $H \in \mathfrak{F}$, then $Z_{\pi\mathfrak{F}}(G)H \in \mathfrak{F}$.

Proof

(a) Let F be the canonical local satellite of $\mathfrak{F}$. First suppose that $\mathfrak{F}$ is hereditary. Let

$$1 = Z_0 < Z_1 < \ldots < Z_t = Z_{\pi\mathfrak{F}}(G)$$

be a chief series of G below $Z_{\pi\mathfrak{F}}(G)$ and $C_i = C_G(Z_i/Z_{i-1})$. Let q be a prime divisor of

$$|Z_i \cap H/Z_{i-1} \cap H| = |Z_{i-1}(Z_i \cap H)/Z_{i-1}|.$$

Suppose that q divides $|Z_i \cap H/Z_{i-1} \cap H|$. Then q divides $|Z_i/Z_{i-1}|$, so $G/C_i \in F(q)$ by Proposition 1.15. Hence, $H/H \cap C_i \simeq C_i H/C_i \in F(q)$. But $H \cap C_i \leq C_H(Z_i \cap H/Z_{i-1} \cap H)$. Hence, $H/C_H(Z_i \cap H/Z_{i-1} \cap H) \in F(q)$ for all primes q dividing $|Z_i \cap H/Z_{i-1} \cap H|$. Thus, $Z_{\pi\mathfrak{F}}(G) \cap H \leq Z_{\pi\mathfrak{F}}(H)$ by Lemma 1.14 and Theorem 2.6 (k).

But then

$$Z_{\pi\mathfrak{F}}(A) \cap H = Z_{\pi\mathfrak{F}}(A) \cap (H \cap A) \leq Z_{\pi\mathfrak{F}}(H \cap A).$$

Similarly, one may prove the second assertion of (a).

(b) This assertion is evident.

(c) Since $H \in \mathfrak{F}$ we have

$$H Z_{\pi\mathfrak{F}}(G)/Z_{\pi\mathfrak{F}}(G) \simeq H/H \cap Z_{\pi\mathfrak{F}}(G) \in \mathfrak{F}$$

and

$$Z_{\pi\mathfrak{F}}(G) \leq Z_{\pi\mathfrak{F}}(Z_{\pi\mathfrak{F}}(G)H)$$

by (a). Hence, $H Z_{\pi\mathfrak{F}}(G) \in \mathfrak{F}$ by (b).

Our next goal is to prove the following result.

Theorem 2.8 (Guo, Skiba [200]). *Let $\mathfrak{F}$ be any formation, E a normal subgroup of G.*

(i) If $F^*(E)$ is $\mathfrak{F}\Phi^*$-hypercentral in G, then E is also $\mathfrak{F}\Phi^*$-hypercentral in G.

(ii) If $F^*(E)$ is $\mathfrak{F}$-hypercentral in G, then E is also $\mathfrak{F}$-hypercentral in G.

(iii) If E is soluble and $F(E)$ is $\mathfrak{F}\Phi$-hypercentral in G, then E is also $\mathfrak{F}\Phi$-hypercentral in G.

In this theorem, $F^*(E)$ is the generalized Fitting subgroup of E, that is, the product of all normal quasinilpotent subgroup of E (see [250, Chap. X]). The following result collects some known properties about this subgroup which will be used in many places of this book.

Proposition 2.9 ([250, Chap. X]). *Let G be a group. Then*

(1) If N is a normal subgroup of G, then $F^*(N) \leq F^*(G)$.

(2) If N is a normal subgroup of G and $N \leq F^*(G)$, then $F^*(G)/N \leq F^*(G/N)$.

(3) $F(G) \leq F^*(G) = F^*(F^*(G))$.

(4) $F^*(G) = F(G)E(G)$, $F(G) \cap E(G) = Z(E(G))$ and $[F(G), E(G)] = 1$, where $E(G)$ is the layer of G.

(5) $C_G(F^*(G)) \leq F(G)$.

(6) If $N \leq Z(G)$, then $F^*(G)/N = F^*(G/N)$.

(7) If $F(G) = 1$, then $F^*(G)$ is the direct product of some non-abelian minimal normal subgroups of G.

(8) If $F^*(G)$ is soluble, then $F^*(G) = F(G)$.

(9) $F^*(G)$ is quasinilpotent.

Lemma 2.10 *$F(G) \leq C_G(H/K)$ for any chief factor H/K of a group G, $O_{p'p}(G) \leq C_G(H/K)$ for any chief factor H/K of G divisible by prime p and $F^*(G) \leq C_G(H/K)$ for any abelian chief factor H/K of G.*

Proof The first two assertions are well known (see, for example, [135, Theorem 1.8.13]).

Now let H/K be an abelian chief factor of G. In view of Lemma 2.5, we may suppose that $H \leq F(G)$. Since $F(G) \leq C_G(H/K)$ and, on the other hand, $F^*(G) = E(G)F(G)$ and $[E(G), F(G)] = 1$ by Proposition 2.9(4), where $E(G)$ is the layer of G, we obtain $F^*(G) \leq C_G(H/K)$.

A group is called semisimple provided it is either identity or the direct product of some simple non-abelian groups.

Lemma 2.11 *Let N be a non-abelian minimal normal subgroup of G and $C = C_G(N)$. Let E be a normal quasinilpotent subgroup of G. If $N \leq E$, then $E = N \times (C \cap E)$.*

Proof Let $F = F(E)$. Since F is characteristic in E, it is normal in G. Hence, NF/F is G-isomorphic to N, and so $C = C_G(NF/F)$. On the other hand, by Lemma 2.10, $F \leq C$, which implies that $C_{G/F}(NF/F) = C/F$. By [250, X, 13.6], E/F is semisimple. Hence,

$$E/F = (C_{G/F}(NF/F) \cap (E/F))(NF/F) = ((C/F) \cap (E/F))(NF/F) = (C \cap E)N/F.$$

This implies that $E = N \times (C \cap E)$.

Lemma 2.12 *Let M and N be normal subgroups of a group G.*

(a) *If $G = MN$, then every chief factor H/K of G is either G-isomorphic to some G-chief factor of M or G-isomorphic to some G-chief factor of N*

(b) *If $M \cap N = 1$, then every chief factor H/K of G is G-isomorphic to some chief factor T/L of G such that either $M \leq L$ or $N \leq L$.*

Proof

(a) It follows from the G-isomorphism $MN/M \simeq N/N \cap M$ and Lemma 2.5.

(b) This follows from Lemma 2.5 and Lemma 1.10.11 in [135].

Lemma 2.13 (see [135, 3.2.3]). *Let H be a π-group for some set of primes π, a subgroup A of $Aut(H)$ stabilizers a normal series of H. Then A is a π-group.*

Lemma 2.14 *If a π-subgroup A of G is subnormal in G, then $A \leq O_\pi(G)$.*

Proof See [449] or the proof of Corollary (3.20) below.

A factor H/K of G is said to be a pd-factor of G if the prime p divides $|H/K|$. A factor H/K of G is said to be a p-factor of G if $|H/K|$ is a power of prime p.

Lemma 2.15 (see [135, Lemma 1.7.11]). *If H/K is a pd-chief factor of a group G, then $O_p(Aut_G(H/K)) = 1$.*

Proposition 2.16 (Vedernikov). $\mathfrak{G}_{cp}$ *is a Fitting formation.*

Proof It is clear that $\mathfrak{G}_{cp}$ is a formation. Now let $M \in \mathfrak{G}_{cp}$ and $N \in \mathfrak{G}_{cp}$ are normal subgroups of a group G. We show that $MN \in \mathfrak{G}_{cp}$. Without loss of the generality, we may assume that $MN = G$. Let H/K be a p-chief factor of G (that is, the order of H/K is p-power) and $C = C_G(H/K)$. We need to show that $C = G$.

Assume that $M \cap N \neq 1$. Then $G/M \cap N = (M/M \cap N)(N/M \cap N)$, where $M/M \cap N$ and $N/M \cap N$ belong to $\mathfrak{G}_{cp}$ since it is a formation. Hence, by induction, $G/M \cap N \in \mathfrak{G}_{cp}$. If H/K is G-isomorphic to some G-chief factor between $M \cap N$ and G, then $C = G$. Now suppose that H/K is G-isomorphic to some G-chief factor of $M \cap N$. In this case, we may assume that $H \leq M \cap N$ and that $K = 1$. Let

$$1 = H_0 < H_1 < \ldots < H_t = H, \tag{1.4}$$

where H_i/H_{i-1} is a chief factor of M for all $i = 1, \ldots, t$. Let $C_i = C_M(H_i/H_{i-1})$ and $C = C_1 \cap \ldots \cap C_t$. Then $M/C \cap M$ stabilizes series (1.4). Hence, $M/C \cap M \simeq MC/C$ is a p-group by Lemma 2.13. Similarly, NC/C is a p-group. Hence, G/C is a p-group, which implies that $C = G$ by Lemma 2.15. Now assume that $M \cap N = 1$. In this case, since $M, N \in \mathfrak{G}_{cp}$, $C = G$ by Lemma 2.12(b).

Finally, if $G \in \mathfrak{G}_{cp}$ and M is a normal subgroup of G, then $M \in \mathfrak{G}_{cp}$ by Lemma 2.5. Therefore, $\mathfrak{G}_{cp}$ is a Fitting formation.

Proposition 2.17 $C^p(G) = G_{\mathfrak{G}_{cp}}$, *for any group* G.

Proof The inclusion $C^p(G) \leq G_{\mathfrak{G}_{cp}}$ is evident. In order to prove the inverse inclusion, in view of Lemma 2.5, it is enough to show that $G_{\mathfrak{G}_{cp}} \leq C_G(H/K)$ for any G-chief factor H/K of $G_{\mathfrak{G}_{cp}}$ of p-power order. Without loss of generality, we may suppose that $K = 1$. Let $C = C_G(H)$. Since H is a p-group and $H \leq G_{\mathfrak{G}_{cp}}$, $G_{\mathfrak{G}_{cp}}/C \cap G_{\mathfrak{G}_{cp}}$ is a p-group (see the proof of Proposition 2.16). Since

$$G_{\mathfrak{G}_{cp}}/C \cap G_{\mathfrak{G}_{cp}} \simeq CG_{\mathfrak{G}_{cp}}/C \leq O_p(G/C) = 1.$$

We obtain that $G_{\mathfrak{G}_{cp}} \leq C$. The lemma is proved.

Lemma 2.18 *Let* N *and* E *be normal subgroups of a group* G.

(1) $C^p(G) \cap N = C^p(N)$.
(2) *If* $p \notin \pi(Com(N))$, *then* $N \leq C^p(G)$ *and* $C^p(G/N) = C^p(G)/N$.
(3) *If* $N \leq E \cap \Phi^*(G)$, *then* $C^p(E/N) = C^p(E)/N$.

Proof

(1) By Proposition 2.17, $C^p(G) = G_{\mathfrak{G}_{cp}}$. On the other hand, by Proposition 2.16, the class $\mathfrak{G}_{cp}$ is a Fitting formation. Hence, we have (1) by Lemma 1.1(2).
(2) This statement is evident.
(3) Let $C^p/N = C^p(E/N)$. Since $N \subseteq \Phi^*(G) \subseteq F(G)$, $N \subseteq F(E)$ and so $N \subseteq C^p(E)$ by Lemma 2.10. Then, clearly, $C^p(E)/N \leq C^p(E/N)$. Hence, we only need to show that $C^p/N = C^p(E/N) \leq C^p(E)/N$.

First we suppose that $E = G$. By Proposition 2.17 we have $C^p/N \in \mathfrak{G}_{cp}$. Hence, by Proposition 2.16, $C^p/\Phi^*(G) \in \mathfrak{G}_{cp}$, so $C^p/\Phi^*(G) \leq C^p(G/\Phi^*(G))$ by Proposition 2.17. But by [89, IV, 4.11], we have $C^p(G)/\Phi^*(G) = C^p(G/\Phi^*(G))$. Then from the definition of the subgroup $C^p(G)$ and Lemma 2.5, we get that $C^p/N = C^p(G/N) \leq C^p(G)/N$.

Now by (1), $C^p/N = C^p(E/N) = C^p(G/N) \cap (E/N) = (C^p(G) \cap N)/N$. Hence, $C^p = C^p(G) \cap E \in \mathfrak{G}_{cp}$. By Lemma 2.15 and Lemma 2.16, $C^p \leq C^p(E)$.

Lemma 2.19 *Suppose that $\mathfrak{F}$ is a solubly saturated formation containing all nilpotent groups. Let $N \leq E$ be normal subgroups of G and $E/N \in \mathfrak{F}$. If either $N \leq \Phi^*(G)$ or $\mathfrak{F}$ is saturated and $N \leq \Phi(G)$, then $E \in \mathfrak{F}$.*

Proof Suppose that $N \leq \Phi^*(G)$. In view of Theorem 2.1(c), we may suppose without loss of generality that N is a minimal normal subgroup of G. Then N is a p-group for some prime p. Let F be the canonical composition satellite of $\mathfrak{F}$. In view of Theorem 1.6 and Lemma 2.18(3) we only need to prove that $E/C^p(E) \in F(p)$. But from $E/N \in \mathfrak{F}$ we get $(E/N)/C^p(E/N) \in F(p)$. On the other hand, by Lemma 2.18, we have $C^p(E)/N = C^p(E/N)$. Thus, $E/C^p(E) \in F(p)$.

The second statement of the lemma may be proved similarly.

Proof of Theorem 2.8

(i) Suppose that this assertion is false and consider a counterexample (G, E) for which $|G| + |E|$ is minimal. Put $F^* = F^*(E)$ and $F = F(E)$.

Let $1 = N_0 < N_1 < \ldots < N_t = F^*$ be a chief series of G below F^* such that every non-solubly-Frattini factor N_i/N_{i-1} of this series is $\mathfrak{F}$-central in G. Let $N = N_1$ and $C = C_E(N)$. Note that in the case where N is $\mathfrak{F}$-central in G we have $G/C_G(N) \in \mathfrak{F}$. Hence, every chief factor of G between $C_G(N)$ and G is $\mathfrak{F}$-central in G. Hence, from the G-isomorphism $E/C \simeq C_G(N)E/C_G(N)$ we deduce that every chief factor of G between C and E is $\mathfrak{F}$-central in G.

Suppose N is non-abelian. Let $D = N \times C$. Since $C = C_G(N) \cap E$, D is normal in G. Moreover, $F^* = F^*(D)$. Indeed, in view of Lemma 2.11, $F^* \leq F^*(D)$. On the other hand, $F^*(D) \leq F^*$ by Proposition 2.9. Hence, $F^* = F^*(D)$ and so the hypothesis holds for D. Suppose that $D \neq E$. Then D is $\mathfrak{F}\Phi^*$-hypercentral in G by the choice of $|G| + |E|$. On the other hand, from the above we know that E/D is $\mathfrak{F}$-hypercentral in G/D. Hence, E is $\mathfrak{F}\Phi^*$-hypercentral in G, a contradiction. Therefore, $D = E$. We now show that in this case the hypothesis holds for $(G/N, E/N)$. Let $W/N = F^*(E/N)$. In view of Proposition 2.9, $F^*/N \leq W/N$. On the other hand, $W = N \times (W \cap C)$, where $W \cap C \simeq W/N$ is quasinilpotent, so W is quasinilpotent by [250, X, 13.3(d)]. Hence, $F^*/N = W/N$. This implies that $F^*(E/N)$ is $\mathfrak{F}\Phi^*$-hypercentral in G/N by Theorem 2.6 (1). The choice of $|G| + |E|$ implies that E/N is $\mathfrak{F}\Phi^*$-hypercentral in G/N and so E is $\mathfrak{F}\Phi^*$-hypercentral in G since N is $\mathfrak{F}$-central in G. This contradiction shows that N is abelian.

Assume that $N \leq \Phi^*(G)$. Since the class $\mathfrak{N}^*$ of all quasinilpotent groups is a saturated formation (see Theorem 3.7 below), in view of Lemma 2.19, $F^*(E/N) = F^*(E)/N$ and so E/N is $\mathfrak{F}\Phi^*$-hypercentral in G/N by the choice of (G, E), which implies that E is $\mathfrak{F}\Phi^*$-hypercentral in G, a contradiction. Hence, $N \nleq \Phi^*(G)$, and consequently N is $\mathfrak{F}$-central in G. Now we show that the hypothesis is still true for $(G/N, C/N)$. Indeed, clearly $N \leq Z(C)$. Moreover, $F^* \leq C$ by Lemma 2.10. Therefore, $F^*(C/N) = F^*/N$ by Proposition 2.9(1)(6). This shows that the hypothesis is still true for $(G/N, C/N)$. The choice of (G, E) implies that C/N is $\mathfrak{F}\Phi^*$-hypercentral in G/N. Then since N is $\mathfrak{F}$-central in G, C is $\mathfrak{F}\Phi^*$-hypercentral in G. It follows that E is $\mathfrak{F}\Phi^*$-hypercentral in G. This contradiction completes the proof of (i).

For (ii) and (iii), See the proof of (i).

Corollary 2.20 (Asaad [13], Li [277], Skiba [385]). *Let E be a normal subgroup of G. If every chief factor of G below $F^*(E)$ is cyclic, then every chief factor of G below E is also cyclic.*

Since $F^*(E) = F(E)$ whenever E is soluble by Proposition 2.9(8), we get from Theorem 2.8(ii) the following:

Corollary 2.21 *Let $\mathfrak{F}$ be any formation, E a soluble normal subgroup of G. If $F(E)$ is $\mathfrak{F}$-hypercentral in G, then E is also $\mathfrak{F}$-hypercentral in G.*

Now let A be any simple non-abelian group and let $\mathfrak{F}$ be the class of all groups W such that every composition factor of W is not isomorphic to A. It is clear that $\mathfrak{F}$ is a solubly saturated formation and $F(A) = 1$ is $\mathfrak{F}$-hypercentral in A. Nevertheless, A is not $\mathfrak{F}\Phi$-hypercentral in A. This obvious example shows that in general, if E is not soluble, Theorem 2.8(iii) and Corollary 2.21 are not true.

Among other corollaries of Theorem 2.8, there are the following:

Corollary 2.22 *Let $\mathfrak{F}$ be any solubly saturated formation. If $F^*(G)$ is $\mathfrak{F}\Phi^*$-hypercentral in G, then $G \in \mathfrak{F}$.*

Corollary 2.23 *Let $\mathfrak{F}$ be any solubly saturated formation and E a normal subgroup of G such that $F^*(E)$ is $\mathfrak{F}\Phi^*$-hypercentral in G. If $G/E \in \mathfrak{F}$, then $G \in \mathfrak{F}$.*

Proposition 2.24 (Guo and Skiba [200]). *Let E be a normal non-identity quasinilpotent subgroup of G. If $\Phi(G) \cap E = 1$, then E is the direct product of some minimal normal subgroups of G.*

Proof Let N be a minimal normal subgroup of G contained in E and $C = C_G(N)$. First we show that for some normal subgroup D of G we have $E = N \times D$. If N is non-abelian, then $E = N \times (E \cap C)$ by Lemma 2.11. Hence, in this case we may take $D = E \cap C$. Now suppose that N is an abelian group. Since $\Phi(G) \cap E = 1$, for some maximal subgroup M of G we have $G = N \rtimes M$ and so $E = N \rtimes (E \cap M)$. Let $D = E \cap M$. Note that $N \leq F(E) \leq Z_\infty(E)$ (see Theorem 3.12(2) below). On the other hand, N is the direct product of some minimal normal subgroup of E by [89, A, 4.13(c)], so $N \leq Z(E)$. Therefore, $G = NM \leq N_G(E \cap M)$. Thus, $E = N \times D$, where D is normal in G. If $D \neq 1$, then, by induction, D is the product of some minimal normal subgroups of G. Hence, E is also the product of some minimal normal subgroups of G.

If $\mathfrak{F} = (1)$ is the formation of all identity groups, then $Z_{\mathfrak{F}\Phi}(G) = \Phi(G)$ is the Frattiti subgroup of G. For general case, we prove on the basis of Theorem 2.8 (ii) the following results.

Theorem 2.25 (Guo and Skiba [200]). *Let $\mathfrak{F} \neq (1)$ be a nonempty formation and $\pi = \pi(\mathfrak{F})$. Then*

$$Z_{\mathfrak{F}\Phi}(G)/\Phi(G) = Z_{\mathfrak{F}}(G/\Phi(G)),$$

and $Z_{\mathfrak{F}\Phi}(G) = A \times B$, where $A = O_\pi(Z_{\mathfrak{F}\Phi}(G))$, $B = O_{\pi'}(\Phi(G))$ and $A/A \cap \Phi(G) \leq Z_{\mathfrak{F}}(G/A \cap \Phi(G))$.

Proof It is clear that $\Phi(G) \le Z_{\mathfrak{F}\Phi}(G)$. By Lemma 2.5, every non-Frattini chief factor of G below $Z_{\mathfrak{F}\Phi}(G)$ is $\mathfrak{F}$-central in G. Hence, we have $Z_{\mathfrak{F}}(G/\Phi(G)) \le Z_{\mathfrak{F}\Phi}(G/\Phi(G)) = Z_{\mathfrak{F}\Phi}(G)/\Phi(G)$.

Now let $L/\Phi(G)$ be a minimal normal subgroup of $G/\Phi(G)$ contained in $Z_{\mathfrak{F}\Phi}(G/\Phi(G))$. Then $L/\Phi(G)$ is non-Frattini. Hence, this chief factor of G is $\mathfrak{F}$-central in G. Thus, $\mathrm{Soc}(Z_{\mathfrak{F}\Phi}(G/\Phi(G))) \le Z_{\mathfrak{F}}(G/\Phi(G))$. On the other hand, since $F^*(Z_{\mathfrak{F}\Phi}(G/\Phi(G))$ is characteristic in $Z_{\mathfrak{F}\Phi}(G/\Phi(G))$, it is normal in $G/\Phi(G)$. Hence, in view of Proposition 2.24, $F^*(Z_{\mathfrak{F}\Phi}(G/\Phi(G))$ is the product of some minimal normal subgroups of $G/\Phi(G)$. But then $F^*(Z_{\mathfrak{F}\Phi}(G/\Phi(G))) \le Z_{\mathfrak{F}}(G/\Phi(G))$. Hence, $Z_{\mathfrak{F}\Phi}(G/\Phi(G)) \le Z_{\mathfrak{F}}(G/\Phi(G))$ by Theorem 2.8(ii), which implies that

$$Z_{\mathfrak{F}\Phi}(G)/\Phi(G) = Z_{\mathfrak{F}}(G/\Phi(G)).$$

It follows that $Z_{\mathfrak{F}\Phi}(G)/\Phi(G)$ is a π-group. Hence, $Z_{\mathfrak{F}\Phi}(G)/O_{\pi'}(\Phi(G))$ is a π-group. By the Schur–Zassenhaus theorem, the subgroup $O_{\pi'}(\Phi(G))$ has a complement A in $Z_{\mathfrak{F}\Phi}(G)$ and any two complements of $O_{\pi'}(\Phi(G))$ in $Z_{\mathfrak{F}\Phi}(G)$ are conjugate in $Z_{\mathfrak{F}\Phi}(G)$. Then by the Frattini argument, $G = Z_{\mathfrak{F}\Phi}(G)N_G(A) = O_{\pi'}(\Phi(G))N_G(A) = N_G(A)$. Therefore, $Z_{\mathfrak{F}\Phi}(G) = A \times O_{\pi'}(\Phi(G))$ and $A = O_{\pi}(Z_{\mathfrak{F}\Phi}(G))$. Moreover, in view of the G-isomorphism $A\Phi(G)/\Phi(G) \simeq A/A \cap \Phi(G)$ from the above we deduce that $A/A \cap \Phi(G) \le Z_{\mathfrak{F}}(G/A \cap \Phi(G))$. The theorem is proved.

Lemma 2.26 *Let $\mathfrak{F}$ be a saturated (solubly saturated) formation and F the canonical local (the canonical composition, respectively) satellite of $\mathfrak{F}$. Let E be a normal p-subgroup of G. Then $E \le Z_{\mathfrak{F}}(G)$ if and only if $G/C_G(E) \in F(p)$.*

Proof First suppose that $\mathfrak{F} = CF(F)$ is solubly saturated. Let $1 = E_0 < E_1 < \ldots < E_t = E$ be a chief series of G below E. Let $C_i = C_G(E_i/E_{i-1})$ and $C = C_1 \cap \ldots \cap C_t$. Then $C_G(E) \le C$ and so $C/C_G(E)$ is a p-group by Lemma 2.13. If $E \le Z_{\mathfrak{F}}(G)$, then by Proposition 1.15, $G/C_i \in F(p)$ for all $i = 1, \ldots, t$, so $G/C \in F(p)$. This induces that $G/C_G(E) \in F(p) = \mathfrak{G}_p F(p)$. On the other hand, if $G/C_G(E) \in F(p)$, then $G/C_i \in F(p)$ for all $i = 1, \ldots, t$, so $E \le Z_{\mathfrak{F}}(G)$ by Proposition 1.15.

Now let $\mathfrak{F} = LF(F)$ be a saturated formation. Then $\mathfrak{F}$ is solubly saturated, and if H is the canonical composition satellite of $\mathfrak{F}$, then $H(p) = F(p)$ for all primes p by Lemma 1.6(3). Hence, the second assertion of the lemma is a corollary of the first one. The lemma is proved.

Recall that $\mathrm{Soc}(G)$ denotes the product of all minimal normal subgroups of G whenever $G \ne 1$, and $\mathrm{Soc}(1) = 1$.

Theorem 2.27 (Guo and Skiba [200]). *Let $\mathfrak{F}$ be a normally hereditary formation containing all nilpotent groups and E a normal subgroup of G. Let $D = E \cap Z_{\mathfrak{F}\Phi}(G)$.*

(i) *If $\mathfrak{F}$ is solubly saturated and $E/E \cap Z_{\mathfrak{F}\Phi^*}(G) \in \mathfrak{F}$, then $E \in \mathfrak{F}$. Hence, if $E \le Z_{\mathfrak{F}\Phi^*}(G)$, then $E \in \mathfrak{F}$; in particular, if $G = Z_{\mathfrak{F}\Phi^*}(G)$, then $G \in \mathfrak{F}$.*

(ii) *If $\mathfrak{F}$ is saturated and $E/D \in \mathfrak{F}$, then $E \in \mathfrak{F}$. Hence, if $E \le Z_{\mathfrak{F}\Phi}(G)$, then $E \in \mathfrak{F}$; in particular, if $G = Z_{\mathfrak{F}\Phi}(G)$, then $G \in \mathfrak{F}$.*

(iii) *If $\mathfrak{F}$ is saturated and $\mathrm{Soc}(E/D) \le Z_{\mathfrak{F}\Phi}(E/D)$, then $E \in \mathfrak{F}$.*

Proof

(i) Suppose that this assertion is false and consider a counterexample (G, E) for which $|G| + |E|$ is minimal. Let $K = E \cap Z_{\mathfrak{F}\Phi*}(G)$. Then $K \neq 1$, in particular $Z_{\mathfrak{F}\Phi*}(G) \neq 1$. By Theorem 2.6(c), $Z_{\mathfrak{F}\Phi*}(G)$ is $\mathfrak{F}\Phi*$-hypercentral in G. Hence, there is a minimal normal subgroup L of G such that either $L \leq \Phi*(G)$ or L is $\mathfrak{F}$-central in G.

We shall show that for any minimal normal subgroup N of G we have $N \leq E$ and $E/N \in \mathfrak{F}$. Indeed, first note that $KN/N \leq EN/N$ and

$$(EN/N)/(KN/N) \simeq EN/KN \simeq E/K(E\cap N) \simeq (E/K)/(K(E\cap N)/K) \in \mathfrak{F}.$$

On the other hand, by Lemma 2.6(d), $KN/N \leq Z_{\mathfrak{F}\Phi*}(G/N)$. Therefore,

$$(EN/N)/((EN/N) \cap Z_{\mathfrak{F}\Phi*}(G/N)) \in \mathfrak{F}.$$

The choice of (G, E) implies that $EN/N \in \mathfrak{F}$. If $N \nleq E$, then from the isomorphism $E \simeq EN/N$ we deduce $E \in \mathfrak{F}$, which contradicts the choice of (G, E). Hence, $N \leq E$. It follows that $N = L$ is the unique minimal normal subgroup of G and $L \nleq \Phi*(G)$, otherwise $E \in \mathfrak{F}$ by Lemma 2.19. Thus, L is $\mathfrak{F}$-central in G.

Let $C = C_G(L)$. Then $L \rtimes (G/C) \in \mathfrak{F}$. Suppose that L is non-abelian. Then $C = 1$. Since $\mathfrak{F}$ is normally hereditary and $G \simeq G/C = G/1 \in \mathfrak{F}$, $E \in \mathfrak{F}$. This contradiction shows that L is a p-group for some prime p. In view of Theorem 1.6(2) and Proposition 1.7, $\mathfrak{F} = CLF(F)$, where F is the canonical composition satellite of $\mathfrak{F}$ and $F(p)$ is a normally hereditary formation. In view of Proposition 1.15, $G/C \in F(p)$, hence $E/C \cap E \simeq CE/C \in F(p)$. Thus, $L \leq Z_{\mathfrak{F}}(E)$ by Lemma 2.26. Since $E/L \in \mathfrak{F}$, it follows that $E \in \mathfrak{F}$. This contradiction completes the proof of (i).

(ii) See the proof of (i).

(iii) Let L/D be a minimal normal subgroup of E/D. If $L/D \leq \Phi(E/D)$, then $L/D \leq \Phi(G/D)$ and so $L \leq E \cap Z_{\mathfrak{F}\Phi}(G) = D$, a contradiction. Hence, $L/D \nleq \Phi(E/D)$, which implies that $\Phi(E/D) = 1$. But $\mathrm{Soc}(E/D) \leq Z_{\mathfrak{F}\Phi}(E/D)$, so $\mathrm{Soc}(E/D) \leq Z_{\mathfrak{F}}(E/D)$. Since $\Phi(E/D) = 1$, we also see by Proposition 2.24 that $F*(E/D)$ is the direct product of some minimal normal subgroups of E/D, so $F*(E/D) \leq Z_{\mathfrak{F}}(E/D)$. Therefore, $E/D = Z_{\mathfrak{F}}(E/D)$ by Theorem 2.8(ii) and hence $E/D \in \mathfrak{F}$. Now, arguing as in the proof of (i), one can show that $E \in \mathfrak{F}$.

Corollary 2.28 *Let $\mathfrak{F}$ be a normally hereditary saturated formation containing all nilpotent groups and E a normal subgroup of G. If $E/E \cap \Phi(G) \in \mathfrak{F}$, then $E \in \mathfrak{F}$.*

Another proof of Corollary 2.28 can be found in [113] or in [359].

Corollary 2.29 (Shemetkov [364]). *Let $\mathfrak{F}$ be a normally hereditary solubly saturated formation containing all nilpotent groups and E a normal subgroup of G. If $E/E \cap \Phi*(G) \in \mathfrak{F}$, then $E \in \mathfrak{F}$.*

Corollary 2.30 *If $\mathfrak{F}$ is a normally hereditary solubly saturated formation, then $Z_{\mathfrak{F}}(G) \in \mathfrak{F}$.*

From Corollary 2.30 we get the following well-known result.

Corollary 2.31 (Huppert [89, IV, 6.15]). *If $\mathfrak{F}$ is a normally hereditary saturated formation, then $Z_{\mathfrak{F}}(G) \in \mathfrak{F}$.*

It is well known that the class $\mathfrak{N}^*$ of all quasinilpotent groups is a solubly saturated formation (L.A. Shemetkov). Hence, we have

Corollary 2.32 *Let E a normal subgroup of G. If $E/E \cap Z_{\mathfrak{N}^* \Phi^*}(G)$ is quasinilpotent, then E is also quasinilpotent.*

Recall that for a class $\mathfrak{X}$ of groups, a maximal subgroup M is said to be $\mathfrak{X}$-abnormal ($\mathfrak{X}$-normal, respectively) if $G/M_G \notin \mathfrak{X}$ ($G/M_G \in \mathfrak{X}$, respectively) (see [135, Definition 2.2.3]). Following [359], we use $\Delta^{\mathfrak{F}}(G)$ to denote the intersection of all $\mathfrak{F}$-abnormal maximal subgroups of G.

Theorem 2.33 (Guo and Skiba [200]). *For any formation $\mathfrak{F}$ we have*

$$\Delta^{\mathfrak{F}}(G) = Z_{\mathfrak{F}\Phi}(G).$$

Proof First note that if M is a maximal subgroup of G and H/K is a chief factor of G such that $K \leq M$ and $H \nleq M$, then by Proposition 1.15, M is $\mathcal{F}$-abnormal in G if and only if H/K is $\mathcal{F}$-eccentric in G.

Assume that for some maximal $\mathfrak{F}$-abnormal subgroup M of G we have $Z_{\mathfrak{F}\Phi}(G) \nleq M$. Since by Theorem 2.25,

$$Z_{\mathcal{F}\Phi}(G)/\Phi(G) = Z_{\mathfrak{F}}(G/\Phi(G)),$$

there is a chief factor H/K of G such that $\Phi(G) \leq K \leq M$, $H \nleq M$, and $H/\Phi(G) \leq Z_{\mathfrak{F}}(G/\Phi(G))$. But then H/K is $\mathcal{F}$-central in G, which contradicts the $\mathfrak{F}$-abnormality of M. This contradiction shows that $Z_{\mathfrak{F}\Phi}(G) \leq \Delta^{\mathfrak{F}}(G)$. If the inverse inclusion is not true, then there is a non-Frattini chief factor H/K of G such that $H \leq \Delta^{\mathfrak{F}}(G)$ and H/K is not $\mathcal{F}$-central. Let M be a maximal subgroup of G such that $K \leq M$ and $H \nleq M$. Then M is $\mathcal{F}$-abnormal in G, so $H \leq \Delta^{\mathfrak{F}}(G) \leq M$, a contradiction. Thus, $\Delta^{\mathfrak{F}}(G) \leq Z_{\mathfrak{F}\Phi}(G)$. The theorem is proved.

From Theorems 2.25, 2.27(ii), and 2.33 we get

Corollary 2.34 (Ballester–Balinches [24]). *For any hereditary saturated formation $\mathfrak{F}$, the subgroup $(\Delta^{\mathfrak{F}}(G))^{\mathfrak{F}}$ is nilpotent.*

Corollary 2.35 (Feng and Chang [104]). *Let $\mathfrak{F}$ be a saturated formation, $\Delta = \Delta^{\mathfrak{F}}(G)$ and $\pi = \pi(\mathfrak{F})$. Then*

$$\Delta^{\mathfrak{F}}(G/O_\pi(\Delta)) = \Phi(G/O_\pi(\Delta)) \leq O_{\pi'}(G/O_\pi(\Delta)).$$

Note that if E is a quasinilpotent normal subgroup of G and $E \cap \Phi(G) = 1$, then E is the direct product $E = E_1 \times \ldots \times E_t$ of some minimal normal subgroups

$E_1, \ldots, E_t$ of G by Proposition 2.24. Hence, $C_G(E) = C_G(E_1) \cap \ldots \cap C_G(E_t)$. Therefore, from Theorems 2.25 and 2.33 we get the following well-known result.

Corollary 2.36 (Gaschütz [112]). $\Delta^{\mathfrak{N}}(G)/\Phi(G) = Z(G/\Phi(G))$.

Corollary 2.36 was a motivation for the following well-known result.

Corollary 2.37 (Shemetkov [357], Selkin [344]) *For any solubly saturated formation $\mathfrak{F}$, in particular, for any saturated formation $\mathfrak{F}$, we have*

$$\Delta^{\mathfrak{F}}(G)/\Phi(G) = Z_{\mathfrak{F}}(G/\Phi(G)).$$

From Theorems 2.27 (ii) and 2.33 we also get

Corollary 2.38 (Shemetkov [359, I, 8.12], Selkin [344]). *Let $\mathfrak{F}$ be a normally hereditary saturated formation containing all nilpotent groups and E a normal subgroup of G. If $E/E \cap \Delta^{\mathfrak{F}}(G) \in \mathfrak{F}$, then $E \in \mathfrak{F}$.*

1.3   The Theory of Quasi-$\mathfrak{F}$-Groups

Recall that a group G is said to be quasinilpotent if for every its chief factor H/K and every $x \in G$, x induces an inner automorphism on H/K (see [250, p. 124]). Since for every central chief factor H/K of a group G an element of G induces trivial automorphism on H/K, one can say that a group G is quasinilpotent if for every its *noncentral* chief factor H/K and every $x \in G$, x induces an inner automorphism on H/K. This observation allows us to consider the following analogue of quasinilpotent groups.

Definition 3.1 [188]. Let $\mathfrak{F}$ be a class of groups and G a group. We say that G is a quasi-$\mathfrak{F}$-group if for every $\mathfrak{F}$-eccentric chief factor H/K of G, every automorphism of H/K induced by an element of G is inner.

By analogy with p-quasinilpotent groups [192] we introduce also the following general version of Definition 3.1.

Definition 3.2 Let $\mathfrak{F}$ be a class of groups, G a group, and $\emptyset \neq \pi \subseteq \pi(\mathfrak{F})$. We say that G is a π-quasi-$\mathfrak{F}$-group if for every $\mathfrak{F}$-eccentric chief factor H/K of G of order divisible by at least one prime in π, every automorphism of H/K induced by an element of G is inner.

In this section, we discuss the theory of the quasi-$\mathfrak{F}$-groups and π-quasi-$\mathfrak{F}$-groups.

Base Theorem of the Theory of Quasi-$\mathfrak{F}$-Groups We use $\mathfrak{F}^*$ and $\mathfrak{F}^*_{\pi}$ to denote the set of all quasi-$\mathfrak{F}$-groups and π-quasi-$\mathfrak{F}$-groups, respectively. It is clear that every $\mathfrak{F}$-group is both a $\mathfrak{F}^*$-group and a π-quasi-$\mathfrak{F}$-group as well.

Lemma 3.3 *For any class $\mathfrak{F}$ of groups, the classes $\mathfrak{F}^*$ and $\mathfrak{F}^*_{\pi}$ are nonempty formations.*

Proof It is obvious by use the proof of [250, Lemma X.13.3].

Lemma 3.4 *Let $L \leq K \leq H \leq D \leq N \leq G$ where L, D, N are normal subgroups of a group G and K, H are normal subgroups of N. Suppose that D/L is a chief factor of G and H/K is a chief factor of N. If $\mathfrak{F}$ is a normally hereditary saturated formation and $(D/L) \rtimes Aut_G(D/L) \in \mathfrak{F}$, then $(H/K) \rtimes Aut_N(H/K) \in \mathfrak{F}$.*

Proof Let F be the canonical local satellite of $\mathfrak{F}$. Then by Proposition 1.7 any value $F(p)$ of F is a normally hereditary formation contained in $\mathfrak{F}$. Since $(D/L) \rtimes Aut_G(D/L) \in \mathfrak{F}$, $Aut_G(D/L) = G/C_G(D/L) \in F(p)$ for all primes p dividing $|D/L|$ by Proposition 1.15. Since the formation $\mathfrak{F}$ is normally hereditary by hypothesis, $N/C_N(H/K) \in F(p)$ for all primes p dividing $|H/K|$. Then by Proposition 1.15 again, we obtain that $(H/K) \rtimes (N/C_N(H/K)) \in \mathfrak{F}$.

Lemma 3.5

1) *Let $L \leq K \leq H \leq D \leq N \leq G$ where L, D, N are normal subgroups of a group G and K, H are normal subgroups of N. Suppose that D/L is a chief factor of G and H/K is a chief factor of N. If $x \in N$ and x induces an inner automorphism on D/L, then x induces an inner automorphism on H/K.*

2) *Let H/K be a chief factor of G. If x induces an inner automorphism on H/K for every $x \in G$, then $G/K = (H/K) \rtimes C_{G/K}(H/K)$.*

Proof

1) see the proof of [250, Lemma X.13.1].

2) Let $gK \in G/K$. Since the automorphism on H/K induced by gK is also induced by some element hK of H/K, $gh^{-1}K \in C_{G/K}(H/K)$. Hence, $G/K = (H/K) \rtimes C_{G/K}(H/K)$.

Proposition 3.6 *Let $\mathfrak{F} = LF(F)$ be a saturated formation, where F is the canonical local satellite of $\mathfrak{F}$, and let $\varnothing \neq \pi \subseteq \pi(\mathfrak{F})$. Then*

(1) $\mathfrak{F}_\pi^* = CLF(f_\pi^*)$, where $f_\pi^*(p) = F(p)$ for all $p \in \pi$ and $f_\pi^*(0) = \mathfrak{F}_\pi^* = f_\pi^*(p)$ for all primes $p \notin \pi$.

(2) $\mathfrak{F}^* = CLF(f^*)$, where $f^*(0) = \mathfrak{F}^*$ and $f^*(q) = F(q)$ for all primes q.

Proof

(1) Let f_π^* be a formation function such that $f_\pi^*(p) = F(p)$ for all $p \in \pi$ and $f_\pi^*(0) = \mathfrak{F}_\pi^* = f_\pi^*(p)$ for all primes $p \notin \pi$. Put $\mathfrak{M}_\pi = CLF(f_\pi^*)$. Then we only need to prove that $\mathfrak{F}_\pi^* = \mathfrak{M}_\pi$. Suppose that $\mathfrak{F}_\pi^* \nsubseteq \mathfrak{M}_\pi$ and let G be a group of minimal order in $\mathfrak{F}_\pi^* \setminus \mathfrak{M}_\pi$. Then $R = G^{\mathfrak{M}_\pi}$ is the only minimal normal subgroup of G.

Suppose that R is either a π'-group or a non-abelian group. Then for every $q \in Com(G) \cap \pi$ we have

$$G/C^q(G) \simeq (G/R)/(C^q(G)/R) = (G/R)/C^q(G/R) \in f_\pi^*(q).$$

On the other hand, since $G \in \mathfrak{F}_\pi^*$, $G/G_\mathfrak{S} \in \mathfrak{F}_\pi^* = f_\pi^*(0)$. Thus, $G \in \mathfrak{M}_\pi$. This contradiction shows that R is an abelian p-group for some $p \in \pi$,

Let $C = C_G(R)$. If $R/1$ is $\mathfrak{F}$-eccentric, then $G = RC = C$ by Lemma 3.5 since $G \in \mathfrak{F}_\pi^*$. This means that $R/1$ is $\mathfrak{F}$-central since $p \in \pi$. This contradiction shows that R is $\mathfrak{F}$-central. If $C = R$, then $R = C^p(G)$ and so

$$G/C = G/C^p(G) \in F(p) = f_\pi^*(p).$$

It follows that $G \in \mathfrak{M}_\pi$, a contradiction. Hence, $R \neq C$. Since R is $\mathfrak{F}$-central, $T = R \rtimes (G/C) \in \mathfrak{F} \subseteq \mathfrak{F}_\pi^*$. But since $|T| < |G|$, $T \in \mathfrak{M}_\pi$ by the choice of G. Hence, $G/C \in F(p) = f_\pi^*(p)$. Since $R = G^{\mathfrak{M}_\pi}$, $(G/R)/C^p(G/R) \in f_\pi^*(p)$. But obviously $C \cap C^p(G/R) = C^p(G)$. Hence, $G/C^p(G) \in f_\pi^*(p)$. This implies that $G \in \mathfrak{M}_\pi$. This contradiction shows that $\mathfrak{F}_\pi^* \subseteq \mathfrak{M}_\pi$.

Next suppose that $\mathfrak{M}_\pi \not\subseteq \mathfrak{F}_\pi^*$ and let G be a group of minimal order in $\mathfrak{M}_\pi \setminus \mathfrak{F}_\pi^*$. Then $R = G^{\mathfrak{F}_\pi^*}$ be the only minimal normal subgroup of G. If $R/1$ is $\mathfrak{F}$-central or is a π'-group, then every $\mathfrak{F}$-eccentric chief factor of G of order divisible by at least one prime in π is above R. Since $G/R \in \mathfrak{F}_\pi^*$, every element of G/R induces an inner automorphism on each $\mathfrak{F}$-eccentric chief factors of order divisible by at least one prime in π. Hence, $G \in \mathfrak{F}_\pi^*$. This contradiction shows that the factor $R/1$ is $\mathfrak{F}$-eccentric of order divisible by some $p \in \pi$. Suppose that R is non-abelian. Then $G_\mathfrak{S} = 1$. Since $G \in \mathfrak{M}_\pi$, $G \simeq G/G_\mathfrak{S} \in f_\pi^*(0) = \mathfrak{F}_\pi^*$. This contradiction shows that R is an abelian p-group. Let $C = C_G(R)$. By Proposition 1.13, $T = R \rtimes (G/C) \in \mathfrak{M}_\pi$. Suppose that $R \neq C$. Then $|T| = |R \rtimes (G/C)| < |G|$. Hence, $T \in \mathfrak{F}_\pi^*$ by the choice of G. Obviously $C_T(R) = R$ and so $R \rtimes (T/C_T(R)) \simeq T = R \rtimes (G/C)$. Then since $R/1$ is $\mathfrak{F}$-eccentric in G, it is also $\mathfrak{F}$-eccentric in T. Thus, $T = CR = C$ by Lemma 3.5. It follows that $R/1$ is $\mathfrak{F}$-central in T since $p \in \pi \leq \pi(\mathfrak{F})$. This contradiction shows that $R = C$. Consequently, $R = C^p(G)$. Since $G \in \mathfrak{M}_\pi$, $G/R \in F(p) = f_\pi^*(p)$. Hence, $G \in \mathfrak{N}_p F(p) \subseteq \mathfrak{F} \subseteq \mathfrak{F}_p^*$. This contradiction shows that $\mathfrak{F}_\pi^* = \mathfrak{M}_\pi$.

(2) This is a special case of (1) when $\pi = \mathbb{P}$.

It is known that the class $\mathfrak{N}^*$ of all quasinilpotent groups is a solubly saturated formation and this formation is normally hereditary (see [250, Theorem X.13.3]).

The following theorem shows that the classes $\mathfrak{F}^*$ and $\mathfrak{F}_\pi^*$ have the same properties.

Theorem 3.7 *Suppose that $\mathfrak{F}$ is a normally hereditary saturated formation, and let $\varnothing \neq \pi \subseteq \pi(\mathfrak{F})$. Then the class $\mathfrak{F}_\pi^*$ is normally hereditary solubly saturated formations.*

Proof By Theorem 1.6 and Proposition 3.6, $\mathfrak{F}_\pi^*$ is a solubly saturated formations.

We now prove that the formation $\mathfrak{F}_\pi^*$ is normally hereditary. Let N be a normal subgroup of the π-quasi-$\mathfrak{F}$-group G. If $L \leq K \leq H \leq D \leq N$, where D/L is a chief factor of G and H/K is a $\mathfrak{F}$-eccentric chief factor of N of order divisible by at least one prime $p \in \pi$, then by Lemma 3.4, D/L is an $\mathfrak{F}$-eccentric chief factor of G of order divisible by p. Hence, by hypothesis, any element $n \in N$ induces an inner automorphism on D/L. It follows from Lemma 3.5 that n induces an inner automorphism on H/K. Therefore, N is π-quasi-$\mathfrak{F}$-group. This completes the proof.

Corollary 3.8 *Suppose that $\mathfrak{F}$ is a normally hereditary saturated formation containing all nilpotent groups. Then both classes $\mathfrak{F}^*$ and $\mathfrak{F}_p^*$ are normally hereditary solubly saturated formations.*

Therefore, from Theorems 2.27 and 3.7 we get

Corollary 3.9 *Let $\mathfrak{F}$ be a normally hereditary saturated formation containing all nilpotent groups and E a normal subgroup of G. If $E/E \cap Z_{\mathfrak{F}^*\Phi^*}(G) \in \mathfrak{F}^*$, then $E \in \mathfrak{F}^*$.*

Corollary 3.10 *Let E a normal subgroup of G. If $E/E \cap Z_{\mathfrak{U}^*\Phi^*}(G)$ is quasisupersoluble, then E is quasisupersoluble.*

We use $Z^*(G)$ to denote the *quasicenter* of G, that is, the largest normal subgroup of G of the form $A \times A_1 \times \ldots \times A_t$, where $A \leq Z(G)$ and A_i is a normal simple non-abelian subgroup of G such that $G = A_i C_G(A_i)$ $(i = 1, \ldots t)$.

Corollary 2.36 is also a motivation for the following result.

Corollary 3.11 (Guo, Skiba, [200]). $\Delta^{\mathfrak{N}^*}(G)/\Phi(G) = Z^*(G/\Phi(G))$.

Proof Let $\Delta = \Delta^{\mathfrak{N}^*}(G)$. In view of Theorems 2.25 and 2.33, $\Delta/\Phi(G) = Z_{\mathfrak{N}^*}(G/\Phi(G))$. Hence, $\Delta/\Phi(G)$ is a normal quasinilpotent subgroup of $G/\Phi(G)$ by Theorem 2.27(i). Since $\Phi(G/\Phi(G)) = 1$, it follows (see the remark before Corollary 2.36) that $\Delta/\Phi(G) = (N_1/\Phi(G)) \times \ldots \times (N_t/\Phi(G))$, where $N_i/\Phi(G)$ is a minimal normal subgroup of $G/\Phi(G)$. Now, the corollary follows from the well-known fact (see Theorem 13.6 in [250, Chap. X]) that a chief factor H/K of G is $\mathfrak{N}^*$-central in G if and only if H/K is simple and $G/K = (H/K)(C_G(H/K))$.

Structure Theorems The structure of quasinilpotent groups is well known: The group G is quasinilpotent if and only if $G/Z_\infty(G)$ is semisimple [250, X,13.6].

Our first application of the base Theorem 3.7 is the following result which describe the structure of quasi-$\mathfrak{F}$-groups and π-quasi-$\mathfrak{F}$-groups.

Theorem 3.12 *Let $\mathfrak{F}$ be a normally hereditary saturated formation, and let $\varnothing \neq \pi \subseteq \pi(\mathfrak{F})$. Then:*

(1) G is π-quasi-$\mathfrak{F}$-group if and only if $G/Z_{\pi\mathfrak{F}}(G)$ is semisimple and the order of each composition factor of $G/Z_{\pi\mathfrak{F}}(G)$ is divisible by at least one prime $p \in \pi$.

(2) G is quasi-$\mathfrak{F}$-group if and only if $G/Z_{\mathfrak{F}}(G)$ is semisimple.

Proof

(1) The proof is analogous to the proof of [250, X, Theorem 13.6]. In fact, we only need to prove that if G is a π-quasi-$\mathfrak{F}$-group, then $G/Z_{\pi\mathfrak{F}}(G)$ is semisimple and the order of each its composition factor is divisible by at least one prime in π. Let $Z = Z_{\pi\mathfrak{F}}(G)$. If $Z \neq 1$, then the inductive hypothesis may be applied to G/Z by Lemma 3.3 and so the assertion holds. Now assume that $Z = 1$. Let R be a minimal normal subgroup of G and $C = C_G(R)$. Obviously, R is $\mathfrak{F}$-eccentric. Hence, there exists a prime $p \in \pi$ which divides $|R|$. Since $Z(G) = 1$, $C \neq G$. By Theorem 3.7, C is a π-quasi-$\mathfrak{F}$-group. Hence, by induction, $C/Z_{\mathfrak{F}_\pi}(C)$ is

semisimple and the order of each its composition factor is divisible by p. Since $Z = 1$, R is $\mathfrak{F}$-eccentric. Therefore, $G = RC$ by Lemma 3.5 and thereby $R \cap C \leq Z(G) = 1$. It follows that $G = R \times C$. Hence, R is non-abelian and $Z_{\mathfrak{F}\pi}(C) = 1$. This induces that G is semisimple and the order of each its composition factor is divisible by at least one prime $p \in \pi.$.

(2) This follows from (1).

Corollary 3.13 *Let $\mathfrak{F}$ be a normally hereditary saturated formation containing all nilpotent groups and p a prime. Then:*

(1) G is p-quasi-$\mathfrak{F}$-group if and only if $G/Z_{\mathfrak{F}_p}(G)$ is semisimple and the order of each composition factor of $G/Z_{\mathfrak{F}_p}(G)$ is divisible by p.
(2) G is quasi-$\mathfrak{F}$-group if and only if $G/Z_{\mathfrak{F}}(G)$ is semisimple.

Now using Theorem 3.7 we prove the following result which shows that in the definition of quasi-$\mathfrak{F}$-groups we may only consider the chief factors between $\Phi(F(G))$ and $F^*(G)$.

Theorem 3.14 (Guo, Skiba [188]). *Let $\mathfrak{F}$ be a saturated formation and G a group. Then G is quasi-$\mathfrak{F}$-group if and only if for every $\mathfrak{F}$-eccentric G-chief factor H/K between $\Phi(F(G))$ and $F^*(G)$, every automorphism of H/K induced by an element of G is inner.*

Proof Let $F = F(G)$. We only need to show that if for every $\mathfrak{F}$-eccentric G-chief factor H/K between $\Phi(F)$ and $F^*(G)$ every automorphism of H/K induced by an element of G is inner, then $G \in \mathfrak{F}^*$. Suppose that this is false and let G be a counterexample of minimal order. By Lemma 2.19 and Theorem 3.7, we have $F^*(G/\Phi(F)) = F^*(G)/\Phi(F)$. Hence, the hypothesis holds for $G/\Phi(F)$. The minimal choice of G implies that $G/\Phi(F) \in \mathfrak{F}^*$. Then by Theorem 3.7 again, $G \in \mathfrak{F}^*$. The contradiction shows that $\Phi(F) = 1$. Therefore, for every $\mathfrak{F}$-eccentric G-chief factor H/K of $F^*(G)$, every automorphism of H/K induced by an element of G is inner. Then since $\mathfrak{N} \subseteq \mathfrak{F}$, we have that every chief factor of F is $\mathfrak{F}$-central.

Let $\mathfrak{F} = LF(F)$, where F is the canonical local satellite of $\mathfrak{F}$. Then by Proposition 3.6, $\mathfrak{F}^* = CLF(f^*)$, where $f^*(0) = \mathfrak{F}^*$ and $f^*(q) = F(q)$ for all primes q. Hence, $G/C_G(H/K) \in f^*(p)$ for every G-chief factor H/K of F with $p \in \pi(H/K)$. On the other hand, if H/K is a chief factor of G between F and $F^*(G)$, then $C_{G/K}(H/K)(H/K) = (C_G(H/K)/K)(H/K) = G/K$ by hypothesis and Lemma 3.5. Hence, $G/C_G(H/K) \simeq H/K$ is a simple non-abelian group. Consequently $G/C_G(H/K) \in \mathfrak{F}^* = f^*(0)$. Thus, $G/C_G(F^*(G)) \in \mathfrak{F}^*$ by the well known Schmid–Shemetkov's Theorem on $\mathfrak{F}^*$-stable automorphism groups (see [135, Theorem 3,2,6]). But by Proposition 2.9(5), $C_G(F^*(G)) \leq F$. Hence, $G \in \mathfrak{F}^*$. This contradiction completes the proof.

Corollary 3.15 *If for every non-central G-chief factor H/K between $\Phi(F)$ and $F^*(G)$, every automorphism of H/K induced by an element of G is inner, then G is quasinilpotent.*

Products of Normal Quasi-$\mathfrak{F}$-Subgroups   Our next goal is to study conditions under which the product of two normal quasi-$\mathfrak{F}$-subgroups is a quasi-$\mathfrak{F}$-subgroup itself.

Theorem 3.16   (Guo and Skiba [188]). *Let $\mathfrak{F} = LF(F) \subseteq \mathfrak{S}$ be a saturated Fitting formation containing all nilpotent groups, where F is the canonical local satellite of $\mathfrak{F}$. Then $\mathfrak{F}^*$ is a solubly saturated Fitting formation and $\mathfrak{F}^* = CLF(f^*)$, where f^* is a composition satellite of $\mathfrak{F}^*$ such that $f^*(p) = F(p)$ for all primes p and $f^*(0)$ is the class of all semisimple groups.*

Proof   Let $\mathfrak{M} = CLF(f^*)$. We first prove that $\mathfrak{F}^* = \mathfrak{M}$. Suppose that $\mathfrak{F}^* \not\subseteq \mathfrak{M}$ and let G be a group of minimal order in $\mathfrak{F}^* \setminus \mathfrak{M}$. Then $R = G^{\mathfrak{M}}$ is the only minimal normal subgroup of G. Since every $\mathfrak{F}$-group is soluble and $G \in \mathfrak{F}^*$, $G/Z_{\mathfrak{F}}(G)$ is semisimple by Theorem 3.12 and $Z_{\mathfrak{F}}(G) \subseteq G_{\mathfrak{S}}$. Suppose that R is non-abelian. Then $Z_{\mathfrak{F}}(G) = 1$ and so $G = R$ is a simple non-abelian group. But then $G \in \mathfrak{M}$, a contradiction. Therefore, R is abelian and by analogy to the proof of Proposition 3.6, we see that $G \in \mathfrak{M}$, a contradiction. This contradiction shows that $\mathfrak{F}^* \subseteq \mathfrak{M}$. Now suppose that $\mathfrak{M} \not\subseteq \mathfrak{F}^*$ and let G be a group of minimal order in $\mathfrak{M} \setminus \mathfrak{F}^*$. Then $R = G^{\mathfrak{F}^*}$ is the only minimal normal subgroup of G.

Analogy to the proof of Proposition 3.6, we may assume that R is non-abelian. Then $G_{\mathfrak{S}} = 1$. But since $G \in \mathfrak{M}$, $G \simeq G/G_{\mathfrak{S}} \in f^*(0)$. Hence, G is a semisimple group. Consequently $G \in \mathfrak{F}^*$ by Corollary 3.13. This contradiction shows that $\mathfrak{M} = \mathfrak{F}^*$.

Finally, we prove that $\mathfrak{F}^*$ is a Fitting class. Since $\mathfrak{F}$ is a Fitting class, $\mathfrak{F}^*$ is a normally hereditary solubly saturated formation by Theorem 3.7. Thus, it remains to prove that if $G = AB$ where A, B are normal subgroup of G and A, B are quasi-$\mathfrak{F}$-group, then G is also a quasi-$\mathfrak{F}$-group.

Let $C^p = C^p(G)$. By Propositions 2.16 and 2.17, $C^p(A) = C^p \cap A$ and $C^p(B) = C^p \cap B$ for any prime p. Since $A \in \mathfrak{F}^*$, $A/C^p(A) \in f^*(p) = F(p)$. Analogously $B/C^p(B) \in F(p)$. Hence, $C^p A/C^p, C^p B/C^p \in F(p)$. By Proposition 1.7(b), $F(p)$ is a Fitting formation. Hence, $G/C^p = (C^p A/C^p)(C^p B/C^p) \in F(p) = f^*(p)$. Since the class of all semisimple groups is evidently also a Fitting formation, we can analogously obtain that $G/G_{\mathfrak{S}} \in f^*(0)$. Thus, $G \in \mathfrak{F}^*$. This completes the proof.

As immediate consequences of Theorem 3.16, we have the following

Corollary 3.17   *The product of two normal quasisoluble subgroups of a group is a quasisoluble subgroup.*

Corollary 3.18   [250, Theorem X.13.11]. *The product of two normal quasinilpotent subgroups of a group is a quasinilpotent subgroup.*

Corollary 3.19   *The product of two normal quasimetanilpotent subgroups of a group is a quasimetanilpotent subgroup.*

Corollary 3.20   *Let H be a subnormal subgroup of a group G. Suppose that H is quasisoluble. Then $H \leq G_{\mathfrak{S}^*} \in \mathfrak{S}^*$.*

Proof   Let $H = G_0 \leq G_1 \leq \ldots \leq G_t = G$, where the subgroup G_{i-1} is normal in G_i and $G_{i-1} \neq G_i$, for all $i = 1, 2, \ldots, t$. We shall prove by the induction on t

that $H \leq G_{\mathfrak{S}^*}$. If $t = 1$, it is clear. Now assume that $t > 1$. Then by induction we have $H \leq (G_{t-1})_{\mathfrak{S}^*}$. On the other hand, since $(H_{t-1})_{\mathfrak{S}^*}$ is a characteristic subgroup of H_{t-1}, $(H_{t-1})_{\mathfrak{S}^*}$ is normal in G and so $H \leq (H_{t-1})_{\mathfrak{S}^*} \leq G_{\mathfrak{S}^*}$. Finally, since $\mathfrak{S}^*$ is a Fitting class by Theorem 3.7 and Corollary 3.17, we have $G_{\mathfrak{S}^*} \in \mathfrak{S}^*$.

Similarly, the following corollaries may be proved.

Corollary 3.21 *Let H be a subnormal subgroup of a group G. Suppose that H is quasimetanilpotent. Then $H \leq G_{\mathfrak{N}^{2*}} \in \mathfrak{N}^{2*}$.*

Corollary 3.22 [250, Theorem X.13.10]. *Let H be a subnormal subgroup of a group G. Suppose that H is quasinilpotent. Then $H \leq G_{\mathfrak{N}^*} \in \mathfrak{N}^*$.*

A Criterion for the Product of Two Quasisupersoluble Groups to be Quasisupersoluble It is well known that if $G = AB$ is the product of two normal supersoluble subgroups A, B of G, G is not necessarily supersoluble. Nevertheless, such product is supersoluble under the additional condition that $|G : A|$ and $|G : B|$ are coprime. By using Theorem 3.12, it is easy to see that $\mathfrak{U}^* \cap \mathfrak{S} = \mathfrak{U}$. This shows that the product of two normal quasisupersoluble is not necessarily quasisupersoluble. In this part, our goal is to give a condition under which the product of two normal quasisupersoluble groups is a quasisupersoluble group.

Lemma 3.23 *Let $\mathfrak{F} = \mathfrak{M}\mathfrak{H}$ be a formation, where $\mathfrak{M}$ is a hereditary formation with $\mathfrak{M}\mathfrak{M} = \mathfrak{M}$ and $\mathfrak{H}$ is some nilpotent formation. Suppose that $G = AB$ where A, B are normal subgroups of G. If $A, B \in \mathfrak{F}$ and $Com(G/A) \cap Com(G/B) = \emptyset$, then $G \in \mathfrak{F}$.*

Proof Suppose that this lemma is false and G be a counterexample of minimal order. It is clear that the hypothesis still holds on any quotient of G. Then, since $\mathfrak{F}$ is a formation, G has a unique minimal normal subgroup $H = G^{\mathfrak{F}}$. If $G_{\mathfrak{M}} \neq 1$, then $H \leq G_{\mathfrak{M}}$ and so $G \in \mathfrak{M}(\mathfrak{M}\mathfrak{H}) = (\mathfrak{M}\mathfrak{M})\mathfrak{H} = \mathfrak{M}\mathfrak{H} = \mathfrak{F}$, which contradicts the choice of G. Thus, $G_{\mathfrak{M}} = 1$ and consequently $A_{\mathfrak{M}} = 1$ since $A_{\mathfrak{M}}$ is a characteristic subgroup of A. This shows that $A \in \mathfrak{H}$. Analogously, we have $B \in \mathfrak{H}$. Hence, G is nilpotent as it is a product of two normal nilpotent groups. Since $Com(G/A) \cap Com(G/B) = \emptyset$, every Sylow subgroup of G is contained at least in one of the subgroups A and B. But since A, B are nilpotent, every Sylow subgroups of G is in $\mathfrak{H}$ and normal in G. This induces that $G \in \mathfrak{H} \subseteq \mathfrak{F}$. The contradiction completes the proof.

Theorem 3.24 (Guo and Skiba [189]). *Let $\mathfrak{F} \subseteq \mathfrak{N}^2$ be a saturated formation containing all nilpotent groups. Let G be a group and $G = AB$ where A and B are normal quasi-$\mathfrak{F}$-subgroups of G. If $Com(G/A) \cap Com(G/B) = \emptyset$, then G is a quasi-$\mathfrak{F}$-group.*

Proof Let F be the canonical local satellite of $\mathfrak{F}$. Then by Proposition 3.6, $\mathfrak{F}^* = CLF(f^*)$ where $f^*(0) = \mathfrak{F}^*$ and $f^*(p) = F(p)$ for all primes p. Note that by Theorem 1.2, $F(p) = \mathfrak{S}_p\mathfrak{F}(p)$ for all primes p. Suppose that this theorem is false and G be a counterexample of minimal order. It is clear that the hypothesis still holds on every quotient of G. So G has a unique minimal normal subgroup $H = G^{\mathfrak{F}^*}$. Obviously $H \leq A \cap B$. Assume that H is a p-group for some prime p. Let $C^p = C^p(G)$, $C_1 = C^p(A)$, and $C_2 = C^p(B)$. By the hypothesis, $A/C_1 \in f^*(p) =$

$F(p) = \mathfrak{G}_p\mathfrak{F}(p)$ and $B/C_2 \in f^*(p) = F(p) = \mathfrak{G}_p\mathfrak{F}(p)$. Hence, $AC_1C_2/C_1C_2 \simeq A/A \cap C_1C_2 = A/C_1(A \cap C_2) \in \mathfrak{G}_p\mathfrak{F}(p)$. Analogously, $BC_1C_2/C_1C_2 \in \mathfrak{N}_p\mathfrak{F}(p)$. Since every group in $\mathfrak{F}$ is metanilpotent, $\mathfrak{F}(p) \subseteq \mathfrak{N}$ for all primes p by Theorem 1.2. Then by Lemma 3.23, we have $G/C_1C_2 \in \mathfrak{G}_p\mathfrak{F}(p) = f^*(p)$. However by Proposition 2.17, we have that $C_1, C_2 \le C^p$. Thus, $G/C^p \in f^*(p) \subseteq \mathfrak{F}^*$. Since $H = G^{\mathfrak{F}^*}$ is an abelian group, $G/C_G(H) \simeq (G/C^p)/(C_G(H)/C^p) \in f^*(p)$. Hence, G is a quasi-$\mathfrak{F}$-group. This contradiction shows that H is non-abelian. Let $Z = Z_{\mathfrak{F}}(A) \neq 1$, then Z char $A \trianglelefteq G$ and so $1 \neq Z \trianglelefteq G$. Since $\mathfrak{F} \subseteq \mathfrak{G}$, $H \subseteq Z$ is soluble. This contradiction shows that $Z_{\mathfrak{F}}(A) = 1$. Analogously $Z_{\mathfrak{F}}(B) = 1$. Hence, A, B are semisimple by Corollary 3.12. Let $A = A_1 \times \cdots \times A_s$ and $B = B_1 \times \cdots \times B_t$, where A_i, B_i are non-abelian simple groups. Since $H \le A \cap B$, there exist A_i and B_j such that $A_i = B_j \trianglelefteq AB = G$. Without lose of generality, we may assume that $H = A_1 = B_1$ is non-abelian simple group. If $H \rtimes (A/C_A(H)) \in \mathfrak{F}$, then $H \in \mathfrak{G}$ since $\mathfrak{F} \subseteq \mathfrak{G}$, a contradiction. This shows that H is $\mathfrak{F}$-eccentric in A. Analogously H is $\mathfrak{F}$-eccentric in B. It follows from Lemma 3.5 that $A = HC_A(H)$ and $B = HC_B(H)$. Then $G = AB = HC_A(H)C_B(H) = HC_G(H) = H \times C_G(H)$. But since H is a unique minimal normal subgroup of G, $G = H$ and so G is quasi-$\mathfrak{F}$-group. This completes the proof.

The following corollaries are immediate from Theorem 3.24.

Corollary 3.25 *Let* $G = AB$ *where* A *and* B *are normal quasisupersoluble subgroups of* G. *If* $Com(G/A) \cap Com(G/B) = \emptyset$, *then* G *is quasisupersoluble.*

Corollary (3.26) *Let* $\mathfrak{F}$ *be a saturated formation containing all nilpotent groups and every group in* $\mathfrak{F}$ *have a nilpotent commutator subgroup. If a group* $G = AB$ *where* A *and* B *are normal quasi-$\mathfrak{F}$-subgroups of* G *and* $Com(G/A) \cap Com(G/B) = \emptyset$, *then* G *is quasi-$\mathfrak{F}$-group.*

The following known results now follows immediately from Theorem 3.24.

Corollary 3.27 (Friesen [108]). *Let* G *be a group and* $G = AB$, *where* A *and* B *are normal subgroups of* G. *If* $|G : A|$ *and* $|G : B|$ *are coprime and* A *and* B *are supersoluble, then* G *is supersoluble.*

Corollary 3.28 [450, Chap. 4, Theorem 3.5]. *Let* $G = AB$ *where* A *and* B *are normal subgroups of* G. *If* $|G : A|$ *and* $|G : B|$ *are coprime and* A' *and* B' *are nilpotent, then* G' *is nilpotent.*

Lemma 3.29 *Let* H/K *be an abelian chief factor of a group* G *with* $|H/K| = p^n$. *Then:*

(1) If $Aut_G(H/K)$ *is abelian, then* $Aut_G(H/K)$ *is cyclic and* $|Aut_G(H/K)|$ *divides* $p^n - 1$ *and* n *is the smallest positive integer such that* $|Aut_G(H/K)|$ *dividers* $p^n - 1$ *([248, Chap. VI, Lemma 8.1] and [359, Chap. I, Lemma 4.1]).*

(2) $|H/K| = p$ *if and only if* $Aut_G(H/K)$ *is an abelian group of exponent dividing* $p - 1$. *([450, Chap. I, Theorem 1.4]).*

Theorem 3.30 (Guo and Skiba [192]). *A group* G *is quasisupersoluble if and only if* $G = AB$ *where* A *and* B *are normal quasisupersoluble subgroups of* G *and either* $G' \le F^*(G)$ *or* $Com(G/A) \cap Com(G/B) = \emptyset$.

Proof First suppose that $G = GG$ is a quasisupersoluble group. We shall show that $G' \leq F^*(G)$. Let $Z = Z_{\mathfrak{U}}(G)$ and $1 = Z_0 \leq Z_1 \ldots \leq Z_t = Z$, where Z_i/Z_{i-1} is a chief factor of G for all $i = 1, 2, \ldots, t$. Let $C_i = C_G(Z_i/Z_{i-1})$, $i = 1, 2, \ldots, t$, and $C = C_1 \cap C_2 \cap \ldots \cap C_t$. If $x \in C$, then x induces the trivial automorphism on every factor Z_i/Z_{i-1}. Besides, by Theorem 3.12, G/Z is a semisimple group and so x induces an inner automorphism on every chief factor of G above Z. Now by Lemma 2.5, we see that x induces an inner automorphism on every chief factor of G. Hence, $x \in F^*(G)$. Therefore, $C \leq F^*(G)$ and so $G' \leq F^*(G)$ since G/C_i is cyclic.

Now assume that $G = AB$, where A and B are normal quasisupersoluble subgroups. Suppose that $G' \leq F^*(G)$. We shall prove that G is quasisupersoluble. Assume that this is false and G is a counterexample of minimal order. Then $A \neq G \neq B$ and $A \cap B \neq 1$ since the class of all quasisupersoluble groups is a formation. Hence, $A \neq 1 \neq B$. Let N be a minimal normal subgroup of G. We show that the hypothesis holds for G/N. In fact, since $F^*(G)N/N \simeq F^*(G)/F^*(G) \cap N$, $(GN/N)' = G'N/N \leq F^*(G)N/N \leq F^*(G/N)$. On the other hand, since $\mathfrak{U}^*$ is a formation, AN/N and BN/N are both quasisupersoluble. Thus, the hypothesis holds for G/N. This implies that N is the only minimal normal subgroup of G and $N = G^{\mathfrak{U}^*}$. It follows that $N \leq A \cap B$. Moreover, N is a p-group for some prime p (see the proof of Theorem 3.24). Hence, $N \leq F(G) \leq F^*(G)$. Let $C = C_G(N)$. Then $F^*(G) \leq C$ by Lemma 2.10. Hence, G/C is abelian and so it is cyclic by Lemma 3.29. Let

$$1 = N_0 \leq N_1 \leq \ldots \leq N_t = N, \tag{1.5}$$

where N_i/N_{i-1} is a chief factor of A and let $C_i = C_A(N_i/N_{i-1})$ for all $i = 1, 2, \ldots, t$. Let $D = C_1 \cap C_2 \cap \ldots \cap C_t$. Then $C \cap A \leq D$, $D/C \cap A$ is isomorphic to some subgroup of $Aut(N)$ and $D/C \cap A$ stabilizes the Series (1.5). Hence, by Lemma 2.13, $D/C \cap A$ is a p-group. On the other hand, A/D is a cyclic group of exponent dividing $p-1$. Hence, $AC/C \simeq A/C \cap A = A_1 \times P_1$, where A_1 is a cyclic group of exponent dividing $p - 1$ and P_1 is a cyclic p-group. Similarly, $BC/C \simeq B/C \cap B = B_1 \times P_2$, where B_1 is a cyclic group of exponent dividing $p - 1$ and P_2 is a cyclic p-group. Since $G/C = (AC/C)(BC/C)$ is cyclic, every Sylow subgroup of G/C is contained at least in one of the subgroups AC/C or BC/C. Hence, $G/C = V \times W$, where V is a cyclic p-group and W is a cyclic group of exponent dividing $p - 1$. Since $O_p(G/C) = 1$ by Lemma 2.15, $V = 1$. Hence, $|N| = p$ by Lemma 3.29(2). But since G/N is quasisupersoluble, G is quasisupersoluble, a contradiction.

Finally, if $Com(G/A) \cap Com(G/B) = \emptyset$, then G is a quasisupersoluble group by Corollary 3.25. The contradiction completes the proof.

Characterizations of Quasisoluble, Quasisupersoluble and Quasimetanilpotent Groups A subgroup A of a group G is called a *CAP-subgroup* of G [89, p. 37] (or A with the cover-avoid property (see [450, p. 211] if A either covers H/K (that is, $H \leq AK$) or A avoids H/K (that is, $A \cap H \leq K$) for each chief factor H/K of G.

It is known that a group G is soluble if and only if every maximal subgroup of G is a CAP-subgroup of G (cf. [34]). On the other hand, a group G is soluble if the index $|G : M|$ of every maximal subgroup M of G is a prime or the square of a prime (Hall). Now we prove the following

Theorem 3.31 (Guo and Skiba [188]). *Let G be a A_4-free group. Suppose that G has a normal subgroup X such that G/X is semisimple and every maximal subgroup M of X is a CAP-subgroup of G and $|X : M|$ is a prime or the square of a prime. Then G is quasisoluble.*

Proof Suppose that this theorem is false and G be a counterexample of minimal order. Then since a semisimple group is quasinilpotent (and thereby it is quasisoluble) and X is soluble by the hypothesis, $1 \neq X \neq G$. Obviously, the hypothesis still holds on every quotient of G, and so G has a unique minimal normal subgroup, H say. This means that $H = G^{\mathfrak{S}^*} \leq X$. Since X is soluble and the class of all quasisoluble groups $\mathfrak{S}^*$ is a solubly saturated formation by Corollary 3.8, $H \nsubseteq \Phi(X)$.

Let $C = C_G(H)$. We first claim that $CX \neq G$. Indeed, if $G = CX$, then $G/C \simeq X/X \cap C$ is soluble and so $H \leq Z_{\mathfrak{S}}(G)$. But since $G/H \in \mathfrak{S}^*$, $(G/H)/Z_{\mathfrak{S}}(G/H)$ is semisimple by Theorem 3.12. Let $L/H = Z_{\mathfrak{S}}(G/H)$. Then since $H \subseteq Z_{\mathfrak{S}}(G)$, $L \leq Z_{\mathfrak{S}}(G)$ and so $G/Z_{\mathfrak{S}}(G)$ is semisimple. It follows from Theorem 3.12 that G is quasisoluble. This contradiction shows that $CX \neq G$.

Since $H \nsubseteq \Phi(X)$, $X = HM$ for some maximal subgroup M of X. By the hypothesis, M either covers or avoids the factor $H/1$ and $|X : M|$ is p or p^2 for some prime p. Hence, $X = H \rtimes M$, H is a minimal normal subgroup of X and $|H| = |X : M|$. If $|H| = p$, then G/C is isomorphic with some subgroup of the cyclic group $Aut(H)$ and consequently $H \leq Z_{\mathfrak{S}}(G)$, which is impossible. Thus, $|H| = p^2$. In this case, G/C is isomorphic with some subgroup D of $GL(2, p)$. Since $HX \neq G$, D has a normal subgroup N such that D/N is a non-trivial semisimple group. Since $GL(2, p)/SL(2, p)$ is cyclic and $DSL(2, p)/NSL(2, p) \simeq D/N(D \cap SL(2, p)) \simeq (D/N)/(N(D \cap SL(2, p)/N))$ is semisimple, $DSL(2, p) = NSL(2, p)$ and thereby $D = N(D \cap SL(2, p))$. Therefore, $(SL(2, p) \cap D)/(SL(2, p) \cap N) \simeq N(D \cap SL(2, p))/N = D/N$ is a semisimple group. Let $Z = Z(SL(2, p))$. If $(SL(2, p) \cap D)Z = (SL(2, p) \cap N)Z$, then $SL(2, p) \cap D = (SL(2, p) \cap N)(Z \cap (SL(2, p) \cap D))$. It follows that $(SL(2, p) \cap D)/(SL(2, p) \cap N) = ((SL(2, p) \cap N)(Z \cap (SL(2, p) \cap D))/(SL(2, p) \cap N) \simeq (Z \cap SL(2, p) \cap D)/(Z \cap SL(2, p) \cap N)$ is an abelian group. This contradiction shows that $(SL(2, p) \cap D)Z \neq (SL(2, p) \cap N)Z$ and so $(SL(2, p) \cap D)Z/(SL(2, p) \cap N)Z \simeq (SL(2, p) \cap D)/(SL(2, p) \cap N)(SL(2, p) \cap D \cap Z)$ is a non-trivial semisimple section of $SL(2, p)/Z = PSL(2, p)$. In this case, we have $p > 2$. Then, by Dikson's theorem [248, Theorem II.8.27], $(SL(2, p) \cap N)Z = Z$ and either $(SL(2, p) \cap D)Z/Z = PSL(2, p)$ or $(SL(2, p) \cap D)Z/Z \simeq A_5$. But then D has a section which is isomorphic with A_4. This contradiction completes the proof.

Recall that a subgroup H of a group G is said to be *c-normal* in G if G has a normal subgroup T such that $T \cap H \leq H_G$ and $HT = G$ (Wang [435]).

We need the following generalization of this concept.

Definition 3.32 (Guo and Skiba [189]). *Let H be a subgroup of a group G. We say that H is nearly normal in G if G has a normal subgroup T such that $T \cap H \leq H_G$ and $HT = H^G$.*

The following lemma can be proved by direct calculations.

Lemma 3.33 *Let G be a group and $H \leq K \leq G$.*

(1) Suppose that H is normal in G. Then K/H is nearly normal (c-normal) in G/H if and only if K is nearly normal (c-normal, respectively) in G.

(2) If H is nearly normal (c-normal) in G, then H is nearly normal (c-normal, respectively) in K.

(3) Suppose that H is normal in G. Then HE/H is nearly normal (c-normal) in G/H for every nearly normal (c-normal, respectively) in G subgroup E satisfying $(|H|, |E|) = 1$.

Lemma 3.34 *Let X be a normal subgroup of a group G. Suppose that every maximal subgroup of X is nearly normal in G. Then X is soluble.*

Proof We proceed the proof by induction on $|G|$. Let N be a minimal normal subgroup of G contained in X. Then X/N is soluble. Indeed, if $N = X$, it is clear. Otherwise, by Lemma 3.33(1) the hypothesis is still true for G/N and so by induction X/N is soluble. If G has a minimal normal subgroup $R \neq N$ contained in X, then $X \simeq X/1 = X/N \cap R$ is soluble. Therefore, we may assume that N is the only minimal normal subgroup of G contained in X.

Let p be a prime dividing $|N|$ and N_p be a Sylow p-subgroup of N. Then $N_p = N \cap P$ for some Sylow p-subgroup P of X. Obviously, $P \leq N_X(N_p)$. Also, by the Frattini argument, $X = NN_X(N_p)$. Suppose that $N \neq N_p$. Then for some maximal subgroup M of X we have $N_X(N_p) \leq M$. Hence, $N \nsubseteq M$ and p does not divide $|X : M|$. It follows that $M_G = 1$. By hypothesis M is nearly normal in G. Let T be a normal subgroup of G such that $X = MT = M^G$ and $T \cap M \leq M_G = 1$. Then $X = T \rtimes M$ and $N \leq T$. It is also clear that T is a minimal normal subgroup of X. Hence, $N = T$ and so p divides $|X : M| = |N|$. This contradiction shows that N is a p-group. Consequently, X is soluble.

Theorem 3.35 (Guo and Skiba [189]). *A group G is quasisoluble if and only if G has a normal subgroup X such that G/X is semisimple, every maximal subgroup of X is nearly normal in G and for every $x \in G$ and for every G-chief factor H/K of X, the automorphism of H/K induced by x is also induced by some element of X.*

Proof To demonstrate the sufficiency of the theorem, we proceed by induction $|G|$. Let N be a minimal normal subgroup of G contained in X. We claim that G/N is quasisoluble. Indeed, if $N = X$, it is clear. Otherwise, by Lemma 3.33(2) the hypothesis is still true for G/N and so by induction G/N is quasisoluble. By Lemma 3.34, X is soluble and so N is a p-group for some prime p. Now let $C = C_G(N)$ and $g \in G$. Then by the hypothesis the automorphism of N induced by g is also induced by some element x of X. Hence, $gx^{-1} \in C$ and so $G = CX$. Then $G/C \simeq X/X \cap C$ is soluble. This means that the factor $N/1$ is $\mathfrak{S}$-central. Since G/N is quasisoluble, by Corollary 3.13, we obtain that G is quasisoluble.

Now we prove that necessity part by induction on $|G|$. Let $X = Z_{\mathfrak{S}}(G)$ be the $\mathfrak{S}$-hypocenter of the quasisoluble group G. Then by Corollary 3.13, G/X is a semisimple group. Moreover, for any G-chief factor H/K of X the group $(H/K) \rtimes (G/C_G(H/K))$ is soluble. Hence, $G/C_G(H/K)$ is soluble and so $XC_G(H/K) = G$. It follows that for every $x \in G$ the automorphism of H/K induced by x is also induced by some element of X. Finally, we prove that every maximal subgroup M of X is nearly normal in G. Suppose that $M_G \neq 1$. Then by induction M/M_G is nearly normal in G/M_G. Hence, by Lemma 3.33(1), M is nearly normal in G. Now, let $M_G = 1$ and N be a minimal normal subgroup of G such that $NM = X$. Let $D = N \cap M$. Since $N \subseteq Z_\mathfrak{G}(G)$, N is abelian. Hence, D is normal in X. On the other hand, $C_G(N) \leq N_G(D)$. This implies that D is normal in $G = XC_G(N)$ and so $D \leq M_G = 1$. Thus, M is nearly normal in G. The theorem is proved.

Corollary 3.36 *A group G is soluble if and only if every its maximal subgroup is nearly normal in G.*

Corollary 3.37 (Wang [435]). *A group G is soluble if and only if for every maximal subgroup M of G there is a normal subgroup T such that $G = MT$ and $M \cap T \leq M_G$.*

Proof Suppose that G is soluble and M is a maximal subgroup of G. If M is normal in G, then it is clear. Now assume that M is not normal in G and let T/M_G be a chief factor of a soluble group G. Then T/M_G is abelian. Hence, $T \cap M \leq M_G$ and $MT = G$. The converse is obvious since by hypothesis, we have $M^G = M(M^G \cap T)$.

Lemma 3.38 *Let P be a normal p-subgroup of a group G. If P is elementary and every maximal subgroup of P is nearly normal in G, then every minimal normal subgroup of G contained in P has prime order.*

Proof Let N be any minimal normal subgroup of G contained in P. Suppose that $|N| > p$ and let M be a maximal subgroup of P such that $NM = P$. Then M is not normal in G and so $M^G = P$. Let T be a normal subgroup of G such that $P = MT$ and $T \cap M \leq M_G$. Suppose that $M_G \neq 1$. By Lemma 3.33 the hypothesis holds for G/M_G and so $|NM_G/M_G| = |N| = p$ by induction, a contradiction. Hence, $M_G = 1$ and thereby $T \cap M = 1$. This implies that $|T| = p$ and $T \neq N$. But by induction, we have also that $|TN/T| = |N| = p$. The contradiction completes the proof.

Lemma 3.39 *Let G be a group and p a prime such that $(p - 1, |G|) = 1$. If the Sylow p-subgroups of G are cyclic, then G is p-nilpotent.*

Proof See the proof of Theorem 10.1.9 in [337].

Lemma 3.40 *Suppose that every maximal subgroup M of every noncyclic Sylow subgroup of a group G is nearly normal in G. Then G is soluble.*

Proof Suppose that this lemma is false and let G be a counterexample with minimal order. Let P be a Sylow p-subgroup of G, where p is the smallest prime dividing $|G|$. Then $p = 2$ by Feit–Thompson theorem on groups of odd order. By Lemma 3.39, P is not cyclic. Suppose that for some maximal subgroup V of P we have $V_G \neq 1$. Then by Lemma 3.33 the hypothesis holds on G/V_G and so G/V_G is soluble, which implies the solubility of G. Therefore, $V_G = 1$ for all maximal subgroups V of P. Let $P = V_1 V_2$ for some maximal subgroups V_1 and V_2 of P. By hypothesis G has a normal subgroup T_i such that $D_i = V_i^G = V_i T_i$ and $T_i \cap V_i \leq (V_i)_G = 1$. Clearly, $P \cap T_i$ is a Sylow 2-subgroup of T_i. But since $T_i \cap V_i = 1$, we have $|T_i \cap P| \leq 2$. Hence, T_i is soluble and so V_i^G is soluble. It follows that $D = V_1^G V_2^G$ is soluble and therefore G is soluble since G/D is a $2'$-group. This contradiction completes the proof.

Theorem 3.41 (Guo and Skiba [189]) *The following statements are equivalent.*

(1) *G is quasisupersoluble.*
(2) *G has a normal subgroup E such that G/E is semisimple and every maximal sugroup M of every Sylow subgroup of $F^*(E)$ is nearly normal in G.*
(3) *G has a normal subgroup E such that G/E is quasisupersoluble and every maximal subgroup M of every Sylow subgroup of $F^*(E)$ is nearly normal in G.*

Proof

(1) $\Rightarrow$ (2) Let $E = Z_{\mathfrak{U}}(G)$. Then by Corollary 3.13, G/E is semisimple and E is supersoluble. Then $F^*(E) = F(E)$ by Proposition 2.9(8). Let M be a maximal subgroup of some Sylow subgroup P of $F(E)$. Since $F(E)$ is characteristic in E and P is characteristic in $F(E)$, P is normal in G. We now prove that M is nearly normal in G. If M is normal in G, then it is clear. We may, therefore, assume that $M \neq M^G$. If $\Phi = \Phi(G) \cap P \neq 1$. Then M/Φ is a maximal subgroup of the Sylow subgroup P/Φ and $F(E)/\Phi = F^*(E)/\Phi = F^*(E/\Phi)$. By induction M/Φ is nearly normal in G/Φ and so M is nearly normal in G by Lemma 3.33. Now suppose that $\Phi = 1$. Then by Proposition 2.24, P is a product of some minimal normal subgroups $N_1, N_2, \ldots, N_t$ of G. Clearly, for some i we have $N_i \nsubseteq M$. Since $N_i \subseteq Z_{\mathfrak{U}}(G)$, $|N_i|$ is a prime. Hence, $N_i \cap M = 1$ and $MN_i = M^G = P$. This shows that M is nearly normal in G.

(2) $\Rightarrow$ (3) This is obvious since a semisimple group is clearly quesisupersoluble.

(3) $\Rightarrow$ (1) Suppose that G has a normal subgroup E such that G/E is quasisupersoluble and every maximal subgroup M of every Sylow subgroup of $F^*(E)$ is nearly normal in G. We shall prove that G is quasisupersoluble. Suppose that this is false and let G be a counterexample with minimal $|G||E|$. Let p be prime dividing $|F^*(E)|$ and P be a Sylow p-subgroup of $F^*(E)$. In view of Corollary 3.36 and Proposition 2.9(8), we have $F^*(E) = F(E)$.

We first show that every minimal subgroup of P is not normal in G. Indeed, suppose that some minimal subgroup L of P is normal in G and let $C = C_E(L)$. We claim that the hypothesis is true for $(G/L, C/L)$. Indeed, by Theorem 3.7, $G/C = G/(E \cap C_G(L))$ is quasisupersoluble. In addition, since $L \leq Z(C)$ and $F^*(E) = F(G) \leq C$, $L \leq Z(F^*(E))$. Hence, $F^*(C/L) = F^*(E)/L$. Now, by Lemma 3.33, the hypothesis is still true for $(G/L, C/L)$. Hence, G/L is quasisupersoluble and so G is quasisupersoluble, a contradiction. Hence, every minimal subgroup of P is not normal in G.

If $\Phi(P) = 1$, then P is elementary abelian p-group. Hence, by Lemma 3.38 for every minimal normal subgroup L of G contained in P we have $|L| = p$, which is a contradiction. Thus, $\Phi(P) \neq 1$. By Lemma 3.9, Theorem 3.7, and Lemma 2.19, we have $F^*(E/\Phi(P)) = F^*(E)/\Phi(P)$. Then by Lemma 3.33, the hypothesis is still true for $(G/\Phi(P), E/\Phi(P))$. But $|G/\Phi(P)| < |G|$ and hence

$G/\Phi(P)$ is quasisupersoluble by the choice of G. Hence, by Theorem 3.7, G is quasisupersoluble, a contradiction. The theorem is proved.

From Lemma 3.40 and Theorem 3.41, we obtain the following

Corollary 3.42 *A group G is supersoluble if and only if every maximal subgroup of every Sylow subgroup of $F^*(G)$ is nearly normal in G.*

Corollary 3.43 (Ramadan [328]). *Let G be a soluble group. If all maximal subgroups of the Sylow subgroups of $F(G)$ are normal in G, then G is supersoluble.*

Corollary 3.44 (Li and Guo [283]). *Let G be a group and E a soluble normal subgroup of G with supersoluble quotient G/E. If all maximal subgroups of the Sylow subgroups of $F(E)$ are c-normal in G, then G is supersoluble.*

Lemma 3.45 (Guo and Shum [224]). *Let $G = N \rtimes M$, where N is a minimal normal subgroup of G and M a maximal subgroup of G. If M is soluble, then N is abelian.*

Proof Clearly, the hypothesis holds for G/M_G. Hence, in the case when $M_G \neq 1$, $N \simeq NM_G/M_G$ is abelian by induction. Now assume $M_G = 1$. Let T be a minimal normal subgroup of M. Since M is soluble, T is a p-group for some prime p. Since $M_G = 1$, $M = N_G(R)$. Let P be a Sylow p-subgroup of G containing a Sylow p-subgroup of M. Then $T \leq P$ and $N \cap P$ is normal in P. Suppose that $N \cap P \neq 1$. Then $N \cap P \cap Z(P) \neq 1$ and so $M < N_G(T)$. This contradiction shows that $N \cap P = 1$. Consequently, N is a p'-group. Therefore, for every prime q dividing $|N|$, there is a unique T-invariant Sylow q-subgroup Q of N (see [250, Chap. X, Lemma 11.1]). On the other hand, for any $x \in M$ we have $(Q^x)^T = Q^{Tx} = Q^x$. Thus, by the uniqueness of the Sylow q-subgroup Q, we see that $Q = N$ is abelian.

Theorem 3.46 (Guo and Skiba [188]). *Let G be a group. Then the following are equivalent:*

(1) G is quasisupersoluble.
(2) G has a normal subgroup X such that G/X is semisimple and every subgroup of X is a CAP-subgroup of G.
(3) G has a normal subgroup X such that G/X is semisimple and every maximal subgroup of every noncyclic Hall subgroup of X is a CAP-subgroup of G.

Proof

(1) $\Rightarrow$ (2) Let $Z = Z_{\mathfrak{U}}(G)$. By Theorem 3.12, the quotient G/Z is semisimple. So we only need to prove the statement that every subgroup E of Z either covers or avoids every chief factor H/K of G. By Lemma 2.5, we know that either $|H/K| = p$ is a prime or H/K is a non-abelian simple group. If $|H/K| = p$, then the statement is evident. Now assume that H/K is a non-abelian simple group. Since $ZH/ZK \simeq H/K(H \cap Z)$, either $H = K(H \cap Z)$ or $K = K(H \cap Z)$. Since Z is soluble, in the former case, by $H/K = K(H \cap Z)/K \simeq (H \cap Z)/K \cap H \cap Z$, we have H/K is abelian, which is a contradiction. Thus, $H \cap Z \leq K$ and thereby $H \cap E \leq K$.

(2) $\Rightarrow$ (3) It is trivial.

(3) $\Rightarrow$ (1) Suppose that this is false and let G be a counterexample of minimal order. It is clear that the hypothesis holds on every quotient of G and so G has a unique minimal normal subgroup $H = G^{\mathfrak{U}^*}$. If $|H| = p$ is a prime, then by the definition of the quasi-$\mathfrak{U}$-group or by Theorem 3.12, we have that G is quasisupersoluble. Now assume that $|H|$ is not a prime. Let $Z/H = Z_\infty^{\mathfrak{U}}(G/H)$. We claim that $X = Z$. In fact, by Theorem 3.12, we see that $Z \leq X$ and hence X is not cyclic. Suppose that $Z \neq X$ and let E be a normal subgroup of G such that $Z \leq E \leq X$ and X/E is a non-abelian simple group. Let M be a maximal subgroup of X such that $E \leq M$. Then M neither covers nor avoids the chief factor X/E, which contradicts the hypothesis. Thus, $Z = X$ is not cyclic and X/H is a soluble group. Let $\Phi = \Phi(X)$. Suppose that $\Phi \neq 1$. Then $H \leq \Phi$ and hence G/Φ is a quasisupersoluble group. It follows from Theorem 3.7 that G is quasisupersoluble, which contradicts the choice of G. Hence, $\Phi = 1$ and $H \nsubseteq \Phi$. It follows that $X = HM$ for some maximal subgroup M of X. By the hypothesis, M either covers or avoids the factor $H/1$. This implies that $X = H \rtimes M$. Obviously, H is a minimal normal subgroup of X. Hence, H is a p-group for some prime p by Lemma 3.45. Let P be a Sylow p-subgroup of X. Then $H \leq P$. Since $\Phi(X) = 1$, $H \nsubseteq \Phi(P)$ by [248, Lemma III.3.3]. Let E be a maximal subgroup of P such that $P = HE$. Then $|H : H \cap E| = p$. Then, by the hypothesis, E avoids the factor $H/1$. Thus, $|H| = p$. This contradiction completes the proof.

Corollary 3.47 *A group G is quasisupersoluble if it has a normal subgroup X such that G/X is semisimple and every Sylow subgroup of X is cyclic.*

Lemma 3.48 *Let G be a quasisoluble group. If H is a minimal normal subgroup of G with $H \leq Z = Z_{\mathfrak{S}}(G)$, then H is a minimal normal subgroup of Z.*

Proof Let $C = C_G(H)$. Since $H \leq Z$, G/C is soluble. By Theorem 3.12, G/Z is the direct product of some non-abelian simple groups. Hence, $G = ZC$. Let L be a minimal normal subgroup of Z contained in H, then L is normal in G. Hence, $L = H$.

Theorem 3.49 (Miao, Chen, and Guo, [312]). *A group G is metanilpotent if and only if every Sylow subgroup of G is c-normal in G.*

Proof Suppose that every Sylow subgroup P of G is c-normal in G. Then, by the definition of c-normal subgroup, we can see that $G/P_G = G/O_p(G)$ is p-nilpotent. This implies that $G/F(G)$ is p-nilpotent for all prime p. Consequently, $G/F(G)$ is nilpotent. Thus, G is metanilpotent.

Conversely, suppose that G is metanilpotent and let P be a Sylow p-subgroup of G. If $P_G \neq 1$, then by induction P/P_G is c-normal in G. Hence, P is c-normal in G by Lemma 3.33. Now let $P_G = 1$. Then $F(G)$ is a p'-group. Since $G/F(G)$ is nilpotent, there is a normal p'-subgroup T such that $F(G) \leq T$ and $(T/F(G))(PF(G)/F(G)) = G/F(G)$. Then $PT = G$ and $T \cap P = 1$. Hence, P is c-normal in G.

Theorem 3.50 (Guo and Skiba [188]). *A quasisoluble group G is quasimetanilpotent if and only if every Sylow subgroup of $Z_\mathfrak{S}(G)$ is c-normal in $Z_\mathfrak{S}(G)$.*

Proof Put $Z = Z_\mathfrak{S}(G)$. First suppose that G is quasimetanilpotent. We prove by induction on G that every Sylow subgroup of Z is c-normal in Z. It is evident in the case $Z = 1$. So we may assume that $Z \neq 1$. Let $F = F(Z)$ and P be a Sylow p-subgroup of Z. If p divides $|F|$ and F_p is a Sylow p-subgroup of F, then F_p is normal in G. By induction P/F_p is c-normal in Z/F_p and thereby P is c-normal in Z. Now let $(p, |F|) = 1$. Since G is quasisoluble and quasimetanilpotent as well, by Theorem 3.12, we easy see that $Z_\mathfrak{S}(G) = Z_{\mathfrak{N}^2}(G) = Z$. Thus, Z/F is nilpotent and so PF/F has a normal complement T/F in Z/F. It is clear that T is a normal complement of P in Z. Thus, P is c-normal in Z.

Now suppose that every Sylow subgroup of $Z = Z_\mathfrak{S}(G)$ is c-normal in Z. We prove that G is quasimetanilpotent. Suppose that this is false and G be a counterexample of minimal order. It is clear that the hypothesis still holds on $Z = Z_\mathfrak{S}(G)$ and on every quotient of G. Hence, G has a unique minimal normal subgroup $H = G^{(\mathfrak{N}^2)^*}$. If $Z = 1$, then by Theorem 3.12, G is the direct product of some non-abelian simple groups and hence it is quasimetanilpotent, which contradicts the choice of G. Thus, $Z \neq 1$ and $H \leq Z$. If $Z = G$, then G is metanilpotent by Theorem 3.49 and consequently, G is quasimetanilpotent. So we may assume that $Z \neq G$. Then, by the choice of G, we have that Z is quasimetanilpotent. Since, obviously, $(\mathfrak{N}^2)^* \cap \mathfrak{S} = \mathfrak{N}^2$ by Theorem 3.12, Z is metanilpotent and thereby Z is also soluble. Since $H \leq Z$, H is a minimal normal subgroup of Z by Theorem 3.49. Thus, $Z/C_Z(H)$ is nilpotent. Since G is quasisoluble and $H \leq Z$, $(H/1) \rtimes (G/C_G(H/1)) \in \mathfrak{S}$ and consequently, $G/C_G(H)$ is soluble. However, since G/Z is semisimple, $G/ZC_G(H)$ is also semisimple. This induces that $G = ZC_G(H)$. Hence, $G/C_G(H) \simeq Z/Z \cap C_G(H) = Z/C_Z(H)$ is nilpotent. It follows that $H \rtimes G/C_G(H) \in \mathfrak{N}^2$, that is, H is $\mathfrak{N}^2$-centric. On the other hand, since $H = G^{(\mathfrak{N}^2)^*}$, G is quasimetanilpotent. This contradiction completes the proof.

1.4   On Factorizations of Groups with $\mathfrak{F}$-hypercentral Intersections of the Factors

In this section, we examine the following question: Let $G = AB$ be the product of two subgroups A and B of G. What we can say about the structure of G if $A \cap B \leq Z_\mathfrak{F}(A) \cap Z_\mathfrak{F}(B)$ for some class of groups $\mathfrak{F}$?

Lemma 4.1 *Let $\mathfrak{F}$ be a class of groups. Let $G = AB$ and $A \cap B \leq Z_\mathfrak{F}(A) \cap Z_\mathfrak{F}(B)$. Then*

(1) $A \cap B^x \leq Z_\mathfrak{F}(A) \cap Z_\mathfrak{F}(B^x)$, for any $x \in G$.

(2) If N is a normal subgroup of G such that $N \leq A$, then $A/N \cap BN/N \leq Z_\mathfrak{F}(A/N) \cap Z_\mathfrak{F}(BN/N)$.

(3) Let N be a normal subgroup of G such that $N \leq A$. If $A \cap B \leq Z_n(Z_\infty(A)) \cap Z_n(Z_\infty(B))$, then $A/N \cap BN/N \leq Z_n(Z_\infty(A/N)) \cap Z_n(Z_\infty(BN/N))$.

Proof

(1) Let $x = ba$, where $a \in A$ and $b \in B$. Then $A \cap B^x = A \cap B^{ba} = A \cap B^a = (A \cap B)^a \leq (Z_{\mathfrak{F}}(A) \cap Z_{\mathfrak{F}}(B))^a = Z_{\mathfrak{F}}(A) \cap (Z_{\mathfrak{F}}(B))^a = Z_{\mathfrak{F}}(A) \cap Z_{\mathfrak{F}}(B^a) = Z_{\mathfrak{F}}(A) \cap Z_{\mathfrak{F}}(B^x)$.

(2) $A/N \cap BN/N = N(A \cap B)/N \leq N(Z_{\mathfrak{F}}(A) \cap Z_{\mathfrak{F}}(B))/N \leq Z_{\mathfrak{F}}(A/N) \cap Z_{\mathfrak{F}}(BN/N)$ by Theorem 2.6(e). Hence, (2) holds.

(3) It follow from (2) and Theorem 2.6(n).

A group G is called *p-decomposable* if G has a normal Sylow p-subgroup and a normal Hall p'-subgroup.

Lemma 4.2 *Let* $\mathfrak{F} = \mathfrak{N}_p\mathfrak{A}(p - 1)$, E *be a normal p-subgroup of* G, *where* p *is a prime, and* V *a normal p'-subgroup of* G. *Then:*

(1) If $E \leq Z_{\mathfrak{N}^2}(G)$, then $(G/C_G(E))^{\mathfrak{N}} \leq O_p(G/C_G(E))$.

(2) If $E \leq Z_{\mathfrak{X}}(G)$, where $\mathfrak{X}$ is the class of all p-supersoluble groups, then $G/C_G(E) \in \mathfrak{F}$. If $V \leq Z_{\mathfrak{X}}(G)$, then $G/C_G(V)$ is p-supersoluble.

(3) If $E \leq Z_{\mathfrak{U}}(G)$, then $G/C_G(E) \in \mathfrak{F}$.

(4) If $E \leq Z_{\mathfrak{F}}(G)$, then $G/C_G(E) \in \mathfrak{F}$.

(5) If $E \leq Z_{\mathfrak{X}}(G)$, where $\mathfrak{X}$ is the class of all p-closed groups, then $G/C_G(E) \in \mathfrak{N}_p\mathfrak{X}$ and $Z_{\mathfrak{X}}(G)$ is p-closed.

(6) If $E \leq Z_{\mathfrak{X}}(G)$, where $\mathfrak{X}$ is the class of all p-decomposable groups, then $G/C_G(E)$ is a p-group.

Proof Let $1 = E_0 < E_1 < \ldots < E_t = E$ be a chief series of G below E, $C_i = C_G(E_i/E_{i-1})$ and $C = C_1 \cap C_2 \cap \ldots \cap C_t$. Then $C_G(E) \leq C$ and by Lemma 2.13, $C/C_G(E)$ is a p-group.

(1) If $E \leq Z_{\mathfrak{N}^2}(G)$, then $G/C_G(E_i/E_{i-1}) \in \mathfrak{N}$ and so G/C is nilpotent. Hence, $(G/C_G(E))^{\mathfrak{N}} \leq C/C_G(E) \leq O_p(G/C_G(E))$.

(2) If $E \leq Z_{\mathfrak{X}}(G)$, then by Lemma 3.29, $(G/C_G(E))^{\mathfrak{A}(p-1)} \leq O_p(G/C_G(E))$. If follows that $G/C_G(E) \in \mathfrak{F}$. Suppose that $V \leq Z_{\mathfrak{X}}(G)$. Let $1 = V_0 < V_1 < \ldots < V_t = V$ be a chief series of G below E, $C_i^* = C_G(V_i/V_{i-1})$ and $C^* = C_1^* \cap C_2^* \cap \ldots \cap C_t^*$. Then $C_G(V) \leq C^*$. Since $(V_i/V_{i-1}) \rtimes G/C_G(V_i/V_{i-1}) \in \mathfrak{X}$, $G/C^* \in \mathfrak{X}$ and thereby G/C^* is p-supersoluble. On the other hand, by Lemma 2.13, $C^*/C_G(V)$ is a p'-group. Hence, $G/C_G(V)$ is p-supersoluble.

(3), (4) By Lemma 3.29 we see that $\mathfrak{F} \subseteq \mathfrak{U} \subseteq \mathfrak{X}$, where $\mathfrak{X}$ is the class of all p-supersoluble groups. It follows that $Z_{\mathfrak{F}}(G) \leq Z_{\mathfrak{U}}(G) \leq Z_{\mathfrak{X}}(G)$. Now, we see that (3) and (4) follows from (2).

(5) The first assertion follows from the fact that G/C_i is a p'-group for all $1, 2, \ldots, t$. The second assertion of (5) follows from Corollary 2.30.

(6) This assertion follows from the fact that $C_i = G$ for all $i = 1, 2, \ldots, t$.

Lemma 4.3 *Let A, B and E be normal subgroups of G. Suppose that $G = AB$. If $E \leq Z_{\mathfrak{N}^2}(A) \cap Z_{\mathfrak{N}^2}(B)$, then $E \leq Z_{\mathfrak{N}^2}(G)$.*

Proof Assume that the lemma is false and let (G, E) be a counterexample with $|G||E|$ minimal. Then $E \neq 1$. Let L be a minimal normal subgroup of G contained in E. Since $Z_{\mathfrak{N}^2}(A)$ is soluble and $L \subseteq E \subseteq Z_{\mathfrak{N}^2}(A)$, L is a p-group for some prime p. Let $C = C_G(L)$, $C_1 = C \cap A$, and $C_2 = C \cap B$.

Since $L \leq Z_{\mathfrak{N}^2}(A)$, by Lemma 4.2(1), A has a normal subgroup V_1 such that $C_1 \leq V_1$, V_1/C_1 is a p-group and A/V_1 is nilpotent. Similarly, B has a normal subgroup V_2 such that $C_2 \leq V_2$, V_2/C_2 is a p-group and B/V_2 is nilpotent. Since $CA/C \simeq A/C_1$ and $CB/C \simeq B/C_2$, G/C has a normal p-subgroup D/C such that G/D is nilpotent. By Lemma 2.15, $O_p(G/C) = 1$ since L is a p-group. Hence, G/C is nilpotent and thereby $L \leq Z_{\mathfrak{N}^2}(G)$. By Theorem 2.6(d), it is easy to see that the hypothesis is still true for G/L. Hence, $E/L \leq Z_{\mathfrak{N}^2}(G/L) = Z_{\mathfrak{N}^2}(G)/L$ by the choice of (G, E). It follows that $E \leq Z_{\mathfrak{N}^2}(G)$. The contradiction completes the proof.

Theorem 4.4 (Guo and Skiba [194]). *Let p be a prime and $\mathfrak{F}$ the formation of all p-supersoluble group. Let A, B and E be normal subgroups of G. Suppose that $G = AB$ and $E = A \cap B \leq Z_{\mathfrak{F}}(A) \cap Z_{\mathfrak{F}}(B)$. If either $(|G : A|, |G : B|) = 1$ or $G' \leq O_{p'p}(G)$, then $E \leq Z_{\mathfrak{F}}(G)$.*

Proof Assume that this theorem is false and let G be a counterexample with $|G|$ minimal. Then $E \neq 1$. Let L be a minimal normal subgroup of G contained in E and $C = C_G(L)$. Since $L \leq Z_{\mathfrak{F}}(A) \cap Z_{\mathfrak{F}}(B)$, $A/C_A(L)$ and $B/C_B(L)$ are p-supersoluble by Lemma 4.2(2). Note that $E/L = (A \cap B)/L \leq (Z_{\mathfrak{F}}(A) \cap Z_{\mathfrak{F}}(B))/L \leq (Z_{\mathfrak{F}}(A)/L) \cap (Z_{\mathfrak{F}}(B)/L) \leq Z_{\mathfrak{F}}(A/L) \cap Z_{\mathfrak{F}}(B/L)$ by Theorem 2.6(d) and, if $G' \leq O_{p'p}(G)$, then $(G/L)' = G'L/L' \leq O_{p'p}(G)L/L \leq O_{p'p}(G/L)$. Hence, the hypothesis holds for G/L. The minimal choice of G implies that $E/L \leq Z_{\mathfrak{F}}(G/L)$. Then $L \nleq Z_{\mathfrak{F}}(G)$ by the choice of G.

First assume that L is a p'-group. In this case, G/C is not a p-supersoluble group. Since $L \subseteq E \subseteq Z_{\mathfrak{F}}(A) \cap Z_{\mathfrak{F}}(B)$, $AC/C \simeq A/A \cap C = A/C_A(L)$ and $BC/C \simeq B/B \cap C = B/C_B(L)$ are p-supersoluble groups. If $C = 1$, then A and B are p-supersoluble and therefore $G/E = (A/E)(B/E) = (A/E) \times (B/E)$ is p-supersoluble. But since $E/L \leq Z_{\mathfrak{F}}(G/L)$, G/L is p-supersoluble. It follow that G is p-supersoluble since L is a p'-group. The contradiction shows that $C \neq 1$. Since $AC/C \simeq A/A \cap C$ and $BC/C \simeq B/B \cap C$ are p-supersoluble, $V = (AC/C) \cap (BC/C) \leq Z_{\mathfrak{F}}(AC/C) \cap Z_{\mathfrak{F}}(BC/C)$. This shows that the hypothesis is true for G/C. Hence, $V/C \leq Z_{\mathfrak{F}}(G/C)$. On the other hand, $(G/C)/(V/C) = AC/C \times BC/C$ is p-supersoluble. This implies that G/C is p-supersoluble and thereby $L \rtimes (G/C)$ is p-supersoluble. Thus, $L \leq Z_{\mathfrak{F}}(G)$. This induces that G is p-supersoluble and $E \leq Z_{\mathfrak{F}}(G) = G$. The contradiction shows that L is a p-group.

By Lemma 4.2(2), $A/C_A(L) \in \mathfrak{N}_p\mathfrak{A}(p - 1)$ and $B/C_B(L) \in \mathfrak{N}_p\mathfrak{A}(p - 1)$. Assume that $(|G : A|, |G : B|) = 1$. Since $CA/C \simeq A/C_A(L)$ and $CB/C \simeq B/B_B(L)$, $G/C \in \mathfrak{N}_p\mathfrak{A}(p - 1)$ by Lemma 3.23. Then by Lemma 2.15, we have $G/C \in \mathfrak{A}(p-1)$. Hence, $|L| = p$ by Lemma 3.29. Now suppose that $G' \leq O_{p'p}(G)$. Since $O_{p'p}(G) \leq C$, G/C is abelian. Consequently it is cyclic by Lemma 3.29. Let

$$1 = L_0 \leq L_1 \leq \ldots \leq L_t = L, \tag{1.6}$$

where L_i/L_{i-1} is a chief factor of A and let $C_i = C_A(L_i/L_{i-1})$ for all $i = 1, 2, \ldots, t$. Since $L \subseteq Z_{\mathfrak{F}}(A)$ and L is a p-group, $|L_i/L_{i-1}| = p$. Hence, by Lemma 3.29, $A/C_i \in \mathfrak{A}(p-1)$. Let $D = C_1 \cap C_2 \cap \ldots \cap C_t$. Then $C \cap A \leq D$, $D/C \cap A$ is isomorphic to some subgroup of $\text{Aut}(L)$ and $D/C \cap A$ stabilizes Series (1.6). Hence, by Lemma 2.13, $D/C \cap A$ is a p-group. On the other hand, A/D is a cyclic group of exponent dividing $p - 1$. Hence, $AC/C \simeq A/C \cap A = A_1 \times P_1$, where A_1 is a cyclic group of exponent dividing $p - 1$ and P_1 is a cyclic p-group. Similarly, $BC/C \simeq B/C \cap B = B_1 \times P_2$, where B_1 is a cyclic group of exponent dividing $p-1$ and P_2 is a cyclic p-group. Since $G/C = (AC/C)(BC/C)$ is cyclic, every Sylow subgroup of G/C is contained at least in one of the subgroups AC/C or BC/C. Hence, $G/C = V \times W$, where V is a cyclic p-group and W is a cyclic group of exponent dividing $p - 1$. Since $O_p(G/C) = 1$ by Lemma 2.15, $V = 1$. Hence, we have $|L| = p$ again. It follows that $L \leq Z_{\mathfrak{U}}(G)$. This contradiction completes the proof.

Corollary 4.5 *Let A, B and E be normal subgroups of G. Suppose that $G = AB$ and $E \leq Z_{\mathfrak{U}}(A) \cap Z_{\mathfrak{U}}(B)$. If either $(|G : A|, |G : B|) = 1$ or $G' \leq F(G)$, then $E \leq Z_{\mathfrak{U}}(G)$.*

Corollary 4.6 (Baer [23]). *Let $G = AB$ where A, B are normal supersoluble subgroups of G. If $G' \leq F(G)$, then G is supersoluble.*

Proof Let $L = A \cap B$. If $L = 1$, then $G = A \times B$ is supersoluble. If $L \neq 1$, then the result follows from Theorem 4.4 by induction on $|G|$.

Corollary 4.7 (Friesen [108]). *Let $G = AB$ where A, B are normal supersoluble subgroups of G. If $(|G : A|, |G : B|) = 1$, then G is supersoluble.*

Proof Similar to the proof of Corollary 4.6.

Lemma 4.8 *Let $\mathfrak{F}$ be a class of groups such that $A \in \mathfrak{F}$ for every group A with $A = Z_{\mathfrak{F}}(A)$. Suppose that a group G has three subgroups A_1, A_2 and A_3 whose indices $|G : A_1|$, $|G : A_2|$, $|G : A_3|$ are pairwise coprime. If $G \notin \mathfrak{F}$ and $A_i \cap A_j \leq Z_{\mathfrak{F}}(A_i) \cap Z_{\mathfrak{F}}(A_j)$ for all $i \neq j$, then $A_i \cap A_j \neq 1$ for all $i \neq j$.*

Proof Suppose, for example, that $A_1 \cap A_2 = 1$. Then A_1 and A_2 are Hall subgroups of G. Hence, for any prime p dividing $|G : A_3|$, p either divides $|G : A_1|$ or divides $|G : A_2|$. The contradiction shows that $|G : A_3| = 1$, that is, $G = A_3$. Therefore, A_1, A_2 are contained in $Z_{\mathfrak{F}}(G)$. It follows that $G = A_1 A_2 = Z_{\mathfrak{F}}(G) \in \mathfrak{F}$, a contradiction.

Theorem 4.9 (Guo and Skiba [194]). *Suppose that G has three subgroups A_1, A_2, and A_3 whose indices $|G : A_1|$, $|G : A_2|$, $|G : A_3|$ are pairwise coprime. If $A_i \cap A_j \leq Z_{\mathfrak{S}}(A_i) \cap Z_{\mathfrak{S}}(A_j)$ for all $i \neq j$, then G is soluble.*

Proof Assume that it is false and let G be a counterexample with $|G|$ minimal. Then by Lemma 4.8, $A_i \cap A_j \neq 1$ for all $i \neq j$. Since $A_1 \cap A_2 \neq 1$ and $A_1 \cap A_2 \leq Z_{\mathfrak{S}}(A_2)$, for some minimal normal subgroup V of A_2 we have $V \leq Z_{\mathfrak{S}}(A_2)$. Hence, V is a p-group for some prime p. Then either p does not divide $|G : A_1|$ or p does not divide $|G : A_3|$. Assume that p does not divide $|G : A_1|$. Then for some $b \in A_2$, we have $V \leq A_1^b$. Hence, $V = V^{b^{-1}} \leq A_1 \cap A_2 \leq Z_\mathfrak{S}(A_2)$, which implies that $V^G = V^{A_2 A_1} = V^{A_1} \leq A_1$. Then $E = V^G \cap A_2 \leq A_1 \cap A_2 \leq Z_\mathfrak{S}(A_2)$ and E is normal in A_2. Hence, $E^G = E^{A_2 A_1} = E^{A_1} \leq A_1$. It follows that $E^G = E^{A_1} \subseteq Z_\mathfrak{S}(A_1)^{A_1} = Z_\mathfrak{S}(A_1)$ and so E^G is soluble. This shows that G has a soluble normal subgroup $E^G \leq A_1$. Let L be a minimal normal subgroup of G contained in E^G. Then L is a p-group, for some prime p. Hence, either $L \leq A_2$ or $L \leq A_3$. Clearly, the hypothesis holds for G/L. Hence, G/L is soluble by the choice of G, which implies the solubility of G. This contradiction completes the proof.

Corollary 4.10 (Wielandt [447]). *If G has three soluble subgroups A_1, A_2, and A_3 whose indices $|G : A_1|$, $|G : A_2|$, $|G : A_3|$ are pairwise coprime, then G is itself soluble.*

In the following theorem, $c(G)$ denotes the nilpotent class of the nilpotent group G.

Theorem 4.11 (Guo and Skiba [194]). *Suppose that G has three subgroups A_1, A_2, and A_3 whose indices $|G : A_1|$, $|G : A_2|$, $|G : A_3|$ are pairwise coprime. Let p be a prime. Then:*

(1) If $A_i \cap A_j \leq Z_\mathfrak{F}(A_i) \cap Z_\mathfrak{F}(A_j)$ for all $i \neq j$, where $\mathfrak{F}$ is the class of all p-closed groups, then G is p-closed.

(2) If $A_i \cap A_j \leq Z_\mathfrak{F}(A_i) \cap Z_\mathfrak{F}(A_j)$ for all $i \neq j$, where $\mathfrak{F}$ is the class of all p-decomposable groups, then G is p-decomposable.

(3) If $A_i \cap A_j \leq Z_n(Z_\infty(A_i)) \cap Z_n(Z_\infty(A_j))$ for all $i \neq j$, then G is nilpotent and $c(G) \leq n$.

Proof

(1) Assume that it is false and let G be a counterexample with $|G|$ minimal. Then p divides $|G|$. By hypothesis, there exists $i \neq j$ such that $p \nmid |G : A_i|$ and $p \nmid |G : A_j|$. Hence, $p \nmid |G : A_i \cap A_j|$ and G has a Sylow p-subgroup P such that $P \leq A_i \cap A_j \leq Z_\mathfrak{F}(A_i) \cap Z_\mathfrak{F}(A_j)$. By Lemma 4.2(5), P is a characteristic subgroup of $Z_\mathfrak{F}(A_i)$ and hence P is normal in A_i. Similarly, we have $A_j \leq N_{A_j}(P)$. Hence, $G = A_i A_j \leq N_G(P)$. Therefore, G is p-closed, a contradiction.

(2) Assume that this is false and let G be a counterexample with $|G|$ minimal. Then p divides $|G|$. Let P be a Sylow p-subgroup of G contained in $A_i \cap A_j$, for some $i \neq j$. By (1), P is normal in G. Let L be a minimal normal subgroup of G contained in P and $C = C_G(L)$. Then the hypothesis is true for G/L by Lemma 4.1. Thus, G/L is p-decomposable by the choice of G. Since $L \leq Z_\mathfrak{F}(A_i) \cap Z_\mathfrak{F}(A_j)$, $A_i/C_{A_i}(L)$ and $A_j/C_{A_j}(L)$ are p-groups by Lemma 4.2(6). Since $G/C = (A_i C/C)(A_j C/C)$, where $A_i C/C \simeq A_i/A_i \cap C = A_i/C_{A_i}(L)$ and $A_j C/C \simeq A_l/A_j \cap C = A_j/C_{A_j}(L)$, we have that G/C is a p-group. But since L is a p-group and $O_p(G/C) = 1$, we have $G = C$. Since G/L is p-decomposable, G/L has a normal Hall p'-subgroup E/L. By Schur–Zassenhaus Theorem, E has a Hall p'-subgroup V. Since $L \leq Z(E)$, V is characteristic in E and so V is normal G. It follows that G p-decomposable. This contradiction completes the proof of (2).

(3) By (2), G is nilpotent. Let P be a Sylow subgroup of G. Then for some $i \neq j$ we have $P \leq A_i \cap A_j \leq Z_n(Z_\infty(A_i)) \cap Z_n(Z_\infty(A_j))$. Hence, $c(P) \leq n$, which implies $c(G) \leq n$.

The theorem is proved.

Corollary 4.12 (Kegel [264]). *If G has three nilpotent subgroups A_1, A_2, and A_3 whose indices $|G : A_1|$, $|G : A_2|$, $|G : A_3|$ are pairwise coprime, then G is itself nilpotent.*

Corollary 4.13 (Doerk [87]). *If G has three abelian subgroups A_1, A_2, and A_3 whose indices $|G : A_1|$, $|G : A_2|$, $|G : A_3|$ are pairwise coprime, then G is itself abelian.*

A subgroup H of G is said to be abnormal if $x \in \langle H, H^x \rangle$ for all $x \in G$. It is clear that if H is a abnormal in G, then $N_G(H) = H$ (see Chap.2, Lemma 5.16).

The following lemma is well known.

Lemma 4.14 *Let A and B be proper subgroups of a group G such that $G = AB$. Then:*

1) $G_p = A_p B_p$ for some Sylow p-subgroup G_p, A_p and B_p of G, A, B, respectively;
2) $G = A^x B$ and $G \neq AA^x$ for all $x \in G$.

Lemma 4.15 *Let $\mathfrak{H}$ be a formation with $\mathfrak{H} \subseteq \mathfrak{N}$ and $\mathfrak{F} = \mathfrak{G}_p \mathfrak{H}$, where p is a prime. Suppose that G has three subgroups A_1, A_2, and A_3 whose indices $|G : A_1|$, $|G : A_2|$, $|G : A_3|$ are pairwise coprime. Then $G \in \mathfrak{F}$ if one of the following conditions holds:*

(1) $A_1, A_2, A_3 \in \mathfrak{F}$;
(2) $A_1, A_2 \in \mathfrak{F}$, A_1 and A_2 are abnormal in G and a Sylow p-subgroup of G contained in $A_1 \cap A_2$.

Proof Assume that p does not divide $|G|$. Assume first that the condition (1) holds and let Q be any Sylow subgroup of G. Then G is nilpotent by Corollary 4.12 and $Q \leq A_i \leq \mathfrak{H}$ for some i. Hence, $G \in \mathfrak{H} \subseteq \mathfrak{F}$. Now assume that the condition (2) holds. Since $p \nmid |G|$, A_1 and A_2 belong to $\mathfrak{H}$ and so A_1 and A_2 are nilpotent. Then A_1 and A_2 are Carter subgroups of G. By Corollary 4.10, G is soluble. Hence, A_1 and A_2 are conjugate in G. It follows from Lemma 4.14 that $G = A_1 A_2 = A_1 = A_2 \in \mathfrak{F}$.

Now assume that p divide $|G|$. Then by hypothesis, G has a Sylow p-subgroup P such that $P \leq A_i \cap A_j$ for some $i \neq j$. Since $A_i \in \mathfrak{G}_p \mathfrak{H}$ where $\mathfrak{H} \subseteq \mathfrak{N}$, P is normal in A_i. Similarly, $A_j \leq N_G(P)$. Thus, P is normal in $G = A_i A_j$. By induction, we see that $G/P \in \mathfrak{F}$. This implies that $G \in \mathfrak{G}_p \mathfrak{G}_p \mathfrak{H} = \mathfrak{G}_p \mathfrak{H} = \mathfrak{F}$.

Theorem 4.16 (Guo and Skiba [194]). *Suppose that G has four subgroups A_1, A_2, A_3 and A_4 whose indices $|G : A_1|$, $|G : A_2|$, $|G : A_3|$, $|G : A_4|$ are pairwise coprime. If $A_i \cap A_j \leq Z_\mathfrak{U}(A_i) \cap Z_\mathfrak{U}(A_j)$ for all $i \neq j$, then G is supersoluble.*

Proof Assume that this theorem is false and let G be a counterexample with $|G|$ minimal. By Theorem 4.9, G is soluble. Let L be a minimal normal subgroup of G. Then L is a p-group for some prime p. By Lemma 4.1(2), we see that G/L satisfies that hypothesis. Hence, G/L is supersoluble by the choice of G. Obviously, there are different i, j and k such that $L \leq Z_\mathfrak{U}(A_i) \cap Z_\mathfrak{U}(A_j) \cap Z_\mathfrak{U}(A_k)$. Let $C = C_G(L)$. Then for any $h \in \{i, j, k\}$, $CA_h/C \simeq A_h/(C \cap A_h) \in \mathfrak{G}_p\mathfrak{A}(p-1)$ by Lemma 4.2(3). Hence, $G/C \in \mathfrak{G}_p\mathfrak{A}(p-1)$ by Lemma 4.15. By Lemma 2.15, we have $G/C \in \mathfrak{A}(p-1)$. Thus, $|L| = p$ by Lemma 3.29. It follows that G is supersoluble. This contradiction completes the proof.

Lemma 4.17   (see [89, A, 15.2]). *Let G be a primitive group. If G has an abelian minimal normal subgroup N, then $N = C_G(N) = F(G) = O_p(G)$, for some prime p.*

Corollary 4.18   (Doerk [87]). *If G has four supersoluble subgroups A_1, A_2, A_3 and A_4 whose indices $|G : A_1|$, $|G : A_2|$, $|G : A_3|$, $|G : A_4|$ are pairwise coprime, then G is supersoluble.*

Theorem 4.19   (Guo and Skiba [194]). *Suppose that G has three abnormal subgroups A_1, A_2 and A_3 whose indices $|G : A_1|$, $|G : A_2|$, $|G : A_3|$ are pairwise coprime.*

(1) If $A_i \cap A_j \leq Z_{\mathfrak{N}^2}(A_i) \cap Z_{\mathfrak{N}^2}(A_j)$ for all $i \neq j$, then G is metanilpotent.
(2) If $A_i \cap A_j \leq Z_\mathfrak{U}(A_i) \cap Z_\mathfrak{U}(A_j)$ for all $i \neq j$, then G is supersoluble.

Proof

(1) Assume that it is false and let G be a counterexample with $|G|$ minimal. By Theorem 4.9, G is soluble. Let L be a minimal normal subgroup of G. By Lemma 4.1, the hypothesis holds for G/L. The minimal choice of G implies that G/L is metanilpotent. By (1.3), the class of all metanilpotent groups is a saturated formation. Hence, L is the only minimal normal subgroup of G and $L \not\subseteq \Phi(G)$. Then G is a primitive group and so $L = C_G(L)$ by Lemma 4.17. Since G is soluble, L be a p-group for some prime p. By hypothesis, G has a Sylow p-subgroup P such that $P \leq A_i \cap A_j$, for some $i \neq j$. Thus, $L \leq Z_{\mathfrak{N}^2}(A_i) \cap Z_{\mathfrak{N}^2}(A_j)$. By Lemma 4.2(1), $A_i/C_{A_i}(L) = A_i/L \in \mathfrak{G}_p\mathfrak{N}$ and so $A_i \in \mathfrak{G}_p\mathfrak{N}$. Similarly, $A_j \in \mathfrak{G}_p\mathfrak{N}$. Hence, $G \in \mathfrak{G}_p\mathfrak{N} \subseteq \mathfrak{N}^2$ by Lemma 4.15. This contradiction shows that (1) holds.

(2) Assume that this is false and let G be a counterexample with $|G|$ minimal. Then G is soluble by Theorem 4.9. Let L be a minimal normal subgroup of G. Then, similar to the proof of (1), we see that G/L is supersoluble. Hence, L is the only minimal normal subgroup of G, L is a p-group for some prime p, $L \not\subseteq \Phi(G)$ and $L = C_G(L)$. Obviously, there are $i \neq j$ such that $L \leq Z_\mathfrak{U}(A_i) \cap Z_\mathfrak{U}(A_j)$. By Lemma 4.2(3), $A_i/C_{A_i}(L) = A_i/L$ and $A_j/C_{A_j}(L) = A_j/L$ belong to $\mathfrak{G}_p\mathfrak{A}(p-1)$. Hence, $G/L = (A_i/L)(A_j/L)$ belongs to $\mathfrak{G}_p\mathfrak{A}(p-1)$ by Lemma 4.15. Consequently, G is supersoluble, a contradiction. The theorem is proved.

Corollary 4.20   *If G has three abnormal metanilpotent subgroups A_1, A_2 and A_3 whose indices $|G : A_1|$, $|G : A_2|$, $|G : A_3|$ are pairwise coprime, then G is itself metanilpotent.*

Corollary 4.21 (Vasilyev [406]). *If G has three abnormal supersoluble subgroups A_1, A_2 and A_3 whose indices $|G : A_1|$, $|G : A_2|$, $|G : A_3|$ are pairwise coprime, then G is itself supersoluble.*

1.5   The Intersection of $\mathfrak{F}$-maximal Subgroups

Introduction Let $\mathfrak{X}$ be a class of groups. A subgroup U of G is called $\mathfrak{X}$-*maximal* in G provided

(a) $U \in \mathfrak{X}$, and

(b) if $U \leq V \leq G$ and $V \in \mathfrak{X}$, then $U = V$.

We use $\text{Int}_{\mathfrak{X}}(G)$ to denote the intersection of all $\mathfrak{X}$-maximal subgroups of G. Clearly, $\text{Int}_{\mathfrak{X}}(G)$ is a characteristic subgroup of G.

In the case where $\mathfrak{X}$ is a hereditary saturated formation, we have $Z_{\mathfrak{X}}(G) \leq \text{Int}_{\mathfrak{X}}(G)$, and the subgroup $\text{Int}_{\mathfrak{X}}(G)$ has properties similar to properties of the $\mathfrak{X}$-hypercenter $Z_{\mathfrak{X}}(G)$ of G (see Proposition 5.6 below).

It is clear that the intersection of all maximal abelian subgroups of a group G coincides with the center $Z(G)$ of G. In the paper [22], Baer proved that $\text{Int}_{\mathfrak{N}}(G)$ coincides with the hypercenter $Z_{\infty}(G) = Z_{\mathfrak{N}}(G)$ of G.

But, in general, for a saturated formation $\mathfrak{X}$, $Z_{\mathfrak{X}}(G) < \text{Int}_{\mathfrak{X}}(G)$, even when $\mathfrak{X} = \mathfrak{U}$ and G is soluble (see Theorem 5.3 and Remark 5.33 below).

In connection with these observations, the following natural questions arise:

Problem I What are the nonempty hereditary saturated formations $\mathfrak{F}$ with the equality

$$\text{Int}_{\mathfrak{F}}(G) = Z_{\mathfrak{F}}(G), \tag{1.7}$$

for all groups G?

Problem II What are the nonempty hereditary saturated formations $\mathfrak{F}$ satisfying the equality

$$\text{Int}_{\mathfrak{F}}(G) = Z_{\pi \mathfrak{F}}(G) \tag{1.8}$$

holds for all group G and at least one nonempty set of primes $\pi \subseteq \pi(\mathfrak{F})$?

The first problem was posed by L.A. Shemetkov in 1995 at the Gomel Algebraic seminar and was also written in [375, p. 41]).

In this section, we give the answer to Problem II, and as a corollary of this result, in the case $\pi = P$, we also give the answer to Problem I.

Firstly we introduce the following concepts.

Definition 5.1 Let $\mathfrak{X}$ be a nonempty class of groups and $\mathfrak{F} = LF(F)$ be a saturated formation, where F is the canonical local satellite of $\mathfrak{F}$. We say that $\mathcal{F}$ satisfies the π-*boundary condition* (*the boundary condition* if $\pi = \mathbb{P}$) *in* $\mathfrak{X}$ if $G \in \mathfrak{F}$ whenever $G \in \mathfrak{X}$ and G is an $F(p)$-critical group for at least one $p \in \pi$.

We say that $\mathfrak{F}$ satisfies the π-*boundary condition* if $\mathfrak{F}$ satisfies the π-boundary condition in the class of all groups.

If $\mathfrak{F}$ is a nonempty formation with $\pi(\mathfrak{F}) = \varnothing$, then $\mathfrak{F} = (1)$, and therefore for any group G we have $Z_{\mathfrak{F}}(G) = 1 = \text{Int}_{\mathfrak{F}}(G)$. In the other limited case, when $\mathfrak{F} = \mathfrak{G}$ is the class of all groups, we have $Z_{\mathfrak{F}}(G) = G = Int_{\mathfrak{F}}(G)$. Similarly, if $\mathfrak{F} = \mathfrak{S}$, then $Z_{\mathfrak{F}}(G) = G = Int_{\mathfrak{F}}(G)$ for every soluble group G.

For the general case, we shall prove the following.

Theorem 5.2 (Guo and Skiba [198]). *Let $\mathfrak{F}$ be a hereditary saturated formation with* $(1) \neq \mathfrak{F} \neq \mathfrak{S}$. *Let* $\pi \subseteq \pi(\mathfrak{F})$. *Then the equality*

$$Z_{\pi\mathfrak{F}}(G) = Int_{\mathfrak{F}}(G)$$

holds for each group G if and only if $\mathfrak{N} \subseteq \mathfrak{F} = \mathfrak{S}_{\pi'}\mathfrak{F}$ and $\mathfrak{F}$ satisfies the π-boundary condition.

Note that $N(p) = \mathfrak{S}_p$, where N is the canonical local satellite of $\mathfrak{N}$. Hence, every $N(p)$-critical group has prime order. This shows that $\mathfrak{N}$ satisfies the boundary condition and so the above-mentioned Baer's result is a corollary of Theorem 5.2.

Theorem 5.3 (Guo and Skiba [198]). *Let $\mathfrak{F}$ be a hereditary saturated formation of soluble groups with* $(1) \neq \mathfrak{F} \neq \mathfrak{S}$. *Let* $\pi \subseteq \pi(\mathfrak{F})$. *Then the equality*

$$Z_{\pi\mathfrak{F}}(G) = Int_{\mathfrak{F}}(G)$$

holds for each soluble group G if and only if $\mathfrak{N} \subseteq \mathfrak{F} = \mathfrak{S}_{\pi'}\mathfrak{F}$ and $\mathfrak{F}$ satisfies the π-boundary condition in the class of all soluble groups.

If for some classes $\mathfrak{F}$ and $\mathfrak{M}$ of groups we have $\mathfrak{F} \subseteq \mathfrak{M}$, then every $\mathfrak{F}$-maximal subgroup of a group is contained in some its $\mathfrak{M}$-maximal subgroup. Nevertheless, the following example shows that in general, $Int_{\mathfrak{F}}(G) \not\leq Int_{\mathfrak{M}}(G)$.

Example 5.4 Let $\mathfrak{F} = \mathfrak{U}$ and $\mathfrak{M}$ be the class of all p-supersoluble groups, where $p > 2$. Let q be a prime dividing $p - 1$ and $G = P \rtimes (Q \rtimes C)$, where C is a group of order p, Q is a simple $\mathbb{F}_q G$-module which is faithful for C and P is a simple $\mathbb{F}_p G$-module which is faithful for $Q \rtimes C$. Then, clearly, $P = Int_{\mathfrak{F}}(G)$ and $Int_{\mathfrak{M}}(G) = 1$.

This example is a motivation for the following result.

Theorem 5.5 (Guo and Skiba [198]). *Let $\mathfrak{F} \subseteq \mathfrak{M} = LF(M)$ be hereditary saturated formations with $\pi \subseteq \pi(\mathfrak{F})$, where M is the canonical local satellites of $\mathfrak{M}$.*

(a) *Suppose that $\mathfrak{N} \subseteq \mathfrak{M} = \mathfrak{S}_{\pi'}\mathfrak{M}$ and $\mathfrak{F}$ satisfies the π-boundary condition in $\mathfrak{M}$. Then the inclusion*

$$Int_{\mathfrak{F}}(G) \leq Int_{\mathfrak{M}}(G)$$

holds for each group G.

(b) *If every (soluble) $M(p)$-critical group belongs to $\mathfrak{F}$ for every $p \in \pi$, then $\mathfrak{N} \subseteq \mathfrak{M}$ and*

$$Int_{\mathfrak{F}}(G) \leq Z_{\pi\mathfrak{M}}(G)$$

for every (soluble) group G.

We use $\psi_0(N)$ denotes the subgroup of N which is generated by all its cyclic subgroups of prime order and order 4 (if the Sylow 2-subgroups of N are non-abelian).

The proofs of Theorems 5.2, 5.3 and 5.5 are based on the following general facts on the subgroup $Int_{\mathfrak{F}}(G)$.

General Properties of the Subgroup $Int_{\mathfrak{F}}(G)$ The applications of the subgroup $Int_{\mathfrak{F}}(G)$ are based on the following.

Proposition 5.6 (Skiba [386]). *Let $\mathfrak{F}$ be a hereditary saturated formation $\varnothing \neq \pi \subseteq \pi(\mathfrak{F})$. Let H, E be subgroups of G, N a normal subgroup of G and $I = Int_{\mathfrak{F}}(G)$. Then:*

(a) *$Int_{\mathfrak{F}}(H)N/N \leq Int_{\mathfrak{F}}(HN/N)$.*
(b) *$Int_{\mathfrak{F}}(H) \cap E \leq Int_{\mathfrak{F}}(H \cap E)$.*
(c) *If $H/H \cap I \in \mathfrak{F}$, then $H \in \mathfrak{F}$. In particular, if $G/Int_{\mathfrak{F}}(G) \in \mathfrak{F}$, then $G \in \mathfrak{F}$.*
(d) *If $H \in \mathfrak{F}$, then $IH \in \mathfrak{F}$.*
(e) *If $N \leq I$, then $I/N = Int_{\mathfrak{F}}(G/N)$.*
(f) *$Int_{\mathfrak{F}}(G/I) = 1$.*
(g) *If every $\mathfrak{F}$-critical subgroup of G is soluble and $\psi_0(N) \leq I$, then $N \leq I$.*
(h) *If $\mathfrak{S}_{\pi'}\mathfrak{F} = \mathfrak{F}$, then $Z_{\pi\mathfrak{F}}(G) \leq I$. In particular, $Z_{\mathfrak{F}}(G) \leq I$.*

In order to prove Proposition 5.6, we first prove the following

Lemma 5.7 *Let $\mathfrak{F}$ be a hereditary saturated formation. Let $N \leq U \leq G$, where N is a normal subgroup of G.*

(i) *If $G/N \in \mathfrak{F}$ and V is a minimal supplement of N in G, then $V \in \mathfrak{F}$.*
(ii) *If U/N is an $\mathfrak{F}$-maximal subgroup of G/N, then $U = U_0N$ for some $\mathfrak{F}$-maximal subgroup U_0 of G.*
(iii) *If V is an $\mathfrak{F}$-maximal subgroup of U, then $V = H \cap U$ for some $\mathfrak{F}$-maximal subgroup H of G.*

Proof

(i) It is clear that $V \cap N \leq \Phi(V)$. Hence, by $V/V \cap N \simeq VN/N = G/N \in \mathfrak{F}$, we have $V \in \mathfrak{F}$ since $\mathfrak{F}$ is saturated.
(ii) Let V be a minimal supplement of N in U. Then $V \in \mathfrak{F}$ by (i). Let U_0 be an $\mathfrak{F}$-maximal subgroup of G such that $V \leq U_0$. Then $U/N = VN/N \leq U_0N/N \simeq U_0/U_0 \cap N \in \mathfrak{F}$. Hence, $U = U_0N$.
(iii) Let H be an $\mathfrak{F}$-maximal subgroup of G such that $V \leq H$. Then $V \leq H \cap U \in \mathfrak{F}$ since $\mathfrak{F}$ is hereditary, which implies $V = H \cap U$.

Proof of Proposition 5.6

(a) First suppose that $H = G$. If U/N is an $\mathfrak{F}$-maximal subgroup of G/N, then for some $\mathfrak{F}$-maximal subgroup U_0 of G we have $U = U_0N$ by Lemma 5.7(ii). Let $Int_{\mathfrak{F}}(G/N) = U_1/N \cap \ldots \cap U_t/N$, where U_i/N is an $\mathfrak{F}$-maximal subgroup of G/N for all $i = 1,\ldots,t$. Let V_i be an $\mathfrak{F}$-maximal subgroup of G such that $U_i = V_iN$. Then $I \leq V_1 \cap \ldots \cap V_t$, and consequently $IN/N \leq Int_{\mathfrak{F}}(G/N)$. Now let H be any subgroup of G, and let $f : H/H \cap N \to HN/N$ be the canonical isomorphism from $H/H \cap N$ onto HN/N. Then $f(Int_{\mathfrak{F}}(H/H \cap$

$N)) = \text{Int}_{\mathfrak{F}}(HN/N)$ and $f(\text{Int}_{\mathfrak{F}}(H)(H \cap N)/(H \cap N)) = \text{Int}_{\mathfrak{F}}(H)N/N$. But from above we have $\text{Int}_{\mathfrak{F}}(H)(H \cap N)/(H \cap N) \leq \text{Int}_{\mathfrak{F}}(H/H \cap N)$. Hence, $\text{Int}_{\mathfrak{F}}(H)N/N \leq \text{Int}_{\mathfrak{F}}(HN/N)$.

(b) If V is any $\mathfrak{F}$-maximal subgroup of H, then $V = H \cap U$ for some $\mathfrak{F}$-maximal subgroup U of G by Lemma 5.7(iii). Thus, there are $\mathfrak{F}$-maximal subgroups $U_1, \ldots, U_t$ of G such that $\text{Int}_{\mathfrak{F}}(H) = U_1 \cap \ldots \cap U_t \cap H$. This means that $I \cap H \leq \text{Int}_{\mathfrak{F}}(H)$ and $\text{Int}_{\mathfrak{F}}(H) \cap E = \text{Int}_{\mathfrak{F}}(H) \cap (H \cap E) \leq \text{Int}_{\mathfrak{F}}(H \cap E)$.

(c) First suppose that $H = G$. Let U be a minimal supplement of I in G. Then $U \in \mathfrak{F}$ by Lemma 5.7(i). Let V be an $\mathcal{F}$-maximal subgroup of G containing U. Then $G = IU = V \in \mathfrak{F}$. Now we consider the general case. By (b), $I \cap H \leq \text{Int}_{\mathfrak{F}}(H)$. Hence, from $H/H \cap I \in \mathfrak{F}$ we obtain that $H/\text{Int}_{\mathfrak{F}}(H) \in \mathfrak{F}$ and therefore $H \in \mathfrak{F}$.

(d) Since $H \in \mathfrak{F}$, $HI/I \simeq H/H \cap I \in \mathfrak{F}$. By (b), $I \leq \text{Int}_{\mathfrak{F}}(HI)$. Hence, $HI/\text{Int}_{\mathfrak{F}}(HI) \in \mathfrak{F}$. Thus, $HI \in \mathfrak{F}$ by (c).

(e) In view of Lemma 5.7(ii), it is enough to prove that if U is an $\mathfrak{F}$-maximal subgroup of G, then U/N is an $\mathcal{F}$-maximal subgroup of G/N. Let $U/N \leq X/N$, where X/N is an $\mathfrak{F}$-maximal subgroup of G/N. By Lemma 5.7(ii), $X = U_0 N$ for some $\mathfrak{F}$-maximal subgroup U_0 of G. But since $N \leq U_0$, $U/N \leq U_0/N$. This implies that $U = U_0$. Thus, $U/N = X/N$.

(f) This follows from (e).

(g) Suppose that this assertion is false and let G be a counterexample with $|G| + |N|$ minimal. Then there is an $\mathcal{F}$-maximal subgroup U of G such that $N \nleq U$. Let $E = NU$. Then $E/N \simeq U/U \cap N \in \mathfrak{F}$. By (b), $\psi_0(N) \leq I \cap E \leq \text{Int}_{\mathfrak{F}}(E)$. Assume that $E \neq G$. Then $N \leq \text{Int}_{\mathfrak{F}}(E)$ by the choice of (G, N). Hence, $E/\text{Int}_{\mathfrak{F}}(E) \in \mathfrak{F}$. It follows from (c) that $E \in \mathfrak{F}$ and consequently $U = E$. Therefore, $N \leq U$. This contradiction shows that $E = G$. Let M be any maximal subgroup of G. We show that $M \in \mathfrak{F}$. Since $\psi_0(N \cap M) \leq \psi_0(N)$, $\psi_0(N \cap M) \leq I \cap M$. Hence, $\psi_0(N \cap M) \leq \text{Int}_{\mathfrak{F}}(M)$ by (b). The choice of (G, N) implies that $N \cap M \leq \text{Int}_{\mathfrak{F}}(M)$. Note that $M/M \cap N \in \mathfrak{F}$. Indeed, if $N \leq M$, then $M/N \leq G/N \in \mathfrak{F}$. On the other hand, if $N \nleq M$, then $M/M \cap N \simeq NM/N = G/N \in \mathfrak{F}$. Therefore, $M \in \mathfrak{F}$ by (c). Hence, $I = \Phi(G)$ and G is an $\mathfrak{F}$-critical group. Consequently, G is soluble. Since $G/N \in \mathfrak{F}$, $\psi_0(G^{\mathfrak{F}}) \leq \psi_0(N) \leq I$. Thus, for any $x \in G^{\mathfrak{F}} \setminus \Phi(G^{\mathfrak{F}})$ we have $x \in \psi_0(N) \leq I = \Phi(G)$ by Proposition 1.8. This implies that $G^{\mathfrak{F}} \leq \Phi(G)$ and so $I = G \in \mathfrak{F}$, a contradiction. Hence, we have (g).

(h) Let H be a subgroup of G such that $H \in \mathfrak{F}$. Then $HZ_{\pi\mathfrak{F}}(G)/Z_{\pi\mathfrak{F}}(G) \simeq H/H \cap Z_{\pi\mathfrak{F}}(G) \in \mathfrak{F}$ and $Z_{\pi\mathfrak{F}}(G) \leq Z_{\pi\mathfrak{F}}(HZ_{\pi\mathfrak{F}}(G))$ by Theorem 2.7(a). It follows from Theorem 2.7(b) that $HZ_{\pi\mathfrak{F}}(G) \in \mathfrak{F}$. Thus, $Z_{\pi\mathfrak{F}}(G) \leq I$.

Lemma 5.8 *Let $\mathfrak{F} = LF(F)$ be a saturated formation with $p \in \pi(\mathfrak{F})$, where F is the canonical local satellite of $\mathfrak{F}$. Suppose that G is a group of minimal order in the set of all $F(p)$-critical groups G with $G \notin \mathfrak{F}$. Then $O_p(G) = 1 = \Phi(G)$ and $G^{\mathfrak{F}}$ is the unique minimal normal subgroup of G.*

Proof Let N be a minimal normal subgroup of G. We first show that $G/N \in \mathfrak{F}$. Indeed, suppose that $G/N \notin \mathfrak{F}$. Then $G/N \notin F(p)$ since $F(p) \subseteq \mathfrak{F}$. On the other hand, for any maximal subgroup M/N of G/N we have $M/N \in F(p)$ since $F(p)$ is a formation and G is an $F(p)$-critical group. Thus, G/N is an $F(p)$-critical group with $G/N \notin \mathfrak{F}$, which contradicts the minimality of G. Hence, $G/N \in \mathfrak{F}$. Since $\mathfrak{F}$ is a saturated formation, $N = G^{\mathfrak{F}}$ is a unique minimal normal subgroup of G and $\Phi(G) = 1$. Suppose that $N \leq O_p(G)$ and let M be a maximal subgroup of G such that $G = NM$. Then $G/N \simeq M/N \cap M \in F(p) = \mathfrak{G}_p F(p)$. It follows that $G \in F(p) \subseteq \mathfrak{F}$. This contradiction shows that $O_p(G) = 1$. The lemma is proved.

Lemma 5.9 (see [366, Lemma 18.8]). *Suppose that $O_p(G) = 1$. If G has a unique minimal normal subgroup, then there exists a simple $\mathbb{F}_p G$-module L, which is faithful for G, that is, $C_G(L) = 1$.*

Proof of Theorem 5.5

(a) Suppose that this assertion is false and let G be a counterexample with minimal order. Let $I = \mathrm{Int}_{\mathfrak{F}}(G)$ and $I_1 = \mathrm{Int}_{\mathfrak{M}}(G)$. Then $1 < I < G$ and $I_1 \neq G$. Let L be a minimal normal subgroup of G contained in I and $C = C_G(L)$. Then $\pi(L) \subseteq \pi(\mathfrak{F})$.

(1) $IN/N \leq \mathrm{Int}_{\mathfrak{F}}(G/N) \leq \mathrm{Int}_{\mathfrak{M}}(G/N)$ for any non-identity normal subgroup N of G.

Indeed, by Proposition 5.6(a), we have $IN/N \leq \mathrm{Int}_{\mathfrak{F}}(G/N)$. On the other hand, by the choice of G, $\mathrm{Int}_{\mathfrak{F}}(G/N) \leq \mathrm{Int}_{\mathfrak{M}}(G/N)$.

(2) $L \not\leq I_1$; in particular, the order of L is divisible by some prime $p \in \pi$.

Suppose that $L \leq I_1$. Then $I_1/L = \mathrm{Int}_{\mathfrak{M}}(G/L)$ by Proposition 5.6(e). But by (1), $IL/L = I/L \leq \mathrm{Int}_{\mathfrak{F}}(G/L) \leq \mathrm{Int}_{\mathfrak{M}}(G/L)$. Hence, $I/L \leq I_1/L$ an so $I \leq I_1$, a contradiction. Thus, $L \not\leq I_1$. This means that there exists an $\mathfrak{M}$-maximal subgroup M of G such that $L \not\leq M$. Suppose that L is a π'-group. Then $LM \in \mathfrak{G}_{\pi'}\mathfrak{M} = \mathfrak{M}$, which contradicts the maximality of M. Hence, the order of L is divisible by some prime $p \in \pi$.

(3) If $L \leq M < G$, then $L \leq \mathrm{Int}_{\mathfrak{M}}(M)$.

By Proposition 5.6(b), $L \leq I \cap M \leq \mathrm{Int}_{\mathfrak{F}}(M)$. But since $|M| < |G|$, $\mathrm{Int}_{\mathfrak{F}}(M) \leq \mathrm{Int}_{\mathfrak{M}}(M)$ by the choice of G. Hence, $L \leq \mathrm{Int}_{\mathfrak{M}}(M)$.

(4) $G = LU$ for any $\mathfrak{M}$-maximal subgroup U of G not containing L. In particular, $G/L \in \mathfrak{M}$.

Indeed, suppose that $LU \neq G$. Then by (3), $L \leq \mathrm{Int}_{\mathfrak{M}}(LU)$, which implies that $LU \in \mathfrak{M}$ by Proposition 5.6(c). This contradicts the $\mathfrak{M}$-maximality of U. Hence, we have (4).

(5) $C_G(L) \cap U = U_G = 1$ for any $\mathfrak{M}$-maximal subgroup U of G not containing L.

Since $C_G(L)$ is normal in G and $G = LU$ by (4), $U_G = C_G(L) \cap U$. Assume that $U_G \neq 1$. Let $U/U_G \leq W/U_G$, where W/U_G is an $\mathfrak{M}$-maximal subgroup of G/U_G. Then by (1), $LU_G/U_G \leq W/U_G$. Hence, $G = LU \leq W$ by (4), which means that $G/U_G = W/U_G \in \mathfrak{M}$. But by (4), $G/L \in \mathfrak{M}$. Therefore, $G \simeq G/L \cap U_G \in \mathfrak{M}$, and consequently $I = G$, a contradiction. Hence, (5) holds.

The final contradiction for (a).

Since $L \not\leq I_1$ by (2), there is an $\mathfrak{M}$-maximal subgroup M of G such that $L \not\leq M$. But then $G = LM$ by (4). Since $L \leq I$ and $G \notin \mathfrak{F}$, $M \notin \mathfrak{F}$ by Proposition 5.6(d). Let H be an $\mathfrak{F}$-critical subgroup of M, V a maximal subgroup of H. We show that $V \in F(p)$. By Proposition 5.6(d), $D = LV \in \mathfrak{F}$. Hence, $D/O_{p',p}(D) \in F(p)$. First assume that L is a non-abelian group. Then, since p divides $|L|$, $O_{p',p}(D) \cap L = 1$. Hence, $O_{p',p}(D) \leq C_G(L)$ and $O_{p',p}(D) \cap V = 1$ by (5). Since $\mathfrak{F}$ is hereditary, $F(p)$ is hereditary by Proposition 1.7(1). Therefore, $O_{p',p}(D)V/O_{p',p}(D) \simeq V \in F(p)$. Now assume that L is an abelian p-group. Then $L \leq O_{p',p}(D)$ and $O_{p',p}(D) = L(O_{p',p}(D) \cap V)$. Hence, $O_{p'}(D) \leq M \cap C_G(L) = 1$. It follows that $O_{p',p}(D) = O_p(D)$. Therefore, $D/O_p(D) \in F(p) = \mathfrak{G}_p F(p)$, which implies that $D \in F(p)$ and so $V \in F(p)$. Therefore, H is an $F(p)$-critical group. Since $\mathfrak{M}$ is hereditary and $M \in \mathfrak{M}$, $H \in \mathfrak{M}$. But then $H \in \mathfrak{F}$ since $\mathfrak{F}$ satisfies the π-boundary condition in $\mathfrak{M}$ by hypothesis. This contradiction completes the proof of (a).

(b) Suppose that every $M(p)$-critical group G belongs to $\mathfrak{F}$ for every $p \in \pi$. First we show that $\mathfrak{N} \subseteq \mathfrak{M}$. Assume that this is false and let C_q be a group of prime order q with $C_q \notin \mathfrak{M}$. Let $p \in \pi$. Then C_q is $M(p)$-critical and so $C_q \in \mathfrak{F} \subseteq \mathfrak{M}$ by the hypothesis. This contradiction shows that $\mathfrak{N} \subseteq \mathfrak{M}$.

Now we show that $\operatorname{Int}_{\mathfrak{F}}(G) \leq Z_{\pi\mathfrak{M}}(G)$ for every group G. Suppose that this assertion is false and let G be a counterexample with minimal order. Let $I = \operatorname{Int}_{\mathfrak{F}}(G)$ and $Z = Z_{\pi\mathfrak{M}}(G)$. Then $1 < I < G$ and $Z \neq G$. Let N be a minimal normal subgroup of G and L a minimal normal subgroup of G contained in I. Then $\pi(L) \leq \pi(\mathfrak{F})$. We proceed via the following steps.

(1) $IN/N \leq \operatorname{Int}_{\mathfrak{F}}(G/N) \leq Z_{\pi\mathfrak{M}}(G/N)$.

Indeed, by Proposition 5.6(a), we have $IN/N \leq \operatorname{Int}_{\mathfrak{F}}(G/N)$. On the other hand, by the choice of G, $\operatorname{Int}_{\mathfrak{F}}(G/N) \leq Z_{\pi\mathfrak{M}}(G/N)$.

(2) $L \not\leq Z$; in particular, the order of L is divisible by some prime $p \in \pi$.

Suppose that $L \leq Z$. Then, clearly, $Z/L = Z_{\pi\mathfrak{M}}(G/L)$, and $I/L = \operatorname{Int}_{\mathfrak{F}}(G/L)$ by Proposition 5.6(e). But by (1), $\operatorname{Int}_{\mathfrak{F}}(G/L) \leq Z_{\pi\mathfrak{M}}(G/L)$. Hence, $I/L \leq Z/L$. Consequently, $I \leq Z$, a contradiction.

(3) If $L \leq M < G$, then $L \leq Z_{\mathfrak{M}}(M)$.

By Proposition 5.6(b), $L \leq I \cap M \leq \operatorname{Int}_{\mathfrak{F}}(M)$. But since $|M| < |G|$, we have that $\operatorname{Int}_{\mathfrak{F}}(M) \leq Z_{\pi\mathfrak{M}}(M)$ by the choice of G. Hence, $L \leq Z_{\pi\mathfrak{M}}(M)$ and so $L \leq Z_{\mathfrak{M}}(M)$ since the order of L is divisible by some prime $p \in \pi$ by (2).

(4) $L = N$ is the unique minimal normal subgroup of G.

Suppose that $L \neq N$. Then by (1), $NL/N \leq Z_{\pi\mathfrak{M}}(G/N)$. Hence, from the G-isomorphism $NL/N \simeq L$ we obtain $L \leq Z$, which contradicts (2).

(5) $L \not\leq \Phi(G)$.

Suppose that $L \leq \Phi(G)$. Then L is a p-group by (2). Let $C = C_G(L)$ and M be any maximal subgroup of G. Then $L \leq M$. Hence, $L \leq Z_{\mathfrak{M}}(M)$ by (3), so $M/M \cap C \in M(p)$ by Proposition 1.7(1) and Lemma 2.26. If $C \not\leq M$, then $G/C = CM/C \simeq M/M \cap C \in M(p)$. This implies that $L \leq Z$, which contradicts (2). Hence, $C \leq M$ for all maximal subgroups M of G. It follows that C is nilpotent. Then in view of (4), C is a p-group since C is normal in G.

Hence, for every maximal subgroup M of G we have $M \in \mathfrak{G}_p M(p) = M(p)$. But since $M(p) \subseteq \mathfrak{M}$, $G \not\leq M(p)$ (otherwise $G \in \mathfrak{M}$ and so $G = Z$). This shows that G is an $M(p)$-critical group. Therefore, $G \in \mathfrak{F}$ by the hypothesis. But since $\mathfrak{F} \subseteq \mathfrak{M}$, we have $G \in \mathfrak{M}$ and so $G = Z$, a contradiction. Thus, (5) holds.

(6) $C = C_G(L) \leq L$ (This follows from (4), (5) and Theorem 15.6 in [89, Chap. A]).

(7) *If $L \leq M < G$, then $M \in M(p)$.*

First by (3), $L \leq Z_{\mathfrak{M}}(M)$. If $L = C$, then $M/L = M/M\cap C \in M(p)$ by Lemma 2.26, which implies that $M \in \mathfrak{G}_p M(p) = M(p)$ since L is a p-group by (2). Now suppose that L is a non-abelian group. Let $1 = L_0 < L_1 < \ldots < L_n = L$ be a chief series of M below L. Let $C_i = C_M(L_i/L_{i-1})$ and $C_0 = C_1 \cap \ldots \cap C_n$. Since $L \leq Z_{\mathfrak{M}}(M)$, $M/C_i \in M(p)$ for all $i = 1, \ldots, n$. It follows that $M/C_0 \in M(p)$. Since $C = 1$ by (4) and (6), for any minimal normal subgroup R of M we have $R \leq L$. Suppose that $C_0 \neq 1$ and let R be a minimal normal subgroup of M contained in C_0. Then $R \leq L$ and $R \leq C_M(H/K)$ for each chief factor H/K of M. Thus, $R \leq F(M)$ is abelian and hence L is abelian. This contradiction shows that $C_0 = 1$. Consequently, $M \in M(p)$.

(8) *There exists a subgroup U of G such that $U \in \mathfrak{F}$ and $LU = G$.*

Indeed, suppose that every maximal subgroup of G not containing L belongs to $M(p)$. Then by (7), G is an $M(p)$-critical group. Hence, $G \in \mathfrak{F}$ by the hypothesis. But then $I = G$, a contradiction. Hence, there exists a maximal subgroup M of G such that $G = LM$ and $M \notin M(p)$. Take an $M(p)$-critical subgroup U of M. Then in view of (7), $LU = G$ and $U \in \mathfrak{F}$ by the hypothesis.

(9) *The final contradiction for (b).*

Since $L \leq I$ and $G/L = UL/L \simeq U/U \cap L \in \mathfrak{F}$ by (8), it follows from Proposition 5.6(c) that $G \in \mathfrak{F}$ and so $G = I$. The final contradiction shows that $\operatorname{Int}_{\mathcal{F}}(G) \leq Z_{\pi \mathfrak{M}}(G)$ for every group G. The second assertion of (b) can be proved similarly. The theorem is proved.

Proof of Theorem 5.2 and 5.3 Since $Z_{\mathfrak{F}}(G) \leq \operatorname{Int}_{\mathfrak{F}}(G)$ by Proposition 5.6(h), the sufficiency is a special case, when $\mathfrak{F} = \mathfrak{M}$, of Theorem 5.5(b). Now suppose that the equality $Z_{\pi \mathfrak{F}}(G) = \operatorname{Int}_{\mathfrak{F}}(G)$ holds for each (soluble) group G.

First we show that $\mathfrak{N} \subseteq \mathfrak{F}$. Let F be the canonical local satellite of $\mathfrak{F}$. Suppose that for some group C_q of prime order q we have $C_q \notin \mathfrak{F}$. Let $p \in \pi$ and $G = PC_q$, where P is a simple $\mathbb{F}_p C_q$-module P which is faithful for C_q. Then $P = \operatorname{Int}_{\mathfrak{F}}(G)$ and $Z_{\mathfrak{F}}(G) = 1$ since $F(p) \subseteq \mathfrak{F}$. This contradiction shows that $\mathfrak{N} \subseteq \mathfrak{F}$.

Now we show that $\mathfrak{G}_{\pi'} \mathfrak{F} = \mathfrak{F}$ ($\mathfrak{S}_{\pi'} \mathfrak{F} = \mathfrak{F}$, respectively). The inclusion $\mathfrak{F} \subseteq \mathfrak{G}_{\pi'} \mathfrak{F}$ is evident. Suppose that $\mathfrak{G}_{\pi'} \mathfrak{F} \not\subseteq \mathfrak{F}$ ($\mathfrak{S}_{\pi'} \mathfrak{F} \not\subseteq \mathfrak{F}$) and let G be a group of minimal order in $\mathfrak{G}_{\pi'} \mathfrak{F} \setminus \mathfrak{F}$ (in $\mathfrak{S}_{\pi'} \mathfrak{F} \setminus \mathfrak{F}$, respectively). Then $G^{\mathfrak{F}}$ is the unique minimal normal subgroup of G and $G^{\mathfrak{F}}$ is a π'-group. Hence,

$$G^{\mathfrak{F}} \leq Z_{\pi \mathfrak{F}}(G) = \operatorname{Int}_{\mathfrak{F}}(G).$$

It follows from Proposition 5.6(c) that $G \in \mathfrak{F}$. This contradiction shows that $\mathfrak{G}_{\pi'} \mathfrak{F} = \mathfrak{F}$ ($\mathfrak{S}_{\pi'} \mathfrak{F} = \mathfrak{F}$, respectively).

Finally, we show that $\mathfrak{F}$ satisfies the π-boundary condition (the π-boundary condition in the class $\mathfrak{S}$, $\mathfrak{F} \mathfrak{S}$, respectively). Suppose that this is false. Then for some

$p \in \pi$, the set of all (soluble) $F(p)$-critical groups A $A \notin \mathfrak{F}$ is nonempty. Let us choose in this set a group A with minimal $|A|$. Then by Lemma 5.8, $A^{\mathfrak{F}}$ is the unique minimal normal subgroup of G and $O_p(A) = 1 = \Phi(A)$. Hence, by Lemma 5.9, there exists a simple $\mathbb{F}_p A$-module P which is faithful for A. Let $G = P \rtimes A$ and M be any maximal subgroup of G. If $P \not\leq M$, then $M \simeq G/P \simeq A \notin \mathfrak{F}$. On the other hand, if $P \leq M$, then $M = M \cap PA = P(M \cap A)$, where $M \cap A$ is a maximal subgroup of A. Hence, $M \cap A \in F(p)$ and so $M \in \mathfrak{G}_p F(p) = F(p) \subseteq \mathfrak{F}$. Therefore, P is contained in the intersection of all $\mathfrak{F}$-maximal subgroups of G. Then $P \leq Z_{\pi\mathfrak{F}}(G)$ by our assumption on $\mathcal{F}$. It follows that $A \simeq G/P = G/C_G(P) \in F(p) \subseteq \mathfrak{F}$ by Proposition 1.15 and Lemma 2.26. But this contradicts the choice of A. Therefore, $\mathfrak{F}$ satisfies the π-boundary condition ($\mathfrak{F}$ satisfies the π-boundary condition in the class $\mathfrak{G}$). The theorems are proved.

In view of Theorems 5.2, 5.3, and 5.5 we have

Corollary 5.10 *Let $\mathfrak{F}$ be a hereditary saturated formation with* $(1) \neq \mathfrak{F} \neq \mathfrak{G}$. *Then the equality $Int_{\mathfrak{F}}(G) = Z_{\mathfrak{F}}(G)$ holds for each group G if and only if $\mathfrak{N} \subseteq \mathfrak{F}$ and $\mathfrak{F}$ satisfies the boundary condition.*

Corollary 5.11 *Let $\mathfrak{F}$ be a hereditary saturated formation of soluble groups with* $(1) \neq \mathfrak{F} \neq \mathfrak{G}$. *Then the equality $Int_{\mathfrak{F}}(G) = Z_{\mathfrak{F}}(G)$ holds for each soluble group G if and only if $\mathfrak{N} \subseteq \mathfrak{F}$ and $\mathfrak{F}$ satisfies the boundary condition in the class $\mathfrak{G}$.*

For a class $\mathfrak{X}$ of groups, a subgroup H of a group G is said to be $\mathfrak{X}$-*subnormal* (in the sense of Kegel [265]) or K-$\mathfrak{X}$-*subnormal* in G (see p. 236 in [34]) if either $H = G$ or there exists a chain of subgroups

$$H = H_0 < H_1 < \ldots < H_t = G$$

such that either H_{i-1} is normal in H_i or $H_i/(H_{i-1})_{H_i} \in \mathfrak{X}$ for all $i = 1, \ldots, t$.

It is clear that a subnormal subgroup of G is a K-$\mathfrak{X}$-subnormal in G.

Lemma 5.12 *Let $\mathfrak{F}$ be a formation, H and E be subgroups of a group G, where H is K-$\mathfrak{F}$-subnormal in G. Then:*

(i) *If $\mathfrak{F}$ is hereditary, then $H \cap E$ is K-$\mathfrak{F}$-subnormal in E (see Theorem 6.1.7(2) in [34]).*

(ii) *If E is normal in G, then HE/E is K-$\mathfrak{F}$-subnormal in G/E (see Theorem 6.1.6(3) in [34]).*

Lemma 5.13 *Let $\mathfrak{F}$ be a hereditary saturated formation. Let $N \leq U \leq G$, where N is a normal subgroup of G.*

(i) *If U/N is a non-K-$\mathfrak{F}$-subnormal $\mathfrak{F}$-maximal subgroup of G/N, then $U = U_0 N$ for some non-K-$\mathfrak{F}$-subnormal $\mathfrak{F}$-maximal subgroup U_0 of G.*

(ii) *If V is a non-K-$\mathfrak{F}$-subnormal $\mathfrak{F}$-maximal subgroup of U, then $V = H \cap U$ for some non-K-$\mathfrak{F}$-subnormal $\mathfrak{F}$-maximal subgroup H of G.*

Proof

(i) By Lemma 5.7(ii), there is an $\mathfrak{F}$-maximal subgroup U_0 of G such that $U = U_0N$. Since U/N is non-K-$\mathfrak{F}$-subnormal in G/N, U_0 is not K-$\mathfrak{F}$-subnormal in G by Lemma 5.12(ii).

(ii) By Lemma 5.7(iii), for some $\mathfrak{F}$-maximal subgroup H of G we have $V = H \cap U$. Since V is a non-K-$\mathfrak{F}$-subnormal in U, H is not K-$\mathfrak{F}$-subnormal in G by Lemma 5.12(i).

For any group G, we write $\mathrm{Int}^*_{\mathfrak{F}}(G)$ to denote the intersection of all non-K-$\mathfrak{F}$-subnormal $\mathcal{F}$-maximal subgroups of G. The following theorem shows that for any hereditary saturated formation $\mathfrak{F}$ with $\mathfrak{N} \subseteq \mathfrak{F}$, the intersection of all non-K-$\mathfrak{F}$-subnormal $\mathfrak{F}$-maximal subgroups of a group G coincides with $\mathrm{Int}_{\mathfrak{F}}(G)$.

Theorem 5.14 (Guo and Skiba [198]). *Let $\mathfrak{F}$ be a hereditary saturated formation containing all nilpotent groups. Then the equality*

$$Int^*_{\mathfrak{F}}(G) = Int_{\mathfrak{F}}(G)$$

holds for each group G.

Proof We will prove the theorem by induction on $|G|$. If $G \in \mathfrak{F}$, then

$$\mathrm{Int}^*_{\mathfrak{F}}(G) = G = \mathrm{Int}_{\mathfrak{F}}(G).$$

We may, therefore, assume that $G \notin \mathfrak{F}$. Let $I = \mathrm{Int}_{\mathfrak{F}}(G)$, $I^* = \mathrm{Int}^*_{\mathfrak{F}}(G)$ and N be a minimal normal subgroup of G. Then $I \leq I^*$. Hence, we may assume that $I^* \neq 1$.

(1) $I^*N/N \leq Int^*_{\mathfrak{F}}(G/N)$.

If U/N is a non-K-$\mathfrak{F}$-subnormal $\mathfrak{F}$-maximal subgroup of G/N, then for some non-K-$\mathfrak{F}$-subnormal $\mathcal{F}$-maximal subgroup U_0 of G we have $U = U_0N$ by Lemma 5.13(i). Let

$$\mathrm{Int}^*_{\mathfrak{F}}(G/N) = U_1/N \cap \ldots \cap U_t/N,$$

where U_i/N is a non-K-$\mathfrak{F}$-subnormal $\mathfrak{F}$-maximal subgroup of G/N for all $i = 1, \ldots, t$. Let V_i be a non-K-$\mathfrak{F}$-subnormal $\mathfrak{F}F$-maximal of G such that $U_i = V_iN$. Then $I^* \leq V_1 \cap \ldots \cap V_t$. Hence, $I^*N/N \leq \mathrm{Int}^*_{\mathfrak{F}}(G/N)$.

(2) *If $N \leq I^*$, then $Int^*_{\mathfrak{F}}(G/N) = I^*/N$.*

By Lemma 5.13(i), it is enough to prove that if U is a non-K-$\mathfrak{F}$-subnormal $\mathfrak{F}$-maximal subgroup of G, then U/N is a non-K-$\mathfrak{F}$-subnormal $\mathfrak{F}$-maximal subgroup of G/N. Let $U/N \leq X/N$, where X/N is a non-K-$\mathfrak{F}$-subnormal $\mathfrak{F}$-maximal subgroup of G/N. By Lemma 5.13(i), $X = U_0N$ for some non-K-$\mathfrak{F}$-subnormal $\mathfrak{F}$-maximal subgroup U_0 of G. But since $N \leq U_0$, $U/N \leq U_0/N$ and so $U = U_0$. Thus, $U/N = X/N$.

(3) *If $I^* \cap H \leq Int^*_{\mathfrak{F}}(H)$ for any subgroup H of G.*

Let V be an arbitrary non-K-$\mathfrak{F}$-subnormal $\mathcal{F}$-maximal subgroup of H. Then $V = H \cap U$ for some non-K-$\mathfrak{F}$-subnormal $\mathfrak{F}$-maximal subgroup U of G by Lemma 5.13(ii). Thus, there are non-K-$\mathfrak{F}$-subnormal $\mathcal{F}$-maximal subgroups $U_1, \ldots, U_t$ of G such that

$$\mathrm{Int}^*_{\mathfrak{F}}(H) = U_1 \cap \ldots \cap U_t \cap H.$$

This induces that $I^* \cap H \leq \mathrm{Int}^*_{\mathfrak{F}}(H)$.

(4) *If $E = DV$ for some normal subgroup D of G contained in I^* and some K-$\mathfrak{F}$-subnormal subgroup $V \in \mathfrak{F}$ of G, then $E \in \mathfrak{F}$.*

First note that $R \leq \mathrm{Int}_{\mathfrak{F}}^*(E)$ by (3). On the other hand, by Lemma 5.12(i), V is a K-$\mathfrak{F}$-subnormal subgroup of E. Hence, we need only consider the case when $G = E$. Assume that $G \notin \mathfrak{F}$. Let R be any minimal normal subgroup of G. Then $(DR/R)(VR/R) = G/R$, where $DR/R \leq \mathrm{Int}_{\mathfrak{F}}^*(G/R)$ by (1), and $VR/R \simeq V/V \cap R \in \mathfrak{F}$ is a K-$\mathfrak{F}$-subnormal subgroup of G/R. Hence, by induction we have $G/R \in \mathfrak{F}$. This implies that R is the only minimal normal subgroup of G and so $R = G^{\mathfrak{F}} \leq I^*$. Let W be a minimal supplement of R in G. Then $W \in \mathfrak{F}$ by Lemma 5.7(i). Let $W \leq M$, where M is an $\mathfrak{F}$-maximal subgroup of G. If M is not K-$\mathfrak{F}$-subnormal in G, then $R \leq M$. Thus, $G = RW = RM = M \in \mathfrak{F}$, a contradiction. This shows that M is K-$\mathfrak{F}$-subnormal in G. But then there is a proper subgroup X of G such that $M \leq X$ and either X is normal in G or $R = G^{\mathfrak{F}} \leq X$. In both of this cases, we have that $G = RM = RX = X < G$, a contradiction. Hence, we have (4).

Conclusion.

Let R be a minimal normal subgroup of G contained in I^*. If $R \leq I$, then $I/R = \mathrm{Int}_{\mathfrak{F}}(G/R)$ by Proposition 5.6(e), and $I^*/R = \mathrm{Int}_{\mathfrak{F}}^*(G/R)$ by (2). Therefore, by induction, $\mathrm{Int}_{\mathfrak{F}}^*(G/R) = \mathrm{Int}_{\mathfrak{F}}(G/R)$. It follows that $I = I^*$.

Finally, suppose that $R \not\leq I$. Then $R \not\leq U$ for some $\mathcal{F}$-maximal subgroup U of G. Let $E = RU$. Then $R \leq \mathrm{Int}_{\mathfrak{F}}^*(E)$ by (3). On the other hand, it is clear that U is a K-$\mathfrak{F}$-subnormal subgroup of G. Hence, by (4), $E \in \mathfrak{F}$. But then $E = U$, which implies $R \leq U$, a contradiction. The theorem is proved.

Some Applications of Theorems 5.2, 5.3, 5.5 and 5.14   We say that $\mathfrak{F}$ *satisfies the p-boundary condition* if $\mathfrak{F}$ satisfies the $\{p\}$-boundary condition in the class of all groups.

Lemma 5.15   *Let $\mathfrak{F} = LF(F)$, where F is the canonical local satellite of $\mathfrak{F}$. Suppose that for some prime p we have $F(p) = \mathfrak{F}$. Then $\mathfrak{F}$ does not satisfy the p-boundary condition.*

Proof   Indeed, in this case every $\mathfrak{F}$-critical group is also $F(p)$-critical.

A group G is called π-closed if G has a normal Hall π-subgroup.

Proposition 5.16   *The formation $\mathfrak{F}$ of all π-closed groups satisfies the π'-boundary condition, but $\mathfrak{F}$ does not satisfy the p-boundary condition for any $p \in \pi$.*

Proof   Let $\mathfrak{F} = \mathfrak{G}_\pi \mathfrak{G}_{\pi'}$ be the formation of all π-closed groups, F the canonical satellite of $\mathfrak{F}$. Then $F(p) = \mathfrak{F}$ for all $p \in \pi$, and $F(p) = \mathfrak{G}_{\pi'}$ for all primes $p \in \pi'$ by 3.1.20 in [135]. Hence, $\mathfrak{F}$ satisfies the π'-boundary condition and does not satisfy the p-boundary condition for any $p \in \pi$ by Lemma 5.15.

Lemma 5.17   *Let $\{\mathfrak{F}_i \mid i \in I\}$ be any set of nonempty saturated formations and $\mathfrak{F} = \cap_{i \in I} \mathfrak{F}_i$.*

(1) If for each $i \in I$, $\mathfrak{F}_i$ satisfies the p-boundary condition, then $\mathfrak{F}$ satisfies the p-boundary condition.

(2) *Suppose that* $I = \{1, 2\}$, F_i *is the canonical local satellite of* $\mathfrak{F}_i$ *and that there is a set* π *of primes satisfying the following conditions:*

(a) $\mathfrak{F}_1$ *satisfies the* π-*boundary condition, and for any* $p \in \pi$, *we have* $F_1(p) \subseteq \mathfrak{F}_2 = F_2(p)$ *and every* $F_1(p)$-*critical group belongs to* $\mathfrak{F}_2$.
(b) $\mathfrak{F}_2$ *satisfies the* π'-*boundary condition, and for any* $p \in \pi'$, *we have* $F_2(p) \subseteq \mathfrak{F}_1 = F_1(p)$ *and every* $F_2(p)$-*critical group belongs to* $\mathfrak{F}_1$.
   Then $\mathfrak{F}$ *satisfies the boundary condition.*

Proof

(1) Let F_i be the canonical local satellite of $\mathfrak{F}_i$ and F the canonical local satellite of $\mathfrak{F}$. If $f(p) = \cap_{i \in I} F_i(p)$, then $F(p) = \mathfrak{G}_p f(p)$ by Proposition 3.8 in [89, IV]. Now let G be any $F(p)$-critical group, $i \in I$. Since $F(p) \subseteq F_i(p)$, all maximal subgroup of G belongs to $F_i(p)$. Hence, $G \in \mathfrak{F}_i$ since $F_i(p) \subseteq \mathfrak{F}_i$ and $\mathfrak{F}_i$ satisfies the p-boundary condition. This implies that $G \in \mathfrak{F}$ and therefore $\mathfrak{F}$ satisfies the p-boundary condition.
(2) In this case, $F(p) = F_1(p)$ for all $p \in \pi$ and $F(p) = F_2(p)$ for all $p \in \pi'$, where F is the canonical local satellite of $\mathfrak{F}$. Hence, if $p \in \pi$ and G is an $F(p)$-critical groups, then $G \in \mathfrak{F}$ by hypothesis (a). This shows that $\mathfrak{F}$ satisfies the π-boundary condition. Similarly, we see that $\mathfrak{F}$ satisfies the π'-boundary condition.

Corollary 5.18 *The formation of all* p-*decomposable groups satisfies the boundary condition.*

Proof Let $\mathfrak{F}$ be the formation of all p-decomposable groups. Then $\mathfrak{F} = \mathfrak{G}_{p'}\mathfrak{G}_p \cap \mathfrak{G}_p\mathfrak{G}_{p'}$. Hence, the assertion follows from Proposition 5.16 and Lemma 5.17.
   From Corollary 5.18 and Theorem 5.2 we get

Corollary 5.19 *Let* D *be the intersection of all maximal* p-*decomposable subgroups of* G. *Then* D *is the largest normal subgroup of* G *satisfying* $D = O_{p'}(D) \times O_p(D)$, *and* G *induces the trivial automorphisms group on every chief factor of* G *below* $O_p(D)$ *and a* π'-*group of automorphisms on every chief factor of* G *below* $O_{p'}(D)$.

   Since a p-nilpotent group is p'-closed, the following result directly follows from Proposition 5.16.

Corollary 5.20 *The formation of all* p-*nilpotent groups satisfies the* p-*boundary condition.*

   From Corollary 5.20 and Theorem 5.2, we have

Corollary 5.21 *Let* D *be the intersection of all maximal* p-*nilpotent subgroups of a group* G. *Then* D *is the largest normal subgroup of* G *satisfying* $O_{p'}(D) = O_{p'}(G)$, *and* $D/O_{p'}(G) \leq Z_\infty(G/O_{p'}(G))$.

Lemma 5.22 *Let* $\mathfrak{F} = LF(F)$ *be a nonempty saturated formation, where* F *is the canonical local satellite of* $\mathfrak{F}$.

(1) If $\mathfrak{F} = \mathfrak{G}_p\mathfrak{F}$ for some prime p, then $F(p) = \mathfrak{F}$.
(2) If $\mathfrak{F} = \mathfrak{N}\mathfrak{H}$ for some nonempty formation $\mathfrak{H}$, then $F(p) = \mathfrak{G}_p\mathfrak{H}$ for all primes
   p.

Proof

(1) Since $F(p) \subseteq \mathfrak{F}$, we need only prove that $\mathfrak{F} \subseteq F(p)$. Suppose that this is
   false and let A be a group of minimal order in $\mathfrak{F} \setminus F(p)$. Then $A^{F(p)}$ is a unique
   minimal normal subgroup of A since $F(p)$ is a formation. Moreover, $O_p(A) = 1$
   since $F(p) = \mathfrak{G}_p F(p)$. Let $G = C_p \wr A = K \rtimes A$ where K is the base group of
   the regular wreath product G. Then $K = O_{p',p}(G)$ and $G \in \mathfrak{F} = \mathfrak{G}_p\mathfrak{F}$. Hence,
   $A \simeq G/K = G/O_{p',p}(G) \in F(p)$, a contradiction. Thus, $F(p) = \mathfrak{F}$.
(2) The inclusion $F(p) \subseteq \mathfrak{G}_p\mathfrak{H}$ is evident. The inverse inclusion can be proved
   similarly as the inclusion $\mathfrak{F} \subseteq F(p)$ in the proof of (1).

Proposition 5.23   (Guo and Skiba [198]). *Let $\mathfrak{F} = \mathfrak{N}\mathfrak{L}$.*

(i) *If $\mathfrak{L}$ is a hereditary saturated formation satisfying the boundary condition in the
   class of all soluble groups, then $\mathfrak{F}$ satisfies the boundary condition in the class
   of all soluble groups.*
(ii) *If $\mathfrak{L}$ is a formation of nilpotent groups with $\pi(\mathfrak{L}) = \mathbb{P}$, then $\mathfrak{F}$ satisfies the
   boundary condition.*

Proof Let F be the canonical local satellite of $\mathfrak{F}$. Then by Lemma 5.22(2), $F(p) = \mathfrak{G}_p\mathfrak{L}$ for all primes p. Assume that $\mathfrak{F}$ does not satisfy the boundary condition ($\mathfrak{F}$ does not satisfy the boundary condition in the class of all soluble groups). Then for some prime p, the set of all $F(p)$-critical (soluble) groups A with $A \notin \mathfrak{F}$ is nonempty. Let G be a group of minimal order in this set. Then $L = G^{\mathfrak{F}}$ is the unique minimal normal subgroup of G and $O_p(G) = 1 = \Phi(G)$ by Lemma 5.8.

First suppose that G is soluble. Then $L = C_G(L)$ is a q-group for some prime $q \neq p$ and $G = L \rtimes M$ for some maximal subgroup M of G with $O_q(M) = 1$ by Proposition 1.4 and Lemma 4.17. Let M_1 be any maximal subgroup of M. Then $LM_1 \in F(p)$, and so $LM_1 \in \mathfrak{L}$ since $L = C_G(L)$ is a q-group.

Suppose that $\mathfrak{L}$ satisfies the condition of (i) and let L be the canonical local satellite of $\mathfrak{L}$. Since $LM_1 \in \mathfrak{L}$,

$$LM_1/O_{q',q}(LM_1) = LM_1/O_q(LM_1) = LM_1/LO_q(M_1)$$
$$\simeq M_1/M_1 \cap LO_q(M_1 \in L(q) = \mathfrak{G}_pL(q).$$

Hence, every maximal subgroup of M belongs to $L(q)$. Since $\mathfrak{L}$ satisfies the boundary condition in the class of all soluble groups, we obtain that $M \in \mathfrak{L}$ and consequently $G \in \mathfrak{F}$. This contradiction completes the proof of (i).

Suppose that $\mathfrak{L}$ satisfies the condition (ii). Let M_1 be a normal maximal subgroup of M. Since $LM_1 \in \mathfrak{L}$ and $L = C_G(L)$, we have $M_1 = 1$. This implies that $|M|$ is prime. Hence, $M \in \mathfrak{L}$ since $\pi(\mathfrak{L}) = \mathbb{P}$. But then $G \in \mathfrak{F} = \mathfrak{N}\mathfrak{L}$. Therefore, G is not soluble.

Let $q \neq p$ be any prime divisor of $|G|$. Suppose that G is not q-nilpotent. Then G has a q-closed $\mathfrak{N}$-critical subgroup $H = Q \rtimes R$ by Proposition 1.10, where Q is a Sylow q-subgroup of H, R is a cyclic Sylow r-subgroup of H. Since G is not soluble, $H \neq G$. Hence, $H \leq M \in F(p) = \mathfrak{G}_p \mathfrak{L}$ for some maximal subgroup M of G. Since $M \in \mathfrak{F}G_p\mathfrak{L}$, $M^{\mathfrak{L}} \leq O_p(M)$ and hence $H^{\mathfrak{L}} \leq Q \cap O_p(H) = 1$. This shows that H is nilpotent. This contradiction shows that G is q-nilpotent for all primes $q \neq p$. This induces that $G^{\mathfrak{N}}$ is a p-subgroup of G and thereby G is soluble. This contradiction completes the proof of (ii).

Remark 5.24 The condition "$\mathfrak{L}$ is a hereditary saturated formation satisfying the boundary condition in the class of all soluble groups" can not be omitted in Proposition 5.23(i). Indeed, let $\mathcal{F} = \mathfrak{N}\mathfrak{U}$ and $G = P \rtimes A_4$, where P is a simple $\mathbb{F}_3 A_4$-module which is faithful for A_4. Let F be the canonical local satellite of $\mathfrak{F}$. Then $F(2) = \mathfrak{G}_2 \mathfrak{U}$ by Lemma 5.22(2). Therefore, G is an $F(2)$-critical group and $G \notin \mathfrak{F}$. Thus, G does not satisfy the boundary condition in the class of all soluble groups.

Corollary 5.25 *Let $\mathfrak{F}$ be the class of all groups with $G' \leq F(G)$. Then $\mathfrak{F}$ satisfies the boundary condition.*

From Corollary 5.25 and Theorem 5.2, we obtain

Corollary 5.26 *Let D be the intersection of all maximal subgroups H of G with the property $H' \leq F(H)$. Then D is the largest normal subgroup of G such that $D' \leq F(D)$ and G induces an abelian group of automorphisms on every chief factor of G below D.*

Note that Corollary 5.26 can be also found in [47, Corollary 7 and Remark 4].

Following [89, VII, Definitions 6.9] we write $l(G)$ to denote the nilpotent length of the group G. Recall that $\mathfrak{N}^r$ is the product of r copies of $\mathfrak{N}$; $\mathfrak{N}^0$ is the class of groups of order 1 by definition. It is well known that $\mathfrak{N}^r$ is the class of all soluble groups G with $l(G) \leq r$. It is also known that $\mathfrak{N}^r$ is a hereditary saturated formation by (1.3). Hence, from Proposition 5.23 we get.

Corollary 5.27 *Let $\mathfrak{F} = \mathfrak{N}^r \mathfrak{L}$ ($r \in \mathbb{N}$), where $\mathfrak{L}$ is a subformation of the formation of all abelian groups with $\pi(\mathfrak{L}) = \mathbb{P}$. Then $\mathfrak{F}$ satisfies the boundary condition in the class of all soluble groups.*

From Proposition 5.23 and Theorem 5.2 we get

Corollary 5.28 *Let D be the intersection of all maximal metanilpotent subgroups of G. Then D is the largest normal subgroup of G such that D is metanilpotent and G induces a nilpotent group of automorphisms on every chief factor of G below D.*

It is clear that every subnormal subgroup is a K-$\mathfrak{F}$-subnormal subgroup as well. On the other hand, in the case when $\mathfrak{N} \subseteq \mathfrak{F}$, every K-$\mathfrak{F}$-subnormal subgroup of any soluble subgroup G is $\mathfrak{F}$-subnormal in G. Hence, from Lemma 5.12 and the above corollaries we get

Corollary 5.29 *Let $\mathfrak{F}$ be the class of all groups G with $G' \leq F(G)$. Then:*

(i) *The subgroup $Z_{\mathfrak{F}}(G)$ may be characterized as the intersection of all non-subnormal $\mathfrak{F}$-maximal subgroups of G, for each group G.*

(ii) *The subgroup $Z_{\mathfrak{F}}(G)$ may be characterized as the intersection of all non- $\mathfrak{F}$-subnormal $\mathfrak{F}$-maximal subgroups of G, for each soluble group G.*

Corollary 5.30   *Let $\mathfrak{F}$ be one of the following formations:*

(1) the class of all nilpotent groups (Baer [22]);
(2) the class of all groups G with $G' \le F(G)$ (Skiba [387]);
(3) the class of all p-decomposable groups (Skiba [386]).
   Then for each group G, the equality $Int_{\mathfrak{F}}(G) = Z_{\mathfrak{F}}(G)$ holds.

Corollary 5.31   (Sidorov [375]). *Let $\mathfrak{F}$ be the class of all soluble groups G with $l(G) \le r$ ($r \in \mathbb{N}$). Then for each soluble group G, the equality $Z_{\mathfrak{F}}(G) = Int_{\mathfrak{F}}(G)$ holds.*

It is well known that every supersoluble group is dispersive (in the sense of Ore). As an application of Theorem 5.5(a) we have the following result.

Proposition 5.32   *Let $\mathfrak{F}$ be the formation of all supersoluble groups and $\mathfrak{M}$ be the formation of all dispersive groups. Then $Int_{\mathfrak{F}}(G) \le Int_{\mathfrak{M}}(G)$ for any group G.*

Proof   Let F be the canonical local satellite of $\mathfrak{F}$. Then $F(p) = \mathfrak{G}_p\mathfrak{A}(p - 1)$ for primes p (see (1.3)). Let G be any $F(p)$-critical dispersive group. We show that $G \in \mathfrak{F}$. Let q be the largest prime dividing $|G|$, P the Sylow q-subgroup of G. If $G = P$, then clearly $G \in \mathfrak{F}$. Let $P \ne G$. Then every Sylow subgroup of G belongs to $F(p)$. Hence, $q = p$ and if E is a Hall p'-subgroup of G, then $E \in \mathfrak{A}(p - 1)$. But then $G \in F(p)$, a contradiction. Therefore, $G \in \mathfrak{F}$. This shows that $\mathfrak{F}$ satisfies the boundary condition in $\mathfrak{M}$. Hence, in view of Theorem 5.5(a), we have $Int_{\mathfrak{F}}(G) \le Int_{\mathfrak{M}}(G)$.

Remark 5.33   (I) If $\pi \ne \{2\}$, then the formation $\mathfrak{F}$ of all π-supersoluble groups does not satisfy the π-boundary condition in the class of all soluble groups. Indeed, let F be the canonical local satellite of $\mathfrak{F}$. Then $F(p) = \mathfrak{N}_p\mathfrak{A}(p - 1)$, where $\mathfrak{A}(p - 1)$ is the formation of all abelian groups of exponent dividing $p - 1$, for all $p \in \pi$ and $F(p) = \mathfrak{F}$ for all primes $q \notin \pi$ by (1.3). Let $2 \ne p \in \pi$ and q any prime with q divides $p - 1$. Let $G = Q \rtimes C$, where C is a group of order p and Q is an $\mathbb{F}_q C$-module which is faithful for C. Then G is soluble, but G is a non-supersoluble $F(p)$-critical groups.

(II) If $\mathfrak{F} \subseteq \mathfrak{M}$ be hereditary saturated formations, then for every group G we have $Z_{\mathfrak{F}}(G) \le Z_{\mathfrak{M}}(G)$ and $Z_{\mathfrak{F}}(G) \le Int_{\mathfrak{F}}(G)$ by Proposition 5.6(h). Therefore, if $\mathfrak{F}$ satisfies the boundary condition, then $Int_{\mathfrak{F}}(G) \le Int_{\mathfrak{M}}(G)$ for every group G. But, by (I), $\mathfrak{F}$ does not necessarily satisfy the boundary condition. Hence, we can not deduce Proposition 5.32 from Theorem 5.2.

On Problem of Agrawal   Recall that a subgroup H of a group G is said to be permutable (s-permutable) or quasinormal (s-quasinormal, respectively) in G if $HP = PH$ for all subgroups (for all Sylow subgroups, respectively) P of G.

The following lemma collects some known properties about quasinormal and s-quasinormal subgroups which will be used in the proofs. All the assertions in the following lemma may be found in Chap. 1 of the book [450] (see also [85], [263] and [342]).

Lemma 5.34 *Let G be a group, H be a subgroup of G and N be a normal subgroup of G.*

(1) If H is s-quasinormal in G, then H is subnormal in G and H/H_G is nilpotent.

(2) If H is s-quasinormal in G, then HN/N is s-quasinormal in G/N

(3) If H/N is s-quasinormal in G/N, then H is s-quasinormal in G.

(4) If H is s-quasinormal (quasinormal) in G, then $H \cap S$ is s-quasinormal (quasinormal, respectively) in S for any subgroup S of G. In particular, If H is s-quasinormal in G and $H \leq K$, then H is s-quasinormal in K.

(5) If H is an s-quasinormal nilpotent subgroup of G, then every Sylow subgroup of H is s-quasinormal in G and every characteristic subgroup of H is s-quasinormal in G.

(6) Let V a p-subgroup of G for some prime p. Then V is s-quasinormal in G if and only if $O^p(G) \leq N_G(V)$.

(7) If H, K are s-quasinormal in G, then $H \cap K$ and $< H, K >$ are s-quasinormal in G.

Lemma 5.35 *Let B be a subgroup of G, $A \leq K \leq G$ and $B \leq G$.*

(1) Suppose that A is normal in G. Then K/A is subnormal in G/A if and only if K is subnormal in G [89, A, 14.1].

(2) If A is subnormal in G, then $A \cap B$ is subnormal in B [89, A, 14.1].

(3) If A is a subnormal Hall subgroup of G, then A is normal in G [449].

(4) If A is subnormal in G and B is a Hall π-subgroup of G, then $A \cap B$ is a Hall π-subgroup of A [449].

(5) If A is subnormal in G and B is a minimal normal subgroup of G, then $B \leq N_G(A)$ [89, A, 14.3].

(6) If A is subnormal in G and A is a π-subgroup of G, then $A \leq O_\pi(G)$ [449].

(7) If A is subnormal in G and A is p-nilpotent (p-soluble), then A is contained in some p-nilpotent normal subgroup of G (A is contained in some p-soluble normal subgroup of G, respectively) [449].

(8) If A and B are subnormal in G, then $A \cap B$ and $\langle A, B \rangle$ is subnormal in G [89, A, 14.4].

(9) If A is subnormal in G and the index $|G : A|$ is a p'-number, then A contains all Sylow p-subgroups of G [449].

(10) If A and B are subnormal in G and $A = A'$ and $B = B'$, then $AB = BA$ [447].

(11) If A is subnormal in K and $|K : A|$ is a p-number, then $O^p(K) \leq A$ (see [452, VIII, (1.2)].

We may use $H \ sn \ G$ denotes that H is a subnormal subgroup of G.

The hyper-generalized-center $genz^*(G)$ of a group G coincides with the largest term of the chain of subgroups

$$1 = Q_0(G) \leq Q_1(G) \leq \ldots \leq Q_t(G) \leq \cdots$$

where $Q_i(G)/Q_{i-1}(G)$ is the subgroup of $G/Q_{i-1}(G)$ generated by the set of all cyclic s-quasinormal subgroups of $G/Q_{i-1}(G)$ (see [450, p. 21]). It is known that $genz^*(G)$ is supersoluble.

In the paper [2], Agrawal proved that $genz^*(G)$ is contained in every maximal supersoluble subgroup of the group G and posed the following question: Whether $genz^*(G) = \mathrm{Int}_\mathfrak{U}(G)$ for any group G? (see [2, p. 19] or [450, p. 22])

Yun Fan had found a concrete group G for which $genz^*(G) \neq \mathrm{Int}_\mathfrak{U}(G)$ (see [468, pp. 82–86] (in Chinese). Furthermore, other extremely simple examples in this trend were found by Beidleman and Heineken in [47] and by Skiba in [386]. Now we give the following general result.

Theorem 5.36 (Guo, Skiba [201]). *For any group A, there is a group G such that $A \leq G$ and $genz^*(G) \neq Int_\mathfrak{U}(G)$.*

Proof Let A be any group. Then there is a prime p such that $p > |A|$ and $p - 1$ is not a prime power. Indeed, if p is a prime with $p > |A|$, then, by the classical Chebyshev's theorem, there is a prime q such that $p < q < 2p$. It is also clear that at lest one of the numbers p and q is not of form $r^n + 1$, where r is a prime.

Let C_p be a group of prime order p. Then $|\pi(Aut(C_p))| > 1$. Let R and L be Hall subgroups of $Aut(C_p)$ such that $Aut(C_p) = R \times L$ and for any $r \in \pi(R)$ and $q \in \pi(L)$ we have $r < q$. Let $B = (C_p \rtimes R) \wr L = K \rtimes L$ be the regular wreath product of $C_p \rtimes R$ with L, where K is the base group of G. Let $P = C_p^\natural$ (see [89, p. 63]). Then by Proposition 18.5 in [89, Chap. A], B is a primitive group and P is a unique minimal normal subgroup of B. Hence, $P = F(B) = C_B(P)$. Moreover, by Lemma 18.2 in [89, Chap. A], $B = P \rtimes M$, where $M \simeq U := R \wr L = D \rtimes L$, where D is the base group of U.

It is clear that D is an abelian Hall subgroup of U and L is a cyclic subgroup of U such that $r < q$ for any $r \in \pi(D)$ and $q \in \pi(L)$. Moreover, since $|Aut(C_p)| = p - 1$, D and L are groups of exponent dividing $p - 1$.

First we show that every supersoluble subgroup W of U is nilpotent. Suppose that this is false and let H be an $\mathfrak{N}$-critical subgroup of W. Then $1 < D \cap H < H$, where $D \cap H$ is a normal Hall subgroup of H. By Proposition 1.9, there are primes r and q such that $H = H_r \rtimes H_q$, where H_r is a Sylow r-subgroup of H, H_q is a cyclic Sylow q-subgroup of H. Hence, $D \cap H = H_r$. Since $H \leq W$, H is supersoluble and hence $r > q$. But $Q \simeq H/D \cap H \simeq HD/D$ isomorphic to some subgroup of L, so $r < q$. This contradiction shows that W is nilpotent.

Now we show that $P \leq \mathrm{Int}_\mathfrak{U}(B)$. Let V be any supersoluble subgroup of B and W a Hall p'-subgroup of V. Then $PV = PW$. It is clear that M is a Hall p'-subgroup of B. Hence, for some $x \in B$ we have $W \leq M^x \simeq U^x$. Hence, W is nilpotent. It is clear that the Sylow subgroups of W are abelian. Hence, W is an abelian group of exponent dividing $p - 1$. It follows from (1.3) that PV is supersoluble. Therefore, $P \leq \mathrm{Int}_\mathfrak{U}(B)$.

Next we show that $genz^*(B) = 1$. Indeed, suppose that $genz^*(B) \neq 1$. Then B has a non-identity cyclic s-quasinormal subgroup, say V. The subgroup V is subnormal in B by Lemma 5.34(1). Therefore, $V \leq F(B) = P$ by Lemma 5.35(6). Moreover, if Q is a Sylow q-subgroup of B, where $q \neq p$, then V is subnormal in VQ and so $Q \leq N_B(V)$. Hence, V is normal in B and therefore $V = P$ is cyclic. But then $|P| = p = |C_p{}^\natural|$, a contradiction. Thus, $genz^*(B) = 1$.

Finally, let $G = A \times B$ and $I = \mathrm{Int}_\mathfrak{U}(G)$. We will prove that $I \neq genz^*(G)$. First we show that $P \leq I$. Let V be a supersoluble subgroup of G. By Proposition 5.6(a), $AP/A \leq \mathrm{Int}_\mathfrak{U}(G/A)$. On the other hand, $VA/A \simeq V/V \cap A$ is supersoluble. Hence,

$$(PA/A)(VA/A) = PVA/A \simeq PV/PV \cap A$$

is supersoluble by Proposition 5.6(d). Moreover,

$$PVB/B \simeq PV/PV \cap B = PV/P(V \cap B) = P(V \cap B)V/P(V \cap B) = V/V \cap B$$

is supersoluble. Therefore, $PV \simeq PV/1 = PV/(PV \cap A) \cap (PV \cap B)$ is supersoluble. This implies that $P \leq I$.

Now let

$$1 = Q_0 \leq Q_1 \leq \ldots \leq Q_t = genz^*(G),$$

where Q_i/Q_{i-1} is the subgroup of G/Q_{i-1} generated by the set of all cyclic s-quasinormal subgroups of G/Q_{i-1}. Suppose that $I = genz^*(G)$. Then $P \leq genz^*(G)$ and so $t > 0$. Since P is a minimal normal subgroup of B, P is also a minimal normal subgroup of G. Hence, there is a number i such that $P \nleq Q_{i-1}$ and $P \leq Q_i$. Let H/Q_{i-1} be any cyclic s-quasinormal subgroup of G/Q_{i-1}. Then H/Q_{i-1} is subnormal in G/Q_{i-1}. Hence, $H/Q_{i-1} \leq O_\pi(G/Q_{i-1})$, where $\pi = \pi(H/Q_{i-1})$ by Lemma 5.35(6). It follows that some subgroup V/Q_{i-1} of G/Q_{i-1} of order p is s-quasinormal in G/Q_{i-1}. Since PQ_{i-1}/Q_{i-1} is a Sylow p-subgroup of G/Q_{i-1} and P is abelian, V/Q_{i-1} is normal in G/Q_{i-1} by Lemma 5.34(6). By the Schur-Zassenhaus theorem, $V = Q_{i-1} \rtimes C_p$ for some subgroup C_p of V of order p. Since $I = genz^*(G)$, V is supersoluble. On the other hand, by the choice of p, p is the largest prime dividing $|V|$. Hence, C_p is normal in V and so C_p is characteristic in V. It follows that C_p is normal in G. Consequently $C_p \leq genz^*(B) = 1$. This contradiction completes the proof.

1.6 Additional Information and Some Problems

I. The theory of the $\mathfrak{F}$-hypercenter of a group for a saturated formation $\mathfrak{F}$ were represented in many books (see, for example, [34, 135, 359]). In Sect. 2 we give some general properties of the generalized $\mathfrak{F}$-hypercenter of groups which will be need in all next chapters of the book.

II. Examples 2.3 and 2.4 were given in [368] . Here we correct some places in Example 2.4.

III. Theorems 3.7 and 3.12 are main results of Sect. 3. These theorems developed
   related results in [188, 189, 192], where, in particular, were involved only saturated formations containing all nilpotent groups. In connection with Theorem
   3.14 the following question arises;

Problem 6.1 To obtain versions of Theorem 3.14 for the π-quasi-$\mathfrak{F}$-groups and for
p-quasi-$\mathfrak{F}$-groups.

IV. It is easy to see that $Z_{\pi\mathfrak{M}}(G) \leq Z_{\pi\mathfrak{F}}(G)$ and $Z_{\mathfrak{M}\Phi}(G) \leq Z_{\mathfrak{F}\Phi}(G)$ for any group
   G and for any two formations $\mathfrak{M}$ and $\mathfrak{F}$ with $\mathfrak{M} \subseteq \mathfrak{F}$. The following example
   shows that related equality for $\mathrm{Int}_{\mathfrak{M}}(G)$ and $\mathrm{Int}_{\mathfrak{F}}(G)$ is not true, in general.

Example 6.2 Let p and q be primes, where q divides $p-1$. Let C_p be a group of
order p, $A = Q \rtimes C_p$, where Q is a simple $\mathbb{F}_q C_p$-module which is faithful for C_p,
and $G = P \rtimes A$, where P is a simple $\mathbb{F}_p C_p$-module which is faithful for A. Let
$\mathfrak{M} = \mathfrak{U}$ be the formation of all supersoluble groups and $\mathfrak{F}$ be the formation of all
p-supersoluble groups. Then $\mathrm{Int}_{\mathfrak{F}}(G) = 1$ and $\mathrm{Int}_{\mathfrak{M}}(G) = P$.

At the same time, for any hereditary saturated formation $\mathfrak{M}$ satisfying the bound-
ary condition and for every hereditary saturated formation $\mathfrak{F}$ with $\mathfrak{M} \subseteq \mathfrak{F}$ we
have

$$\mathrm{Int}_{\mathfrak{M}}(G) = Z_{\mathfrak{M}}(G) \leq Z_{\mathfrak{F}}(G) \leq \mathrm{Int}_{\mathfrak{F}}(G)$$

by Theorem 5.3. Hence, $\mathrm{Int}_{\mathfrak{M}}(G) \leq \mathrm{Int}_{\mathfrak{F}}(G)$.

These observations force us to the following question.

Problem 6.3 What are the nonempty hereditary saturated formation $\mathfrak{M}$ with the
property that for every hereditary saturated formation $\mathfrak{F}$ with $\mathfrak{M} \subseteq \mathfrak{F}$ and for each
group G, the inclusion

$$\mathrm{Int}_{\mathfrak{M}}(G) \leq \mathrm{Int}_{\mathfrak{F}}(G)$$

holds?

Problem 6.4 What are the nonempty hereditary saturated formations $\mathfrak{F}$ with the
property that for every hereditary saturated formation $\mathfrak{M}$ with $\mathfrak{M} \subseteq \mathfrak{F}$ and for each
group G, the inclusion

$$\mathrm{Int}_{\mathfrak{M}}(G) \leq \mathrm{Int}_{\mathfrak{F}}(G)$$

holds?

V. Maier and Schmid [307] proved that if H is a quasinormal subgroup of G,
   then $H^G/H_G \leq Z_\infty(G/H_G)$. In the class of all soluble groups this result was
   generalized by P. Schmid as follows.

Theorem 6.5 [342]. *Let H be an s-quasinormal subgroup of a soluble group G
and H permutes with some system normalizer of G. Then $H/H_G \leq Z_\infty(G/H_G)$.*

As a development, L. M. Ezquerro and X. Soler-Escrivà [99] proved the following
result:

Theorem 6.6 [99]. *Let $\mathfrak{F}$ be a saturated formation containing all nilpotent groups.
Let H be an s-quasinormal subgroup of a soluble group G. Then H permutes with
some $\mathfrak{F}$-normalizer (see [135]) of G if and only if $H^G/H_G \leq Z_{\mathfrak{F}}(G/H_G)$.*

VI. Corollary 5.11 follows also directly from [47, Main Theorem].

Lemma 6.7 *Let $\mathfrak{F} = LF(F)$ be a saturated formations, where F is the canonical local satellites of $\mathfrak{F}$. Let $\mathfrak{M} = \mathfrak{F} \cap \mathfrak{S}$ and M the canonical local satellites of $\mathfrak{M}$. Then $M(p) = F(p) \cap \mathfrak{S}$ for all primes p.*

Proof It is clear that $\mathfrak{M} = LF(F_1)$, where $F_1(p) = F(p) \cap \mathfrak{S}$ for all primes p. On the other hand, for any prime p we have $F_1(p) \subseteq \mathfrak{M}$, and $F_1(p) = \mathfrak{G}_p F_1(p)$ since

$$\mathfrak{G}_p F_1(p) = \mathfrak{G}_p(F(p) \cap \mathfrak{S}) \subseteq F(p) \cap \mathfrak{S}.$$

Hence, $F_1 = M$ is the canonical local satellite of $\mathfrak{M}$.

Lemma 6.8 *Let $(1) \neq \mathfrak{F} = LF(F)$ be a saturated formations and $\mathfrak{M} = \mathfrak{F} \cap \mathfrak{S}$. Let $\pi \subseteq \pi(\mathfrak{F})$. Then $\mathfrak{F}$ satisfies the boundary condition in the class S of all soluble groups if and only if $\mathfrak{M}$ satisfies the boundary condition in $\mathfrak{S}$.*

Proof Let F and M be the canonical local satellite of the formations $\mathfrak{F}$ and $\mathfrak{M}$, respectively. Let $p \in \pi$. Then $M(p) = F(p) \cap \mathfrak{S}$ by Lemma 6.7. Hence, a soluble group G is $F(p)$-critical group if and only if G is $M(p)$-critical. On the other hand, the soluble group G belongs to $\mathfrak{F}$ if and only if G belongs to $\mathfrak{M}$.

VII. In view of Lemmas 6.7 and 6.8, one may prove the following general form of Theorem 5.3.

Theorem 6.9 *Let $\mathfrak{F}$ be a saturated formation such that $\mathfrak{M} = \mathfrak{F} \cap \mathfrak{S}$ is hereditary and $(1) \neq \mathfrak{M} \neq \mathfrak{S}$. Let $\pi \subseteq \pi(\mathfrak{F})$. Then the equality*

$$Z_{\pi\mathfrak{F}}(G) = Int_{\mathfrak{F}}(G)$$

holds for each soluble group G if and only if $\mathfrak{N} \subseteq \mathfrak{F} = \mathfrak{S}_{\pi'}\mathfrak{F}$ and $\mathfrak{F}$ satisfies the π-boundary condition in the class of all soluble groups.

VIII. A. Ballester-Bolinches, L. M. Ezquerro and A. N. Skiba proved the following result.

Theorem 6.10 [38]. *Let $\mathfrak{F}$ be a Baer-local formation. Given a group G and a normal subgroup E of G, let $Z_{\mathfrak{F}}(G)$ contain a p-subgroup A of E which is maximal being abelian and of exponent dividing p^k, where k is some natural number, $k \neq 1$ if $p = 2$ and the Sylow 2-subgroups of E are non-abelian. Then $E/O_{p'}(E) \leq Z_{\mathfrak{F}}(G)$.*

IX. Let A be a subgroup of a group G. We say that A is a generalized CAP-subgroup of G if for each chief factor H/K of G either A avoids H/K or the following hold:

(1) If H/K is non-abelian, then $|H : (A \cap H)K|$ is a p'-number for every $p \in \pi((A \cap H)K/K)$;
(2) If H/K is a p-group, then $|G : N_G(K(A \cap H))|$ is a p-number.

It is easy to see that every CAP-subgroup of G is a generalized CAP-subgroup of G, and every s-quasinormally embedded subgroup A of G (see Chap. 3, Sect. 3.1) is also a generalized CAP-subgroup of G.

We said that a subgroup H of G is $\mathfrak{U}$-*embedded* in G if for some K-$\mathfrak{U}$-subnormal subgroup T of G we have that $HT = G$ and $H \cap T \le S \le H$, where S is a generalized CAP-subgroup of G.

In [209], the authors proved the following

Theorem 6.11   (Guo and Skiba, Yang, [209]). *Let E be a normal subgroup E of G. Then $E \le Z_{\mathfrak{U}}(G)$ if and only if for any non-cyclic Sylow subgroup P of E and for some integer $k = k(P)$, every k-intermediate subgroup of P (see p. 147 below) and every cyclic subgroup of P of order 4 (if $k = 1$ and P is a non-abelian 2-group) are $\mathfrak{U}$-embedded in G.*

Problem 6.12 (Guo and Skiba). Let E be a normal subgroup of G. Suppose that for any non-cyclic Sylow subgroup P of E and, for some integer $k = k(P)$, every k-intermediate subgroup of P and every cyclic subgroup of P of order 4 ($k = 1$ and P is a non-abelian 2-group) having no supersoluble supplements in G are $\mathfrak{U}$-embedded in G. Does it true then that $E \le Z_{\mathfrak{U}}(G)$?

Problem 6.13 (Guo and Skiba). Let E be a normal subgroup of G and P a Sylow p-subgroup of E with $(p - 1, |E|) = 1$. Suppose that $|P| > p$ and, for some integer k, every k-intermediate subgroup of P and every cyclic subgroup of P of order 4 (if $k = 1$ and P is a non-abelian 2-group) having no p-supersoluble supplements in G are $\mathfrak{U}$-embedded in G. Does it true then that $E/O_{p'}(E) \le G/O_{p'}(E)$?

Problem 6.14 Is it true that the statement of Theorem 6.11 also holds if the condition of every chief factor in the generalized CAP-subgroup is replaced by every factor in some chief series?

Problem 6.15 Is it true that Theorem 6.11 holds if the condition "K-$\mathfrak{U}$-subnormal subgroup" is replaced by "subgroup" in the definition of $\mathfrak{U}$-embedded subgroup.

X. Recently, there are some researches about the intersection of normalizers of some subgroups of a finite group, see, for example, [68, 116, 290, 371, 399].

Chapter 2
Groups With Given Systems of X-Permutable Subgroups

2.1 Base Concepts

X-Permutable and c-Permutable Subgroups A subgroup A of a group G is said to be permutable with a subgroup B if $AB = BA$. A subgroup A is said to be a permutable or a quasinormal subgroup of G if A is permutable with all subgroups of G. But we often meet the situation $AB \neq BA$, nevertheless there exists an element $x \in G$ such that $AB^x = B^x A$, for instance, we have the following cases:

1) Let $G = AB$ be a group. If A_p and B_p are Sylow p-subgroups of A and of B respectively, then $A_p B_p \neq B_p A_p$ in general, but G has an element x such that $A_p B_p^x = B_p^x A_p$.

2) If A and B are Hall subgroups of a soluble group G, then there exists an element $x \in G$ such that $AB^x = B^x A$. (see [89, Theorem (I, 4.11)])

3) If A and B are normally embedded subgroups (see [89, Definition (I, 7.1)]) of a soluble group, then A is permutable with some conjugate of B. (see [89, Theorem (I, 7.11)])

4) If $|G : A| = p^\alpha$ is a prime power, then for every Sylow subgroup Q of G, there is $x \in G$ such that $AQ^x = Q^x A$.

Based on the above observations, we give the following definitions.

Definition 1.1 Let A and B be subgroups of a group G, and let $\varnothing \neq X \subseteq G$. Then we say:

(1) A is *X-permutable* with B in G if there exists some $x \in X$ such that $AB^x = B^x A$;

(2) A is *completely X-permutable* (or *hereditary X-permutable*) with B in G if there exists some $x \in X \cap \langle A, B \rangle$ such that $AB^x = B^x A$;

(3) A is *conditionally permutable* (or in brevity, *c-permutable*) with B in G provided A is G-permutable with B;

(4) A is *completely c-permutable* or (*hereditary c-permutable*) with B in G provided A is complete G-permutable with B in G.

© Springer-Verlag Berlin Heidelberg 2015
W. Guo, *Structure Theory for Canonical Classes of Finite Groups*,
DOI 10.1007/978-3-662-45747-4_2

By Chap. 1, Lemma 5.34(1), every s-quasinormal subgroup is subnormal. The following examples show that a subgroup of a group G which is c-permutable with all Sylow subgroups of G is not necessarily s-quasinormal in general even in the case when it is subnormal.

Example 1.2 Fixing some odd prime p, put $A = \langle x, y | x^{p^2} = y^p = 1, x^y = x^{1+p} \rangle$ and $L = \langle y \rangle$. Take some involution g in $\text{Aut} L$, and put $B = L \rtimes \langle g \rangle$. Consider a transitive permutation representation $\alpha : B \longrightarrow Sym(p)$ of degree p. Take the wreath product $G = A \wr_\alpha B = K \rtimes B$ of A and B with respect to α, where K is the base of $A \wr_\alpha B$. Using the terminology of [89], put $R = L^\natural$, and consider $N = N_G(R)$. It is clear that $B \subseteq N$ and $N \cap K = (N_A(L))^\natural$. Since $|A| = p^3$ and $N_A(L) \neq A$, $N_A(L)$ is an abelian group, and so $N \cap K$ is also abelian. It is clear that R is G-permutable with all Sylow subgroups of G and that R is subnormal in G. Suppose that R is permutable with all Sylow 2-subgroups of G. Then for each $x \in G$ we have $\langle g \rangle^x \subseteq N$. Consequently, the normal closure $\langle g \rangle^G$ of the subgroup $\langle g \rangle$ in G satisfies $L \subseteq \langle g \rangle^B = B \subseteq \langle g \rangle^G \subseteq N$; thus, $B^G \subseteq N$. Suppose further that $M = \{(a_1, \ldots, a_p) | a_i \in A, a_1 \ldots a_p \in A'\}$. Then $B^G = MB$ by [89, A, (18.4)]. Consequently, $M \subseteq N$. However, if $a_1 = \cdots = a_p$, then $a_1^p \subseteq A'$. Hence M has a subgroup which is isomorphic to A. This means that $N \cap K$ is not abelian. This contradiction shows that R is not permutable with some Sylow 2-subgroup of G.

Example 1.3 Let M be a subgroup of a soluble group G. Suppose that $|G : M| = p$ is a prime. Then

(i) *M completely c-permutes with all subgroups of G.* Indeed, let $T \leq G$. Let $M_1, \ldots, M_t$ and $T_1, \ldots, T_t$ be some Sylow systems of the groups M and T, respectively. Then G has Sylow systems $\Sigma = \{P_1, \ldots, P_t\}$ and $\Sigma_1 = \{Q_1, \ldots, Q_t\}$ such that $M_i = P_i \cap M$ and $T_i = Q_i \cap T$ for all $i = 1, \ldots, t$ (see Sect. 2 in [248, Chap. VI]). Moreover, the systems Σ and Σ_1 are conjugate, that is, G has an element x such that $Q_i^x = P_i$ for all $i = 1, \ldots, t$. Without loss of generality, we may assume that P_1 is a Sylow p-subgroup of G. Then $M_2 = P_2, \ldots, M_t = P_t$. Assume that $T_1^x \subseteq M_1$. Then we have

$$T^x \subseteq M_1 P_2 \ldots P_2 = M,$$

and so $T^x M = M = MT^x$.
Now let $T_1^x \nsubseteq M_1$. Since $|G : M| = p$, we have $|P_1 : M_1| = p$ and $P_1 = T_1^x M_1$. Hence,

$$T^x M = T_2^x \ldots T_t^x T_1^x M_1 M_2 \ldots M_t = T_2^x \ldots T_t^x P_1 P_2 \ldots P_t = G = MT^x.$$

(ii) *If G is supersoluble, then M is $F(G)$-permutable with every subgroup T of G.* Indeed, if $F(G) \leq M$, then $G' \leq M$. Hence M is normal in G and so it is quasinormal in G. Now assume that $F(G) \nleq M$. Then by (i), there exists an element x of G such that $MT^x = T^x M$. Then $G = MF(G)$ and hence $x = mf$ for some $m \in M$ and $f \in F(G)$. Therefore $MT^x = MT^{mf} = MT^f = T^f M$.

These examples are a motivation for introducing the following concepts.

Definition 1.4 Let A be a subgroup of a group G and $\varnothing \neq X \subseteq G$. Then we say that:

(1) A is *(completely) X-quasinormal* or *(completely) X-permutable in G* if A is (completely, respectively) X-permutable with all subgroups of G.

(2) A is *(completely) X-s-permutable in G* if A is *(completely, respectively) X-permutable with all Sylow subgroups of G. In particular, if $X = 1$, then an X-s-permutable subgroup is said to be s-permutable (or s-quasinormal) in G.*

(3) A is *(completely) c-permutable in G* if A is *(completely, respectively) c-permutable with all subgroups of G.*

(4) A is *(completely) s-c-permutable in G* if A is *(completely, respectively) c-permutable with all Sylow subgroups of G.*

In the following lemma we give the general properties of X-permutability.

Lemma 1.5 *Let A, B, X be subgroups of G and $K \trianglelefteq G$. Then the following statements hold:*

(1)   *If A is (completely) X-permutable with B, then B is (completely) X-permutable with A.*

(2)   *If A is (completely) X-permutable with B, then A^x is (completely) X^x-permutable with B^x for all $x \in G$.*

(3)   *If A is (completely) X-permutable with B, then AK/K is (completely) XK/K-permutable with BK/K in G/K.*

(4)   *Suppose that $K \leq A$. Then A/K is (completely) XK/K-permutable with BK/K if and only if A is (completely) X-permutable with B.*

(5)   *If $A, B \leq M \leq G$ and A is completely X-permutable with B, then A is completely $(X \cap M)$-permutable with B.*

(6)   *If A is (completely) X-permutable with B and $X \leq M \leq G$, then A is (completely) M-permutable with B.*

(7)   *If A is X-permutable with B and $X \leq N_G(A)$, then A is permutable with B.*

(8)   *If F is a quasinormal subgroup of G and A is (completely) X-permutable with B, then AF is (completely) X-permutable with B.*

(9)   *If $A \leq T$, where T is a subnormal subgroup of a (solvable) group G, and A is G-permutable with all Sylow (Hall) subgroups of G, then A is T-permutable with all Sylow (Hall) subgroups of T.*

(10) *Suppose that $G = AT$ and T_1 is a subgroup of T. If A is (completely) G-permutable with T_1, then A is (completely) T-permutable with T_1.*

(11) *If A is a maximal subgroup of G, T is a minimal supplement to A in G and A is c-permutable with all subgroups of T, then $T = \langle a \rangle$ is a cyclic p-group for some prime p, and $a^p \in A$.*

Proof (1)–(3) and (5)–(8) are obvious.

(4) Suppose that A/K is (completely) XK/K-permutable with KB/K in G/K. Then there exists an element xK of XK/K (an element of $(XK/K) \cap \langle A/K, KB/K \rangle$,

respectively) such that

$$(A/K)(BK/K)^{xK} = (BK/K)^{xK}(A/K).$$

This implies that $AB^xK = AB^x = B^xA$. Clearly, $xK = hK$ for some $h \in X$. We may, therefore, assume that $x \in X$ (respectively $x \in X \cap \langle A, KB \rangle = X \cap \langle A, B \rangle$). This means that A is (completely) X-permutable with B in G. On the other hand, if A is (completely) X-permutable with B in G, then by (3), A/K is (completely) X-permutable with BK/K in G/K.

(9) Take some Sylow p-subgroup T_p of T and some Sylow subgroup G_p of G containing T_p. Pick $x \in G$ such that $AG_p^x = G_p^xA$. Then AG_p^x is a subgroup of G. Hence $AG_p^x \cap T = A(G_p^x \cap T) = (G_p^x \cap T)A$ is a subgroup of T. Since T is subnormal in G, $G_p^x \cap T$ is a Sylow p-subgroup of T. Take some element t in T such that $(G_p^x \cap T)^t = T_p$. Then $A(G_p^x \cap T) = AT_p^{t^{-1}} = T_p^{t^{-1}}A$. Similarly we can prove the second claim.

(10) Suppose that A is completely G-permutable with T_1. Then there exists some element $x \in \langle A, T_1 \rangle$ such that $AT_1^x = T_1^xA$. Since $G = AT$, $x = at$ for some $a \in A$ and $t \in T$. Hence

$$AT_1^x = AatT_1t^{-1}a^{-1} = aAtT_1t^{-1}a^{-1} = a(AT_1^t)a^{-1}$$

is a subgroup of G. Hence, $AT_1^t = T_1^tA$, where $t \in T \cap \langle A, T_1 \rangle$ (because $x \in \langle A, T_1 \rangle$). This shows that A is completely T-permutable with T_1. If A is G-permutable with T_1, then similarly we can show that A is T-permutable with T_1.

(11) Take a maximal subgroup M of T. By (10) for some $t \in T$ we have $AM^t = M^tA$. Since T is a minimal supplement to A in G, $AM \neq G$ and so $AM^t \neq G$. Since A is a maximal subgroup of G, $M^t \leq A$. Suppose that T has some maximal subgroup M_1 which is not conjugate to M. Then as above we see that $M_1^{t_1} \leq A$ for some $t_1 \in T$. It is clear that $M^t \neq M_1^{t_1}$ and $T = \langle M^t, M_1^{t_1} \rangle \leq A$. This implies that $G = AT = A$, a contradiction. Therefore T is a primary cyclic group and $M \leq A$.

Lemma 1.6 *Suppose that $G = HT$, where H is a completely c-permutable proper subgroup of G and T is a nilpotent subgroup of G. Then G has a chain of subgroups*

$$H = T_0 \leq T_1 \leq \ldots \leq T_{t-1} \leq T_t = G$$

such that $|T_i : T_{i-1}|$ is a prime for all $i = 1, \ldots, t$.

Proof We may assume that $G \neq DH$ for any proper subgroups D of T. Let T_1 be a maximal subgroup of T. Suppose that $T_1 \leq H$. Then $|G| = \frac{|T||H|}{|T \cap H|} = \frac{|T||H|}{|T_1|}$. Since T is nilpotent, $|G : H| = |T : T_1|$ is a prime.

Now assume that $T_1 \nleq H$. By hypothesis, for some element $x \in G$, we have $HT_1^x = T_1^xH$. Since $G = HT$, $x = th$, where $t \in T$ and $h \in H$. Hence $T_1^x = T_1^h$. Since $T_1 \nleq H^{h^{-1}} = H$, $T_1^h \nleq H$. Moreover, because $T_1^hH \leq G$, we have $(T_1^hH)^{h^{-1}} = T_1H \leq G$. It is clear that $T_1H \neq G$. Assume that $T \cap H \nleq T_1$. Then

$T = T_1(T \cap H) \subseteq T_1H$, and so $G = TH \subseteq T_1H$, which is impossible. Hence $T \cap H \subseteq T_1$ and so $|T \cap H| = |T_1 \cap H|$. It follows that

$$|TH : T_1H| = \frac{|T||H|}{|T \cap H|} \cdot \frac{|T_1 \cap H|}{|T_1||H|} = |T : T_1|$$

is a prime. Since $|T_1H| < |G|$ and by hypothesis H is completely c-permutable in G, by induction on $|G|$, T_1H has a chain of subgroups

$$H = D_0 \leq D_1 \leq \ldots \leq D_{n-1} \leq D_n = HT_1$$

such that $|D_i : D_{i-1}|$ is a prime for all $i = 1, \ldots, n$. This completes the proof.

Proposition 1.7 [207]. *Let A be a proper group of a group G.*

(1) *If G is soluble and A is a completely c-permutable subgroup of G, then G has a chain of subgroups*

$$A = T_0 \leq T_1 \leq \ldots \leq T_{t-1} \leq T_t = G$$

such that $|T_i : T_{i-1}|$ is a prime for all $i = 1, \ldots, t$.

(2) *If A is subnormal in G and A c-permutes with all Sylow subgroup of G, then A/A_G is soluble.*

Proof

(1) Suppose the claim false and take a counterexample G of minimal order. Let L be any minimal normal subgroup of G. Since G is soluble, L is abelian. Suppose that $G = LA$. Then Lemma 1.6 implies (1), which contradicts the choice of G. Now assume that $LA \neq G$. Since $|LA| < |G|$ and A is a completely c-permutable subgroup of LA by Lemma 1.5(5), the choice of G implies that LA has a series
$$A = T_0 \leq T_1 \leq \ldots \leq T_{t-1} \leq T_t = LA$$
such that $|T_i : T_{i-1}|$ is a prime for all $i = 1, \ldots, t$.
Consider G/L. By Lemma 1.5(3), AL/L is a completely c-permutable subgroup of G/L, and so G/L has se series
$$AL/L = T_0/L \leq T_1/L \leq \ldots \leq T_{t-1}/L \leq T_t/L = G/L$$
such that $|T_i/L : T_{i-1}/L| = |T_i : T_{i-1}|$ is a prime for all $i = 1, \ldots, t$. Hence G has a series
$$A = T_0 \leq T_1 \leq \ldots \leq T_{t-1} \leq T_t = G$$
of subgroups with prime indexes. This completes the proof of Claim (1).

(2) By Lemma 1.5(4) we may assume that $A_G = 1$. Let $R = A^{\mathfrak{S}}$. Then, obviously, $R = R'$. Assume that A is not soluble. Then $R \neq 1$. Let $p \in \pi(G)$ and G_p be a Sylow p-subgroup of G such that $D = G_pA = AG_p$. Let Q be a Sylow q-subgroup of A, where $q \neq p$. Then evidently Q is a Sylow q-subgroup of

D. Since A is subnormal in G, A is subnormal in D and $Q^x \cap A$ is a Sylow q-subgroup of A, for all $x \in D$. Hence $L_q = \langle Q^x \mid x \in D \rangle \subseteq A$. Clearly, $L_q \trianglelefteq D$. Let L be the product of all L_q, where q runs through all prime divisors of the order of A which are different from p. Then $L \trianglelefteq D$. Since $LR/L \simeq R/L \cap R$ and D/L is a p-group, we have $R \subseteq L$. Let R_1 be the smallest normal subgroup of L with a soluble quotient. Then R_1 char $L \trianglelefteq A$, and so $R_1 \trianglelefteq A$. Since L/R_1 is a soluble group, $R_1 \subseteq R$. But since $R' = R$, $R_1 = R$ char $L \trianglelefteq D$. Therefore $R \trianglelefteq D$. Consequently, $G_p \subseteq N_G(R)$ for any $p \in \pi(G)$. It follows that $R \trianglelefteq G$. Therefore $A_G \neq 1$. This contradiction completes the proof.

X-semipermutable Let A, B be subgroups of a group G. If $AB = G$, then B is called a supplement of A in G.

Definition 1.8 Let A and B be subgroups of a group G, and let $\varnothing \neq X \subseteq G$. Then we say that A is (completely) X-semipermutable in G if A is (completely) X-permutable with all subgroups of some supplement T of A in G.

We use $X(A)$ $(X_c(A))$ to denote the set of all supplements T of A in a group G such that A is (completely) X-permutable in G with all subgroups of T. Thus A is (completely) X-semipermutable in G if and only if $X(A) \neq \varnothing$ $(X_c(A) \neq \varnothing$, respectively).

Example 1.9 Let $G = A_5 \times C_7$, where C_7 is a group of order 7 and A_5 is the alternating group of degree 5. Let C_5 be a Sylow 5-subgroup of A_5. Let $A \simeq A_4$ be a subgroup of G with $|G : A| = 5$ and $T = C_5 \times C_7$. Then $AT = G$ and evidently A permutes with all subgroups of T. Hence A is 1-semipermutable in G. On the other hand, A is not c-permutable in G. Indeed, let P be a Sylow 3-subgroup of A_5. Then $|A_5 : N_{A_5}(P)| = 10$. Hence A is not c-permutable with $N_{A_5}(P)$.

Lemma 1.10 *Let A and X be subgroups of G. Then the following statements hold:*

(1) *If N is a permutable subgroup of G and A is X-semipermutable in G, then NA is a X-semipermutable subgroup of G.*
(2) *If $N \trianglelefteq G$, A is X-semipermutable in G and $T \in X(A)$, then AN/N is XN/N-semipermutable in G/N and $TN/N \in (XN/N)(AN/N)$.*
(3) *If A/N is XN/N-semipermutable in G/N and $T/N \in (XN/N)(A/N)$, then A is X-semipermutable in G and $T \in X(A)$.*
(4) *If A is X-semipermutable in G and $A \leq D \leq G$, $X \leq D$, then A is X-semipermutable in D.*
(5) *If A is a maximal subgroup of G, T is a minimal supplement of A in G and $T \in G(A)$, then $T = \langle a \rangle$ is a cyclic p-group, for some prime p and $a^p \in A$.*
(6) *If $T \in X(A)$ and $A \leq N_G(X)$, then $T^x \in X(A)$, for all $x \in G$.*
(7) *If A is X-semipermutable in G and $X \leq D$, then A is D-semipermutable in G.*

Proof

(1) This part follows directly from Lemma 1.5(8).
(2) It is obvious that TN/N is a supplement of AN/N in G/N. If T_1/N is a subgroup of TN/N, then $T_1/N = (T_1 \cap N)T/N = N(T_1 \cap T)/N$ and so

AN/N is XN/N-permutable with T_1/N in G/N by Lemma 1.5(3). Hence,
$TN/N \in (XN/N)(AN/N)$.

(3) The proof of this part is the same as the proof in (2).

(4) This part is evident.

(5) See Lemma 1.5(11).

(6) Obviously, T^x is a supplement of A in G for any $x \in G$. Let T_1 be a subgroup of T^x. We need to prove that A is X-permutable with T_1. Since $G = AT$, we have $x = ta$, for some $a \in A, t \in T$. Hence $T^x = T^a$. Note that $T_1^{a^{-1}} \leq T$ and $A = A^{a^{-1}}$. By hypothesis, for some $d \in X$, we have $A(T_1^{a^{-1}})^d = (T_1^{a^{-1}})^d A = A^a(d^{-1})^a(T_1^{a^{-1}})^a d^a = AT_1^{d^a} = T_1^{d^a} A$, where $d^a \in X$ since $A \in N_G(X)$. This shows that $T^x \in X(A)$.

(7) This part is evident.

X_m-semipermutable subgroups.

Definition 1.11 Let A be a subgroup of a group G and X a nonempty subset of G. Then we say that:

(1) A is X_m-*permutable in* G if A is X-permutable with all maximal subgroups of all Hall subgroups of G.

(2) A is X_m-*semipermutable in* G if A is X-permutable with all maximal subgroups of all Hall subgroups of some minimal supplement of A in G.

In particular, if A is 1_m-permutable (1_m-semipermutable) in G, then we say that A is m-permutable (respectively, m-semipermutable) in G.

Example 1.12 Let A be a p-group with p an odd prime and $B = D_m = \langle x, y | x^{2^{m-1}} = y^2 = 1, x^y = x^{-1} \rangle$ a dihedral group of order 2^m, where $m > 2$. Let $G = A \times B$ and $L = \langle y \rangle$. Since G is nilpotent, every maximal subgroup of any Hall subgroup of G is normal in G. Hence L is m-permutable in G. On the other hand, L is clearly not permutable with $\langle y^x \rangle$. Thus the class of the X_m-permutable subgroups is in general a broader class than the class of the X-permutable subgroups.

Example 1.13 Let B and L be the groups as in Example 1.12. Then, it is easy to see that there is a 2-group P such that $B \leq P'$ and so $B \leq \Phi(P)$. Therefore P is the only minimal supplement of L in P. This shows that L is X_m-semipermutable but not X-semipermutable in P. Thus the class of the X_m-semipermutable subgroups is in general a broader class than the class of the X-semipermutable subgroups.

We will use $X_m(A)$ to denote the set of all minimal supplement T of A in a group G such that A is X-permutable with all maximal subgroups of any Hall subgroup of T. Thus A is X_m-semipermutable in G if and only if $X_m(A) \neq \varnothing$.

Lemma 1.14 *Let* A, T, X *be subgroups of* G *and* H *be a minimal normal subgroup of* G. *Then:*

(1) *If either* $H \leq A$ *or* $H \leq T$ *and if* T *is a minimal supplement of* A *in* G, *then* TH/H *is a minimal supplement of* AH/H *in* G/H.

(2) *Suppose that A is X-permutable with all maximal subgroups of any Hall subgroup of T. Assume that or H is abelian or $(|H|, |T|) = 1$ or T is soluble. Then AH/H is XH/H-permutable with all maximal subgroups of any Hall subgroup of TH/H.*

Proof

(1) Suppose that $H \leq A$ and let E/H be a supplement of A/H in G/H such that $E/H \leq TH/H$. Then $E = E \cap TH = H(E \cap T)$ and so $G = AE = A(E \cap T)$. Hence $T \leq E$ and $E/H = TH/H$ is a minimal supplement of A/H in G/H. On the other hand, if $H \leq T$ and E/H is a subgroup of T/H such that $G/H = (AH/H)(E/H)$, then $G = AE$ and so $E/H = T/H$ is a minimal supplement of AH/H in G/H.

(2) Let E/H be a Hall π-subgroup of TH/H and M/H be any maximal subgroup of E/H. We prove that AH/H is XH/H-permutable with M/H. We first note that $E = E \cap TH = H(E \cap T)$, $M = M \cap TH = H(M \cap T)$ and so $|T : E \cap T|$ is a π'-integer. Therefore, if H is a π-group, then E is also a π-group and so $E \cap T$ is a Hall π-subgroup of T. On the other hand, if $(|H|, |T|) = 1$, then $H \cap E \cap T = 1$ and so in this case $E \cap T$ is also a Hall π-subgroup of T. Now we show that $M \cap T$ is a maximal subgroup of $E \cap T$. Clearly $M \cap T \neq E \cap T$. Assume that for some subgroup D of G we have $M \cap T \leq D \leq E \cap T$. Then $M = H(M \cap T) \leq HD \leq H(E \cap T) = E$ and hence or $M = DH$ or $DH = E$. If $M = DH$, then

$$D = D \cap H(M \cap T) = (M \cap T)(D \cap H) = M \cap T.$$

If $DH = E$, then,

$$D = D \cap E \leq (E \cap T) \cap H(E \cap T) \leq (E \cap T)(E \cap T \cap H) = E \cap T.$$

Therefore, $M \cap T$ is maximal in $E \cap T$ and so by the hypothesis A is X-permutable with $M \cap T$. It follows from Lemma 1.5(4) that AH/H is XH/H-permutable with $M/H = (M \cap T)H/H$.

Finally, let either T be soluble or H be a p-group where $p \in \pi'$. Then for some Hall π-subgroup T_π of T, we have $T_\pi \leq E \cap T$. Indeed, the result is evident if T is soluble. For the second case, by using $|E : H| = |(E \cap T)H : H| = |E \cap T : (E \cap T) \cap H|$ and the well known Schur-Zassenhaus Theorem, we see that $E \cap T$ has a Hall π-subgroup T_π. Since $|T : E \cap T|$ is a π'-integer, T_π is a Hall subgroup of T. Hence $T_\pi H/H = E/H = H(E \cap T_\pi)/H$ and $M/H = H(M \cap T_\pi)/H$. As above, we may prove that $M \cap T_\pi$ is a maximal subgroup of $E \cap T_\pi$ and so again AH/H is XH/H-permutable with M/H. The Lemma is proved.

Lemma 1.15 *Let A and X be subgroups of G, $H \trianglelefteq G$. Suppose that $T \in X_m(A)$ and either $H \leq A$ or $H \leq T$. Suppose also that or H is an abelian minimal normal subgroup of G or $(|H|, |T|) = 1$ or T is soluble. Then $TH/H \in (XH/H)_m(AH/H)$.*

Proof This Lemma is a direct consequence of Lemma 1.14.

2.2   Criterions of Existence and Conjugacy of Hall Subgroups

A group G is said to be π-separable if G has a normal series

$$1 = G_0 \le G_1 \le \ldots \le G_{t-1} \le G_t = G, \tag{2.1}$$

where each index $|G_i : G_{i-1}|$ is either a π-number or a π'-number.

A group G is said to be:

  (I) E_π-group provided G has a Hall π-subgroup;
 (II) C_π-group provided G is a E_π-group and every two Hall π-subgroups of G are conjugate;
(III) D_π-group provided G is a C_π-group and every π-subgroup of G contained in some Hall π-subgroup of G.

The famous Schur-Zassenhaus Theorem asserts that: *If G has a normal Hall π-subgroup A, then G is an $E_{\pi'}$-group. Moreover, if either A or G/A is soluble, then A is a $C_{\pi'}$-subgroup.*

In 1928, Hall [228] proved that: *A soluble group is a D_π-group for any nonempty set π of primes.*

The most important result of the theory of π-separable groups is the following generalization of the above Hall result.

Theorem 2.1 (P. Hall [233], Čunihin [77]). *If G is a π-separable group, then G is a D_π-group.*

It is well known that the above Schur-Zassenhaus theorem, Hall theorem and Hall–Čunihin's theorem are truly fundamental results of group theory. In connection with these important results, the following two problems have naturally arisen:

Problem 2.2   Can we weaken the condition of normality for the Hall subgroup A of G so that the conclusion of the Schur-Zassenhaus Theorem is still true?

Problem 2.3   Whether we can replace the condition of normality for the members of series (2.1) by some weaker condition, for example, by permutability of the members of series (2.1) with some systems of subgroups of G.

In this section we give positive answers to the above two Problems.

A generalization of the Schur-Zassenhaus theorem.

Lemma 2.4   *Let N be a normal C_π-subgroup of G.*

  (i) *If G/N is a C_π-group, then G is a C_π-group (Čunihin [78]).*
 (ii) *If G/N is an E_π-group, then G is an E_π-group (Čunihin [78]).*
(iii) *If G has a nilpotent Hall π-subgroup, then G is a D_π-group [445].*
 (iv) *If G has a Hall π-subgroup with cyclic Sylow subgroups, then G is a D_π-group (S. A. Rusakov [339]).*

Lemma 2.5 *Let N be a normal C_π-subgroup of G and N_π a Hall π-subgroup of N.*

(i) *If G is a C_π-group, then G/N is a C_π-group ([233]).*
(ii) *If every Sylow subgroup of N_π is cyclic and G/N is a D_π-group, then G is a D_π-group* (Shemetkov [356] or [359, IV, Theorem 18.17]).
(iii) *G is a D_π-group if and only if G/N and N are D_π-groups* (See [334]).

Lemma 2.6 (Kegel [262]). *Let A and B, be the subgroups of G such that $G \neq AB$ and $AB^x = B^x A$, for all $x \in G$. Then G has a proper normal subgroup N such that either $A \leq N$ or $B \leq N$.*

The following lemma is obvious.

Lemma 2.7 *If N is normal in G and T is a minimal supplement of N in G, then $N \cap T \leq \Phi(T)$.*

Lemma 2.8 (Knyagina, Monakhov [266]). *If H, K, and N be pairwise permutable subgroups of G and H is a Hall subgroup of G, then*

$$N \cap HK = (N \cap H)(N \cap K).$$

Now we prove the following generalization of the Schur-Zassenhaus theorem.

Theorem 2.9 (Guo, Skiba [202]). *Let X be a normal C_π-subgroup of G and A a subgroup of G such that $|G : A|$ is a π-number. Suppose that A has a Hall π-subgroup A_0 such that either A_0 is nilpotent or every Sylow subgroup of A_0 is cyclic. Suppose that A X-permutes with every Sylow p-subgroup of G for all primes $p \in \pi$ or for all primes $p \in \pi \setminus \{q\}$ for some prime q dividing $|G : A|$. Then G is a C_π-group.*

Proof Assume that this proposition is false and let G be a counterexample of minimal order. Then $|\pi \cap \pi(G)| > 1$.

(1) *G/R is a C_π-group for any nonidentity normal subgroup R of G.*
   In order to prove this assertion, in view of the choice of G, it is enough to show that the hypothesis is still true for $(G/R, AR/R, XR/R)$. First note that $|G/R : AR/R| = |G : AR|$ is a π-number, and $A_0 R/R$ is a Hall π-subgroup of AR/R since

$$|AR/R : A_0 R/R| = |AR : A_0 R| = |A : A \cap A_0 R| = |A : A_0(A \cap R)|.$$

   On the other hand, $XR/R \simeq X/X \cap R$ is a C_π-group by Lemma 2.5(i), and either $A_0 R/R \simeq A_0/R \cap A_0$ is nilpotent or every Sylow subgroup of $A_0 R/R$ is cyclic. Finally, let P/R be a Sylow p-subgroup of G/R, where $p \in \pi \setminus q$. Then for some Sylow p-subgroup G_p we have $G_p R/R = P/R$. Hence AR/R XR/R-permutes with P/R by Lemma 1.5(3). Therefore the hypothesis holds for $(G/R, AR/R, XR/R)$.

(2) *$X = 1$.*
   Indeed, if $X \neq 1$, then G/X is a C_π-group by (1). Hence G is C_π-group by Lemma 2.4(i), a contradiction.

(3) *G has a proper nonidentity normal subgroup N.*

Let $p \in \pi \cap \pi(G)$, where $p \neq q$. Let P be a Sylow p-subgroup of G. First assume that $AP = G$. Since $|\pi \cap \pi(G)| > 1$, there is a prime $r \in \pi \cap \pi(G)$ such that $r \neq p$, so r does not divide $|G : A|$. Let R be a Sylow r-subgroup of G. Then for any $x \in G$ we have $AR^x = R^x A$. Hence $R \leq A_G$. Since G is not a C_π-group, $A \neq G$ by Lemma 2.4(iii)(iv). Hence $1 \neq A_G \neq G$. Now suppose that $AP \neq G$. By (2), $P^x A = AP^x$ for all $x \in G$. Hence we have (3) by Lemma 2.6.

(4) *N is a C_π-group.*

In view of the choice of G it is enough to prove that the hypothesis holds for (N, A_1), where $A_1 = A \cap N$. Since $|AN : A| = |N : A \cap N|$, $|N : A_1|$ is a π-number. On the other hand, $A_0 \cap N$ is a Hall π-subgroup of N since

$$|A \cap N : A_0 \cap N| = |A_0(A \cap N) : A_0|.$$

It also clear that either $A_0 \cap N$ is nilpotent or every Sylow subgroup of $A_0 \cap N$ is cyclic. Now let N_r be any Sylow r-subgroup of N, where $r \in \pi \setminus \{q\}$. Then for some Sylow r-subgroup G_r of G we have $N_r = G_r \cap N$ and

$$N \cap AG_r = (A \cap N)(N \cap G_r) = A_1 N_r = N_r A_1$$

by Lemma 2.8. Therefore the hypothesis holds for (N, A_1).

Finally, in view of (1) and (4), G is a C_π-group by Lemma 2.4(i), which contradicts the choice of G.

Lemma 2.10 *Let $U \leq B \leq G$ and $G = AB$, where the subgroup A permutes with U^b for all $b \in B$. Then A permutes with U^x for all $x \in G$.*

Proof Since $G = AB$, $x = ab$ for some $a \in A$ and $b \in B$. Hence

$$AU^x = AU^{ab} = Aa(U^b)a^{-1} = a(U^b)a^{-1}A = U^x A.$$

Corollary 2.11 *Let X be a normal C_π-subgroup of G and A a subgroup of G such that $|G : A|$ is a π-number. Suppose that A has a Hall π-subgroup A_0 such that either A_0 is nilpotent or every Sylow subgroup of A_0 is cyclic. Suppose that G has a subgroup T such that $|G : T|$ is a π'-number, $G = AT$ and A X-permutes with every Sylow p-subgroup of T for all primes $p \in \pi(G) \setminus \{q\}$, for some prime q dividing $|G : A|$. Then G is a C_π-group.*

Proof Let $p \in \pi(G) \setminus \pi$, $p \neq q$, and P be a Sylow p-subgroup of G. Since $G = AT$ and $|G : T|$ is a π'-number, every Sylow p-subgroup T_p of T is a Sylow p-subgroup of G. Hence for some $x \in G$ we have $T_p = P^x$, so A permutes with P by Lemma 2.10. Therefore Lemma 2.11 follows from Theorem 2.9.

Corollary 2.12 (Guo, Shum, Skiba [180]). *Let X be a normal nilpotent subgroup of G and A a Hall π-subgroup of G. Suppose $G = AT$ and A X-permutes with every subgroup of T. Then G is a $C_{\pi'}$-group.*

Corollary 2.13 (Guo, Skiba [193]). *Let A be a Hall subgroup of a group G and T a minimal supplement of A in G. Suppose that A permutes with all Sylow subgroups of T and with all maximal subgroups of any Sylow subgroup of T. Then G is a $C_{\pi'}$-group.*

Corollary 2.14 (Guo, Skiba in [197]). *Let H be a Hall π-subgroup of G. Let $G = HT$ for some subgroup T of G, and q a prime. If H permutes with every Sylow p-subgroup of T for all primes $p \neq q$, then T contains a complement of H in G and any two complements of H in G are conjugate.*

Proof It is clear that $T \cap H$ is a Hall π-subgroup of T. Moreover, if P is a Sylow subgroup of T and $HP = PH$, then $HP \cap T = P(H \cap T) = (H \cap T)P$. Hence by Theorem 2.9, T is a C_π-group. Consequently, Corollary 2.14 holds.

Corollary 2.15 (Foguel [107]). *Let A be a Hall π-subgroup of G, $G = AT$ and A permutes with every subgroup of T. Then G is an $E_{\pi'}$-group.*

Some classes of E_π-group Recall that a subgroup H of G is said to be a supplement of a subgroup A in G if $AH = G$. Let

$$1 = H_0 \leq H_1 \leq \ldots \leq H_{t-1} \leq H_t = G \tag{2.2}$$

be some subgroup series of G. We say that a subgroup series

$$1 = T_t \leq T_{t-1} \leq \ldots \leq T_1 \leq T_0 = G$$

is a supplement of series (2.2) in G if T_i is a supplement of H_i in G for all $i = 0, 1, \ldots, t$.

Theorem 2.16 (Guo, Skiba [197]). *Let X be a normal π-separable subgroup of G. Suppose that G has a subgroup series*

$$1 = H_0 \leq H_1 \leq \ldots \leq H_{t-1} \leq H_t = G$$

and a supplement

$$1 = T_t \leq T_{t-1} \leq \ldots \leq T_1 \leq T_0 = G$$

of this series in G such that H_i X-permutes with every Sylow subgroup of T_i for all $i = 1, 2, \ldots, t$. If each index $|H_{i+1} : H_i|$ is either a π-number or a π'-number, then G is an E_π-group and an $E_{\pi'}$-group. Moreover, if each π-index $|H_{i+1} : H_i|$ is a prime power, then G has a soluble Hall π-subgroup.

Proof Suppose that this theorem is false and let G be a counterexample of minimal order. Without loss of generality, we may assume that $H_1 \neq 1$. We proceed the proof by proving the following claims:

(1) *The assertions of the theorem hold for every nontrivial quotient G/N of G.*

We consider the series

$$1 = H_0 N/N \leq H_1 N/N \leq \ldots \leq H_{t-1}N/N \leq H_t N/N = G/N \tag{2.3}$$

and its supplement

$$1 = T_t N/N \le T_{t-1} N/N \le \ldots \le T_1 N/N \le T_0 N/N = G/N$$

in G/N. By Lemma 1.5(3), $H_i N/N$ is XN/N-permutable with any Sylow subgroup of $T_i N/N$ for all $i = 1, 2, \ldots, t$. On the other hand, since $|H_{i+1} N/N :$ $H_i N/N| = |H_{i+1} N : H_i N| = |H_{i+1} : H_i| : |N \cap H_{i+1} : N \cap H_i|$, every index of the series (2.3) is either π-number or π'-number (a π-prime power or π'-number). Moreover, obviously, $XN/N \simeq X/(X \cap N)$ is π-separable. This shows that the hypothesis holds on G/N. Hence in the case $N \ne 1$, the assertions of the theorem hold for G/N by the choice of G.

(2) $O_{\pi'}(G) = 1 = O_\pi(G)$.

Suppose that $D = O_{\pi'}(G) \ne 1$. Then by (1), G/D has a Hall π'-subgroup A/D and a Hall π-subgroup B/D. Then, obviously, A is a Hall π'-subgroup of G. By the Schur-Zassenhaus Theorem, D has a complement V in B, which, clearly, is a Hall π-subgroup of G. Hence G is an E_π-group and an $E_{\pi'}$-group. Besides, if for every π-index of series (2.3) is a prime power, then B/D has a soluble Hall π-subgroup B/D. It follows that V is also soluble. This contradiction shows that $O_{\pi'}(G) = 1$. Analogously, we may prove that $O_\pi(G) = 1$.

(3) $X = 1$.

Indeed, if N is a minimal normal subgroup of G contained in X, then N is either a π'-group or a π.-group, which contradicts (2).

(4) $T_1 \ne G$.

Suppose that $T_1 = G$. Then by hypothesis and (3), H_1 permutes with all Sylow subgroups of G. It follows from Chap. 1, Lemma 5.34 that H_1 is subnormal in G. Since H_1 is either a π-group or a π'-group, $H_1 \le O_\pi(G)$ or $H_1 \le O_{\pi'}(G)$ by Chap. 1, Lemma 2.14. It follows from (2) that $H_1 = 1$, which contradicts $H_1 \ne 1$. Hence (4) holds.

(5) *The assertions of the theorem hold for T_1.*

We consider the series

$$1 = H_0 \cap T_1 \le H_1 \cap T_1 \le \ldots \le H_{t-1} \cap T_1 \le H_t \cap T_1 = T_1. \qquad (2.4)$$

Then the series
$$1 = T_t \le T_{t-1} \le \ldots \le T_1$$

is a supplement of the series (2.4) in T_1 since $(H_i \cap T_1)T_i = H_i T_i \cap T_1 = G \cap T_1 = T_1$. Since $H_{i+1} = H_i T_1 \cap H_{i+1} = H_i(H_{i+1} \cap T_1)$, $|H_{i+1} : H_i| = |H_{i+1} \cap T_1 : H_i \cap T_1|$, for all $i = 1, 2, \ldots, t-1$ and $|H_1 \cap T_1 : H_0 \cap T_1| = |H_1 \cap T_1| \le |H_1 : 1|$, we see that every index of the series (2.4) is either π-number or π'-number. Moreover, if every π-index of the series (2.3) is a prime power, then every π-index of the series (2.4) is a prime power. Now let E be a Sylow subgroup of T_i. By (3) and the hypothesis, $H_i E = E H_i$. Hence $H_i E \cap T_1 = E(H_i \cap T_1) = (H_i \cap T_1)E$. This shows that the hypothesis holds for T_1. The minimal choice of G implies that (5) holds.

Final contradiction.

Let $(T_1)_\pi$ and $(T_1)_{\pi'}$ are a Hall π-subgroup and a Hall π'-subgroup of T_1, respectively. By (3) and the hypothesis, H_1 permutes with all Sylow subgroups of $(T_1)_\pi$. Hence H_1 permutes with $(T_1)_\pi$. Similarly, H_1 permutes with $(T_1)_{\pi'}$. By hypothesis, H_1 is either a π-group or a π'-group. Assume that H_1 is a π-group. Since $G = H_1 T_1$, we see that $G_\pi = H_1(T_1)_\pi$ is a Hall π-subgroup of G and $(T_1)_{\pi'}$ is a Hall π'-subgroup of G. If H_1 is a π'-group, then $G_{\pi'} = H_1 T_{\pi'}$ is a Hall π'-subgroup of G and T_π is a Hall π-subgroup of G. Finally, we prove that if every π-index of the series (2.3) is a prime power, then G has a soluble Hall π-subgroup. In fact, by (5), we see that $(T_1)_\pi$ is soluble. If H_1 is a π'-group, then $(T_1)_\pi$ is a soluble Hall π-subgroup of G since $G = H_1 T_1$. If H_1 is a p-group, then $H_1(T_1)_\pi$ is a Hall π-subgroup of G. Since $(T_1)_\pi$ is soluble and H_1 permutes with every Sylow subgroup of $(T_1)_\pi$, we see that $H_1(T_1)_\pi$ is soluble by Hall's theorem (see [337, (9.1.7), (9.1.8)]). The contradiction completes the proof.

In the case when $X = 1$ we obtain from Theorem 2.16 the following

Theorem 2.17 (Guo, Skiba [197]). *Suppose that G has a subgroup series*

$$1 = H_0 \le H_1 \le \ldots \le H_{t-1} \le H_t = G$$

and a supplement

$$1 = T_t \le T_{t-1} \le \ldots \le T_1 \le T_0 = G$$

of this series in G such that H_i permutes with every Sylow subgroup of T_i for all $i = 1, 2, \ldots, t$. If each index $|H_{i+1} : H_i|$ is either a π-number or a π'-number, then G is an E_π-group and an $E_{\pi'}$-group. Moreover, if each π-index $|H_{i+1} : H_i|$ is a prime power, then G has a soluble Hall π-subgroup.

Corollary 2.18 *Suppose that G has a subgroup series*

$$1 = H_0 \le H_1 \le \ldots \le H_{t-1} \le H_t = G$$

and a supplement

$$1 = T_t \le T_{t-1} \le \ldots \le T_1 \le T_0 = G$$

of this series in G such that H_i permutes with all Sylow subgroups of T_i for all $i = 1, 2, \ldots, t$. If each index $|H_{i+1} : H_i|$ $(i = 0, 1, \cdots, t-1)$ is a prime power, then G is an E_π-group, for any set π of primes.

Corollary 2.19 *Suppose that G has a subgroup series*

$$1 = H_0 \le H_1 \le \ldots \le H_{t-1} \le H_t = G$$

and a supplement

$$1 = T_t \le T_{t-1} \le \ldots \le T_1 \le T_0 = G$$

of this series in G such that H_i permutes with all Sylow subgroups of T_i for all $i = 1, 2, \ldots, t$. If each index $|H_{i+1} : H_i|$ is a prime power, then G is soluble.

Proof By Theorem 2.17, for any nonempty set π of primes, G has a Hall π-subgroup. Hence in the case, where $|\pi(G)| > 2$, G is soluble by Chap. 1, Corollary 4.10. Finally, in the case, where $|\pi(G)| = 2$, G is soluble by Burnside's $p^a q^b$-theorem.

Some classes of C_π-group. Now, based on Theorem 2.17, we prove the following result.

Lemma 2.20 ([89, A, 1,6]). *Let H, K, and N be subgroups of G. If $HK = KH$ and $HN = NH$, then $H\langle K, N\rangle = \langle K, N\rangle H$.*

Theorem 2.21 (Guo, Skiba [202]). *Let X be a normal C_π-subgroup of G. Suppose that G has a subgroup series*

$$1 = H_0 < H_1 \le \ldots \le H_{t-1} \le H_t = G,$$

where $|H_{i+1} : H_i|$ is either a π-number or a π'-number for all $i = 1, \ldots, t$. Suppose that G has a subgroup T such that $H_1 T = G$ and $|G : T|$ is a π'-number. If H_i is X-permutable with each Sylow subgroup of T, for all $i = 1, \ldots, t$, then G is a C_π-group.

Proof Suppose that this theorem is false and let G be a counterexample of minimal order. By Theorem 2.17, G has a Hall π-subgroup S. Hence some Hall π-subgroup S_1 of G is not conjugated with S. Without loss of generality, we may assume that $H_1 \ne 1$. Since $|G : T|$ is π'-number, every Sylow p-subgroup P of T, where $p \in \pi$, is also a Sylow p-subgroup of G. We proceed the proof via the following steps.

(1) *G/N is a C_π-group for every nontrivial quotient G/N of G.*
   We consider the subgroup series

$$1 = H_0 N/N \le H_1 N/N \le \ldots \le H_{t-1} N/N \le H_t N/N = G/N \qquad (2.5)$$

in G/N. Then $(H_1 N/N)(TN/N) = G/N$ and $|G/N : TN/N| = |G : TN|$ is a π'-number. Moreover, by Lemma 1.5, $H_i N/N$ is XN/N-permutable with every Sylow subgroup of TN/N for all $i = 1, \ldots, t$. On the other hand, since

$$|H_{i+1} N/N : H_i N/N| = |H_{i+1} N : H_i N| = |H_{i+1} : H_i| : |N \cap H_{i+1} : N \cap H_i|,$$

every index of the series (2.5) is either π-number or π'-number. Moreover, $XN/N \simeq X/(X \cap N)$ is a C_π-group by Lemma 2.5(i). All these show that the hypothesis holds for G/N. Hence in the case, where $N \ne 1$, G/N is a C_π-group by the choice of G.

(2) *$O_{\pi'}(G) = 1 = O_\pi(G)$.*
   Suppose that $D = O_{\pi'}(G) \ne 1$. Then by (1), there is an element $x \in G$ such that $S_1{}^x D = SD$. But by the Schur-Zassenhaus theorem, $S_1{}^x$ and S are conjugate in SD, which implies that S_1 and S are conjugate in G. This contradiction shows that $O_{\pi'}(G) = 1$. Analogously, one can prove that $O_\pi(G) = 1$.

(3) $X = 1$ (This follows from (1), Lemma 2.4(i) and the choice of G).

(4) $T \ne G$.
   Suppose that $T = G$. Then by hypothesis and (3), H_1 permutes with all Sylow subgroups of G. It follows from Chap. 1, Lemma 5.34 that H_1 is subnormal in

G. Since H_1 is either a π-group or a π'-group, $H_1 \leq O_\pi(G)$ or $H_1 \leq O_{\pi'}(G)$ by Chap. 1, Lemma 5.35 (6). Hence by (2), $H_1 = 1$, which contradicts to our assumption about H_1.

(5) *The hypothesis holds for T.*

Consider the subgroup series

$$1 = H_0 \cap T \leq H_1 \cap T \leq \ldots \leq H_{t-1} \cap T \leq H_t \cap T = T \qquad (2.6)$$

of the group T. Since

$$H_{i+1} = H_i T \cap H_{i+1} = H_i(H_{i+1} \cap T)$$

we have

$$|H_{i+1} : H_i| = |H_{i+1} \cap T : H_i \cap T|,$$

for all $i = 1, \ldots, t - 1$ and $|H_1 \cap T : H_0 \cap T| = |H_1 \cap T|$ divides $|H_1 : 1|$, we see that every index of the series (2.6) is either π-number or π'-number. Now let E be a Sylow subgroup of T. By (3) and the hypothesis, $H_i E = E H_i$. Hence

$$H_i E \cap T = E(H_i \cap T) = (H_i \cap T)E.$$

This shows that the hypothesis holds for T.

(6) *T is a C_π-group.*

Since $T \neq G$ by (4), and the hypothesis holds for T by (5), the minimal choice of G implies that (6) holds.

(7) *T is a $E_{\pi'}$-subgroup.*

This follows from (3), (5), and Theorem 2.17.

(8) *Let $T_{\pi'}$ be a Hall π'-subgroup of T and D a normal subgroup of G. Then $T_{\pi'} \neq 1$, and if either $H_1 \leq D$ or $T_{\pi'} \leq D$, then $D = G$.*

Suppose that $T_{\pi'} = 1$. Then H_1 is a Hall π'-subgroup of G. Therefore G is a C_π-group by Corollary 2.11, a contradiction. Hence $T_{\pi'} \neq 1$.

We show that the hypothesis holds for D. Consider the subgroup series

$$1 = D_0 \leq D_1 \leq \ldots \leq D_{t-1} \leq D_t = D,$$

where $D_i = H_i \cap D$ for all $i = 1, \ldots, t$. Let $T_0 = D \cap T$. First we show that $D_1 T_0 = D$. If $H_1 \leq D$, then

$$D = H_1(D \cap T) = H_1 T_0 = D_1 T_0.$$

Now suppose that $T_{\pi'} \leq D$. In view of (3), H_1 permutes with $T_{\pi'}$. Since $H_1 T = G$ and $T \neq G$ with $|G : T|$ is a π'-number, we have that H_1 is a π'-group. Hence $H_1 T_{\pi'}$ is a Hall π'-subgroup of G. Therefore

$$D = (D \cap H_1 T_{\pi'})(D \cap T_\pi) = T_{\pi'}(D \cap H_1)(D \cap T_\pi)$$

$$= (D \cap H_1)(T_{\pi'}(D \cap T_\pi)) = (D \cap H_1)(D \cap T) = D_1 T_0.$$

It is also clear that $|D : T_0|$ is a π'-number. Now let $p \in \pi(T_0)$ and P be a Sylow p-subgroup of T_0. Then for some Sylow p-subgroup T_p of T we have $P = T_p \cap D$. Hence in view of Lemma 2.8,

$$D \cap H_i T_p = (D \cap H_i)(D \cap T_p) = D_i P = P D_i.$$

Thus D_i is permutable with every Sylow subgroup of T_0 for all $i = 1, \ldots, t$. Finally, since the number

$$|D_i : D_{i-1}| = |(D \cap H_i)H_{i-1} : H_{i-1}|$$

divides $|H_i : H_{i-1}|$, each index $|D_i : D_{i-1}|$ is either a π-number or a π'-number. Therefore the hypothesis holds for D. Suppose that $D \neq G$. Then D is a C_π-group by the choice of G. Since either $1 \neq H_1 \leq D$ or $1 \neq T_{\pi'} \leq D$, G/D is a C_π-group by (1). It follows from Lemma 2.4(i) that G is a C_π-group, which contradicts the choice of G. Hence, (8) holds.

Final contradiction. Since $G = H_1 T$ and H_1 permutes with all Sylow subgroups of T by (3),

$$H_1(T_{\pi'})^x = (T_{\pi'})^x H_1$$

for all $x \in G$ by Lemmas 2.10 and 2.20. Therefore by Lemma 2.6, either $H_1{}^G \neq G$ or $(T_{\pi'})^G \neq G$. But in view of (8) both these cases are impossible. The contradiction completes the proof of the result.

Corollary 2.22 (Guo, Skiba [197]). *Let X be a normal π-separable subgroup of G. Suppose that G has a subgroup series*

$$1 = H_0 < H_1 \leq \ldots \leq H_{t-1} \leq H_t = G,$$

where $|H_{i+1} : H_i|$ is either a π-number or a π'-number for all $i = 1, \ldots, t$. Suppose that G has a subgroup T such that $H_1 T = G$ and $|G : T|$ is a π'-number. If H_i is X-permutable with each subgroup of T, for all $i = 1, \ldots, t$, then G is a C_π-group.

A generalization of Hall-Čunihin's theorem.

Theorem 2.23 (Guo, Skiba [202]). *Let X be a normal E_π-subgroup of G and X_π a Hall π-subgroup of X. Suppose that G has a subgroup series*

$$1 = H_0 < H_1 \leq \ldots \leq H_{t-1} \leq H_t = G,$$

where $|H_{i+1} : H_i|$ is either a π-number or a π'-number for all $i = 1, \ldots, t$. Suppose that G has a subgroup T such that $H_1 T = G$ and $|G : T|$ is a π'-number.

(i) *Suppose that the Sylow subgroups of X_π are cyclic. If H_i is X-permutable with each cyclic subgroup H of T of prime power order, for all $i = 1, \ldots, t$, then G is a D_π-group.*

(ii) *Suppose that X is a D_π-subgroup. If H_i is X-permutable with each cyclic subgroup H of T of prime power order, for all $i = 1, \ldots, t$, then G is a D_π-group.*

Proof (i) Suppose that this assertion is false and let G be a counterexample of minimal order. Since G is not a D_π-group, there is a π-subgroup U of G such that for any Hall π-subgroup E of G we have $U \nleq E$.

(1) G/N is a D_π-group for any nonidentity normal subgroup N of G.

Consider the subgroup series

$$1 = H_0 N/N \leq H_1 N/N \leq \ldots \leq H_{t-1} N/N \leq H_t N/N = G/N.$$

It is clear that $(H_1 N/N)(T N/N) = G/N$, $|G/N : T N/N|$ is a π'-number and $|H_{i+1} N/N : H_i N/N|$ is either a π-number or a π'-number for all $i = 1, \ldots, t$ (see (1) in the proof of Theorem 2.21). Now let H/N be a cyclic subgroup of $T N/N$ of prime power order $|H/N|$. Then $H = N(H \cap T)$. Let W be a group of minimal order with the properties that $W \leq H \cap T$ and $NW = H$. If $N \cap W \not\leq \Phi(W)$, then for some maximal subgroup S of W we have $(N \cap W)S = W$. Hence $H = NW = N(N \cap W)S = NS$, a contradiction. Hence $N \cap W \leq \Phi(W)$. Since $W/W \cap N \simeq H/N$ is a cyclic group of prime power order, W is also a cyclic group of prime power order. Hence H_i is X-permutable with W. Thus $H_i N/N$ is XN/N-permutable with $WN/N = H/N$ by Lemma 1.5(3). Therefore the hypothesis holds for G/N. But since, $N \neq 1$, $|G/N| < |G|$, so G/N is a D_π-group by the choice of G.

(2) $X = 1$, and so G is a C_π-group by Theorem 2.21 and Corollary 2.22.

Suppose that $X \neq 1$. Then G/X is a D_π-group by (1). On the other hand, by Lemma 2.4(iv), X is a D_π-group. Hence G is a D_π-group by Lemma 2.5(iii), a contradiction. Thus we have (2).

(3) H_1 permutes with every subgroup of U.

Let Z be any cyclic subgroup of U of prime power order p^n. Then $p \in \pi$ and $Z \leq G_p$ for some Sylow p-subgroup G_p of G. Since $|G : T|$ is a π'-number, there is an element $x = ht$, where $h \in H_1$, $t \in T$, such that $(G_p)^x \leq T$. Hence $H_1 Z = Z H_1$ by Lemma 2.10, which in view of Lemma 2.20 implies that $H_1 U = U H_1$.

(4) $T \neq G$ (see the proof of (4) in the proof of Theorem 2.21).

(5) T is a D_π-group.

The hypothesis holds for T (see (5) in the proof of Theorem 2.21), so by (4) and the minimal choice of G, we have (5).

(6) $V = H_1 U$ is a C_π-group.

Indeed, V is a group by (3), and since $T \neq G$, H_1 is a Hall π'-subgroup of V. Therefore we have (6) by (2) and Corollary 2.11.

(7) $|V : T \cap V|$ is a π'-number and $T \cap V$ is a C_π-group.

First note that $|G : T| = |V : T \cap V|$ is a π'-number. But H_1 is a Hall π'-subgroup of V. Hence $V = H_1(T \cap V)$, which implies that

$$|V : H_1| = |T \cap V : H_1 \cap T \cap V|$$

is a π-number. Hence $A = H_1 \cap T \cap V$ is a Hall π'-subgroup of $T \cap V$. If W is any π-subgroup of $T \cap V$, then $H_1 W = W H_1$ by (3). Therefore

$$AW = (H_1 \cap T \cap V)W = H_1 W \cap (T \cap V) = WA.$$

Hence $T \cap V$ is a C_π-group by Theorem 2.9.

Final contradiction for (i). In view of (7), $|V : T \cap V|$ is a π'-number and $T \cap V$ is a C_π-group. Hence in view of (6), there is an element $x \in V$ such that $U^x \leq T \cap V \leq T$. But by (5), T is a D_π-group. Hence for some Hall π-subgroup T_π of T we have $U \leq T_\pi$, which is a contradiction since T_π is clearly a Hall π-subgroup of G.

(ii) See the proof of (i) and use Lemma 2.5 (iii).

The following example shows that, under conditions of Theorems 2.9, 2.21 and 2.23, the group G is not necessary π-separable.

Example 2.24 Let $G = A_5 \times C_7$, where C_7 is a group of order 7 and A_5 is the alternating group of degree 5. Let C_5 be a Sylow 5-subgroup of A_5. Consider the subgroup series

$$1 = H_0 < H_1 < H_2 < H_3 = G, \tag{2.7}$$

where $H_1 = A_4$ and $H_2 = A_5$. Then the series $1 = T_3 < T_2 < T_1 < T_0 = G$, where $T_2 = C_7$ and $T_1 = C_5 \times C_7$, is a supplement of series (2.7) in G. It is clear also that H_i permutes with all subgroups of T_i, for all i. Let $\pi = \{5, 7\}$. Then every index of series (2.7) is either a π-number or a π'-number. However, G is not π-separable.

Groups With Soluble Hall π-Subgroups A group G is said to be:

(a) A E_π^S-*group* (E_π^N-*group*) if G has a soluble (a nilpotent, respectively) Hall π-subgroup;
(b) A C_π^S-*group* (C_π^N-*group*) if G has a soluble (nilpotent, respectively) Hall π-subgroup and G is a C_π-group;
(c) A D_π^S-*group* if G is a C_π^S-group and any π-subgroup of G is contained in some Hall π-subgroup of G.

Our next results are new criteria for a group to be an E_π^S-group.

Lemma 2.25 (See [233] or [135, 1.10.3]). *Let N be a normal E_π^N-subgroup of G. If G/N is a D_π^S-group, then G is a D_π^S-group*

Theorem 2.26 (Guo, Skiba [202]). *Let X be a normal E_π^N-subroup of G. Suppose that G has a subgroup series*

$$1 = H_0 < H_1 \leq \ldots \leq H_{t-1} \leq H_t = G,$$

where $|H_{i+1} : H_i|$ is divisible by at most one prime in π, for all $i = 1, \ldots, t$. Suppose that G has a subgroup T such that $H_1 T = G$ and $|G : T|$ is a π'-number. If H_i is X-permutable with each Sylow subgroup of T, for all $i = 1, \ldots, t$, then G is an E_π^S-group.

Proof Suppose that this theorem is false and let G be a counterexample of minimal order.

(1) *Every nontrivial quotient G/N of G is an E_π^S-group* (See (1) in the proof of Theorem 2.21).
(2) $X = 1$.

Suppose that $X \neq 1$. Then G/X of G is an E_π^S-group by (1). Let E/X be a soluble π-Hall subgroup of G/X. Then, by Lemma 2.25, E is a D_π^S-group and if U is a soluble Hall subgroup of E, then U is a soluble Hall subgroup of G. Hence G is a E_π^S-group, a contradiction. Thus we have (2).

(3) $T \neq G$.

Suppose that $T = G$. Then by hypothesis and (2), H_1 permutes with all Sylow subgroups of G. Hence H_1 is subnormal in G by Chap. 1, Lemma 5.34(1), so

$$1 < H_1 \leq O_{\pi' \cup \{p\}}(G)$$

for some $p \in \pi$ by Chap.1, Lemma 5.35(6). Therefore $O_{\pi' \cup \{p\}}(G)$ is a non-identity normal D_π^N-subgroup of G. Then by Lemma 2.25, G is a D_π^S-group, a contradiction. Thus (3) holds.

Final contradiction. In view of (2), the hypothesis is true for T (see (5) in the proof of Theorem 2.21). Hence T is an E_π^S-group by (3) and the choice of G. But since $|G : T|$ is a π'-number, any Hall π-subgroup of T is a Hall π-subgroup of G as well. Hence G is an E_π^S-group, a contradiction. The theorem is proved.

Theorem 2.27 (Guo, Skiba [202]). *Let X be a normal E_π^N-subgroup of G. Suppose that G has a subgroup series*

$$1 = H_0 < H_1 \leq \ldots \leq H_{t-1} \leq H_t = G,$$

where $|H_{i+1} : H_i|$ is divisible by at most one prime in π, for all $i = 1, \ldots, t$. Suppose that G has a subgroup T such that $H_1 T = G$ and $|G : T|$ is a π'-number. If H_i is X-permutable with each cyclic subgroup of T of prime power order, for all $i = 1, \ldots, t$, then G is a D_π^S-group.

Proof Suppose that this theorem is false and let G be a counterexample of minimal order. Then $|\pi \cap \pi(G)| > 1$. Without loss of generality, we may suppose that $H_1 \neq 1$ and $H_{t-1} \neq G$.

(1) *Every nontrivial quotient G/N of G is a D_π^S-group* (See (1) in the proof of Theorem 2.23).

(2) *If N is a normal E_π^N-subgroup of G, then $N = 1$. In particular, $X = 1$.*
   Suppose that $N \neq 1$ Then the factor group G/N of G is a D_π^S-group by (1). Hence G is a D_π^S-group by Lemma 2.25, which contradicts the choice of G. Hence we have (2).

(3) $T \neq G$ (See (3) in the proof of Theorem 2.26).

(4) *The hypothesis holds for H_{t-1} and for T.*
   Since

$$H_{t-1} = H_1(H_{t-1} \cap T)$$

and

$$|H_{t-1} : H_{t-1} \cap T| = |G : T|,$$

the hypothesis holds for H_{t-1} by (2). The second assertion of (4) may be proved as (5) in the proof of Theorem 2.21.

(5) G is a C_π^S-group.

Since $|G : H_{t-1}|$ is divisible by at most one prime in π and $|\pi \cap \pi(G)| > 1$, there is a prime $p \in \pi \cap \pi(G)$ such that for a Sylow p-subgroup P of G we have $P \le H_{t-1}$, so in view of (2), Lemmas 2.10 and 2.20, we have $1 < P \le (H_{t-1})_G$ (In fact, for $p \in \pi$, clearly, every Sylow p-subgroup T_p of T is also a Sylow p-subgroup of G. Hence by (2), Lemmas 2.10 and 2.20, H_{t-1} is permutable with T_p^x for all $x \in G$. Therefore $P \subseteq T_p^x$. Since $p \nmid |G : H_{t-1}|$, $T_p^x \subseteq H_{t-1}$ for any $x \in G$, which implies that $P = T_p \subseteq (H_{t-1})_G$). Since $H_{t-1} \ne G$ and the hypothesis holds for H_{t-1}, H_{t-1} is a D_π^S-group by the choice of G. Hence $(H_{t-1})_G$ is a C_π^S-group by Lemma 2.5(iii), so G is a C_π^S-group by Lemma 2.4(i) and Theorem 2.26.

(6) *For any i, H_i permutes with every π-subgroup of G* (see (3) in the proof of Theorem 2.23).

(7) $V = H_1 U$ *is a D_π^S-group for any π-subgroup U of G.*

By (6), V is a subgroup of G. Since $T \ne G$, H_1 is a π'-group. By Theorem 2.9, V is a C_π-group. We show that the hypothesis holds for V. Let V_π be a Hall π-subgroup of V. Then $V = H_1 V_\pi$ and

$$V_i = V \cap H_i = H_1(V_\pi \cap H_i),$$

for all $i = 1, \ldots, t$. Let W be any subgroup of V_π. Then by (2), (5) and Lemma 2.20,

$$V_i W = (H_1(V_\pi \cap H_i))W = H_1(V_\pi \cap H_i W) = H_1 W(V_\pi \cap H_i)$$
$$= W(H_1(V_\pi \cap H_i)) = W V_i.$$

It is also clear that $|V_{i+1} : V_i|$ is divisible by at most one prime in π, for all $i = 0, \ldots, t$. Hence the hypothesis holds for V. Suppose that $G = V$. In this case, in view of (5), we may suppose that V_π is a soluble Hall π-subgroup of G. Let L be a minimal normal subgroup of V_π. Then

$$L^G = L^{V_\pi H_1} = L^{H_1} \le LH_1 \cap L^{H_1} = L(L^G \cap H_1)$$

and L is a q-group for some $q \in \pi$. It follows from (2) that for some prime $p \in \pi$ with $q \ne p$ we have $p \in \pi(L^G \cap H_1)$ and a Sylow p-subgroup P of $L^G \cap H_1$ is also a Sylow subgroup of L^G. For some Sylow p-subgroup G_p of G we have $P = L^G \cap G_p$, so by Lemma 2.8,

$$L^G \cap H_1 G_p = (L^G \cap H_1)(L^G \cap G_p) = (L^G \cap H_1)P = P(L^G \cap H_1).$$

Hence $P \le (L^G \cap H_1)_{L^G}$. Therefore G has a nonidentity subnormal subgroup $R = (L^G \cap H_1)_{L^G}$ of order divisible by at most one prime in π, which contradicts (2). Hence $V \ne G$, so V is a D_π^S-group by the choice of G.

Final contradiction. Let U be any π-subgroup of G and $V = H_1 U$. Then V is a D_π^S-subgroup of G by (7), and $|G : T| = |V : V \cap T|$ is a π'-number. By (3) and (4), T is a D_π-group. Hence for some Hall π-subgroup V_π of V we have $U \le V_\pi$, and

$V_\pi \leq T_\pi$. But since $|G : T|$ is a π'-number, T_π is a Hall π-subgroup of G. Therefore G is a D_π^S-group by (5).

Corollary 2.28 *Suppose that G has a subgroup series*

$$1 = H_0 < H_1 \leq \ldots \leq H_{t-1} \leq H_t = G,$$

where $|H_{i+1} : H_i|$ is divisible by at most one prime in π, for all $i = 1, \ldots t$. Suppose also that G has a subgroup T such that $H_1 T = G$ and $|G : T|$ is a π'-number. If H_i is quasinormal in T, for all $i = 1, \ldots, t$, then G is a D_π^S-group.

Corollary 2.29 (Čunihin [76]). *If G has a chief subgroup series*

$$1 = H_0 < H_1 \leq \ldots \leq H_{t-1} \leq H_t = G,$$

where $|H_{i+1} : H_i|$ is divisible by at most one prime in π, for all $i = 1, \ldots, t$, then G is a D_π^S-group.

2.3   On the Groups in Which Every Subgroup Can be Written as an Intersection of Some Subgroups of Prime Power Indices

A number of authors have studied the groups G in which, for all subgroups H and all divisors d of $|G : H|$, G has a subgroup K such that $H \leq K$ and $|K : H| = d$ (see Sect. 6 in [450]). In particular, it was proved that such groups are just the groups in which every subgroup can be written as an intersection of subgroups of pairwise coprime prime power indexes (see Theorem 6.1 in [450, Chap. 6]). In connection with this, the following questions have naturally arisen:

Problem 3.1 What is the structure of the groups G in which every subgroup can be written as an intersection of subgroups of prime power indexes?

Problem 3.2 What is the structure of the groups G in which every subnormal subgroup can be written as an intersection of subnormal subgroups of prime power indexes?

The solutions of these problems are connected with the following two concepts.

Definition 3.3 A subgroup H of a group G is said to be meet-irreducible (or primitive) if H is a proper subgroup of the intersection of all subgroups of G containing H as a proper subgroup.

Definition 3.4 A subnormal subgroup H of a group G is said to be subnormally meet-irreducible (or subnormally primitive) if H is a proper subgroup of the intersection of all subnormal subgroups of G containing H as a proper subgroup.

It is not difficult to show that a (subnormal) subgroup H of G can be written as an intersection of (subnormal) subgroups of prime power indexes if and only if H is a (subnormally) meet-irreducible subgroup of G of prime power index.

Therefore the above problems are equal to the following problems:

Problem 3.5 What is the structure of the groups G in which every meet-irreducible subgroup has prime power index?

Problem 3.6 What is the structure of the groups G in which every subnormally meet-irreducible subgroup has prime power index?

On Problem 3.5, in 1971, Johnson [256] proved that a group G is supersoluble if every of its meet-irreducible subgroup has prime power index in G. Late, this result was generalized in various directions. In particular, A.N. Skiba [376] in 1978 first gave a generalized version of this theorem by using the theory of formations. A nice local version of Johnson's theorem was also given by M. S. Chen [62]. In 2001, N.S. Kosenok and V. N. Ryzhih [270] have also established an analogous theorem of Johnson for subnormally meet-irreducible subgroups. Some other interesting results on Problem 3.5 were given in the book [450]. In this section, we give the full solutions to both Problems 3.5 and 3.6.

Lemma 3.7 *Let G be a group, $H \leq K \leq G$. Then the following statements hold:*

(1) *Every proper (subnormal) subgroup H of G is the intersection of some (subnormally) meet-irreducible subgroups of G.*
(2) *Let $H \leq T \leq G$, where H, T are (subnormal) subgroups of G and H is a (subnormally) meet-irreducible subgroup of T. Then $H = T \cap X$, for some (subnormally) meet-irreducible subgroup X of G.*
(3) *Let $K \leq H \leq G$ and $K \lhd G$. Then H is a (subnormally) meet-irreducible subgroup of G if and only if H/K is a (subnormally) meet-irreducible subgroup of G/K.*
(4) *Let H be a (subnormally) meet-irreducible subgroup of G. If P and Q are normal subgroups of G, and P and Q have coprime orders, then either $P \leq H$ or $Q \leq H$.*
(5) *If H is a (subnormally) meet-irreducible subgroup of G and $H_G = 1$, then $F(G) = O_p(G)$ for some prime p.*

Proof

(1) If H is a (subnormally) meet-irreducible subgroup of G, then (1) clearly holds. Otherwise, $H = H_1 \cap \ldots \cap H_t$, where $\{H_1, \ldots, H_t\}$ is the set of all (subnormal) subgroups of G such that $H < H_i < G$. Then by induction on $|G : H|$, we have $H_i = T_{i_1} \cap \ldots \cap T_{i_r}$, where T_{ij} is a (subnormally) meet-irreducible subgroup of G. Hence (1) holds.
(2) If H is a (subnormally) meet-irreducible subgroup of G, then the statement (2) clearly holds. Otherwise, let $\{X_1, \ldots, X_t\}$ be a set of (subnormally) meet-irreducible subgroups of G such that $H = X_1 \cap \ldots \cap X_t$. Then

$$H = X_1 \cap \ldots \cap X_t = (X_1 \cap \ldots \cap X_t) \cap T$$
$$= (X_1 \cap T) \cap \ldots \cap (X_t \cap T).$$

Since H is a (subnormally) meet-irreducible subgroup of T, there exists some $i \in \{1, \ldots, t\}$ such that $H = T \cap X_i$.
(3) This assertion is obvious.

(4) Assume that $P \nsubseteq H$ and $Q \nsubseteq H$. Let $n = |PH : H| = |P : H \cap P|$ and
$m = |QH : H| = |Q : H \cap Q|$. Then $n \neq 1$, $m \neq 1$, and $(n, m) = 1$. Hence
$H = PH \cap QH$, which is impossible since H is meet-irreducible in G.

(5) This directly follows from (4).

Lemma 3.8 *Let G be a group. Then the following statements hold:*

(i)   *If G is supersoluble, then $G' \subseteq F(G)$ and G is a dispersive group (see [135,
        Theorem 1.9.9]);*

(ii)   *If $L \trianglelefteq G$ and $G/\Phi(L)$ is supersoluble (a dispersive group), then G is super-
        soluble (respectively, G is a dispersive group) (see (1.3) and [135, Theorem
        1.8.1])*

(iii)   *G is supersoluble if and only if $|G : M|$ is a prime for every maximal subgroup
        M of G (see [135, Theorem 1.9.6]).*

Proposition 3.9 (Yang, Guo [457]). *Let G be a soluble group. If every (subnormally) meet-irreducible subgroup of G not having a nilpotent Hall supplement in G
is normal in G, then $G = D \rtimes M$ is a supersoluble group, where D and M are nilpotent Hall subgroups of G, $D = G^{\mathfrak{N}}$ is the $\mathfrak{N}$-residual of G and $G = DN_G(D \cap X)$,
for every (subnormally) meet-irreducible subgroup X of G. In particular, if each
subnormally meet-irreducible subgroup of G has a nilpotent Hall supplement in G,
then G is nilpotent.*

Proof Suppose that the proposition is false and let G be a counterexample of minimal
order. We proceed the proof by proving the following claims.

(1) *The hypotheses hold for every factor group of G.*

Let H be a normal subgroup of G and T/H a (subnormally) meet-irreducible
subgroup of G/H. Then T is a (subnormally) meet-irreducible subgroup of G
by Lemma 3.7(3). Hence, by the hypotheses, either T is normal in G or $G = TX$
for some nilpotent Hall subgroup X of G. In the former case, T/H is normal in
G/H. In the second case, $XH/H \simeq X/X \cap H$ is a nilpotent Hall subgroup of
G/H and $(T/H)(XH/H) = G/H$. Thus (1) holds.

(2) *The hypotheses hold for every normal subgroup N of G.*

Let H be a (subnormally) meet-irreducible subgroup of N. Then, by
Lemma 3.7(2), G has a (subnormally) meet-irreducible subgroup X such that
$H = N \cap X$. If X is normal in G, then H is normal in N. Assume now that
X has a nilpotent Hall supplement T in G. If $X \leq N$, then $N = N \cap XT =
X(N \cap T) = H(N \cap T)$ and $N \cap T$ is a nilpotent Hall subgroup of N since
N is normal in G. Now assume that $X \nsubseteq N$ and let π be the set of all prime
divisors of $|XN : X| = |N : N \cap X| = |N : H|$. Then clearly each $p \in \pi$ is a
divisor of $|T|$. Hence, by Theorem 2.1, there is a Hall π-subgroup G_π of G such
that $G_\pi \leq T$. Therefore $G_\pi \cap N$ is a nilpotent Hall π-subgroup in N such that
$N = H(G_\pi \cap N)^x$ for some element $x \in M$. This shows that the hypotheses
hold for N.

(3) *G is a supersoluble group.*

Assume that G is not supersoluble. Then, by claim (1), G has a unique minimal normal subgroup H such that G/H is supersoluble, and so $G^{\mathfrak{U}} = H = C_G(H) = F(G) = O_p(G) \nsubseteq \Phi(G)$ for some prime p and $|H| \neq p$.

Let M be a maximal normal subgroup of G. Then $H \leq M$. Let q be the largest prime divisor of $|M|$ and M_q a Sylow q-subgroup of M. Then by claim (2) and the choice of G, M is supersoluble. Hence $M_q \trianglelefteq M$ by Lemma 3.8. If $q \neq p$, then $M_q \leq C_G(H) = H$, which is impossible. Hence $q = p$. Since $M \trianglelefteq G$ and M_p is a characteristic subgroup of M, $M_p \trianglelefteq G$. Obviously, $H \leq M_p$ and so $M_p/H \leq O_p(G/H)$. Because $H = C_G(H)$, $O_p(G/H) = 1$ by Chap. 1, Lemma 2.15. It follows that $M_p = H$. Assume that $|G : M| = p$. Then p is the largest prime divisor of $|G|$. Since G/H is supersoluble, $O_p(G/H) \neq 1$, a contradiction. Hence $|G : M| = q \neq p$. This leads to that H is a Sylow p-subgroup of G. Let H_1 be a maximal subgroup of H. Then H_1 is (subnormally) meet-irreducible in H. Hence, by Lemma 3.7(2), $H_1 = X \cap H$ for some (subnormally) meet-irreducible subgroup X of G. Let T be a nilpotent Hall subgroup of G such that $G = XT$. Since $p \mid |G : X|$, $H \leq T$. It follows from $C_G(H) = H$ that $T = H$. Thus, $G = HX$ and so $H_1 \trianglelefteq G$. This contradiction shows that (3) holds.

(4) *Let $D = G^{\mathfrak{N}}$ be the $\mathfrak{N}$-residual of G. Then D is a nilpotent Hall subgroup of G.*

Since G is supersoluble, $G' \subseteq F(G)$ by Lemma 3.8. Obviously, $D \leq G'$. Hence D is nilpotent. We now prove that D is a Hall subgroup of G. Assume that it is false. Then $D \neq 1$.

We first suppose that G has two minimal normal subgroups H and R such that H is a p-group and R is a q-group with $p \neq q$. Without loss of generality, we may assume that $H \leq D$. By Chap. 1, Lemma 1.1, $(G/R)^{\mathfrak{N}} = (G^{\mathfrak{N}})R/R = DR/R$, and so DR/R is a Hall subgroup of G/R by the choice of G. Let D_p be a Sylow p-subgroup of D. Then RD_p/R is a Sylow p-subgroup of DR/R, and so it is also a Sylow p-subgroup of G/R. It follows that D_p is a Sylow p-subgroup of G. Assume that $D_p \neq D$ and let D_r be a Sylow r-subgroup of D with $r \neq p$. Then as above we see that D_r is a Sylow r-subgroup of G. Thus D is a Hall subgroup of G.

Next assume that all minimal normal subgroups of G are p-groups. In this case, $F(G) = O_p(G)$ is a Sylow p-subgroup of G and so $D \leq O_p(G)$. Let H be a minimal normal subgroup of G. If $H \neq D$, then by using the same arguments as above, we see that D is a Sylow p-subgroup of G. Hence we may assume $H = D$. We claim that $\Phi = \Phi(O_p(G)) = 1$. Indeed, if $\Phi \neq 1$, then, by the choice of G, we know that $\Phi D/\Phi = \Phi G^{\mathfrak{N}}/\Phi = (G/\Phi)^{\mathfrak{N}}$ is a Hall subgroup of G/Φ. If $H \leq \Phi$, then G/Φ is a nilpotent group. But $O_p(G) \trianglelefteq G$ and consequently $\Phi \leq \Phi(G)$. This leads to G is nilpotent and thereby $H = G^{\mathfrak{N}} = 1$, which contradicts the choice of G. Hence $H \nleq \Phi$ and $H\Phi/\Phi$ is a nonidentity p-group. Since $(G/\Phi)^{\mathfrak{N}} = G^{\mathfrak{N}}\Phi/\Phi = H\Phi/\Phi$ is a Hall subgroup of G/Φ, $H\Phi = O_p(G)$ and so $D = H = O_p(G)$ is a Hall subgroup of G. This contradiction shows that $\Phi(O_p(G)) = 1$.

We now prove that every proper subgroup T of $O_p(G)$ is normal in G. Assume that T is a maximal subgroup of $O_p(G)$. Then T is (subnormally) meet-irreducible in $O_p(G)$. It follows from Lemma 3.7(2) that $T = O_p(G) \cap X$, for some (subnormally) meet-irreducible subgroup X of G. If X is normal in G, then $T = O_p(G) \cap X$ is normal in G. So we may assume that X is not normal in G. Hence, by hypothesis, G has a nilpotent Hall subgroup X_0 such that $G = XX_0$. Let q be a prime divisor of $|G : X|$. Then $q \mid |X_0|$. Assume that $q \neq p$. Since $O_p(G) \cap X = T \neq O_p(G)$, $p \mid |G : X|$, and so $p \mid |X_0|$. Hence $O_p(G) \leq X_0$. Since X_0 is nilpotent, $X_0 \leq C_G(O_p(G))$, which contrary to Chap. 1, Proposition 2.9(5)(8). Hence $|G : X| = p$. But $X \cap O_p(G) \trianglelefteq X$, so $T \trianglelefteq G$. Thus every maximal subgroup of $O_p(G)$ is normal in G. Let T be a non-maximal subgroup of $O_p(G)$. In order to prove that T is normal in G, we only need to prove that T is the intersection of all maximal subgroups T_i of $O_p(G)$ containing T. Since $\Phi(O_p(G)) = 1$, $O_p(G)$ is an elementary abelian p-group. Hence $O_p(G)/T$ is also an elementary abelian group and consequently $\Phi(O_p(G)/T) = 1$. This shows that T is the intersection of all maximal subgroups of $O_p(G)$ containing T.

Since $\Phi(O_p(G)) = 1$, we may assume that $O_p(G) = \langle a \rangle \times \langle a_2 \rangle \times \ldots \times \langle a_t \rangle$, where $\langle a_i \rangle$ is a minimal normal subgroup of G, $\langle a \rangle = H$ and $|a| = |a_i| = p$. Let $a_1 = aa_2 \ldots a_t$. Then since $\langle a_1 \rangle \cap \langle a_2 \rangle \ldots \langle a_t \rangle = 1$, we have $O_p(G) = \langle a_1 \rangle \times \langle a_2 \rangle \times \ldots \times \langle a_t \rangle$. Because G is not nilpotent, $O_p(G) \nsubseteq Z(G)$. Hence there exists an index i such that $a_i \notin Z(G)$. Since G is not a p-group, G has an element g such that $(|g|, p) = 1$ and $g \notin C_G(a_i)$. Let $y = [[a_i, y_1], \ldots, y_n]$, where $y_1 = \ldots = y_n = g$ and $n = |G|$. Since $G/H = G/G^{\mathfrak{N}}$ is nilpotent, $y \in H = D$. On the other hand, as $\langle a_i \rangle \trianglelefteq G$, we have $y \in \langle a_i \rangle$. Since $y_i = g \notin C_G(a_i)$, $y \neq 1$. Hence $\langle a_i \rangle = \langle y \rangle = H$, a contradiction. This completes the proof of (4).

(5) *For every (subnormally) meet-irreducible subgroup X of G, we have $G = DN_G(X \cap D)$.*

Let π_1 be the set of all prime divisors of $|D|$. Assume that for some $p \in \pi_1$, there is a Sylow p-subgroup G_p of G which is contained in X. Since the hypothesis holds for G/G_p, by induction, we have

$$(D/G_p)N_{G/G_p}((X/G_p) \cap (D/G_p)) = (D/G_p)N_{G/G_p}((X \cap D)/G_p) = G/G_p.$$

Since $N_{G/G_p}((X \cap D)/G_p) = N_G(X \cap D)/G_p$, we obtain that $G = DN_G(X \cap D)$. Suppose that $p \mid |G : X|$ for all $p \in \pi_1$. Then, by Lemma 3.7(4), $|\pi_1| = 1$. Let $\pi_1 = \{p\}$. Then $D = G_p$. Let π_2 be the set of all prime divisors of $|G : X|$. Then $p \in \pi_2$. By the hypothesis, either G has a nilpotent Hall subgroup X_0 such that $G = XX_0$ or X is normal in G. Assume that we have the former case. Then $G_p \subseteq X_0$ and if Q is a Sylow q-subgroup of X_0 with $q \neq p$, then Q is a Sylow q-subgroup of G such that $Q \subseteq C_G(X \cap G_p)$. Hence for all $q \in \pi_2 \setminus \{p\}$, we have $q \nmid |G : N_G(X \cap G_p)|$. Since $G_p \cap X \trianglelefteq X$, $r \nmid |G : N_G(X \cap G_p)|$ for each prime $r \notin \{p\} \cup \pi_2$. This shows that $|G : N_G(X \cap G_p)| = p^\alpha$ for some $\alpha \in \{0\} \cup \mathbb{N}$, and so $G = G_p N_G(X \cap G_p)$. Finally, suppose that X is normal in G. Then $N_G(X \cap D) = DN_G(X \cap D)$.

(6) *The conclusion*

Since $D = G^{\mathfrak{N}}$ is a normal Hall subgroup of G, by Shur-Zassenhaus Theorem, we know that D has a complement M in G. Since $G/D \simeq M$ is nilpotent, M is a nilpotent Hall subgroup of G such that $G = D \rtimes M$.

Finally, suppose that every subnormally meet-irreducible subgroup of G has a nilpotent Hall supplement in G. We prove that G is nilpotent. Suppose that this is false and let G be a counterexample of minimal order. Then, as in the proof of (3), it may be proved that G has a unique minimal normal subgroup H such that G/H is nilpotent and $H = F(G)$. But then 1 is a subnormally meet-irreducible subgroup of G and so G is nilpotent.

The theorem is proved.

Lemma 3.10 *Let X be a meet-irreducible subgroup of a group G. Then X is c-permutable with all Sylow subgroups of G if and only if $|G : X| = p^a$ for some prime p.*

Proof Suppose that X is c-permutable with all Sylow subgroups of G but the index $|G : X|$ has a least two different prime divisors p and q. Then, for some Sylow p-subgroup P and some Sylow q-subgroup Q of G, we have $X < XP = PX$ and $X < XQ = QX$. It follows that $X = XP \cap XQ$, which is impossible since X is a meet-irreducible subgroup of G. Thus, $|G : X| = p^a$ for some prime p. Conversely, suppose that $|G : X| = p^a$ for some prime p and let Q be a Sylow q-subgroup of G. If $q = p$, Then $G = XQ^x$ for some $x \in G$. On the other hand, if $q \neq p$, then $Q^x \leq X$ for some $x \in G$. Hence X is c-permutable with Q. Lemma is proved.

Lemma 3.11 *If every meet-irreducible subgroup of a group G not having a nilpotent Hall supplement in G is normal in G, then G is soluble.*

Proof Suppose that this lemma is false and let G be a counterexample of minimal order. It is clear that the hypothesis still holds on every subnormal subgroup of G and on every quotient of G (see the proof of Proposition 3.9). Hence G is a simple non-abelian group. Let p be the smallest prime divisor of $|G|$ and P a Sylow p-subgroup of G. Let M be a maximal subgroup of P and $N = N_G(M)$. Since P/M is a cyclic group, N/M is p-nilpotent by Chap. 1, Lemma 3.39. So N has a maximal subgroup T such that $M \leq T \trianglelefteq N$ and $|N : T| = p$. Since T is meet-irreducible in N, $T = N \cap X$, for some meet-irreducible subgroup X in G by Lemma 3.7(2). Now, by hypothesis, either X is normal in G or G has a nilpotent Hall subgroup T_0 such that $G = XT_0$. In the former case, G is not simple, a contradiction. Hence we have the second case. If $q \mid |G : X|$ for some prime $q \neq p$, then $q \mid |T_0|$. We claim that $p \mid |T_0|$. Indeed, if $p \nmid |T_0|$, then $p \nmid |G : X|$, and so $P^x \subseteq X$ for some $x \in G$. Because $T \subseteq X$, we have $N \subseteq X$ and so $T = N \cap X = N$, a contradiction. Therefore, $p \mid |T_0|$. Let Q be a Sylow p-subgroup of T_0 and x an element of G such that $Q^x = P$. Since T_0^x is a nilpotent group, $T_0^x \subseteq N$. Hence $q \nmid |G : N|$ and so $q \nmid |G : X|$. This contradiction shows that $|G : X| = p^\alpha$ for some $\alpha \in \mathbb{N}$. But since $M \subseteq X$, $|G : X| = p$ and so $X \trianglelefteq G$ since p is the smallest prime divisor of $|G|$ (see [452, II, 4.6]). It is clear that the hypothesis of the lemma still holds for every quotient of G and every normal subgroup of G. Hence G is soluble by induction on $|G|$.

The following theorem give the answer to Problem 3.5.

Theorem 3.12 (Yang, Guo [457]). *Let G be a group. The following are equivalent.*

(a) *Each subgroup of G can be written as an intersection of subgroups of prime power indices in G.*

(b) *Every meet-irreducible subgroup of G has prime power index.*

(c) *Every meet-irreducible subgroup of G not having a nilpotent Hall supplement in G is normal in G.*

(d) *$G = D \rtimes M$ is a supersoluble group, where $D = G^{\mathfrak{N}}$ and M are nilpotent Hall subgroups of G and $G = DN_G(D \cap X)$ for every meet-irreducible subgroup X of G.*

(e) *Every meet-irreducible subgroup X of G is $G^{\mathfrak{N}}$-permutable with all Sylow subgroups of G.*

Proof It is clear that (a) and (b) are equivalent (see p. 132 in [450]). It is also clear that (c) follows from (b), and (d) follows from (c) in view of Proposition 3.9 and Lemma 3.11. Therefore in view of Lemma 3.10, we only need to prove that (e) follows from (d). Suppose that this is false and let G be a group of minimal order. Then G has a meet-irreducible subgroup X such that X does not $G^{\mathfrak{N}}$-permute with some Sylow subgroup Q of G. First suppose that $X_G \neq 1$. Then the hypothesis holds on G/X_G by Lemma 1.5 and $(G/N_G)^{\mathfrak{N}} = G^{\mathfrak{N}}X_G/X_G = DX_G/X_G$. Hence X/X_G is DX_G/X_G-permutable with QX_G/X_G in G/X_G. But then, by Lemma 1.5(4), X is D-permutable with Q in G. This contradiction shows that $X_G = 1$. Hence by Lemma 3.7(5), $D = F(G) = O_p(G)$ is the Sylow p-subgroup of G for some prime p.

Let $N = N_G(X \cap D)$. Then, by hypothesis, $G = DN$ and so $|G : N| = p^\alpha$ for some $\alpha \in \{0\} \cup \mathbb{N}$. Therefore, without loss of generality, we may assume that $Q \leq N$. Since $D \trianglelefteq G$, $X \subseteq N$. Let $X_{p'}$ be a Hall p'-subgroup in X. Since G is soluble, by Theorem 2.1 we see that for some Hall p'-subgroup $N_{p'}$ of the group N, we have $X_{p'} \subseteq N_{p'}$. Since $G/D \simeq M$ is a nilpotent group, $N_{p'}$ is nilpotent. Now

$$XQ = ((X \cap D)X_{p'})Q = Q(X \cap D)X_{p'} = QX.$$

This contradiction completes the proof of the implication (d) $\Rightarrow$ (e).

Corollary 3.13 (Johnson [256]). *If every meet-irreducible subgroup of a group G has prime-power index, then G is supersoluble.*

Corollary 3.14 *If every meet-irreducible subgroup of a group G has a Hall nilpotent supplement in G, then G is supersoluble.*

Corollary 3.15 (Guo, Shum, and Skiba [178]). *Let G be a group. Then the following are equivalent:*

(a) *Every meet-irreducible subgroup of G has a nilpotent Hall supplement in G.*

(b) *$G = D \rtimes M$ is a supersoluble group, where D and M are nilpotent Hall subgroups of G, $D = G^{\mathfrak{N}}$ is the $\mathfrak{N}$-residual of G and $G = DN_G(D \cap X)$ for every meet-irreducible subgroup X of G.*

(c) *Every meet-irreducible subgroup of G has a prime-power index.*

The following theorem give the answer to Problem 3.6.

Theorem 3.16 (Yang, Guo [457]). *Let G be a group. Then the following are equivalent:*

(a) *Every subnormal subgroup can be written as an intersection of subnormal subgroups of prime power indexes.*
(b) *Every subnormally meet-irreducible subgroup of G has prime power index.*
(c) *G is nilpotent.*

Proof Items (a) and (b) are equivalent by Lemma 3.7(1). It is clear that (b) follows from (c). Finally, from Proposition 3.9 and Lemma 3.11 we have (b) $\Rightarrow$ (c).

2.4   Criteria of Supersolubility and p-Supersolubility for Products of Groups

A well-known Fitting's theorem says that any group G which is a product of normal nilpotent subgroups of G is nilpotent. However, supersoluble groups do not have such property (see Theorem I, 1.13 in the book [450]). It is natural to ask: *Under what additional conditions, will the product of two supersoluble groups still be supersoluble?* In the literature, we know, for example, that the product $G = AB$ of two normal supersoluble subgroups A and B is supersoluble if either G' is nilpotent (see R. Baer [23]) or the subgroups A and B have coprime indices in G [59]. An interesting approach for solving the supersolubility problem in finite groups was proposed by M. Asaad and A. Shaalan in 1989 [20]. They obtained the following nice result: *Assume that $G = AB$ is the product of two supersoluble subgroups A and B. If every subgroup of A is permutable with every subgroup of B, then G is supersoluble.* In addition, they have also generalized the above mentioned result of Baer by replacing the condition of normality of A, B in G by using the weaker condition: *A permutes with all subgroups of B and B permutes with all subgroups of A.* Their results in [20] were further developed and applied by many authors (see, for example, the book [34]). In this section, we develop the above results on base of the concept of c-permutability.

The following result is well known.

Lemma 4.1 *Let $A, B \leq G$ and $G = AB$. Then $P = A_p B_p$ for some $P \in \mathrm{Syl}_p(G)$, $A_p \in \mathrm{Syl}_p(A)$, and $B_p \in \mathrm{Syl}_p(B)$.*

Lemma 4.2 (see [359, Chap. 1]). *The class of all dispersive groups is a hereditary saturated formation.*

Lemma 4.3 (see [135, 2.4.4]). *Let M_1, M_2 be maximal subgroups of a soluble group G such that $(M_1)_G = (M_2)_G$. Then M_1 and M_2 are conjugate.*

Lemma 4.4 (Čunihin). *Let $G = AB$ be the product of the subgroups A and B. If L is a normal subgroup of A and $L \leq B$, then $L \leq B_G$.*

Proof $L^G = L^{AB} = L^B \leq B_G$.

Theorem 4.5 (Liu, Guo, Shum [298]). *Let $G = AB$ be the product of dispersive subgroups A and B. If every subnormal subgroup of A is completely c-permutable in G with every subnormal subgroup of B, then G is a dispersive group.*

Proof Assume that this Theorem is not true and let G be a counterexample of minimal order. Without loss of generality, we may assume that for every proper subnormal subgroup A_1 of A and every proper subnormal subgroup B_1 of B, we have $A_1 B \neq G$ and $G \neq AB_1$. We proceed the proof via the following steps.

(a) *Let $a \in A$. Then $G = AB^a$ and every subnormal subgroup of A is completely c-permutable with every subnormal subgroup of B^a.*
   By Chap. 1, Lemma 4.14 we have $G = AB^a$. Now let $H \lhd\lhd A$, $T \lhd\lhd B^a$ and $\langle H, T \rangle = D \leq G$. Then $H^{a^{-1}} \lhd\lhd A$, $T^{a^{-1}} \lhd\lhd B$ and $\langle H^{a^{-1}}, T^{a^{-1}} \rangle = D^{a^{-1}}$. By hypothesis, for some $d \in D$, we have $H^{a^{-1}}(T^{a^{-1}})^{d^{a^{-1}}} = (T^{a^{-1}})^{d^{a^{-1}}} H^{a^{-1}}$. Then $(aHa^{-1})(ad^{-1}a^{-1})(aTa^{-1})(ada^{-1}) = aHd^{-1}Tda^{-1} = ad^{-1}TdHa^{-1}$. This implies that $HT^d = T^d H$.

(b) *G has an abelian minimal normal subgroup.*
   Let L be a minimal normal subgroup of A. Then, by hypothesis, G has an element x such that $LB^x = B^x L$. Assume that $L \subseteq B^x$. By Chap. 1, Lemma 4.14, $G = AB^x$, and by Lemma 4.4, $L^G \subseteq B^x$. But since B^x is a dispersive group, a minimal normal subgroup of G contained in L^G is abelian. Hence we may suppose that L is not contained in B^x. Assume that $LB^x \neq G$ and let M be a maximal subgroup of G such that $LB^x \subseteq M$. Let $x = ba$, where $a \in A$, $b \in B$. Then $B^x = B^a$. By Chap. 1, Lemma 4.14 again, $G = AB^a$. In view of (a) and Lemma 1.5, M satisfies the hypothesis. The choice of G implies that M is a dispersive group. Since $L \subseteq A \cap M$, $L^G \subseteq M$ by Lemma 4.4. This shows that a minimal normal subgroup of G contained in L^G is abelian. Now, we let $G = LB^a$. Since $L \subseteq A$, by (a) every subnormal subgroup of L is completely c-permutable with every subnormal subgroup of B^a. Let R be a minimal normal subgroup of B^a. Then by using the same argument as above, we come to the case that $G = LR$. Since L and R are abelian groups, G has an abelian minimal normal subgroup.

(c) *G/L is a dispersive group for any nonidentity normal subgroup L of G.*
   Obviously, $G/L = (AL/L)(BL/L)$. Let $H/L \lhd\lhd AL/L$ and $T/L \lhd\lhd BL/L$, and let $D = \langle H, T \rangle$. By the hypothesis, $(H \cap A)(H \cap B)^x = (H \cap B)^x(H \cap A)$ for some $x \in D$. Hence

$$(H/L)(T/L)^{xL} = (L(A \cap H)/L)(L(T \cap B)/L)^{xL} = L(A \cap H)(T \cap B)^x/L$$

$$= ((T \cap B)^x L/L)(L(A \cap H)/L)$$

$$= ((T \cap B)L/L)^{xL}(L(A \cap H)/L) = (T/L)^{xL}(H/L),$$

where $xL \in D/L$. This shows that G/L is a product of the dispersive subgroups AL/L and BL/L such that every subnormal subgroup of AL/L is completely

c-permutable with every subnormal subgroup of BL/L in G/L. Since $|G/L| < |G|$, we conclude that G/L is a dispersive group.

(d) *G has a unique minimal normal subgroup $L = O_p(G) = C_G(L)$, for some prime p, and $G = L \rtimes M$, where M is a maximal subgroup of G with $O_p(M) = 1$ and $|L| \neq p$.*

Let L be a minimal normal subgroup of G. Then by Lemma 4.2, L is the unique minimal normal subgroup of G and $L \nsubseteq \Phi(G)$. Let M be a maximal subgroup of G not containing L and $C = C_G(L)$. Then $M_G = 1$ and so G is a primitive group. Hence $L = O_p(G) = C_G(L)$ for some prime p by I. Chap. 1, Lemma 4.17, and $M \simeq G/L$ is a dispersive group with $O_p(M) = 1$. Finally, we claim that $|L| \neq p$. Indeed, if $|L| = p$, then $G/L = G/C_G(L)$ is a cyclic group with order $p - 1$ and so G is a dispersive group, a contradiction.

(e) *p is the largest prime divisor of $|G|$.*

Assume that q is the largest prime divisor of the order of G with $q \neq p$. Let A_1 be a maximal subgroup of A such that $A_1 \trianglelefteq A$ and $|A : A_1| = r$, where r is the least prime divisor of $|A|$. Analogously, we let B_1 be a maximal subgroup of B such that $B_1 \trianglelefteq B$ and $|B : B_1| = s$ where s is the least prime divisor of $|B|$. Let $x \in A$ and $y \in B$ such that $T_1 = AB_1^x = B_1^x A$ and $T_2 = A_1^y B = BA_1^y$. We now claim that $|G : T_1| = s$. Indeed, $|G : T_1| = |AB| : |AB_1| = (|B : B_1||A \cap B_1|) : |A \cap B| = s$. Analogously we see that $|G : T_2| = r$. Hence both T_1 and T_2 are maximal subgroups of G. By Chap. 1, Lemma 4.14, T_1 and T_2 are not conjugate in G. Since all maximal subgroups of G not containing L are conjugate in G by Lemma 4.3, either T_1 contains L or T_2 contains L. Without loss of generality, we may assume that $L \leq T_1$. Let Q be a Sylow q-subgroup of T_1. Now we prove that T_1 is a dispersive group. Indeed, by Dedekind Law, we have $T_1 = T_1 \cap AB_1^x = A(T_1 \cap B_1^x)$. Hence T_1 is a product of the dispersive group subgroups A and $T_1 \cap B_1^x$. Besides, by (a), every subnormal subgroup of A is completely c-permutable in T_1 with every subnormal subgroup of B_1^x. This shows that T_1 is a dispersive group since $|T_1| < |G|$. Similarly one can prove that T_2 is a dispersive group. But then Q is normal in T_1 and so $Q \subseteq C_G(L) = L$. Since $p \neq q$, this shows that $Q = 1$ and hence $|G : T_1| = s = q$ is the order of a Sylow q-subgroup G_q of G. Moreover, since s is the least prime divisor of B, $|B| = q$ and so $L \subseteq A$. It follows that p is the largest prime divisor of $|A|$. Assume that $L \neq A$. Then $L \subseteq A_1$ and hence $L \subseteq T_2$. However, since T_2 is a dispersive group, $B \subseteq C_G(L) = L$. This contradiction shows that $L = A$ and so $LB = G$. Let L_1 be a maximal subgroup of L. Then for some $x \in G$, we have $L_1^x B = L_1^x B$. Hence $L \cap L_1^x B = L_1^x \trianglelefteq L_1^x B$ and consequently $L_1^x \trianglelefteq G$. By (d), we see that $L_1 \neq 1$. This contradiction shows that p is the largest prime divisor of $|G|$.

To complete the proof.

First we claim that L is a Sylow subgroup of G. Indeed, assume it is not true. Then $p \mid |G : L|$. This means that $p \mid |M|$, and so by (c), we have that $O_p(M) \neq 1$. This contradicts (d). Hence, L is a Sylow p-subgroup of G. Now from (c) and (e), we obtain that G is a dispersive group. This contradiction completes the proof.

Theorem 4.6 (Liu, Guo, Shum [298]). *A group G is supersoluble if and only if $G = AB$ is a product of supersoluble subgroups A and B such that every subnormal subgroup of A is completely c-permutable with every subgroup of B in G and every subnormal subgroup of B is completely c-permutable with every subgroup of A in G.*

Proof The necessity part is evident. We only need to prove the sufficiency part. Suppose that it is false and let G be a counterexample with minimal order. Then G has a unique minimal normal subgroup L such that $L = C_G(L) = O_p(G)$, for some prime p, and $|L| \neq p$ (see the proof of Theorem 4.5). Also, by Theorem 4.5, G is a dispersive group. Hence L is a Sylow subgroup of G. Without loss of generality, we may assume that $p \mid |A|$. It is clear that $L \cap A \trianglelefteq A$ and $L \cap A$ is a Sylow p-subgroup of A. Since A is supersoluble, we know that A has a normal subgroup Z_p of order p contained in $L \cap A$. Let $B_{p'}$ be a Hall p'-subgroup of B and x an element of G such that $Z_p B_{p'}^x = B_{p'}^x Z_p$. Then, $Z_p = L \cap Z_p B_{p'}^x \trianglelefteq Z_p B_{p'}^x$. Consequently, $B_{p'}^x \subseteq N_G(Z_p)$. It is clear that a Sylow p-subgroup of B is contained in $N_G(Z_p)$. Thus $Z_p \trianglelefteq G$ and so $Z_p = L$. It follows that G is supersoluble.

Corollary 4.7 (Guo, Shum and Skiba [177]). *A group G is supersoluble if and only if $G = AB$ is a product of supersoluble subgroups A and B such that every subgroup of A is completely c-permutable in G with every subgroup of B in G.*

Lemma 4.8 (Arad, Michael [7]). *A group G is soluble if and only if every two of its Sylow subgroups are G-permutable.*

Lemma 4.9 (see Theorems 4.11 and 4.14 in [248, Chap. VI]). *If $G = AB$, where A and B are nilpotent subgroups of G, then G is soluble.*

Lemma 4.10 *Let p be a prime number and G a p-soluble group. If $O_{p'}(G) = 1$, then the following statements are equivalent:*

  (i) G is p-supersoluble;
 (ii) G is supersoluble;
(iii) $G/O_p(G)$ is an abelian group of exponent dividing $p - 1$.

Proof (i)$\Longrightarrow$(ii). Since G is p-supersoluble, for every chief p-factor H/K of G, we have $|H/K| = p$ and so by Chap.1, Lemma 3.29, $G/C_G(H/K)$ is an abelian group of exponent dividing $p - 1$. Since $O_{p'}(G) = 1$, the intersection of the centralizers of all such factors contains the subgroup $O_{p'p}(G) = O_p(G)$ by Chap.1, Lemma 2.10. Hence G is supersoluble by Chap. 1, (1.3). By using the same arguments, we can also prove that (ii)$\Longrightarrow$(iii) and (iii)$\Longrightarrow$(i).

Theorem 4.11 (Guo, Shum and Skiba [177]). *A group G is supersoluble if and only if $G = AB$, where A, B are nilpotent subgroups of G and A has a chief series*

$$1 = A_0 \leq A_1 \leq \ldots \leq A_{t-1} \leq A_t = A \tag{2.8}$$

such that every A_i is completely c-permutable (permutable) with all subgroups of B, for all $i = 1, \ldots, t$.

Proof Assume that G is a supersoluble group. Then, by Lemma 3.8, $G' \subseteq F(G)$. Let $A = F(G)$ and B be a minimal supplement of A in G. Then, evidently, $A \cap B \subseteq$

$\Phi(B)$. Since $AB/A \simeq B/A \cap B$, $B/A \cap B$ is nilpotent and so B is a nilpotent group. Now considering a chief series of G below $F(G)$

$$1 = A_0 \leq A_1 \leq \ldots \leq A_{t-1} \leq A_t = A = F(G).$$

We can see immediately that this series is also a chief series of A (since $|A_i/A_{i-1}|$ is a prime for all $i = 1, \ldots, t$) and that A_i is permutable with all subgroups of B for all $i = 1, \ldots, t$.

Now we assume that $G = AB$, where A, B are nilpotent subgroups of G and A has a chief series $1 = A_0 \leq A_1 \leq \ldots \leq A_{t-1} \leq A_t = A$ such that every term of which is completely c-permutable with all subgroups of B. We prove that G is a supersoluble group. Suppose that it is not true and let G be a counterexample of minimal order. Without loss of generality, we may assume that $A_{t-1}B \neq G$ and $G \neq AB_1$ for every proper subgroup B_1 of B. First of all we note that by Corollary 4.7, G is soluble since it is a product of two nilpotent groups. We now proceed via the following steps:

(a) *G/N is supersoluble for every normal subgroup $N \neq 1$ of G.*
Clearly, $G/N = (AN/N)(BN/N)$, where $AN/N \simeq A/A \cap N$ and $BN/N \simeq B/B \cap N$ are nilpotent groups. Consider the series

$$1 = A_0N/N \leq A_1N/N \leq \ldots \leq A_{t-1}N/N \leq A_tN/N = AN/N \quad (2.9)$$

of AN/N. Without loss of generality, we may assume that all terms of this series are distinct. Obviously, every term of series (2.9) is completely c-permutable with all subgroups of the group BN/N by Lemma 1.5. Since $A \subseteq N_G(A_iN)$, $A_iN/N \trianglelefteq AN/N$. Since $|A_i/A_{i-1}|$ is a prime, $|A_iN/N : A_{i-1}N/N|$ is also a prime. Hence the series (2.9) is a chief series of AN/N. Thus our hypothesis is true for G/N. But $|G/N| < |G|$, so G/N is supersoluble.

(b) *G has a unique minimal normal subgroup H such that $H = C_G(H) = O_p(G)$, for some prime p, and $|H| \neq p$* (see the proof of Theorem 4.5).

(c) *The orders of A and B are not prime.*
Indeed, if $|A| = q$ for some prime q, then G is the product of two supersoluble groups A and B. By Corollary 4.7, G is supersoluble. This contradicts the choice of G, and hence $|A|$ is not a prime. Next, we assume that $|B| = q$ is a prime. If $q \neq p$, then $H \subseteq A$. Since A is nilpotent, A is a p-group by (b). We now prove that $H = A$. Assume that $p > q$. Then $A/H = G_p/H \lhd G/H$ since G/H is supersoluble. But $H = C_G(H)$, and by Chap. 1, Lemma 2.15, we have $O_p(G/C_G(H)) = 1$. Hence $H = A$. On the other hand, suppose that $q > p$. In this case, let $x \in G$ such that $T = A_{t-1}B^x = B^xA_{t-1}$. Since $A_{t-1} \trianglelefteq A$, $A_{t-1} \subseteq (A_{t-1}B^x)_G$ by Theorem 4.5. Hence $H \subseteq T$. It is clear that the hypothesis still holds for T. This means that the group T is supersoluble, and hence $B^x \trianglelefteq T$. It follows that $B^x \subseteq C_G(H) = H$, a contradiction. Therefore $A = H$. This shows that H is a Sylow p-subgroup of G and B is a maximal subgroup of G. By the hypothesis, there exists some $x \in G$ such that $BA_1^x = A_1^xB$. Since B is a maximal subgroup of G and $A_1^x \nsubseteq B$, $G = BA_1^x$. This contradicts our assumption on

G. Hence $q = p$. By our hypothesis again, we have $A_1 B^x = B^x A_1$, for some $x \in G$. Hence $G = AB^x = A(A_1 B^x)$. Now by using Theorem 4.5, we see that $H \subseteq A_1 B^x$ and so $H = A_1 B^x$ since the order of H is not a prime. It follows that $A_1 \trianglelefteq A(A_1 B^x) = G$, which contradicts (b). Thus (c) is proved.

(d) *For every $x \in G$ and all $i = 1, \ldots, t$, the subgroup A_i is completely c-permutable with all subgroups of B^x* (see the proof of Theorem 4.5)

(e) *H is a Sylow p-subgroup of G.*

Assume that the assertion is not true and let q be the largest prime divisor of $|G|$. Then, we see that $p \neq q$, and by (b) and $O_p(G/C_G(H)) = 1$. Let B_1 be a maximal subgroup of B. Let $x, y \in G$ such that $A_{t-1} B^x = B^x A_{t-1}$ and $AB_1^y = B_1^y A$. In view of (d), the hypothesis also holds for the groups $A_{t-1} B^x$ and AB_1^y. By (c), A_{t-1} and B_1^y are nonidentity groups. Since A, B are nilpotent, $A_{t-1} \trianglelefteq A$ and $B_1^y \trianglelefteq B^y$. Now, by Theorem 4.5, we have $A_{t-1} \subseteq (A_{t-1} B^x)_G$ and $B_1^y \subseteq (AB_1^y)_G$. It follows that $H \subseteq A_{t-1} B^x \cap AB_1^y$. It is clear that either $q \mid |A_{t-1} B^x|$ or $q \mid |AB_1^y|$. Suppose that the first case holds and let Q be a Sylow q-subgroup of $A_{t-1} B^x$. Then, by Lemma 3.8, we have $Q \trianglelefteq A_{t-1} B^x$, and so $Q \subseteq C_G(H) = H$, a contradiction. The second case can be similarly considered. Thus (e) holds.

(f) *$H \nsubseteq A$ and $H \nsubseteq B$.*

Assume that $H \subseteq A$. Because A is nilpotent, A is a p-group, and so by (e), $A = H$ is a Sylow p-subgroup of G. Clearly, $H \nsubseteq B$ and $H \cap B \trianglelefteq G$. Hence $H \cap B = 1$. Let $x \in G$ such that $A_1^x B = BA_1^x$. It is clear that $1 \neq A_1^x = H \cap A_1^x B \trianglelefteq A_1^x B$. But then $A_1^x \trianglelefteq G$ and so $A_1^x = H = A$. This contradicts (b). Hence $H \nsubseteq A$. Analogously, we can show that $H \nsubseteq B$.

(g) *The final step.*

Let $B_{p'}$ be a Hall p'-subgroup of B. Then we can easily see that $B_{p'} \neq 1$. Now, let x be an element of G such that $T = AB_{p'}^x = B_{p'}^x A$. Since $B_{p'}^x \trianglelefteq B^x$, $B_{p'}^x \subseteq (B_{p'}^x)^G \subseteq AB_{p'}^x$. Hence $H \subseteq AB_{p'}^x$, and so $H \subseteq A$, which contrary to (f). This contradiction completes the proof.

Theorem 4.12 (Guo, Shum, Skiba [177]). *Assume that $G = AB$, where A, B are supersoluble subgroups of a group G. Assume further that either G' is nilpotent or A and B have coprime orders. If A is completely c-permutable with every subgroup of B and B is completely c-permutable with every subgroup of A, then G is supersoluble.*

Proof We first prove that G is supersoluble whenever $G' \subseteq F(G)$. Assume that the assertion is not true and let G be a counterexample of minimal order. Since $G' \subseteq F(G)$, G is soluble. By using the same arguments as in the proof of Theorem 4.5, one can show that $G = H \rtimes M$, where H is the only minimal normal subgroup of G. Moreover, we can see that $H = O_p(G) = C_G(H) = F(G)$, for some prime p. Since $G' \subseteq F(G)$, G/H is abelian. But then G/H is a cyclic group because G/H is an irreducible automorphism group of H. Now, by Chap. 1, Lemma 2.15, H is a Sylow p-subgroup of G. It is also clear that $|H| \neq p$.

Let G_q be a Sylow q-subgroup of G, where $q \neq p$. Then G_q is a cyclic group. But by Lemma 4.1, $G_q = (A_q)^x (B_q)^y$, for some Sylow q-subgroup A_q of A, Sylow

q-subgroup B_q of B and some $x, y \in G$. Hence we have either $G_q = A_q^x$ or $G_q = B_q^x$. Assume that $H \subseteq A$ and $H \subseteq B$, and let, for example, $G_q \subseteq A$. Since $O_{p'}(A) = 1$, we have $H = O_p(A) = F(A)$. Since A is supersoluble by our hypothesis, we have $exp(A/H)|(p-1)$ by Chap. 1, Lemma 3.29. Hence $|G_q| \mid (p-1)$. If $H \subseteq A \cap B$, we can deduce that $|G/H||(p-1)$. This shows that G/H is an abelian group with exponent dividing $p-1$, and by Lemma 4.10, G is supersoluble. This contradiction shows that either $H \not\subseteq A$ or $H \not\subseteq B$. Assume that $H \not\subseteq B$. Then, $H \cap A \neq 1$. Since A is supersoluble, A has a minimal normal subgroup $L \subseteq H$ with $|L| = p$.

Assume that $p||B|$. Let $A_{p'}$ be a Hall p'-subgroup of A. By hypothesis, for some $x \in G$, we have $T = (A_{p'})^x B = B(A_{p'})^x$. We already know from above that if Q is a Sylow q-subgroup of G with $q \neq p$, then either $Q^y \subseteq A$ or $Q^y \subseteq B$ for some $y \in G$. Hence, $|G : T| = p^\alpha$, for some $\alpha \in \mathbb{N}$ and so $G = TH$. Let $B_p = B \cap H$. Then, $1 \neq B_p \neq H$ and $B_p = H \cap T \trianglelefteq G$, which is impossible. Consequently $B \cap H = 1$.

Let $D = LA_{p'}$ and $F = BD^x = DB^x$ for some $x \in G$. In this case, by using the same arguments as above, we can prove that $L = H \cap F \trianglelefteq G$. This contradiction completes the proof of the first case.

Now we will prove that G is supersoluble whenever A and B have coprime indices in G. Assume that the assertion is not true and let G be a counterexample with minimal order. Without loss of generality, we may suppose that $A_1 B \neq G \neq AB_1$ for all proper subgroups A_1 of A and B_1 of B. We proceed the proof as follows:

(a) *Every subgroup of A is completely c-permutable in G with all subgroups of B^a for all $a \in A$* (see the proof of Theorem 4.5).

(b) *G has an abelian minimal normal subgroup.*
Let L be a minimal normal subgroup of the supersoluble subgroup A. Then, we have $|L| = p$, for some prime p. By hypothesis, $T = LB^a = B^a L$ for some $a \in A$. In view of (a), every subgroup of B^a is completely c-permutable in T with all subgroups of L. Hence by Theorem 4.6, T is supersoluble, and so by the choice of G we have $T \neq G$. Using Lemma 4.4, we see that $L \subseteq T_G$. Therefore (b) holds.

(c) *G has a unique minimal normal subgroup H such that G/H is supersoluble, moreover, $H = O_p(G) = C_G(H)$, for some prime p and $|H| \neq p$* (see the proof of Theorem 4.5).

(d) *The final step.*

Since $(|A|, |B|) = 1$, we have either $H \subseteq A$ or $H \subseteq B$. Without loss of generality, we may assume that $H \subseteq A$. Let L be a minimal normal subgroup of A contained in H. Let $x \in G$ such that $T = L^x B = BL^x$. Then by hypothesis, $p \nmid |B|, L^x = H \cap T \trianglelefteq T$, and so $L^x \trianglelefteq A^x B = G$. Thus $H = L^x$ is a group of order p, which contradicts (c). This contradiction completes the proof.

Some applications of Theorems 4.6, 4.11 and 4.12.

Lemma 4.13 (Burnside $p^a q^b$-theorem) (see [248, V, 7.3]) *If $|G| = p^a q^b$ for some prime p and q, then G is soluble.*

On the base of Theorem 4.6 we now prove the following

Theorem 4.14 (Guo, Shum, Skiba [177]). *Let p be a prime number and $G = AB$ a product of two p-supersoluble groups A and B. If every subgroup of A is completely c-permutable with every subgroup of B in G, then G is p-supersoluble.*

Proof Assume that the assertion is false and let G be a counterexample of minimal order. Since the hypothesis of the theorem holds for every factor group of G, we may put $O_{p'}(G) = 1$. Also, we assume that for every proper subgroup A_1 of A and every proper subgroup B_1 of B, we have $A_1 B \neq G$ and $G \neq AB_1$. We proceed the proof as follows:

(a) *G has a nontrivial normal subgroup which is p-soluble.*

By using the same arguments as in the proof of Theorem 4.5, we obtain that both subgroups A and B are simple groups.

It is clear that if both subgroups A, B are either p'-groups or p-groups, then G is p-supersoluble, which contradicts the choice of G. Hence we may assume that A is a p'-group and B is a p-group of G. Let $q \neq q$ be any prime dividing $|G|$ and Q a Sylow q-subgroup of A. Then $E := BQ^x = Q^x B$ for some $x \in G$. Suppose that $O_{p'}(E) \neq 1$. Then $G = AE$ and $O_{p'}(E) \leq A \cap E$. Hence by Lemma 4.4, $(O_{p'}(E))^G \leq A$. Hence $(O_{p'}(E))^G$ is p-soluble. Now assume that for any prime $q \neq p$ and for any Sylow q-subgroup Q of G such that $E = BQ^x = Q^x B$, for some $x \in G$, we have $O_{p'}(E) = 1$. Obviously, every subgroup of Q^x is completely c-permutable in E with every subgroup of B. If $E = G$, then G is soluble by Theorem 4.6. We may, therefore, assume that $E < G$. Then E is p-supersoluble by the choice of G. But $O_{p'}(E) = 1$, so E is supersoluble by Lemma 4.10. Hence B is normal in E. This implies that for every prime $q \neq p$, q does not divides $|G : N_G(B)|$. Hence B is normal in G. (a) is proved.

(b) *For every nonidentity normal subgroup D of G, the quotient G/D is p-supersoluble* (see the proof of Theorem 4.5).

(c) *G has a unique minimal normal subgroup $L = O_p(G) = C_G(L)$ and $|L| \neq p$.*

Let L be a minimal normal subgroup of G. Then G/L is p-supersoluble by (b). Hence L is not a p'-group, and by Chap. 1 (1.3), L is the unique minimal normal subgroup of G and $L \not\subseteq \Phi(G)$. Thus G is a primitive group and by (a), L is a p-group. Hence $L = O_p(G) = C_G(L)$ by Chap.1, Lemma 4.17. It is also clear that $|L| \neq p$.

(d) *If $L \leq A$, then $L \leq B$ and conversely.*

Assume that $L \leq A$. Assume that $L \not\leq B$. Let $T = LB$ and L_1 be some maximal subgroup of L. Then in view (c), $L_1 \neq 1$. By hypothesis, there exists $x \in G$ such that $L_1^x B = B L_1^x$. Assume that $T = G$. Then $T = L \rtimes B$ and B is a maximal subgroup of G and so $|G : B| = |L| = |L_1|$, which is impossible. Hence $T \neq G$. It is clear that the hypothesis is true for T. Hence T is p-supersoluble by the choice of G. Since $O'_p(T) \subseteq C_G(L) = L$, $O'_p(T) = 1$. Thus, by Lemma 4.10, T is supersoluble. Analogously one can see that A is a supersoluble group. Hence G is supersoluble by Corollary 4.7. This contradiction shows that (d) holds.

(e) $O_{p'}(A) = 1 = O_{p'}(B)$.

Assume that $O_{p'}(A) \neq 1$. Then $L \not\subseteq A$ and so $L \not\subseteq B$ by (d). But by Lemma 4.4, $L \subseteq O_{p'}(A)B^x$, for some $x \in G$. So $L \subseteq B^x$. It follows that $L \subseteq B$, a contradiction. Hence (e) is proved.

Final contradiction. By (e) and Lemma 4.10, A and B are supersoluble. Hence G is supersoluble by Theorem 4.6. The proof is completed.

Corollary 4.15 (Carocca [59]). *Let p be a prime number. Assume that $G = AB$, where A and B are p-supersoluble subgroups of G. If every subgroup of A permutes with every subgroup of B, then G is p-supersoluble.*

Now, we apply Theorem 4.11 to prove the following characterization theorem for p-supersoluble groups.

Theorem 4.16 (Guo, Shum, Skiba [177]). *Let p be a prime and G a soluble group. Then G is p-supersoluble if and only if $G = AB$, where A is p-nilpotent, B is nilpotent and A has a chief series*

$$1 = A_0 \leq A_1 \leq \ldots \leq A_n = O_{p'}(A) \leq A_{n+1} \leq \ldots \leq A_{t-1} \leq A_t = A$$

such that A_i is completely c-permutable (permutable) with all subgroups of B, for all $i = n, \ldots, t$.

Proof First, we assume that G is p-supersoluble. Then by Lemma 4.10, $G/O_{p'}(G)$ is supersoluble and $G/O_{p'p}(G)$ is an abelian group. Let $A = O_{p'p}(G)$ and B be a subgroup of G such that $AB = G$ and $B_1 A \neq G$ for all proper subgroups B_1 of B. Then B is nilpotent. Since $O_{p'}(A) \text{char} A \trianglelefteq G$, we have $O_{p'}(A) \trianglelefteq G$. Hence G has a chief series below A

$$1 = A_0 \leq A_1 \leq \ldots \leq A_n = O_{p'}(A) \leq A_{n+1} \leq \ldots \leq A_{t-1} \leq A_t = A$$

passing through $O_{p'}(A)$. This proves the necessity part of the theorem. The sufficiency part can be proved by using Theorem 4.11 and the arguments adopted in the proof of Theorem 4.14. We omit the details.

Using Theorem 4.12 and the same arguments in the proof of Theorem 4.14, we can also prove the following theorem for p-supersoluble groups.

Theorem 4.17 (Guo, Shum, Skiba [177]). *Let p be a prime and $G = AB$ the product of p-supersoluble groups A and B. Assume that A is completely c-permutable with all subgroups of B and B is completely c-permutable with all subgroups of A. If either $G' \subseteq O_{p'p}(G)$ or A and B have coprime orders, then G is p-supersoluble.*

2.5 Characterizations of Classes of Groups in Terms of X-Permutable Subgroups

A subgroup H of a group G is said to be have non-primary index in G if $|G : M|$ is divisible by at least two distinct prime numbers. A subgroup H of G is said to be a 2-maximal subgroup of G if H is a maximal subgroup of some maximal subgroup of G.

Two Characterizations of Solubility The classical theorem of F. Hall on the existence of Hall systems in soluble groups admits the following interpretation in terms of X-permutable subgroups.

Lemma 5.1 *A group G is soluble if and only if every two Hall subgroups of G are c-permutable in G.*

Proof If G is soluble, and A and B are some Hall subgroups of G, then for some $x \in G$ we have $AB^x = B^x A$. Because A and B are the elements of some Hall systems of G and every two of Hall systems are conjugate (see Theorem 4.11 in [89, Chap. I]). On the other hand, if every two Hall subgroups are c-permutable, then G has a Hall p'-subgroup for each prime divisor p of the order of G. Thus, G is soluble by Theorem 3.5 in [89, Chap. I].

The following theorem gives two new characterization of solubility of groups in terms of X-permutability.

Theorem 5.2 (Guo, Skiba, Shum [207]). *Let G be a group, $X = G^{\mathcal{N}^*}$ and $Y = R(G^{\mathcal{N}^*})$ (that is, the largest solvable normal subgroup of $G^{\mathcal{N}^*}$). Then the following are equivalent:*

(1) *G is soluble.*
(2) *G has a soluble maximal subgroup M, and every Sylow subgroup of G is X-permutable with M and with every Sylow subgroup of G.*
(3) *Every maximal subgroup of G is Y-permutable with all Sylow subgroups of G.*

Proof (2) $\Rightarrow$ (1) Suppose the claim false and take a counterexample G of minimal order. Pick a prime divisor p of the index $|G : M|$ and a Sylow p-subgroup P of G. By hypothesis, $MP^x = P^x M$ for some $x \in X$. It is clear that $G = MP^x$, and so for some $a \in \mathbb{N}$ we have $|G : M| = p^a$. Take some minimal normal subgroup H of M. Since M is solvable by hypothesis, H is a q-group for some prime q. Pick some Sylow q-subgroup M_q of M. If $q \neq p$, then M_q is a Sylow q-subgroup of G. Thus, by hypothesis, there exists an element $y \in X$ such that $M_q^y P^x = P^x M_q^y$. Consequently, $G = MP^x M_q^y$. Then, by Chap. 1, Lemma 4.14, we have that

$$G = M(P^x M_q^y)^{y^{-1}} = M(M_q^y P^x)^{y^{-1}} = M(M_q P^{xy^{-1}}).$$

Since $H \trianglelefteq M$, $H \subseteq M_q \subseteq M_q P^{xy^{-1}}$. Therefore, $H^G = H^{MM_q P^{xy^{-1}}} = H^{M_q P^{xy^{-1}}}$, and so $H \subseteq (M_q P^{xy^{-1}})G$. Since each biprimary group is solvable, G has a minimal normal abelian subgroup. If $q = p$, then for some Sylow p-subgroup G_p of G we have $H \leq G_p$. Clearly, $MG_p^x = G_p^x M = G$ for some $x \in G$. It follows from Chap. 1, Lemma 4.14 that $G = MG_p$. Hence, $H^G \leq H^{MG_p} = H^{G_p} \leq G_p$. Consequently, in any case, G has a minimal normal abelian subgroup, say R. If $R \not\subseteq M$, then $G/R = RM/R \simeq M/R \cap M$ is a solvable group, and thereby G is solvable, which contradicts the choice of G. Thus $R \leq M$. By Lemma 1.5(3), each Sylow subgroup of G/R is (XR/R)-permutable with M/R and with each Sylow subgroup of G/R. In addition, by Chap. 1, Lemma 1.1, $(G/R)^{\mathcal{N}^*} = G^{\mathcal{N}^*} R/R = XR/R$. This shows that the hypothesis holds for G/R. Since $|G/R| < |G|$, G/R is soluble by the choice of G. It follows that G is soluble, a contradiction.

(3)$\Rightarrow$(1) Suppose the implication is false and let G be a counterexample of minimal order. Take some minimal normal subgroup L of G. Since $R(G^{\mathcal{N}^*})$ char $G^{\mathcal{N}^*}$ and $(G/N)^{\mathcal{N}^*} = G^{\mathcal{N}^*}N/N$, $YN/N \leq R((G/N)^{\mathcal{N}^*})$. Hence, by Lemma 1.5(3), the hypothesis holds for the quotient group G/L. The choice of G implies that G/L is solvable. Then L is the unique minimal normal subgroup of G and $Y = 1$. Pick a prime divisor p of $|L|$ and some Sylow p-subgroup L_p of L. Let P be a Sylow subgroup of G containing L_p, and put $N = N_G(L_p)$. Then $P \leq N$, and $N \neq G$ since L is not abelian. Suppose that $N \leq M$, where M is a maximal subgroup of G. Frattinis Lemma implies that $G = LN$, and so $L \nsubseteq M$. Hence $M_G = 1$. Let q be a prime divisor of $|G : M|$ and Q a Sylow q-subgroup of G. Then $Q \nsubseteq M$ and $MQ = QM = G$. Now take some Sylow r-subgroup R of M for $r \neq q$. Then R is a Sylow subgroup of G. Thus, by hypothesis, $R^x \leq M$ for all $x \in G$ and consequently $M_G \neq 1$. This contradiction shows that G is soluble.

(1) $\Rightarrow$ (2), (3) Suppose that G is soluble. Then $X = Y = G^{\mathfrak{N}}$, where $\mathfrak{N}$ denotes the class of all nilpotent groups. Let M be a maximal subgroup of G. We show that M is X-permutable with all Sylow subgroups of G and every two Sylow subgroups of G are X-permutable. Write $|G : M| = p^a$ and pick some Sylow q-subgroup Q of G. If $p = q$, then $Q^x M = MQ^x = G$ for some $x \in G$. It follows that $QM = MQ$. Now assume that $q \neq p$ and pick some Sylow q-subgroup M_q of M. It is clear that $M_q = Q^x$ for some $x \in G$. Then $M = MQ^x = Q^x M$. By [89, I,(4.11)], for some system normalizer N of G we have $N \subseteq N_G(Q)$. Using [89, I, (5.6)], we also see that $G = NX$, and so $x = nd$ for some $n \in N$ and $d \in X$. Then $MQ^d = Q^d M$. This means that given a prime divisor q of $|G|$, every two pairs of Sylow q-subgroups of G are conjugate by some element of X. Consequently, Q is X-permutable with all Sylow subgroups of G.

New characterizations of supersolubility and p-supersolubility

Lemma 5.3 Let $E \leq M \leq G$, where M is a maximal subgroup of a group G and E is a maximal subgroup of M. Suppose that $T \in 1_m(E)$ is a supersoluble group. Then $|G : M|$ is a prime and either $|G : E|$ is a prime power or $|M : E|$ is a prime.

Proof We first note that for any maximal subgroup T_1 of T, the subgroup ET_1 has prime index in G. Indeed, since T is supersoluble, $|T : T_1| = p$ is a prime. Hence $|G : ET_1| = |ET : ET_1| = p$ since $E_1T \neq G$. Clearly, $T \nsubseteq M$ and so $T \cap M \neq T$. Let T_1 be a maximal subgroup of T such that $T \cap M \leq T_1$. Then $M = E(T \cap M) \leq ET_1$. Since T is a minimal supplement of E in G, $ET_1 \neq G$ and by the maximality of M, we have $M = ET_1$. This induces that $|G : M|$ is a prime. Now suppose that $|G : E|$ is not a prime power. Then, we can let $|G : M| = p$ and $q \neq p$ is a prime dividing $|M : E|$. Thus, q divides $|T|$ and T has a maximal subgroup T_1 such that $|T : T_1| = q$ since T is supersoluble. From the above result, we know that $|G : ET_1| = q \neq p$ and hence ET_1 contains some Sylow p-subgroup P of G. But since $|G : M| = p$, $G = MP$ and hence $G = MET_1$. It follows that $|M : M \cap ET_1| = q$ and so $E = M \cap ET_1$ by the maximality of E, consequently $|M : E| = q$.

Theorem 5.4 (Guo, Shum, Skiba [183]). *Let G be a group and $X = G'$ the commutator subgroup of G. Then the following statements are equivalents:*

(1) *Every maximal subgroup of G is G_m-semipermutable in G.*
(2) *G is supersoluble.*
(3) *Every maximal subgroup of G is X-permutable in G.*
(4) *Every maximal subgroup of G is 1_m-semipermutable in G.*

Proof (1) $\Rightarrow$ (2) Let M be a maximal subgroup of G and $T \in G_m(M)$. Let E be any maximal subgroup of T. Then by the hypothesis, $ME^x = E^x M$ for some $x \in G$. Hence, by the maximality of M, we have either $ME^x = G$ or $E^x \leq M$. In the former case, $ME = G$ by Chap. 1, Lemma 4.14, which contradicts the fact that T is a minimal supplement of M in G. This shows that $E^x \leq M$. Since $G = MT$, $G = MT^x$ by Chap. 1, Lemma 4.14 and clearly $E^x = M \cap T^x$. Therefore $|T^x : E^x| = |G : M|$. But $|T^x : E^x| = |T : E|$. Hence for any maximal subgroup E of T we have $|T : E| = |G : M|$. It follows that T is p-group for some prime p and so $|T : E| = |G : M| = p$. Therefore, every maximal subgroup of G has prime index and consequently G is supersoluble by the well known Huppert's Theorem on supersoluble groups (see [135, Theorem 1.9.6]).

(2) $\Rightarrow$ (3), (2) $\Rightarrow$ (4) Let M be a maximal subgroup of G with $|G : M| = p$ and T any subgroup of G. First we show that for some $a \in X$ we have $MT^a = T^a M$. Let $M_1, \ldots, M_t$ and $T_1, \ldots, T_t$ be some Sylow systems of the groups M and T, respectively. Then, by [248, VI, Theorems 2.2, 2.3], G has Sylow systems $\Sigma = \{P_1, \ldots, P_t\}$ and $\Sigma_1 = \{Q_1, \ldots, Q_t\}$ such that $M_i = P_i \cap M$ and $T_i = Q_i \cap T$ for all $i = 1, \ldots, t$. Moreover, the systems Σ and Σ_1 are conjugate, i.e., G has an element x such that $Q_i^x = P_i$ for all $i = 1, \ldots, t$. Without loss of generality, we may assume that P_1 is a Sylow p-subgroup of G. Then $M_2 = P_2, \ldots, M_t = P_t$.

Assume that $T_1^x \subseteq M_1$. Then we have

$$T^x \subseteq M_1 P_2 \ldots P_2 = M,$$

and hence $T^x M = M = MT^x$.

Now assume that $T_1^x \nsubseteq M_1$. Then by $|G : M| = p$, we have $|P_1 : M_1| = p$ and $P_1 = T_1^x M_1$. Thus, we derive that

$$T^x M = T_2^x \ldots T_t^x T_1^x M_1 M_2 \ldots M_t = T_2^x \ldots T_t^x P_1 P_2 \ldots P_t = G = MT^x.$$

Now let $N = N_G(\Sigma_1)$. By [248, VI, Theorem 11.10], N covers all central chief factors of G and so $G = XN$. This leads to $x = na$ for some $a \in X$ and $n \in N$, and hence $MT^a = T^a M$.

Next we show that M is m-semipermutable in G. Let P be a Sylow p-subgroup of G. Then by $|G : M| = p$, we have $G = MP$. This shows that a minimal supplement of M contained in P belongs to $1_m(M)$.

Finally, since the implications (3) $\Rightarrow$ (1) and (4) $\Rightarrow$ (1) are evident. The theorem is proved.

Corollary 5.5 *A group G is supersoluble if and only if every its maximal subgroup is G_m-semipermutable in G.*

Theorem 5.4 shows that in any supersoluble group G any of its maximal subgroup is G_m-semipermutable. The following example shows that there exists a supersoluble group in which some 2-maximal subgroup is not G_m-semipermutable.

Remark 5.6 Let E be a 2-maximal subgroup of a supersoluble group G with $|G : E| = rq$, where r, q are distinct primes. Suppose that $|F(G/E_G)| = |O_p(G/E_G)| > p$ (such conditions, for example, in the group $S_3 \times Z_3$, for one of the subgroups of order 3). We show that E is not G_m-semipermutable in G. In view of Lemmas 1.5 and 1.14, we may assume that $E_G = 1$. Suppose that E is G_m-semipermutable in G and let $T \in G_m(E)$. Since $E_G = 1$, $p \in \{r, q\}$. Let $p = r$. Since $|F(G)| = |O_p(G)| > p$, E has a nonidentity Sylow p-subgroup P. If T_1 is a maximal subgroup of T with $|T : T_1| = p$, then by our hypothesis, $V = ET_1^x = T_1^x E$ is a subgroup of G for some $x \in G$. If $V = G$, then by Chap. 1, Lemma 4.14, $G = ET_1$, which contradicts that T is a minimal supplement of E in G. Hence $V \neq G$. Suppose that $V = E$. Then $T_1^x = E \cap T^x$ and so $pq = |G : E| = |T^x : T_1^x|$. But since T_1^x is a maximal subgroup of the supersoluble group T^x, $|T^x : T_1^x|$ is prime. This contradiction shows that $V \neq E$. Hence V is a maximal subgroup of G. It is clear also that $|G : V| = p$. Thus $P = O_p(G) \cap V$ is normal in V and so P is normal in G since P is a maximal subgroup of $O_p(G)$. Hence $E_G \neq 1$. This contradiction shows that E is not G_m-semipermutable in G. On the other hand, if either $|F(G/E_G)|$ is a prime or $|F(G/E_G)|$ has at least two distinct prime divisors, then E is G_m-semipermutable in G (see Lemma 5.8 below).

In view of Remark 5.6, we now call a 2-maximal subgroup E of a group G a 2-*maximal subgroup of special type* if G/E_G is a supersoluble group with $|F(G/E_G)| = |O_p(G/E_G)| > p$ for some prime p.

Lemma 5.7 (see [320, Theorem 3.4].) *A group G is soluble if $G = AB$, where A is a supersoluble subgroup, B is a cyclic subgroup of G of odd order.*

Lemma 5.8 *Let G be a group and X a normal soluble subgroup of G. Then G is soluble if the set $X_m(E)$ contains a supersoluble group, for every 2-maximal subgroup E of G with a non-primary index such that E is not a 2-maximal subgroup of special type.*

Proof Assume that the lemma is false and let G be a counterexample of minimal order. Then:

(1) *G is not a simple group.*

Assume that G is a simple non-abelian group. Then $X = 1$ and G has a non-supersoluble maximal subgroup, say M, by [248, VI, 9.6]. Moreover, it is clear that every 2-maximal subgroup E of G is not of special type.

Assume that M has non-primary index in G and T is a maximal subgroup of M. Then, it is obvious that T is a 2-maximal subgroup of G satisfying the conditions in the Lemma. Hence, by our hypothesis, the set $X_m(E) = 1_m(E)$ contains a supersoluble group. It follows from Lemma 5.3 that M has prime power index. This contradiction shows that $|G : M| = p^a$ for some prime p. Since M is non-soluble, M has a maximal subgroup T such that $(p, |M : T|) = 1$. This shows that T is a 2-maximal subgroup of G satisfying the condition. Hence $1_m(T)$ is not empty. Let $A \in 1_m(T)$ and let A_1 be a proper subgroup of A. Then $G = TA$ and TA_1 is a proper subgroup of G. Let $x \in G$ and $x = at$, where $t \in T$ and $a \in A$. Since $X = 1$, $T(A_1)^a = (A_1)^a T$ and so $((A_1)^a T)^t = (A_1)^x T$ is a subgroup of G. Hence, G is not simple by Lemma 2.6. This contradiction completes the proof of (1).

(2) *For every minimal normal subgroup N of G, the quotient group G/N is soluble.* Indeed, let M/N be a 2-maximal subgroup of G/N of non-primary index such that M/N is not a group of special type. Then, since $G/M_G \simeq (G/N)/(M_G/N) \simeq (G/N)/(M/N)_{G/N}$, M is not of special type. Hence by our hypothesis the set $X_m(M)$ contains a supersoluble group, say T. Now by Lemma 1.14, the supersoluble group TN/N belongs to $(XN/N)_m(M/N)$, where XN/N is a normal soluble subgroup of G/N. This shows that our hypothesis still holds on G/N. Thus, by the choice of G, G/N is soluble.

(3) *G has a unique minimal normal subgroup L* (This is a direct consequence of claim (2)).

Final contradiction.

By claim (2), we only need to prove that L is soluble. Assume that this assertion is not true. Then by claim (3), we have $X = 1$. By claim (1), G is not simple and hence $L \neq G$. By claim (2) again, G has a normal maximal subgroup M such that $L \leq M$ and $|G : M|$ is a prime. Let $|G : M| = p$. We claim that there exists a maximal subgroup T of M such that $M = LT$ and $(|M : T|, p) = 1$. Indeed, if p divides $|L|$, L_p is a Sylow p-subgroup of L and P is a Sylow subgroup of G containing L_p, then $P \leq N = N_G(L_p)$. By using the Frattini argument, we have $G = LN$, and hence $M = M \cap LN = L(M \cap N)$. Since L is not soluble, $N \neq G$. Therefore, $M \cap N \neq M$. Let T be a maximal subgroup of M containing $M \cap N$. Then, $M = LT$ and $(|M : T|, p) = 1$. Next, we assume that $(|L|, p) = 1$. Then, by claim (2), $L \not\leq \Phi(G)$ and so $L \not\leq \Phi(M)$. Hence, there exists a maximal subgroup T of M such that $M = LT$. It follows that p does not divide $|M : T| = |L|/|L \cap T|$. Hence, our claim is established. This shows that T is a 2-maximal subgroup of G having non-primary index. Since $L \not\subseteq T$, $T_G = 1$ and so $G/T_G \simeq G$ is not supersoluble. This shows that T is not of special type. Thus, by the hypothesis, the set $X_m(T)$ contains a supersoluble group, say A. Since $X = 1$, by Lemma 5.3, $|M : T| = q$ for some prime $q \neq p$ and so $|G : T| = pq$. Since $G = TA$ and A is a minimal supplement of T in G, $A \cap T \leq \Phi(A)$. Thus, A is a $\{p, q\}$-group. Let W be a maximal subgroup of A containing a Sylow p-subgroup A_p of A. Since $X = 1$, $D = TW = WT$. Clearly, $|G : D| = q$. Moreover, since $|G : M| = p$ and $M = LT$, $L \not\leq D$. Thus $D_G = 1$, and by considering the permutation representation of G on the right cosets of D, we see that G is isomorphic with some subgroup of the symmetric group S_q of degree q. This shows that D is a Hall q'-subgroup of G and $G = DZ_q$, where Z_q is a subgroup of order q. Now we have seen that every maximal subgroup V of D has non-primary index in G and clearly V is not a group of special type since $F(G) = 1$. Hence $|D : V|$ is a prime by Lemma 5.3. This implies that D is supersoluble, and by Lemma 5.7, G is soluble. This contradiction completes the proof.

Lemma 5.9 (Guo, Skiba [207, Theorem 3.15]). *In any supersoluble group G, any of its 2-maximal subgroup M which is not of special type is G'-semipermutable in G.*

Proof Let T be a minimal supplement of M in G. We prove that M is G'-permutable with all subgroups of T by induction. Take some subgroup T_1 of T.

Assume that $M_G \neq 1$. Then the statement holds for G/M_G. Note that TM_G/M_G is a minimal supplement of M/M_G in G/M_G. Indeed, if $T_0/M_G \leq TM_G/M_G$ and $(T_0/M_G)(M/M_G) = G/M_G$ then $T_0 = T_0 \cap M_GT = M_G(T_0 \cap T)$, and so $T_0M = G = (T_0 \cap T)M$. Consequently, $T \subseteq T_0$. Hence, $T_0/M_G = TM_G/M_G$ is a minimal supplement to M/M_G in G/M_G. Therefore, the subgroup M/M_G is $(G/M_G)'$-permutable with T_1M_G/M_G. However, $(G/M_G)' = G'M_G/M_G$, so by Lemma 1.5 we see that M is G'-permutable with T_1.

Now assume that $M_G = 1$ and $|G : M| = pq$, where $p > q$. Let $\pi = \pi(F(G))$ the set of all prime divisors of $|F(G)|$. If $|\pi| > 2$ and R is a Sylow d-subgroup of $F(G)$, where $q \neq d \neq p$, then $R \leq M_G$, which contradicts the equality $M_G = 1$. Hence, $\pi \subseteq \{p, q\}$. Since the group G is supersolvable, a Sylow r-subgroup is normal in G, where r is the largest prime divisor of $|G|$. Hence, $r = p$.

If $|F(G)| = p$, then $G = F(G) \rtimes E$, for some maximal subgroup E of G, and $C_G(F(G)) = F(G)$. This implies that E is a cyclic group. Without loss of generality, we may assume that $M \leq E$. We show that M is G'-permutable with T_1. If A is some Hall p'-subgroup of T_1, then $T_1 = PA$, where $P = T_1 \cap F(G)$ is a Sylow p-subgroup of T_1. Obviously, for some $x \in F(G) = G'$ we have $A^x \subseteq E$. Hence $MT_1^x = M(T_1 \cap F(G))A^x = (T_1 \cap F(G))A^xM = T_1^xM$.

Suppose that $|\pi| = 2$, and consider a Sylow p-subgroup F_p and a Sylow q-subgroup F_q of $F(G)$ respectively. It is clear that $G = F(G)M$. Take some Sylow r-subgroup R of $F(G)$. Suppose that $|R| > r$. Then $D = R \cap M \neq 1$. Since R is a characteristic subgroup of $F(G)$, R is normal in G. It is also clear that $|R : D| = r$. Consequently, D is normal in G and so $D = 1$ because $M_G = 1$. Therefore, $|F(G)| = pq$. Suppose that q divides the order $|M|$ and q and p divide $|T_1|$. If $\{M_2, \cdots, M_t\}$ is some Sylow base of M and $\{D_1, D_2\}$ is some Sylow base of T_1, then by [89, I, (4.8) and (4.12)], G possesses Sylow bases $\Sigma = \{P_1, \cdots, P_t\}$ and $\Sigma_1 = \{Q_1, \cdots, Q_t\}$ such that $M_i \subseteq P_i$ for all $i = 2, \cdots, t$ and $D_i \subseteq Q_i$ for $i = 1, 2$. Moreover, there is some element $x \in G$ such that $Q_i^x = P_i$ for all $i = 1, \cdots, t$. Clearly, $P_1 = D_1$ is a Sylow p-subgroup of G and $M_3 = P_3, \cdots, M_t = P_t$. If $D_2^x \subseteq M_2$, then $T_1^xM = P_1M = MT^x$. On the other hand, if $D_2^x \not\subseteq M_2$, then the equality $|G : M| = pq$ implies that $|P_2 : M_2| = q$, and so $P_2 = D_2^xM_2$. Hence, $T_1^xM = G = MT_1^x$. Note that $G = G'N_G(\Sigma_1)$ by [89, VI, 11.10]. Hence $x = fn$ where $f \in G'$ and $n \in N_G(\Sigma_1)$. This implies that $MT_1^f = T_1^fM$. Similarly we may consider the cases in which either $(|M|, q) = 1$ or $(|T_1|, p) = 1$.

Finally, note that since G is supersoluble, $G' \leq F(G)$, and so $X = G'$. Consequently, M is X-permutable with all subgroups of T. The theorem is proved.

Theorem 5.10 (Guo, Shum, Skiba [183]). *Let G be a group and $X = F(G) \cap G'$. Then G is supersoluble if and only if for every 2-maximal subgroup E of G of non-primary index, either E is of special type or the set $X_m(E)$ contains a supersoluble group.*

Proof In view of Lemma 5.9, we only need to prove that if for every 2-maximal subgroup E of G with non-primary index, either E is of special type or the set $X_m(E)$ contains a supersoluble group, then G is supersoluble. Suppose that this assertion is false and let G be a counterexample of minimal order. Let L be a minimal normal subgroup of G. Then, by making use of Lemma 1.14, it is not difficult to see that the hypothesis still holds on the quotient group G/L. Hence by the choice of G, G/L is supersoluble. Since the class of all supersoluble groups is a saturated formation, L is a unique minimal normal subgroup of G and $L \not\subseteq \Phi(G)$. Let M be a maximal subgroup of G such that $L \not\subseteq M$. By Lemma 5.8, G is a soluble group and so $G = L \rtimes M$, where $L = C = C_G(L) = F(G) = O_p(G)$ for some prime p. Since G is soluble, $L \subseteq G'$. Hence $L = X$. Since $M \simeq G/L$ is supersoluble, M contains a maximal subgroup E such that $|M : E| = q \neq p$. Since $L \not\subseteq E$, $E_G = 1$ and so $G/E_G \simeq G$ is not supersoluble. Therefore, by the hypothesis, E is L_m-semipermutable in G and the set $L_m(E)$ contains a supersoluble group, say T. Obviously, $E \neq 1$. Let $D = E \cap T$ and D_p be a Sylow p-subgroup of D. Then, since $|G : E| = |L|q$, $|T : D| = p^a q$, where $|L| = p^a$ and $a > 1$. Let r be the largest prime divisor of $|G|$. Assume that $r = p$. Because G/L is a supersoluble group and $O_p(G/L) = 1$, we see that L is a Sylow subgroup of G. It is clear that $L \leq T$. Let L_1 be a maximal subgroup of L. Then by our hypothesis again, we have $A = L_1 E = E L_1$. Evidently, $|G : A| = pq$ and $A_G = 1$. Hence, A is L_m-semipermutable in G. If $T_1 \in L_m(A)$ and Q is maximal subgroup of T_1 containing some Sylow q-subgroup of T_1, then $B = Q^x A = AQ^x$ for some $x \in L$. However, since $|G : B| = p$, we have $LB = G$. This shows that $|L| \neq |L \cap B| \neq 1$ and $L \cap B \trianglelefteq G$, which contradicts to the minimality of L.

Now, without loss of generality, we may assume that $r = q$. Let K be a maximal subgroup of T such that $|T : K| = p$. Because G is soluble, it is clear that T is a $\{p, q\}$-group. Since $p < q$ and T is supersoluble, K is normal in T. Let $D = E \cap T$ and D_p be a Sylow p-subgroup of D. If $D \not\leq K$, then $KD = T$ and so $G = ET = EDK = EK$, which contradicts the minimality of T. Hence $D \leq K$. Let $x \in L$ such that $V = EK^x = K^x E$. Since $G = ET$, $x = te$, where $e \in E$ and $t \in T$. Hence, we obtain that $V^{e^{-1}} = EK^t = EK$. Let $Y = V^{e^{-1}}$ and Y_p be a Sylow p-subgroup of Y and let G_p be a Sylow p-subgroup of G containing Y_p. Then, $|G_p : Y_p| = (|E_p||T_p|/|D_p|) : (|E_p||K_p|/|D_p|) = p$, where E_p is a Sylow p-subgroup of E. On the other hand, since $|G : LE| = q$, we have $|G_p| = |L||E_p|$. Thus, $L \not\subseteq Y$ and $L \cap Y \neq 1$. This induces that $G = LY$ and so $Y \cap L$ is normal in G, which contradicts the minimality of L. The contradiction shows that G is supersoluble.

Lemma 5.11 *Suppose that $G = HM$, where $H \trianglelefteq G$ and M is H-semipermutable in G. If H is abelian and M is supersoluble, then G is supersoluble.*

Proof Let R be a minimal normal subgroup of G. Then $G/R = (HR/R)(MR/R)$, HR/R is abelian and MR/R is supersoluble. By Lemma 1.10(2), we have that MR/R is HR/R-semipermutable in G/R. Hence G/R is supersoluble by induction on $|G|$. If G has another minimal normal subgroup or $R \subseteq \Phi(G)$, then G is supersoluble since the class of all supersoluble groups is a saturated formation. This leads that $R = C_G(R) = F(G)$ and hence $R = H$ is a minimal normal subgroup of G. Let $|H| = p^\alpha$ for some $p \in \mathbb{P}$ and $\alpha \in \mathbb{N}$.

Suppose $T \in H(M)$ be a minimal supplement of M in G. Then T is a p-group since $|G : M| = |H : (M \cap H)|$ is a p-number and T is a minimal supplement of M in G. Let P be a maximal subgroup of T. Then, by the choice of T, M is permutable with P^x for some $x \in H$. By Chap. 1, Lemma 4.14, we see that $G = MT = MT^x$. Hence $|G : MP^x| = |MT^x : MP^x| = \frac{|M||T^x||M \cap P^x|}{|M||P^x||M \cap T^x|} = \frac{|T^x||M \cap P^x|}{|P^x||M \cap T^x|} = p|M \cap P^x|/|M \cap T^x|$. Clearly, $M \cap P^x \subseteq M \cap T^x$. Since MP^x is a proper subgroup of G by the choice of T, we see that $M \cap P^x = M \cap T^x$ and $|G : MP^x| = p$. Therefore, $H \cap MP^x$ is maximal in H because $|H : H \cap MP^x| = |HMP^x : MP^x| = |G : MP^x| = p$. Since $H \cap MP^x \trianglelefteq MP^x$ and H is abelian, we know that $G = MH \subseteq N_G(H \cap MP^x)$ and hence $H \cap MP^x \trianglelefteq G$. Then the minimal normality of H in G shows that $H \cap MP^x = 1$ and H is cyclic. This leads that G is supersoluble.

Theorem 5.12 (Li, Skiba [281]). *Suppose G is a group. Let $\Sigma_G = \{M \mid M$ is a 2-maximal subgroup of G of non-primary index and $M/M_G \cap F(G/M_G) = 1\}$, $\Delta_G = \{M \mid M$ is a 2-maximal subgroup of G of non-primary index and $\Phi(G) < M \cap F(G) < F(G)\}$. Then G is supersoluble if and only if the following statements hold:*

(a) *Every element in Σ_G is $F(G)$-semipermutable in G.*

(b) *Let N be an arbitrary normal subgroup of G. If $\Delta_{G/N} \neq \emptyset$, then there exists a group $M/N \in \Delta_{G/N}$ such that M/N is $F(G/N)$-semipermutable in G/N.*

Proof 1) Firstly, we assume that the statement (a) and (b) are both true. We prove that G is supersoluble. Suppose that the assertion is false and let G be a counterexample of minimal order. Then:

(1) G *is not a non-abelian simple group.*
     Assume that G is a non-abelian simple group. Then $F(G) = 1$. If all maximal subgroups of G are supersoluble, then G is soluble by Chap. 1, Proposition 1.10. But this is impossible because $F(G) = 1$ and so G has a non-supersoluble maximal subgroup M. Assume that M is of non-primary index in G and let T be a maximal subgroup of M. Then $T \neq 1$ and T is also of non-primary index in G. Hence by hypothesis, T is $F(G)$-semipermutable in G. It follows that T is $F(G)$-semipermutable in M and so by Lemma 1.10(5), $|M : T|$ is a prime. This shows that every maximal subgroup of M has prime index, and hence M is supersoluble (see [135, 1.9.6]). This contradiction shows that $|G : M| = p^\alpha$ for some prime p. Obviously, M has a maximal subgroup T such that $(p, |M : T|) = 1$ and so T is $F(G)$-semipermutable in G. Therefore, there exists a minimal supplement A

of T in G such that every subgroup of A is permutable with T since $F(G) = 1$. Let A_1 be a maximal subgroup of A. For any element $x \in G$, since $G = AT$, we have that $x = at$ for some $a \in A$ and $t \in T$. Then $T(A_1)^a = (A_1)^a T$ and so $((A_1)^a T)^t = (A_1)^x T$ is a subgroup of G. In view of Chap. 1, Lemma 4.14, $(A_1)^x T \neq G$ since A is a minimal supplement of T in G. However, in this case, by using Lemma 2.6, we see that G has a proper normal subgroup, a contradiction.

(2) *For every minimal normal subgroup N of G, the quotient group G/N is supersoluble.*

Indeed, let M/N be a 2-maximal subgroup of G/N such that $M/N \in \Sigma_{G/N}$. Then $|G : M|$ is not a prime power and $M/M_G \cap F(G/M_G) \cong (M/N)/(M/N)_{G/N} \cap F((G/N)/(M/N)_{G/N}) = 1$. It follows that $M \in \Sigma_G$. Then by the hypothesis, M is $F(G)$-semipermutable in G, and so M/N is $F(G)N/N$-semipermutable in G/N Lemma 1.10(2). Since $F(G)N/N \leq F(G/N)$, we have that M/N is $F(G/N)$-semipermutable in G/N. The statement b) is also clearly true for G/N. Hence, by the choice of G, we obtain that G/N is supersoluble.

(3) *G has the only minimal normal subgroup H and $\Phi(G) = 1$.*

It directly follows from (2) and the fact that the class of all supersoluble groups is a saturated formation.

(4) *G is soluble.*

In view of (2), we only need to prove that H is soluble. Assume that it is not true. Then in view of (3), $F(G) = 1$. Let $H \leq M$, where M is a maximal subgroup of G. In view of (2), $|G : M| = p$ is a prime. We claim that there exists some maximal subgroup T of M such that $M = HT$ and $(|M : T|, p) = 1$. Indeed, suppose that p divides $|H|$ and let P be a Sylow p-subgroup of G containing some Sylow p-subgroup M_p of M. Then $H_p = H \cap P$ is a Sylow p-subgroup of H since $L \trianglelefteq G$. Clearly, $P \leq N = N_G(H_p)$. By Frattini argument, $G = HN$, and hence $M = M \cap HN = H(M \cap N)$. Since H is not abelian, $N \neq G$. Therefore $M \cap N \neq M$. Let T be a maximal subgroup of M containing $M \cap N$. Then $M = HT$ and $(|M : T|, p) = 1$ since $M_p \leq M \cap N \leq T$. Now assume that $(|H|, p) = 1$. Then, evidently, since $H \nsubseteq \Phi(M)$, there exists a maximal subgroup T of M such that $M = HT$. It is clear that p does not divide $|M : T| = |H|/|H \cap T|$. Thus our claim is true.

Now we see that T is a 2-maximal subgroup of G of non-primary index and hence $T \in \Sigma_G$ since $T \cap F(G) = F(G) = 1$. By the hypothesis, we have that T is $F(G)$-semipermutable in G. By Lemma 1.10(4)(5), we have $|M : T| = q$ for some prime $q \neq p$. Hence $|G : T| = pq$. Let A be a minimal supplement of T such that every its subgroup is $F(G)$-permutable with T. Then $G = TA$ and $A \cap T \leq \Phi(A)$. Thus A is a $\{p, q\}$-group. Let A_p be a Sylow p-subgroup of A. Then $TA_p = A_pT$ since $F(G) = 1$. Let $D = TA_p$. Then $|G : D| = q$. Besides, since $|G : M| = p$ and $M = HT$, we see that $H \nleq D$. Hence $D_G = 1$, and by considering the permutation representation of G on the right cosets of D, we can see that G is isomorphic to some subgroup of the symmetric group S_q of degree q. It follows that D is a Hall q'-subgroup of G and $G = DZ_q$ where Z_q is a subgroup of order q. Now we see that every maximal subgroup of D has non-primary index in G and hence belongs to Σ_G. So, every maximal subgroup of D is $F(G)$-semipermutable in G and hence is $F(G)$-semipermutable in D by Lemma 1.10(4). Therefore, every maximal subgroup of D has prime index in D by Lemma 1.10(5). Hence D is a supersoluble group and therefore, G is soluble by Lemma 5.7. This contradiction completes the proof of (4).

(5) $H = C_G(H) = F(G) = O_p(G)$, *for some prime* $p \neq |H|$.
  It directly follows (2), (3), and (4).

Final contradiction.

In view of (5), $G = H \rtimes E$, for some maximal subgroup E of G where $E_G = 1$. Then by (2), E is supersoluble. Hence E contains a maximal subgroup E_1 such that $|E : E_1| = q \neq p$. Now, by hypothesis, E_1 is H-semipermutable in G. Let $T \in H(E_1)$ be a minimal supplement of E_1 in G. Denote by r the largest prime divisor of $|G|$.

Assume that $r = p$. Then, since G/H is supersoluble and $O_p(G/H) = 1$, we see that H is a Sylow subgroup of G. Obviously, $H \leq T$. Let H_1 be a maximal subgroup of H. Then for some $x \in H$ we have $H_1^x E_1 = E_1 H_1^x$. But H is abelian, so $H_1^x E_1 = H_1 E_1$ is a subgroup of G. Hence $H_1 = H \cap H_1 E_1 \trianglelefteq H_1 E_1$. This means that $E_1 \leq N_G(H_1)$. If E has another maximal subgroup E_2, then we also have that $E_2 \leq N_G(H_1)$, and hence $E = \langle E_1, E_2 \rangle \leq N_G(H_1)$. This means that $H_1 \trianglelefteq G$ and hence $H_1 = 1$ by the minimality of H. So $|H| = p$ and hence G is supersoluble. This is a contradiction. Thus, E has just one maximal subgroup E_1 and so E is cyclic. Let $A = H_1 E_1$. Clearly, A is a maximal subgroup of $E_1 H$ and $A \in \Delta_G$. Hence, by the hypothesis, there exists a 2-maximal subgroup M in G such that $M \in \Delta_G$ and M is H-semipermutable in G. Let $R = M \cap H$. Since $M \in \Delta_G$, we have $1 \neq R \neq H$. Obviously, $R \trianglelefteq MH$, so $G \neq MH$. Suppose $T \in H(M)$ be a minimal supplement of M in G. Since E is a Hall p'-subgroup of G and G is soluble, by Lemma 4.1, there is an element $x \in G$ such that $E^x = (E^x \cap M)(E^x \cap T)$. But E is cyclic and $E^x \not\subseteq M$, so $E^x \leq T$. Hence, by the hypothesis, there is an element $y \in H$ such that ME^{xy} is a subgroup of G. It is clear that R is a Sylow p-subgroup of ME^{xy}, so we have that $R \trianglelefteq ME^{xy}$. That H is abelian shows that $R \trianglelefteq G$. This is impossible by the choice of R and the minimality of H. This contradiction shows $p \neq r$.

Thus, without loss of generality, we may suppose that $r = q$. We claim that the order of the Sylow q-subgroup Q of E is q and E/Q is a p-group. Otherwise, E has a maximal subgroup X of p'-index in E and X is not a q'-group. In fact, if E is not a $\{p, q\}$-group, we may choose X be any maximal subgroup of E containing some Hall $\{p, q\}$-subgroup of E, and if $q^2 \mid |E|$, then q divides $|X|$ for any maximal subgroup X of E, since $|E : X|$ is prime by the supersolubility of E. Clearly, $X \in \Sigma_G$ and hence X is H-semipermutable in G. By Lemma 1.10(4), X is also H-semipermutable in XH. Hence, by Lemma 5.11, we know that XH is supersoluble. Therefore, the unique Sylow q-subgroup of XH is normal in XH and hence it is contained in $C_G(H) = H$. This is a contradiction. Thus, we may assume $|G| = p^\alpha q$, for some $\alpha \in \mathbb{N}$, and hence a Sylow p-subgroup G_p of G is maximal in G. Let P be a maximal subgroup of G_p containing a Sylow p-subgroup of E. Clearly, $P \in \Delta_G$

and hence $\Delta_G \neq \emptyset$. Therefore, there is an element $M \in \Delta_G$ such that M is H-semipermutable in G. If M is not a p-group, then $q \mid |M|$ and hence the index of M in G is a p-number since $|G| = p^\alpha q$. It is contrary to the choice of M. So M is a p-group. Since M is 2-maximal in G, there exists some maximal subgroup U of G such that M is maximal in U. If $H \nsubseteq U$, then $H \cap U = 1$ since $H \cap U \unlhd HU = G$ and H is a minimal normal subgroup of G. Hence $F(G) \cap M = H \cap M = 1$. This is contrary to the choice of M. So $H \leq U$. Then $|G : U|$ is a prime since G/H is supersoluble. Clearly, M is also H-semipermutable in U. By Lemma 1.10(4)(5), M has prime index in U and hence $|G : M| = pq$. Therefore, M is maximal in some Sylow p-subgroup G_p of G and $M, \cap H$ is a maximal subgroup of H since $M \cap H < H$ by the choice of M. Obviously, $G = MHQ$ and every supplement of M in G must contain some conjugate of Q. So MQ^x is a subgroup of G for some $x \in G$. Since $H \cap M = H \cap MQ^x \unlhd MQ^x$ and H is abelian, we have that $M \cap H \unlhd G$. This is contrary to the minimality of H. Thus, the sufficiency holds.

2) Now we prove the necessity.

Assume that G is a supersoluble group, then $G' \leq F(G)$. We prove firstly by induction on $|G|$ that for every subgroup $M \in \Sigma_G$ and for any minimal supplement T of M in G, it is true that M is G'-permutable with all subgroups of T. In fact, suppose $M_G \neq 1$. Then, clearly, TM_G/M_G is a minimal supplement of M/M_G in G/M_G. Indeed, if $T_0/M_G \leq TM_G/M_G$ and $(T_0/M_G)(M/M_G) = G/M_G$, then $T_0 = T_0 \cap M_G T = M_G(T_0 \cap T)$ and so $T_0 M = G = (T_0 \cap T)M$. Hence $T \leq T_0$ and hence $T_0/M_G = TM_G/M_G$ is a minimal supplement of M/M_G in G/M_G. Therefore, for every subgroup T_1 of T, the subgroup M/M_G is $(G/M_G)'$-permutable with $T_1 M_G/M_G$ by induction. But $(G/M_G)' = G'M_G/M_G$, so it is easy to see that M is G'-permutable with T_1.

Assume that $M_G = 1$. Then $F(G) \cap M = 1$ by the choice of M. Since G is supersoluble, we can assume that $|G : M| = pq$, where $p > q$. Hence, evidently, $|F(G)| \in \{p, q, pq\}$. Since G is supersoluble, it has a normal Sylow r-subgroup G_r, where r is the largest prime divisor of $|G|$. Clearly, $G_r \subseteq F(G)$ and so $|G_r| = p$. Thus the case $|F(G)| = q$ is impossible.

Let $|F(G)| = pq$. Then $G = F(G) \rtimes M$. Clearly, $F(G)$ is abelian and hence $C_G(F(G)) = F(G)$. Thus M is isomorphic with some subgroup of $Aut(F(G))$ and therefore M is a Hall $\{p, q\}'$-subgroup of G. Consequently, $F(G)$ is a Hall $\{p, q\}$-subgroup of G. We note that since T is a $\{p, q\}$-subgroup of G, it follows that $T = F(G)$ and so M is permutable with all subgroups of T since every subgroup in $F(G)$ is normal in G by $|F(G)| = pq$.

Let $|F(G)| = p$. Then $G = F(G) \rtimes E$ for some maximal subgroup E of G and $C_G(F(G)) = F(G)$. Hence E is a cyclic p'-group. Without loss of generality, we may assume that $M \subseteq E$. We show that M is G'-permutable with every subgroup T_1 of T. Let $P = T_1 \cap F(G)$. Since $|F(G)| = p$, $P = 1$ or $F(G)$ and hence it is normal in G. Since $F(G)$ is the unique Sylow p-subgroup of G, P is a Sylow p-subgroup of T_1 or T_1 is a p'-group. Let A be a Hall p'-subgroup of T_1. Then $T_1 = PA$. If $G' = 1$, then G is abelian and hence $MT_1 = T_1M$. Suppose $G' \neq 1$. Then $G' = F(G)$ since $|F(G)| = p$ and $G' \subseteq F(G)$ by the supersolubility of G. Therefore, there exists some element $x \in F(G) = G'$ such that $A^x \subseteq E$ and consequently, $MT_1^x = MPA^x = PA^xM = T_1^xM$.

Therefore, M is G'-permutable with all subgroups of T and hence M is G'-semipermutable in G. Since G is supersoluble, $G' \leq F(G)$. Thus, Lemma 1.10(7), the statement (a) holds.

Now we prove the statement (b).

Since G is supersoluble, for any $N \trianglelefteq G$, it is sufficient to prove that if $\Delta_{G/N} \neq \emptyset$, then there exists a subgroup $M/N \in \Delta_{G/N}$ such that M/N is $(G/N)'$-semipermutable in G/N by Lemma 1.10(7). Suppose it is not true and G is a counterexample of minimal order. Let N be a nontrivial normal subgroup of G and $\Delta_{G/N} \neq \emptyset$. Then, since G/N is still supersoluble, by the choice of G, we know that there exists an element $M/N \in \Delta_{G/N}$ such that M/N is $(G/N)'$-semipermutable in G/N. So we only need to consider the case that $\Delta_G \neq \emptyset$.

We claim that $\Phi(G) = 1$. Assume $\Phi(G) \neq 1$. For any element $M \in \Delta_G$, we have $\Phi(G) \leq M$. Hence $M/\Phi(G) \in \Delta_{G/\Phi(G)}$ and so $\Delta_{G/\Phi(G)} \neq \emptyset$. Then, by the choice of G, there is an element $T/\Phi(G) \in \Delta_{G/\Phi(G)}$ such that $T/\Phi(G)$ is $(G/\Phi(G))'$-semipermutable in $G/\Phi(G)$. Since $(G/N)' = G'N/N$, by Lemma 1.10(3), we have that T is G'-semipermutable in G. $T/\Phi(G) \in \Delta_{G/\Phi(G)}$ also shows that $\Phi(G) < T \cap F(G) < F(G)$. Thus $T \in \Delta_G$, a contradiction.

If $G' = 1$, then G is abelian and hence every subgroup of G is normal and certainly G'-semipermutable in G. Now suppose that $G' \neq 1$ and p is the maximal prime divisor of $|G'|$. We prove that the Sylow p-subgroup G_p of G is normal in G. Let $\pi = \{q \in \mathbb{P} \mid q > p\}$ and G_π be a Hall π-subgroup of G. Then $G_\pi \trianglelefteq G$ since G is supersoluble. Therefore, $[G_\pi, G] \leq G_\pi \cap G' = 1$ since $\pi(G') \cap \pi = \emptyset$. Hence $G_\pi \leq Z(G)$ and so $G_\pi G_p$ is p-closed. Since G is supersoluble and p is the maximal prime divisor of $|G/G_\pi|$, we have that $G_\pi G_p \trianglelefteq G$ and therefore, G_p char $G_p G_\pi \trianglelefteq G$. Choose L be a minimal normal subgroup of G contained in $G_p \cap G'$. Clearly, $|L| = p$. Since $L \nleq \Phi(G) = 1$, we may choose E be a complement of L in G. If G is nilpotent, then G is abelian since $\Phi(G) = 1$. This means that every subgroup in G is normal and certainly G'-semipermutable in G. Therefore, G is not nilpotent. Let E_p be the Sylow p-subgroup of E. We may choose a subgroup M such that M is a maximal normal subgroup of E and $|E : M| \neq p$ because $E_p = E \cap G_p$ is normal in E and G is not a p-group. We prove that M is G'-semipermutable in G. In fact, if we put $T = LR$, where R is a minimal supplement of M in E, then L is the Sylow p-subgroup of T and R is a cyclic p'-group. Let A be any subgroup of T. If $L \leq A$, then, since $ML \trianglelefteq G$, we have $MA = MLA$ is a subgroup of G. If $L \nleq A$, then A is a p'-group since $|L| = p$ and L is the unique Sylow p-subgroup of T. By Sylow's Theorem, there exists some element $x \in L \leq G'$ such that $A^x \leq R \leq E$ and hence $MA^x = A^x M$ since $M \trianglelefteq E$. Now, it is easy to see that M is G'-semipermutable in G. Clearly, the index of M in G is not primary and $M \cap F(G) < F(G)$. If $M \cap F(G) \neq 1$, then $M \in \Delta_G$, a contradiction. So $M \cap F(G) = 1$. Since M is of index pq in G, for some prime $q \neq p$, we have that $|F(G)| = p$ or pq. If $|F(G)| = p$, then for every subgroup U of G, we have $U \cap F(G) = 1$ or $U \cap F(G) = F(G)$, and hence $\Delta_G = \emptyset$, a contradiction. Assume $|F(G)| = pq$. Then $1 \neq F(G) \cap E \trianglelefteq G$. Since $G_p \trianglelefteq G$, $G_p \leq F(G)$. This implies that $G_p = L$ and hence E is a p'-group. Therefore, we may choose M to be the maximal normal subgroup of E containing $F(G) \cap E$. Then $M \in \Delta_G$. Thus, the statement (b) holds. This final contradiction completes the proof.

Theorem 5.13 (Li, Skiba [281]). *All 2-maximal subgroups of a group G of non-primary index are $F(G)$-semipermutable in G if and only if the following conditions are satisfied:*

(a) *G is supersoluble.*
(b) *For every $p \in \pi(G)$, if $\bar{P}$ is a maximal subgroup of a Sylow p-subgroup of $\bar{G} = G/O_{p'}(G)$, then $N_{\bar{G}}(\bar{P})$ is not maximal in $\bar{G}$.*

Proof (1) The proof of the "only if" part.

In order to prove G is supersoluble, by Theorem 5.12 we only need to prove that for any normal subgroup N of G, if $\Delta_{G/N}$ defined as in Theorem 5.12 is not empty, then there is an element in $\Delta_{G/N}$ which is $F(G/N)$-semipermutable in G/N. Assume $M/N \in \Delta_{G/N}$, then M is 2-maximal in G with non-primary index. Hence M is $F(G)$-semipermutable in G by the hypothesis. Since $F(G)N/N \leq F(G/N)$, by Lemma 1.10(3)(7), we have that M/N is $F(G/N)$-semipermutable in G/N and hence G is supersoluble by Theorem 5.12.

Assume that $N_{\bar{G}}(\bar{P})$ is maximal in $\bar{G}$. Since G is supersoluble, $\bar{G}$ is also supersoluble. Hence we may suppose that $|\bar{G} : N_{\bar{G}}(\bar{P})| = q$ for some $q \in \mathbb{P}$. Let $\bar{H}$ be a Hall p'-subgroup of $N_{\bar{G}}(\bar{P})$ and $\bar{M} = \bar{P}\bar{H}$. Then $\bar{M}$ is 2-maximal in $\bar{G}$ and $|\bar{G} : \bar{M}| = pq$. Let M be the inverse image of $\bar{M}$ in G. Then M is 2-maximal in G of non-primary index and hence M is $F(G)$-semipermutable in G by the hypothesis. Assume that T is a minimal supplement of M in G and every subgroup of T is $F(G)$-permutable with M. Let Q be a Sylow q-subgroup of T. Then, by hypothesis, there is some element $x \in F(G)$ such that MQ^x is a subgroup of G. If $Q^x \leq M$, then the complement of Q^x in T^x is a supplement of M in G, which contradicts the minimality of T by Chap. 1, Lemma 4.14. Hence MQ^x is maximal in G and $|G : MQ^x| = p$. Since $MQ^x < G$ is supersoluble, we have that p is the maximal divisor of $|MQ^x/O_{p'}(G)|$ and $|G/O_{p'}(G)|$. Since $\bar{P}$ is the Sylow p-subgroup of $\bar{M}$, we have that $\bar{P}$ is the Sylow p-subgroup of $MQ^x/O_{p'}(G)$. Hence $\bar{P} \trianglelefteq MQ^x/O_{p'}(G)$. On the other hand, Since $\bar{P}$ is a maximal subgroup of the Sylow p-subgroup of $\bar{G}$, $\bar{P}$ is normal in the Sylow p-subgroup of $\bar{G}$. Therefore, $\bar{P}$ is normal in $\bar{G}$. This contradiction shows that $N_{\bar{G}}(\bar{P})$ is not maximal in $\bar{G}$.

(2) The proof of the "if" part.

Since G is supersoluble, $G' \leq F(G)$. We only need to prove that any 2-maximal subgroup of G of non-primary index is G'-semipermutable in G. Let M be 2-maximal in G of non-primary index. Since G is supersoluble, we may suppose that $|G : M| = pq$, where $p, q \in \mathbb{P}$, and without loss of generality, we may assume $p > q$. Suppose that r is the maximal divisor of $|G|$. If $p \neq r$, then the Sylow r-subgroup R of G is contained in M. Clearly, G/R also satisfies the hypothesis. Hence, by induction on $|G|$, we have that M/R is $(G/R)'$-semipermutable in G/R and hence M is G'-semipermutable in G by Lemma 1.10(3) since $(G/R)' = G'R/R$. So, we may assume that $p = r$ is the maximal divisor of $|G|$. Since M is supersoluble, the Sylow p-subgroup P of M is normal in M (If M is a p'-group, then we let $P = 1$). We denote the unique Sylow p-subgroup of G by G_p. Obviously, P is maximal in G_p and so $P \trianglelefteq G_p$. Therefore, the maximal subgroup $G_p M$ of G is contained in the normalizer of P. By condition (b), we have $P \trianglelefteq G$. Let A be a p-group of G. Then $A \leq P$ or $AP = G_p$ by the maximality of P. In the former case, $A \leq P \leq M$ and hence $AM = MA$. Suppose $AP = G_p$. Then $AM = APM = G_pM = MG_p = MA$. Hence all p-subgroups of G are permutable with M. Let $G_{p'}$ be a Hall p'-subgroup of G containing some Hall p'-subgroup $M_{p'}$ of M and G_q a Sylow q-subgroup of $G_{p'}$ containing some Sylow q-subgroup M_q of $M_{p'}$. We prove that any subgroup B of G_q is permutable with M. Since $|G : M| = pq$, M_q is maximal in G_q. Therefore, $B \leq M_q$ or $BM_q = G_q$. If $B \leq M_q$, then $B \leq M$ and hence B is permutable with M. Assume $BM_q = G_q$. Then $MB = PM_{p'}B = PG_{p'}$ is a subgroup of G and hence M is permutable with B. Let T be a minimal supplement of M in G. Then, clearly, T is a $\{p, q\}$-group by the minimality of T. Let D be a subgroup of T. Then $D = D_pD_q$, where D_p is a Sylow p-subgroup of D and D_q a Sylow q-subgroup of D. We prove that there is an element $x \in G'$ such that $D_q{}^x \leq G_q$. By Sylow's Theorem, there is an element $y \in G$ such that $D_q \leq G_q{}^y$. Since G/G' is abelian, we have $G'G_q = G'G_q{}^y$. By Sylow's Theorem again, there is some element $x \in G'$ such that $G_q = G_q{}^{yx}$ and hence $D_q{}^x \leq G_q{}^{yx} = G_q$. Therefore, by the proof above, we have that $MD^x = MD_p{}^xD_q{}^x = D_p{}^xMD_q{}^x = D_p{}^xD_q{}^xM = D^xM$. This shows that M is G'-semipermutable in G. This completes the proof.

New characterizations of nilpotent groups. The first two results are based on Theorem 5.10.

Theorem 5.14 *Let G be a group and $X = F(G) \cap G'$. Then G is nilpotent if and only if for every 2-maximal subgroup M of G with non-primary index, the set $X_m(M)$ contains a supersoluble group and every minimal subgroup of G is contained in the hypercenter of its normalizer.*

Proof The necessity part is evident. In order to prove the sufficiency part, we suppose that for every 2-maximal subgroup M of G with a non-primary index, the set $X_m(M)$ contains a supersoluble group and every minimal subgroup of G is contained in the hypercenter of its normalizer. We prove that G is nilpotent. Assume that the assertion is false. Let H be a Schmidt subgroup of G. Then, by Chap. 1, Proposition 1.8, H has the following properties:

a) $|H| = p^\alpha q^\beta$, where p, q are different primes;
b) $H = H_p \rtimes H_q$, where H_p is a Sylow p-subgroup of H, H_q is a Sylow q-subgroup of H;
c) $H_p/\Phi(H_p)$ is a chief factor of H;
d) H_p is the smallest normal subgroup of H with nilpotent quotient;

Invoking Theorem 5.10, we see immediately that G is supersoluble and so its subgroup H is also supersoluble. It follows that $|H_p/\Phi(H_p)| = p$. Consequently, $|H_p| = p$ and hence by the hypothesis, $H_p \leq Z_\infty(H)$. This induces that H is nilpotent. This contradiction completes the proof.

Theorem 5.15 *Let G be a group and $X = F(G) \cap G'$. Then G is nilpotent if and only if, for every 2-maximal subgroup E of non-primary index in G, the set $X_m(E)$ contains a nilpotent group.*

Proof We only need to prove the "if part". Suppose that the assertion is false and let G be a counterexample of minimal order. Then by using the arguments as in the proof of Theorem 5.10, we can show that G has a unique minimal normal subgroup, say L, $G = L \rtimes M$ where M is a nilpotent maximal subgroup of G and $X = L = C_G(L) = F(G) = O_p(G)$, for some prime p. In view of Theorem 5.10, G is supersoluble. Thus L is a Sylow p-subgroup of G and $|L| = p$. By the choice of G, M contains a maximal subgroup E such that $|M : E| = q \neq p$. Then, by the hypothesis, E is L_m-semipermutable in G and the set $L_m(E)$ contains a nilpotent group, say T. Since $|G : E| = pq$, $L \subseteq T$ and hence $T \subseteq L = C_G(L)$. This contradiction completes the proof.

Lemma 5.16 [89, I,(6.20)]. *Let A be an abnormal subgroup of a group G. Then:*

1) If $A \leq T \leq G$, then T is abnormal in G and A is abnormal in T.
2) If H be a normal subgroup of G, then AH/H is abnormal subgroup of G/H.
3) $N_G(A) = A$.

Theorem 5.17 (Guo, Shum, Skiba [183]). *Let A be a nilpotent abnormal subgroup of a group G and $X = F(G) \cap A$. Then G is nilpotent if and only if every Sylow subgroup is X_m-permutable in G.*

Proof The necessary condition of the theorem is evident. To prove the sufficiency condition, we use induction on $|G|$.

Assume that G has an abelian minimal normal subgroup H. Let M/H be a maximal subgroup of some Hall π-subgroup E/H of G/H and P/H a Sylow p-subgroup of G/H. Then for some Sylow subgroup D of P, we have $P = HD$. Let $\pi(H) = q$. If $q \in \pi$, then E is a Hall subgroup of G. If $q \notin \pi$, then by Schur-Zassenhaus Theorem, for some Hall π-subgroup V of E, we have $E = HV$ and for some maximal subgroup Z of V we have $M = HZ$. It is now clear that D is a Sylow p-subgroup of G and V is a Hall π-subgroup of G. By the hypothesis, D is X_m-permutable in G and hence D is X-permutable with Z. Now by Lemma 1.5, we see that P/H is XH/H-permutable with M/H. By Lemma 5.16, AH/H is an abnormal nilpotent subgroup of G/H and $F(G)H/H \leq F(G/H)$. It follows that $XH/H \leq AH/H \cap F(G/H)$ and so P/H is $(AH/H \cap F(G/H))_m$-permutable in G/H. Thus, the hypothesis still holds on G/H and thereby G/H is nilpotent by using induction. This shows that if G has an abelian minimal normal subgroup H, then H is the only abelian minimal normal subgroup of G and $H \nleq \Phi(G)$.

Now we show that G is soluble. If $F(G) \neq 1$, then the result follows from above. Now assume that $F(G) = 1$ and so $X = 1$. Let P be a Sylow subgroup of G, where P is the smallest prime dividing $|G|$. If P is cyclic, then G has a normal Hall p'-subgroup E and G is soluble by the well known Feit–Thompson odd theorem. Therefore, we may assume that P is not cyclic. But then P has at least two distinct maximal subgroup, say A and B. By the hypothesis, P is permutable with A^x and B^x and so P is permutable with $P^x = A^x B^x$ for any $x \in G$. Thus PP^x is a subgroup of G and so $P = P^x$ for all $x \in G$ since P is a Sylow subgroup of G. This means that P is normal in G and consequently $P \leq F(G) = 1$, a contradiction. Hence G is soluble.

Let M be a maximal subgroup of G such that $H \not\subseteq M$. Then $G = H \rtimes M$ and H is the only minimal normal subgroup of G. It follows that $H = C_G(H) = O_p(G) = F(G)$ for some prime p. Since $M \simeq G/H$ is nilpotent, H is a Sylow p-subgroup of G. Since G is soluble and A is abnormal nilpotent subgroup of G, A is a Carter subgroup of G by Lemma 5.16. Obviously, M is also a Carter subgroup. But since all Carter subgroup of G are conjugate by [135, Theorem 2.4.7], without loss of generality, we may suppose that $A = M$. Let Q be any Sylow subgroup of M. Then since $X = F(G) \cap M = 1$, $MQ^x = Q^x M = M$ for all $x \in G$. It follows that $Q \leq M_G \leq C_G(H) = H$. This contradiction completes the proof.

It is well known that a Carter subgroup is nilpotent abnormal (see [89, I.6.19(c)]). Now by Theorem 5.17, we immediately obtain the following result.

Corollary 5.18 *Let A be a Carter subgroup of a soluble group G and $X = F(G) \cap A$. Then G is nilpotent if and only if every its Sylow subgroup is X_m-permutable in G.*

Groups with X_m-semipermutable Hall subgroups.

Lemma 5.19 *Let A be a Hall subgroup of a group G and T a supplement of A in G. Let $p_1 > p_2 > \ldots > p_t$ be the distinct primes dividing $|T|$. Suppose $p > q$ for all primes p and q such that p divides $|A|$ and q divides $|G : A|$. If every Sylow subgroup of T is cyclic, then G has a normal series $G_0 = A < G_1 < \ldots < G_t = G$ such that $|G_i/G_{i-1}|$ is the order of a Sylow p_i-subgroup of G for all $i = 1, 2, \ldots, t$. In particular, G/A is soluble.*

Proof Assume that this Lemma is false and let G be a counterexample of minimal order. Then $A \neq G$. Let p be the smallest prime dividing $|G|$ and P a Sylow p-subgroup of T. Then p divides $|G : A|$ and P is a Sylow p-subgroup of G. Since P is cyclic, G has a normal Hall p'-subgroup E by Chap. 1, Lemma 3.39. Clearly, $E = E \cap AT = A(E \cap T)$ and so A is normal in E by the choice of G. Since A char E, A is normal in G. Now by induction on $|G|$, the Lemma holds.

Theorem 5.20 (Guo, Shum, Skiba [183]). *Let $|G| = p_1^{a_1} p_2^{a_2} \ldots p_t^{a_t}$ where $p_1 > p_2 > \ldots > p_t$. Let $\pi_i = \{p_1, p_2, \ldots, p_i\}$ for all $i = 1, 2, \ldots, t$ and $X = F(G)$. Then G is a dispersive group if and only if for every $i = 1, 2, \ldots, t$, the group G has a Hall π_i-subgroup E_i such that E_i is X_m-semipermutable in G and the set $X_m(E_i)$ contains a dispersive group.*

Proof We need only to prove that if $A = E_i$ and the set $X_m(A)$ contains a dispersive group T, then A is normal in G. We prove this assertion by induction on $|G|$. In view of Lemma 5.19, it suffices to consider the case when at least one of Sylow subgroups of T is noncyclic.

We first claim that $AL \trianglelefteq G$, for any nonidentity normal subgroup L of G. Indeed, our hypothesis still holds for G/L by Lemma 1.14 and so AL/L is normal in G/L by induction. This implies that $AL \trianglelefteq G$. If $E = A \cap T \neq 1$, then E is normal in T since T is a dispersive group. Hence, for the normal closure E^G of E in G, we have $E^G = E^{AT} = E^A \leq A$. It follows that $A = AE \trianglelefteq G$. Now assume that $A \cap T = 1$. Then T is a complement of A in G and so T is a Hall π_i'-group of G.

Let $\{p_1, p_2, \ldots, p_t\}$ be the set of all such distinct primes dividing $|T|$ that P_i is noncyclic Sylow p_i-subgroup of T for all $i = 1, 2, \ldots, t$. Suppose that X has at least two different Sylow subgroups, say P and Q. Then $A = AQ \cap AP$ is normal in G. Thus, we may assume that X is a p-group for some prime p. If $p \in \pi_i$, then $AX = A \trianglelefteq G$. Now, we may assume that $p \in \pi'$ so that $X \leq T$.

Suppose that $X \neq 1$ is cyclic and let L be a minimal normal subgroup of G contained in X. Then A is normal in AL by Lemma 5.19. But then A is normal in G. Hence either $X = 1$ or $X \neq 1$ is a noncyclic group. First, we suppose that $X = 1$. Then by the hypothesis, A is permutable with all maximal subgroups of the noncyclic subgroup P_i of T and so $AP_i = P_i A$ is a subgroup of G. Clearly, the hypothesis still holds for AP_i and hence $P_i \leq N = N_G(A)$. Now let $x \in G$. Since $G = AT$, $x = ta$, where $a \in A$ and $t \in T$. Then $A(P_i)^x = A(P_i)^{ta} = (A(P_i)^t)^a = ((P_i)^t A)^a = P_i^x A$. It follows that $(P_i)^x \leq N$. Thus, the normal closure $(P_i)^G$ of P_i in G is contained in N for all $= 1, 2, \ldots, t$. Now let $D = (P_1)^G (P_2)^G \ldots (P_i)^G$. Then every Sylow subgroup of TD/D is cyclic and hence TD/D is supersoluble. It follows from Lemma 5.19 that AD/D is normal in G/D, and consequently A is normal in G.

Finally, we suppose that $X \neq 1$ is noncyclic and let L be a minimal normal subgroup of G contained in X. Then X is contained in some noncyclic Sylow subgroup of T. Without loss of generality, we may suppose that $X \leq P_1$. Then, by the hypothesis, A is permutable with all maximal subgroup of P_1 and so AP_1 is a subgroup of G. Clearly, the hypothesis still holds for AP_1 and so $P_1 \leq N = N_G(A)$. It follows that A is normal in AL and consequently A is normal in G since from above we have $AL \trianglelefteq G$. The Theorem is proved.

Corollary 5.21 *Let p be the largest prime divisor of a group G and $X = F(G)$. Then G is p-closed if and only if a Sylow p-subgroup of G is X_m-semipermutable in G.*

Theorem 5.22 (Guo, Shum, Skiba [183]). *Let A be a Hall π-subgroup of a group G and $X = F(G)$ the Fitting subgroup of G. If A is X_m-permutable in G and the set $X_m(A)$ contains a nilpotent group T, then G is p-supersoluble for every prime $p \notin \pi$ such that T contains a Sylow p-subgroup P with $|P| > p$.*

Proof Assume that the theorem is false and let G be a counterexample of minimal order. Then G is not p-supersoluble for some prime $p \in \pi$ such that T has a Sylow p-subgroup P with $|P| > p$. We now proceed our proof by proving the following claims:

(1) $O_\pi(G) = 1$.

Suppose that $D = O_\pi(G) \neq 1$. Clearly $D \leq A$ and so by Lemma 1.15 the hypothesis still holds for G/D. Since PD/D is a Sylow subgroup of TD/D and $|P| > p$, $|PD/D| > p$ and so G/D is p-supersoluble by the choice of G. But then G is p-supersoluble since $D = O_\pi(G)$, which contradicts the choice of G. Thus claim (1) is proved.

(2) $T \cap A = 1$.

Suppose on the contrary that $T \cap A \neq 1$ and let Q be a Sylow q-subgroup of $T \cap A$ for some prime q dividing $|T \cap A|$. Since T is nilpotent and A is a Hall subgroup of G, Q is a Sylow subgroup of T and so Q is normal in T. Hence for the normal closure Q^G of Q in G we have $Q^G = Q^{TA} = Q^A \leq A$, which contradicts (1). Hence T is a complement of A in G.

(3) $O_{p'}(G) = 1$ (cf. the proof of claim (1)).

(4) $X \leq P$.

Assume that $X \nleq P$. Then since P is a Sylow p-subgroup of G, $O := O_{p'}(X) \neq 1$. Since O $char$ $F \trianglelefteq G$, $O \trianglelefteq G$ and so $O \subseteq O_{p'}(G)$, this contradicts to our claim (3). Thus claim (4) is established.

(5) A *is permutable with all maximal subgroups of any Hall subgroup of* T (this claim follows directly from the hypothesis and claim (4) above).

(6) G *is not p-soluble.*

Assume that G is p-soluble and let L be a minimal normal subgroup of G. Then by claim (3), L is a p-group and $L \leq P$. Next we note that G/L is p-supersoluble. Indeed, if $|P/L| \leq p$, then the assertion is evident. On the other hand, if $|P/L| > p$, then the hypothesis is still true for G/L by Lemma 1.15. Hence G/L is p-supersoluble by the choice of G. It follows that $|L| > p$ and $L \nleq \Phi(G)$. Then, clearly, $L \nleq \Phi(P)$. Let P_1 be a maximal subgroup of P such that $L \nleq P_1$ and let M be a maximal subgroup of T containing P_1. Then by claim (5), we have $AM = MA$ and so $D = M^G = M^{AT} = M^A \leq AM$. This implies that $D \cap L$ is a nontrivial subgroup of L and $D \cap L$ is normal in G, which contradicts to the minimality of L.

(7) *If G has a minimal normal subgroup H with* $|H| = p$, *then* $|P| = p^2$.

Indeed, if $|P| > p^2$, then the hypothesis is still true for G/H and so G/H is p-supersoluble by the choice of G, which again contradicts to claim (6). Thus, claim (7) is established.

(8) *If H is a normal subgroup of G and $H \cap A \neq A$, then H is p-soluble.*

It is clear that $H = (A \cap H)(T \cap H)$. Write $E = (H \cap A)T$. If V is a maximal subgroup of some Hall subgroup of T, then $AV = VA$ by claim (5), and so by claim (2), we have $AV \cap (A \cap H)T = (A \cap H)(AV \cap T) = (A \cap H)V(A \cap T) = (A \cap H)V = V(A \cap H)$. Thus our hypothesis still holds for E. If $E = G$, then $A = A \cap (H \cap A)T = (H \cap A)(A \cap T) = H \cap A$. This contradiction shows that $E \neq G$. Thus E is p-supersoluble by the choice of G and consequently $H \leq E$ is p-soluble.

(9) *If H is a p-soluble minimal normal subgroup of G, then* $|H| = p$ *and* $H \leq Z(G)$.

Clearly, H is either p'-group or p-group. But the former case is impossible by claim (3) and so $|H| = p^a$ for some natural a. It is clear that $H \neq P$. If $|P/H| > p$, then the hypothesis still holds for G/H, which implies the p-solubility of G, a contradiction. Hence H is a maximal subgroup of P. Assume that $a > 1$. Then for some maximal subgroup E of P, we have $P = HE$. Let M be a maximal subgroup of T containing E and $K = AM$. Then $1 < K \cap H < H$, where $K \cap H$ is clearly normal in G, which contradicts the minimality of H. Therefore, $|H| = p$. Let $C = C_G(H)$. Then $C = (C \cap A)(C \cap T)$ and G/C is cyclic because it is isomorphic to some subgroup of $Aut(H)$. Suppose that $C \neq G$. By claim (7), P is abelian and so $P \leq C$. This implies that $C \cap A \neq A$ and hence C is p-soluble by claim (8). It follows that G is p-soluble, which contradicts claim (6). Hence, claim (9) is established.

(10) *P is not cyclic.*

Suppose on the contrary that P is cyclic. Then by Chap. 1, Lemma 3.39, we see that $p > 2$. Let Z be a maximal subgroup of P. Then, by claim (5), $M = AZ$. It is clear that $D = Z^G = Z^{AT} = Z^A \le M$ and $D = (A \cap D)(T \cap D)$. Assume that either $D \ne M$ or $T \ne P$. Then D is p-soluble. Indeed, in the former case, we have $A \nsubseteq D$ and by claim (8) D is p-soluble. On the other hand, if $A \le D$ and $T \ne P$, then our hypothesis still holds for DP. Since $|DP| < |G|$, DP is p-supersoluble by the choice of G. Now, let H be a minimal normal subgroup of G contained in D. Since D is p-soluble, $|H| = p$ and $H \le Z(G)$ by claim (9). Let $N = N_G(P)$. If $P \le Z(N)$, then G is p-soluble by [248, IV, Theorem 2.6], which contradicts claim (6). Hence $N \ne C_N(P)$. Let $x \in N \backslash C_G(P)$ with $(|x|, |P|) = 1$ and $E = P \rtimes \langle x \rangle$. By [248, III, 13.4], $P = [E, P] \times (P \cap Z(E))$. Since $H \le P \cap Z(E)$ and P is cyclic, $P = P \cap Z(E)$ and so $x \in C_G(P)$. This contradiction shows that $T = P$ and $D = M$. Let $q \ne p$ be a prime dividing $|A|$ and Q any Sylow q-subgroup of A. Let $N = N_G(Q)$. It is clear that Q is a Sylow subgroup of D and so by using Frattini arguments, we have $G = DN$ and so $P = D_p N_p$, for some Sylow subgroup D_p of D and some Sylow subgroup N_p of N. But since P is cyclic, $P = N_p$. Hence $Q^G = Q^{AP} = Q^A \le A$, which contradicts claim (1). Hence the claim (10) is proved.

(11) $|P| \ne p^2$.

Suppose on the contrary that $|P| = p^2$. By claim (10), P is not cyclic. Thus, P has at least three district subgroups, say Z_1, Z_2, Z_3 of order p. Let $N_i = Z_i^G$ be the normal closure of Z_i in G. Then $N_i \le AZ_i$ and so $N_i \cap N_j$ is contained in $O_{p'}(G) = 1$ for any different $i, j \in \{1, 2, 3\}$. Hence, $P \le C_i = C_G(N_i)$ for all i. Assume that $C_i \ne G$ for some i. Then, we can see that C_i is p-soluble by claim (8), and so G is p-soluble since G/C_i is a p'-group. This contradiction shows that $C_i = G$ for all i. It follows that N_1, N_2, and N_3 are abelian groups and hence $N_i = Z_i$ for all i. This implies that P is normal in G and so G is p-soluble, which contradicts claim (6).

(12) $O_p(G) = 1$.

Let $D = O_p(G) \ne 1$ and H a minimal normal subgroup of G contained in D. Then $|H| = p$ by claims (8) and (9) and so $|P| = p^2$ by claim (7) which contradicts claim (11).

(13) $T = P$.

By claim (10), for some two maximal subgroups P_1 and P_2 of P, we have $P = P_1 P_2$. By claim (5), A is permutable with P_1 and with P_2. Hence A is permutable with P. Let $E = AP$. Assume that $T \ne P$ and so $E \ne G$. Clearly, our hypothesis still holds for E and hence E is p-supersoluble by the choice of G. Let M be a maximal subgroup of P. Then $D = M^G = M^{AT} = M^A \le AP$ and hence D is p-supersoluble. It follows that $O_p(G) \ne 1$, which contradicts claim (12). Thus claim (13) is verified.

Final contradiction.

Let $M_1, \ldots, M_t$ be all maximal subgroups of P and $N_i = M_i^G$ the normal closure of M_i in G. Then $N_i \le AM_i$. Suppose that for some i we have $N_i \cap A \ne A$. Then N_i is

p-soluble by claim (8) and hence $O_{p'}(G) \neq 1$ which contradicts (12). Thus $N_i \cap A = A$ for all i. Now we have that $D = N_1 \cap \cdots \cap N_t = A(M_1 \cap \cdots \cap M_t) = A\Phi(P)$. It follows that $D \cap P = \Phi(P)$. Let $q \neq p$ be a prime dividing $|A|$ and Q any Sylow q-subgroup of A. Then $P \leq N_G(Q)$ (see the proof of claim (10)) and hence $Q^G = Q^{AP} = Q^A \leq A$, which contradicts claim (1). Thus, the theorem is finally proved.

Groups with 1-semipermutable. We denote by $l_p(G)$ the p-length of a p-soluble group G (see [337, p.249]; $l_\pi(G)$ denotes the minimum of numbers of π-factors (where $\pi = \pi(G) \setminus \{p\}$) in all normal series of a p-soluble group G.

Recall that if A and B are subgroups of G, then $A^B = \langle A^b | b \in B \rangle$.

Lemma 5.23 ([276, 7.2.5]). *Let G be a group, H and L subgroups of G, and H be permutable with subgroup L^g for all $g \in G$. Then $H^L \cap L^H$ is subnormal in G.*

Recall that a group G is said to be π-separable (π-soluble) if every composition factor of G is either a π-group or a π'-group (a p-group for some prime $p \in \pi$ or a π'-group respectively).

Lemma 5.24 ([117, 6, 3.2]). *If G is a π-separable group, then $C_G(O_{\pi'\pi}(G)) \leq O_{\pi'\pi}(G)$.*

Lemma 5.25 *Let G be a group and p a prime such that $(|G|, p-1) = 1$.*

(1) *If G is p-supersoluble, then G is p-nilpotent.*
(2) *If for some subgroup M of G we have $|G : M| = p$, then M is normal in G.*

Proof

(1) Assume that G is p-supersoluble. Then every chief factor H/K of G is either a cyclic group of order p or a p'-group. When H/K is a cyclic group of order p, we have that $G/C_G(H/K) \in \mathfrak{A}(p-1)$, that is, it is an abelian group with exponent $p-1$. Hence $G/C_G(H/K) = 1$, that is, $H/K \subseteq Z(G/K)$. Therefore, G is a p-nilpotent group.
(2) If $p = 2$, this is clear. Let $p > 2$. Then G is of odd order, so is soluble. Without loss of generality, we may suppose that $M_G = 1$, so G is isomorphic to some subgroup of S_p. Hence M is a Hall p'-subgroup of G, so G is p-supersoluble. Now (2) follows from (1).

The following lemma is evident.

Lemma 5.26 *If G is a p-supersoluble soluble group, then $l_p(G) \leq 1$.*

Theorem 5.27 (Guo [142]). *Let p be a prime number, P a Sylow p-subgroup of G and $\pi = \pi(G) \setminus \{p\}$. If P is 1-semipermutable in G, then the following statements hold.*

1) *The group G is a p-soluble group and $P' \leq O_p(G)$;*
2) *$l_p(G) \leq 2$ and $l_\pi(G) \leq 2$;*
3) *If a π-Hall subgroup of G is q-supersoluble for some $q \in \pi$, then G is q-supersoluble.*

Proof

1) We first prove by induction on the order of G that G is a p-soluble group. By Lemma 1.10, we see that the condition of the theorem is hereditary for all subgroups of G containing P and for all factor group of G. Hence, we may assume that every minimal normal subgroup of G is not p-soluble. Let K be a minimal supplement of the 1-semipermutable subgroup P in G. Then, for every proper subgroup L of K, PL is a proper subgroup of G, and by Lemma 1.10, P is permutable with all subgroup L^g, for all $g \in G$. By Lemma 5.23, $D = P^L \cap L^P$ is a subnormal subgroup of G. By induction, the subgroup PL is p-soluble, and consequently, D is p-soluble. Now by Chap. 1, Lemma 5.35(7), we see that the subgroup D^G is also p-soluble. This induces that $D = 1$ and

$$[P, L] \leq [P^L, L^P] \leq P^L \cap L^P = 1,$$

that is, $L \leq C_G(P)$. If K has two different maximal subgroups L_1 and L_2, then $K = \langle L_1, L_2 \rangle \leq C_G(P)$ and consequently, P is normal in G. In this case, G is p-soluble. If K only has one maximal subgroup, then K is a cyclic q-group for some prime q. Hence, G is a biprimary group (that is, $\pi(G) = \{p, q\}$ where p, q are two distinct primes), and consequently, G is a soluble group by Burnside's $p^a q^b$-Theorem. Therefore, in any case, G is a p-soluble group.

We now prove that $P' \leq O_p(G)$ by induction on the order of G. Assume that $O_p(G) \neq 1$. Then by Lemma 1.10, $P/O_p(G)$ is a 1-semipermutable Sylow p-subgroup of the group $G/O_p(G)$, and by induction

$$(P/O_p(G))' = P'O_p(G)/O_p(G) \leq O_p(G/O_p(G)) = 1.$$

It follows that $P' \leq O_p(G)$. Hence, without loss of generality, we may assume that $O_p(G) = 1$.

Since G is a p-soluble group and $O_p(G) = 1$, we have $O_{p'}(G) \neq 1$. Support that the subgroup $H = PO_{p'}(G)$ is different from G. Then, by induction and Lemma 5.24,

$$P' \leq O_p(H) \leq C_G(O_{p'}(G)) \leq O_{p'}(G).$$

This induces that $P' = 1$. Now, assume that $PO_{p'}(G) = G$ and M is a maximal subgroup of $O_{p'}(G)$. Then, because P is 1-semipermutable, we have that PM is a proper subgroup of G. Since

$$PM \cap O_{p'}(G) = M(P \cap O_{p'}(G)) = M,$$

M is a normal subgroup of PM. Hence, by induction,

$$P' \leq O_p(PM) \leq C_G(M).$$

This shows that P' centralizes every maximal subgroup of $O_{p'}(G)$. If $O_{p'}(G)$ contains two different maximal subgroups M_1 and M_2, then P' centralizes $\langle M_1, M_2 \rangle = O_{p'}(G)$ and so $P' = 1$. If $O_{p'}(G)$ only contains one maximal subgroup, then $O_{p'}(G)$ is a cyclic q-subgroup, for some prime q. However, since the automorphism group of a cyclic group is an abelian group, $G/C_G(O_{p'}(G))$ is abelian. It follows from Lemma 5.24 that P is a abelian group and so $P' = 1$. Therefore, the statement (1) is proved.

2) Since $P' \leq O_p(G)$, every Sylow p-subgroup of $G/O_p(G)$ is abelian. Hence, by [248, VI, 6.6], we obtain that $l_p(G/O_p(G)) \leq 1$ and $l_p(G) \leq 2$.

We now prove $l_\pi(G) \leq 2$ by induction on the order of G. In fact, obviously, $l_\pi(G) = l_\pi(G/O_p(G))$, so we may assume, without loss of generality, that $O_p(G) = 1$. However, in this case, every Sylow p-subgroup of G is abelian and $l_p(G) \leq 1$. It follows that $l_\pi(G) \leq 2$.

3) We prove the statement (3) by induction on the order of G. By 1), we have known that G is a p-soluble group. Hence, G has a p'-Hall subgroup H by Theorem 2.1. Since P is 1-semipermutable, P is permutable with all subgroups of H. Since the condition of the theorem is inheritable for every factor group and the class of all q-supersoluble groups is a saturated formation, it is easy to see that G is a primitive group, and consequently, by Lemma 1.10(6), Chap. 1, Proposition 1.4 and Chap. 1, Lemma 5.34(1), G has a unique minimal normal subgroup N such that N is a r-subgroup for some prime $r \in \pi(G)$ and $N = C_G(N)$. If $r \neq q$, then, since G/N is q-supersoluble by induction, G is q-supersoluble. Thus, we may assume that N is a q-group and hence $N \leq H$. Since H is q-supersoluble, H has a normal subgroup L of prime order such that $L \leq N$. Since P is 1-semipermutable in G, PL is a subgroup of G and $L = PL \cap N$ is normal in PL. Consequently, L is normal in G. This means that $L = N$ and G/L is a cyclic group. Therefore, G is supersoluble. This completes the proof of the theorem.

Corollary 5.28 *Let G be a group and $\pi(G) = \{p, q\}$. If G has a 1-semipermutable Sylow p-subgroup, then G is a q-supersoluble group and $l(G) \leq 3$. Moreover, when $p > q$, then $l(G) \leq 2$.*

Proof By the statement 3) of Theorem 5.27, we know that G is a q-supersoluble group. If $q < p$, then G is q-nilpotent by Lemma 5.25 and $l(G) \leq 2$. In general case, by Lemma 5.26, $l_q(G) \leq 1$, and hence $l(G) \leq 3$. $\quad\square$

Remark The symmetry group S_4 have nilpotent length 3 and its Sylow 2-subgroup is 1-semipermutable. Hence, the estimation of nilpotent length of a group in Corollary 5.28 is exact.

Corollary 5.29 *Let G be a group and $r \in \pi(G)$. If G has a Sylow p-subgroup which is 1-semipermutable in G, for all $p \in \pi(G) \setminus \{r\}$, then G is a soluble r-supersoluble group.*

Proof Let $\pi = \pi(G) \setminus \{r\}$. By Theorem 5.27, G is a p-soluble group for all $p \in \pi$, and consequently, G is π-soluble. Since $|\pi(G) \setminus \pi| = |\{r\}| = 1$, it is easy to see that G is soluble. Now, by using induction on the order of G, we prove that G is a r-supersoluble group. Indeed, if G is a biprimary group, then it is a r-supersoluble by Corollary 5.28. Assume that G is not a biprimary group, $p \in \pi$ and H is a p'-Hall subgroup of G. By induction, the subgroup H is r-supersoluble group, and consequently, by using the statement 3) of Theorem 5.27, we obtain that G is a r-supersoluble group. This completes the proof.

From Corollary 5.29, we see that if all Sylow subgroups of G are 1-semipermutable, then G is p-supersoluble for every $p \in \pi(G)$, and consequently G is a supersoluble group. Hence, we obtain the following known result.

Corollary 5.30 [434]. *If all Sylow subgroups of a group G are 1-semipermutable, then G is supersoluble.*

Corollary 5.31 *Let G be a group and $r, s \in \pi(G)$. If G has a Sylow p-subgroup which is 1-semipermutable in G, for some $p \in \pi(G) \setminus \{r, s\}$, then G is a soluble group.*

Proof Let $\pi = \pi(G) \setminus \{r, s\}$. Then by Theorem 5.27, G is p-soluble group for all $p \in \pi$. It follows that G is π-soluble group. Hence, G has a normal series in which every factor is soluble π-group or a $\{r, s\}$-group. Since a biprimary group is soluble, G is a soluble group.

Corollary 5.32 *If every noncyclic Sylow subgroup of a group G is 1-semipermutable in G, then the fourth derived subgroup $G^{(4)}$ of G is a nilpotent group.*

Proof Let $\pi = \{p \mid$ every Sylow p-subgroup of G is 1-semipermutable in $G\}$ and $\tau = \{q \mid$ every Sylow q-subgroup of G is cyclic$\}$. Then $\pi(G) = \pi \cup \tau$. By Theorem 5.27, G is p-soluble group for all $p \in \pi$. It follows that G is a π-soluble group. Hence, G has a normal series in which every factor is a soluble π-group or a τ-group. Since all Sylow subgroup of every τ-subgroup are cyclic groups, by [248, Theorem IV.2.11], all τ-factors of the series are metacyclic. Consequently, G is a soluble group. Thus, $G = AB$, where A is a π-Hall subgroup and B is a τ-Hall subgroup.

Let $\mathfrak{N}(\mathfrak{A})^4 = \mathfrak{N}\mathfrak{A}\mathfrak{A}\mathfrak{A}\mathfrak{A}$ is the product of formations. We prove that that $G \in \mathfrak{N}(\mathfrak{A})^4$. In fact, since the product of two saturated formations is still a saturated formation, $\mathfrak{N}(\mathfrak{A})^4$ is a saturated formation. If $G \notin \mathfrak{N}(\mathfrak{A})^4$, then by induction, G is a primitive group. It follows from Chap. 1, Proposition 1.4 and Chap.1, Lemma 4.17 that $G = F(G) \rtimes M$, $C_G(F(G)) = F(G)$, and $F(G)$ is a minimal normal subgroup which is a r-subgroup, for some prime r. If $r \in \tau$, then $F(G) \leq B$ and $F(G)$ is cyclic. Then, $|F(G)| = r$ and $G/F(G)$ is a cyclic group of order dividing $r - 1$. Thus, $G \in (\mathfrak{A})^2 \subseteq \mathfrak{N}(\mathfrak{A})^4$. If $r \in \pi$, then $F(G) \leq A$ and by Theorem 5.27, G has a Sylow r-subgroup R such that $R' \leq F(G)$. Since A is supersoluble by Lemma 1.10(4) and Corollary 5.30, its derived subgroup A' is nilpotent. Since $C_G(F(G)) = F(G)$, A' is a r-subgroup. Hence, it is easy to see that $A = A_1 R$, where A_1 is an abelian r'-Hall subgroup of A. On the other hand, by [248, IV, 2.11], we know that $B = B'B_1$, where B' and B_1 are cyclic Hall subgroups. Obviously, the subgroups R and A_1 are permutable with the subgroups B' and B_1 since, by the condition and Lemma 1.10, all Sylow subgroups of A are permutable with every subgroup of B. Therefore

$$G/F(G) = (A_1 F(G)/F(G))(R/F(G))(B' F(G)/F(G))(B_1 F(G)/F(G))$$

is a product of permutable in pairs abelian subgroups of relatively prime orders. By [248, VI, 14.16], the derived length of $G/F(G)$ is not more than 4. Therefore, $G/F(G) \in \mathfrak{A}^4$, and consequently, $G \in \mathfrak{N}(\mathfrak{A})^4$. This shows that the fourth derived subgroup of G is nilpotent. The proof is completed.

2.6 Additional Information and Some Problems

I. The idea of X-permutability was first introduced and applied in [175] as the form c-permutability. The idea and related results in the paper [175] were developed in various directions in the papers of many authors (see, for example, [8, 9, 14, 61, 63, 146, 234, 240, 241, 266, 277, 282, 323, 325]). Hear we cite some related results as follow:

Theorem 6.1 (Arroyo-Jordá, Arroyo-Jordaá, Martinez-Pastor, Pérez-Ramos [8]). *Let $G = AB$, where A and B be supersoluble subgroups of a group G such that:*
  (i) Every normal subgroup of A is c-permutable with all Sylow subgroup of B.
  (ii) Every normal subgroup of B is c-permutable with all Sylow subgroup of A. Then G is supersoluble.

Theorem 6.2 (Huang, Guo [241]). *A group G is p-supersoluble if and only if G has a p-soluble normal subgroup N such that G/N is p-supersoluble and every cyclic p-subgroup of N (or every maximal subgroup of every Sylow p-subgroup of N) is s-c-permutable in G.*

Theorem 6.3 *(Hu, Guo [240]). Let N be a normal subgroup of G. If every maximal subgroup of G not containing N is c-semipermutable in G, the N is supersoluble.*
   II. O. H. Kegel [263] proved that every s-quasinormal subgroup H of a group G is subnormal in G and H/H_G is nilpotent. B. Li and W. Guo [279] proved that if a subgroup H of a group G is X-permutable with all subgroups where X is a nilpotent normal subgroup of G, then H/H_G is supersoluble. In connection with this results, the following questions arises.

Problem 6.4 Let A be a completely c-permutable subgroup of a group G. What we can say about A/A_G? Is A/A_G supersoluble or soluble at least?

Problem 6.5 Let A be a completely c-permutable subnormal subgroup of a group G. Is A/A_G nilpotent then?

Problem 6.6 Let H be a subgroup of a group G and X be a nilpotent normal subgroup of G. Suppose that H is X-permutable with all subgroups of G. Is then H/H_G nilpotent ?
   III. In the alternative group A_5, any involutions generate a c-permutable subgroup. Nevertheless, the following question are still open.

Problem 6.7 Is there a non-abelian simple group in which some nontrivial subgroup is completely c-permutable?

Problem 6.8 (A.N. Skiba and V.N. Tyutyanov [318, Question 17.112 a)]). Which non-abelian simple groups possess a nontrivial c-permutable subgroup?

Problem 6.9 (A.N. Skiba and V.N. Tyutyanov [318, Question 17.112 b]). Which non-abelian simple groups possess a nontrivial complete c-permutable subgroup?

Problem 6.10 Is a group soluble if every its two maximal subgroups are c-permutable?

IV. Section 2.2 is written on the base of results of the papers [197, 202] which were a further development of the methods and ideas of the papers [180, 181, 193].

V. About the existence and conjugacy of Hall subgroups, E. P. Vdovin, D. O. Revin made important contributions. They proved the following results by using the classification of finite simple groups.

Theorem 6.11 (Revin, Vdovin [336]). *Let a composition series*

$$1 = G_0 < G_1 < \ldots < G_n = G$$

of G be a refinement of some chief series of G. Then the following statements are equivalent:

(1) $G \in E_\pi$;
(2) $Aut_G(G_i/G_{i-1}) \in E_\pi$ *for every* $i = 1, \ldots, n$.

Here $Aut_G(G_i/G_{i-1})$ is the image of $N_G(G_i) \cap N_G(G_{i-1})$ in $Aut(G_i/G_{i-1})$ under the homomorphism given by the rule $x \mapsto \varphi_x$ where $\varphi_x : G_{i-1}g \mapsto G_{i-1}x^{-1}gx$ for every $x \in N_G(G_i) \cap N_G(G_{i-1})$ and $g \in G_i$.

Theorem 6.12 (Gross, Revin, Vdovin [121, 334]). *Suppose that a set π of primes is such that $2 \notin \pi$ or $3 \notin \pi$. In this case G is an E_π-group if and only if every composition factor of G is an E_π-group.*

Theorem 6.13 *(Vdovin, Revin [421]). Let π be a set of primes and A a normal subgroup of G. Then $G \in C_\pi$ if and only if $G/A \in C_\pi$, and for any (equivalently, for some) $K/A \in Hall_\pi(G/A)$ its full pre-image $K \in C_\pi$.*

VI. In the 1 hr survey report by Wielandt at the 13th International Congress of Mathematicians in Edinburgh in 1955 [446], the following problem was first stated:

Problem 6.14 Does an extension of a D_π-group A by a D_π-group B always is a D_π-group?

Many mathematicians studied this problem over the following 50 years. This problem is mentioned in [80, 135, 318, 358, 359, 401, 448] and was studied, for example, in [237, 238, 260, 319, 333, 335, 350, 352-356, 359, 404, 405, 420]. Finally, a positive solution of this problem was obtained by Revin and Vdovin in [334] by applying the classification of finite simple groups. In fact, they proved the following theorems:

Theorem 6.15 (Revin, Vdovin [334]). *Let A be a normal subgroup of G and π be a set of primes. Then $G \in D_\pi$ if and only if $A \in D_\pi$ and $G/A \in D_\pi$.*

Problem 6.16 Let π be a set of primes. Find all finite simple D_π-groups in which

a) Every subgroup is a D_π-group (H. Wielandt);
b) Every subgroup possessing a Hall π-subgroup is a D_π-group (Revin and Vdovin).

Problem 6.17 It is true that under the hypothesis in Theorem 2.21 the group G is a D_π-group?

VII. A subgroup H of a group G is called pronormal if H and H^g are conjugate in $\langle H, H^g \rangle$ for every $g \in G$. A subgroup H is said to be strongly pronormal if L^g is conjugate to a subgroup of H in $\langle H, H^g \rangle$ for every $L \leq H$ and $g \in G$.

It is easy to see that every normal subgroup, maximal subgroup, Sylow subgroup, Carter subgroup of a soluble group, Hall subgroup of a soluble group are all pronormal subgroups. In [419], Vdovin shows that the Carter subgroups of any finite group remain pronormal. But the Hall subgroups of a nonsoluble group maybe is not pronormal (see the examples in [422, 424]) or [159, Lemmas 19 and 20]).

By analogy with Hall's notation, we say that a finite group G enjoys *property P_π* or is a *P_π-group* or belongs to the *class P_π* if $G \in E_\pi$ and every Hall π-subgroup of G is pronormal.

It is shown in [423] that the Hall subgroups of finite simple groups are pronormal, and it is established in [424] that the Hall π-subgroups are pronormal in every group in C_π. Thus, $C_\pi \subseteq P_\pi \subseteq E_\pi$ and every simple E_π-group belongs to the class P_π.

The question [422, Problem 7.20; 6, Problem 6] is open: For which π is the inclusion $C_\pi \subseteq E_\pi$ strict? It is known [121, Theorem A] that $E_\pi = C_\pi$ provided that $2 \notin \pi$. At the same time, many examples of sets π with $E_\pi \neq C_\pi$ are available.

Another natural problem is as follows: For each of the inclusions $C_\pi \subseteq P_\pi$ and $P_\pi \subseteq E_\pi$ determine for which π it is strict? The following statement reduces this question to the strictness of the inclusions $C_\pi \subseteq E_\pi$.

Proposition 6.18 (Guo, Revin, [159]). *For every set π, the following are equivalent:*

(1) $C_\pi = E_\pi$;
(2) $C_\pi = P_\pi$;
(3) $P_\pi = E_\pi$.

In the theory of classes of groups and the theory of properties E_π, C_π, and D_π, there are important questions of closedness under the class-theoretic operations S, H, S_n, R_0, N_0, D_0, E, E_Z, E_Φ, and P defined by their actions on a class $\mathfrak{X}$ of finite groups as follows:

$\text{s}\mathfrak{X} = \{G \mid G \text{ is isomorphic to a subgroup of some group } H \in \mathfrak{X}\}$;
$\text{H}\mathfrak{X} = \{G \mid G \text{ is an epimorphic image of some group } H \in \mathfrak{X}\}$;
$\text{s}_n\mathfrak{X} = \{G \mid G \text{ is isomorphic to a subnormal subgroup of some group } H \in \mathfrak{X}\}$;
$\text{R}_0\mathfrak{X} = \{G \mid \exists N_i \trianglelefteq G \text{ such that } G/N_i \in \mathfrak{X} \text{ and } \bigcap_{i=1}^m N_i = 1 \ (i = 1, \ldots, m)\}$;
$\text{N}_0\mathfrak{X} = \{G \mid \exists N_i \text{ sn } G \text{ such that } N_i \in \mathfrak{X} \text{ and } G = \langle N_1, \ldots, N_m \rangle \ (i = 1, \ldots, m)\}$;
$\text{D}_0\mathfrak{X} = \{G \mid G \simeq H_1 \times \cdots \times H_m \text{ for some } H_i \in \mathfrak{X} \ (i = 1, \ldots, m)\}$;
$\text{E}\mathfrak{X} = \{G \mid G \text{ possesses a series } 1 = G_0 \trianglelefteq G_1 \trianglelefteq \cdots \trianglelefteq G_m = G \text{ such that } G_i/G_{i-1} \in \mathfrak{X} \ (i = 1, \ldots, m)\}$;

$E_Z\mathfrak{X} = \{G \mid \exists N \trianglelefteq G$ such that $N \leq Z_\infty(G)$ and $G/N \in \mathfrak{X}\}$;
$E_\Phi\mathfrak{X} = \{G \mid \exists N \trianglelefteq G$ such that $N \leq \Phi(G)$ and $G/N \in \mathfrak{X}\}$;
$P\mathfrak{X} = \{G \mid \forall H < \cdot\, G\ G/H_G \in \mathfrak{X}\}$, where $H < \cdot\, G$ means that H is maximal in G.
In [159], Guo and Revin proved the following

Theorem 6.19 (Guo, Revin [159]). *(A) If* $c \in \{H, R_0, D_0, E_Z, E_\Phi\}$, *then* $cP_\pi = P_\pi$
for every set π *of prime numbers.*
　　(B) If $c \in \{s, s_n, N_0, E\}$, *then* $cP_\pi \neq P_\pi$ *for some set* π *of prime numbers.*
　　(C) If $c \in \{s_n, E\}$ *and* $cP_\pi = P_\pi$ *for some set* π *of prime numbers, then* $cE_\pi = E_\pi$
and $cC_\pi = C_\pi$.

Corollary 6.20 *For* $c \in \{s, H, s_n, R_0, N_0, D_0, E, E_Z, E_\Phi\}$, *the following are equivalent:*
　　(1) $cP_\pi = P_\pi$ *for every set* π *of prime numbers;*
　　(2) $cE_\pi = E_\pi$ *and* $cC_\pi = C_\pi$ *for every set* π *of prime numbers.*

Recall that a $\langle H, P \rangle$-closed class is called a *Schunk class*. It is known that each saturated
formation is a Schunk class.

Corollary 6.21 *For each set* π *of prime numbers, the class* P_π *is a saturated
formation, but it is not a Fitting class.*

Corollary 6.22 *For every set* π *of prime numbers,* $PP_\pi = P_\pi$. *In particular,* P_π *is
a Schunk class.*

Problem 6.23 (Revin, Vdovin). *a) In a finite simple group, are Hall subgroups
always pronormal?
　　b) In a finite simple group, are Hall subgroups always strongly pronormal?
　　c) In a finite group, is a dispersive Hall subgroup always strongly pronormal?
　　Note that an affirmative answer to (a) is known modulo CFSG for Hall subgroups
of odd order (see [121]), and the affirmative answer to (a) was given by E.P.Vdovin,
D.O.Revin in [404].
　　Note also that there exist finite (non-simple) groups with a non-pronormal Hall
subgroup and dispersive Hall subgroups are known to be pronormal.

Problem 6.24 (Revin, Vdovin). Let π be a set of primes. Is it true that in any
D_π-group G there are three Hall π-subgroups whose intersection coincides with
$O_\pi(G)$?

Problem 6.25 (Revin, Vdovin). Is every Hall subgroup of a finite group pronormal
in its normal closure?

Problem 6.26 (Guo, Revin). Is it true that if $sP_\pi = P_\pi$ for some set π of prime
numbers, then $sE_\pi = E_\pi$ and $sC_\pi = C_\pi$?

Problem 6.27 (Guo, Revin). Is it true that if $N_0 P_\pi = P_\pi$ for some set π of prime
numbers, then $N_0 E_\pi = E_\pi$?

Problem 6.28 (Guo, Revin). Is it true that if $s_n C_\pi = C_\pi$ for some set π of prime
numbers, then $s_n P_\pi = P_\pi$?

Problem 6.29 (Guo, Revin). Is it true that for every set π of prime numbers $s_n P_\pi = P_\pi$ if and only if $E_\pi = C_\pi$?

VIII. In 1959, Wielandt [445] formulated the following conjecture.

(6.30) Wielandt Conjecture. Suppose that a set of primes π is a union of disjoint subsets σ and τ, and a finite group G possesses a π-Hall subgroup $H = H_\sigma \times H_\tau$, where H_σ and H_τ are σ- and τ-subgroups of H, respectively. If G satisfies both D_σ and D_τ, then G satisfies D_π.

This conjecture was mentioned in [238, 351, 353, 356] and was investigated in [351, 353, 356]. Recently, Guo, Revin, and Vdovin [160] proved the following theorem which completely confirms Wielandt's conjecture.

Theorem 6.31 (Guo, Revin and Vdovin [160]). *Let a set π of primes be a union of disjoint subsets σ and τ. Assume that a finite group G possesses a π-Hall subgroup $H = H_\sigma \times H_\tau$, where H_σ and H_τ are σ- and τ-subgroups, respectively. Then G satisfies D_π if and only if G satisfies both D_σ and D_τ.*

IX. There are many criterions of supersolubility in terms of 2-maximal subgroups. In fact, Theorem 1.3 in [180] was the first result base on X-semipermutability in which it was given the sufficient and necessary condition for a group to be supersoluble. Theorem 5.10 in Sect. 2.5 is a stronger version of Theorem 1.3 in [180].

X. A subgroup H of G is called semi-normal in G [397] if there exists a subgroup K of G such that $G = HK$ and H permutes with every proper subgroup of K. It is clear that a seminormal subgroup of G is a 1-semipermutable subgroup of G.

XI. A group G is called a *T-group* (respectively *PT-group*, *PST-group*) if normality (respectively permutability, S-permutability) is a transitive relation, that is, if H and K are subgroups of G such that H is normal (respectively permutable, s-permutable) in K and K is normal (resp. permutable, s-permutable) in G, then H is normal (respectively permutable, s-permutable) in G. By Kegel's result [263], a group G is a PST-group if and only if every subnormal subgroup of G is s-permutable in G. Agrawal [1] showed that a group G is a solvable PST-group if and only if the nilpotent residual L of G is a normal abelian Hall subgroup of G upon which G acts by conjugation as power automorphisms. In particular, a solvable PST-group is supersolvable. A subgroup H of a group G is said to be *SS-permutable* or *SS-quasinormal* [289] (respectively *NSS-permutable* [67]) in G if H has a supplement (normal supplement) K in G such that H permutes with every Sylow subgroup of K. We say that a group G is an *SST-group* (resp. *NSST-group*) if SS-permutability (resp. NSS-permutability) is a transitive relation.

In [67], the following theorems were proved.

Theorem 6.32 (Chen, Guo [67]). *Let G be a group. Then the following statements are equivalent:*

(1) *G is solvable and every subnormal subgroup of G is SS-permutable in G.*
(2) *G is solvable and every subnormal subgroup of G is NSS-permutable in G.*
(3) *Every subgroup of $F^*(G)$ is SS-permutable in G.*
(4) *Every subgroup of $F^*(G)$ is NSS-permutable in G.*
(5) *G is a solvable PST-group.*

Theorem 6.33 (Chen, Guo [67]). *Let G be a solvable group. Then the following statements are equivalent:*

(1) *G is an SST-group.*
(2) *G is an NSST-group.*
(3) *Every subgroup of G is SS-permutable in G.*
(4) *Every subgroup of G is NSS-permutable in G.*
(5) *Every subgroup of G of prime power order is SS-permutable in G.*
(6) *Every subgroup of G of prime power order is NSS-permutable in G.*
(7) *Every cyclic subgroup of G of prime power order is SS-permutable in G.*
(8) *Every cyclic subgroup of G of prime power order is NSS-permutable in G.*

XII. In connection with Theorems 4.6, 4.12, 4.14, 4.15, 4.16, 4.17, and 6.1, the following problems seem natural:

Problem 6.34 What is the structure of the non-supersoluble group which is a product of two supersoluble subgroups?

The Problem (6.34) maybe is too difficult since up to now we do not know the answer even to the following its special cases.

Problem 6.35 What is the structure of the minimal non-supersoluble group which is a product of two normal supersoluble subgroups?

Problem 6.36 Can we describe the structure of the non-supersoluble group which is a products of two normal supersoluble subgroups?

More general, we have the following problem.

Problem 6.37 For some classes $\mathfrak{X}$ (for example, the class $\mathfrak{U}_p$ of all p-supersoluble groups, the class of all metanilpotent groups, and so on), can we describe the structure of the non-$\mathfrak{X}$-group which is a products of two (normal) $\mathfrak{X}$-subgroups?

XIII. A classic Hall's result states: a finite group G is soluble if and only if G has a Sylow basis, that is, the Sylow subgroups in some set of Sylow subgroups $\{P_1, P_2, \ldots, P_k\}$ of G are pairwise permutable. Hence it is very interesting to study the structure of finite groups in which some Hall subgroups are permutable.

Chapter 3
Between Complement and Supplement

If $G = HB$, then B is said to be a *supplement* of H in G. Since $HG = G$, it makes sense to consider only the supplements B with some restrictions on B. For example, we often deal with the situation when for a supplement B of H we have $H \cap B = 1$. In this case, B is said to be a *complement* of H in G and H is said to be *complemented* in G; if, in addition, B is also normal in G, then B is said to be a *normal complement* of H in G. Note that if H either is normal in G or has a normal complement in G, then, clearly, H satisfies the following: there are normal subgroups T and S of G such that $G = HT$, $S \leq H$ and $H \cap T \leq S$ (see Ore [324]); in this case we say H satisfies the *Ore supplement condition* in G. It is clear that H satisfies the Ore supplement condition if and only if H/H_G has a normal complement in G/H_G. In the paper [435], the subgroups which satisfy the Ore supplement condition were called *c-normal*.

It was discovered that many classes of groups that are important for applications (for example, the classes of all soluble, supersoluble, nilpotent, p-nilpotent, metanilpotent, dispersive in the sense of Ore [324] groups) may be characterized in the terms of the Ore supplement condition or in the terms of some generalized Ore supplement conditions. It was the main motivation for introducing, studying, and applying the generalized Ore supplement conditions of various type.

The first some generalizations of the Ore supplement condition was considered in [45, 141, 383]: A subgroup H of G is said to be *c-supplemented* [45] in G if G has a subgroup T such that $HT = G$ and $T \cap H \leq S \leq H$ for some normal subgroup S of G; A subgroup H of G is said to be $\mathfrak{F}$-*supplemented* [141] in G provided there is a supplement T/H_G of H/H_G in G/H_G such that $(H/H_G) \cap (T/H_G) \leq Z_{\mathfrak{F}}(G/H_G)$; A subgroup H of G is said to be *weakly s-supplemented* [383] in G if G has a subgroup T such that $HT = G$ and $T \cap H \leq S \leq H$ for some s-quasinormal subgroup S of G. It is clear that if H is either complemented or s-quasinormal in G, then H is weakly s-supplemented in G.

In this chapter, we consider some new generalizations of the Ore supplement condition, which, in fact, unify and generalize the most part of them. Thus the theorems in this chapter cover main result of a long list of publications (see Sect. 3.5 below).

© Springer-Verlag Berlin Heidelberg 2015
W. Guo, *Structure Theory for Canonical Classes of Finite Groups*,
DOI 10.1007/978-3-662-45747-4_3

3.1 Base Concepts and Lemmas

Supplement Conditions I, I* , and I**

Definition 1.1 Let $\mathfrak{X}$ be a non-empty class of groups and $K \leq H$ subgroups of a group G. Then we say that the pair (K, H) satisfies:

$\mathfrak{X}$-*supplement condition* (I) in G if G has a subgroup T such that $G = HT$ and

$$H \cap T \subseteq K \mathrm{Int}_{\mathfrak{X}}(T).$$

$\mathfrak{X}$-*supplement condition* (I^*) in G if G has a subgroup T such that $HT = G$ and

$$H \cap T \subseteq K Z_{\mathfrak{X}}(T).$$

$\mathfrak{X}$-*supplement condition* (I^{**}) in G if G has a K-$\mathfrak{X}$-subnormal subgroup T such that $HT = G$ and

$$H \cap T \subseteq K Z_{\mathfrak{X}}(T).$$

Note that if $\mathfrak{F}$ is a hereditary saturated formation, then $Z_{\mathfrak{F}}(G) \leq Int_{\mathfrak{F}}(G)$.

Lemma 1.2 *Let $\mathfrak{X}$ be a non-empty class of groups, $K \leq H$ subgroups of G and $N \trianglelefteq G$. Suppose that the pair (K, H) satisfies the $\mathfrak{X}$-supplement condition (I^*) in G (the $\mathfrak{X}$-supplement condition (I) in G).*

(1) *If either $N \leq H$ or $(|H|, |N|) = 1$, then the pair $(KN/N, HN/N)$ satisfies the $\mathfrak{X}$-supplement condition (I^*) (the $\mathfrak{X}$-supplement condition (I), respectively) in G/N.*

(2) *If $H \leq E \leq G$ and $\mathfrak{X}$ is a hereditary saturated formation, then the pair (K, H) satisfies the $\mathcal{X}$-supplement condition (I^*) (the $\mathfrak{X}$-supplement condition (I), respectively) in E.*

(3) *If $K \leq V \leq H$, then the pair (V, H) satisfies $\mathfrak{X}$-supplement condition (I^*) (the $\mathfrak{X}$-supplement condition (I), respectively) in G.*

Proof First suppose that the pair (K, H) satisfies the $\mathfrak{F}$-supplement condition (I^*) in G. Let T be a subgroup of G such that $HT = G$ and $H \cap T \subseteq K Z_{\mathfrak{X}}(T)$.

(1) Clearly, $(HN/N)(TN/N) = G/N$. Moreover, $TN \cap HN = (T \cap H)N$. Indeed, if either $N \leq T$ or $N \leq H$, it is clear. But in the case when $N \nleq H$, we have $N \leq T$ since in this case $(|H|, |N|) = 1$ by hypothesis. Hence

$$(HN/N) \cap (NT/N) = (NH \cap NT)/N = N(H \cap T)/N \subseteq N(K Z_{\mathfrak{X}}(T))/N$$

$$= (NK/N)((Z_{\mathfrak{X}}(T)N/N) \leq (KN/N)(Z_{\mathfrak{X}}(TN/N))$$

by Chap. 1, Theorem 2.6. Hence the pair $(KN/N, HN/N)$ satisfies the $\mathfrak{X}$-supplement condition (I^*) in G/N.

(2) Let $T_0 = T \cap E$. Then $E = E \cap HT = H(T \cap E) = HT_0$. Moreover,

$$T_0 \cap H = T \cap H \subseteq K Z_{\mathfrak{X}}(T) \leq K(Z_{\mathfrak{X}}(T) \cap E) \leq K(Z_{\mathfrak{X}}(T \cap E) = K Z_{\mathfrak{X}}(T_0)$$

by Chap. 1, Theorem 2.7(a). This shows that the pair (K, H) satisfies the $\mathfrak{X}$-supplement condition (I^*) in E.

(3) This is evident.

If the pair (K, H) satisfies the $\mathfrak{F}$-supplement condition (I) in G, then all assertions of the lemma may be proved analogously by using Chap. 1, Proposition 5.6.

Lemma 1.3 *Let $\mathfrak{F}$ be a nonempty formation. Let H be a K-$\mathfrak{F}$-subnormal subgroup of a group G.*

(1) *If K is a K-$\mathfrak{F}$-subnormal subgroup of H, then K is K-$\mathfrak{F}$-subnormal in G* (see [34, 6.1.6(1)]).

(2) *If $G^{\mathfrak{F}} \leq K \leq G$, then K is K-$\mathfrak{F}$-subnormal in G* (see [34, 6.1.7(1)]).

(3) *If U/N is K-$\mathfrak{F}$-subnormal in G/N, then U is K-$\mathfrak{F}$-subnormal in G* (see [34, 6.1.6(2)]).

Lemma 1.4 *Let $\mathfrak{X}$ be a nonempty class of groups, $K \leq H$ subgroups of G and $N \trianglelefteq G$. Suppose that the pair (K, H) satisfies the $\mathfrak{X}$-supplement condition (I^{**}) in G.*

(1) *If either $N \leq H$ or $(|H|, |N|) = 1$, then the pair $(KN/N, HN/N)$ satisfies the $\mathfrak{X}$-supplement condition (I^{**}) in G/N.*

(2) *If $H \leq E \leq G$ and $\mathfrak{X}$ is a hereditary saturated formation, then the pair (K, H) satisfies the $\mathfrak{X}$-supplement condition (I^{**}) in E.*

(3) *If $K \leq V \leq H$, then the pair (V, H) satisfies $\mathfrak{X}$-supplement condition (I^{**}) in G.*

Proof See the proof of Lemma 1.2 and use Lemma 1.3 and Chap. 1, Lemma 5.12.

Inductive Subgroup Functors Recall that a *subgroup functor* is a function τ which assigns to each group G a set $\tau(G)$ subgroups of G satisfying $1 \in \tau(G)$ and $\theta(\tau(G)) = \tau(\theta(G))$ for any isomorphism $\theta : G \to G^*$. If $H \in \tau(G)$, then we say that H is a τ-*subgroup* of the group G. If τ_1 and τ_2 are subgroup functors such that for any group G we have $\tau_1(G) \subseteq \tau_2(G)$, then we write $\tau_1 \leq \tau_2$. In the following definition we collect some type of subgroup functors which are the most useful for various applications.

Definition 1.5 Let τ be a subgroup functor. Then we say that τ is:

(1) *Inductive* provided τ enjoys the following properties:

(A) If $N \trianglelefteq G$, then $N \in \tau(G)$.

(B) If $N \trianglelefteq G$ and $H \in \tau(G)$ then $HN/N \in \tau(G/N)$.

(2) *Strongly* inductive provided τ enjoys the following properties:

(C) If $N \trianglelefteq G$ and $H \in \tau(G)$ then $HN \in \tau(G)$.

(3) *Regular* or a *Li-subgroup functor* [277] if for any group G, whenever $H \in \tau(G)$ is a p-group and N is a minimal normal subgroup of G, then $|G : N_G(H \cap N)|$ is a power of p.

(4) *Quasiregular* if for any group G, whenever $H \in \tau(G)$ is a p-group and N is an abelian minimal normal subgroup of G, then $|G : N_G(H \cap N)|$ is a power of p.

(5) Φ-*regular* (Φ-*quasiregular*, respectively) if for any primitive group G, whenever $H \in \tau(G)$ is a p-group and N is a minimal normal subgroup (an abelian minimal normal subgroup, respectively) of G, then $|G : N_G(H \cap N)|$ is a power of p.

Example 1.6 Let $\tau(G)$ be the set of all normal subgroups of G for any group G. Then, clearly, τ is a regular strongly inductive subgroup functor.

Example 1.7 Let $\tau(G)$ be the set of all subnormal subgroups of G for any group G. Then τ is a strongly inductive subgroup functor by Chap. 1, Lemma 5.35(1)(8).

Example 1.8 Let $\tau(G)$ be the set of all s-quasinormal subgroups of G for any group G. It is clear that τ is strongly inductive (see Chap. 1, Lemma 5.34 (2)). Now let a p-subgroup H of G be s-quasinormal in G. Let N be a minimal normal subgroup of G. Assume that $L = H \cap N \neq 1$. Then, by Chap. 1, Lemma 5.34 (1)(4) and Chap. 1, Lemma 5.35 (6), N is a p-group. Now let Q be any Sylow q-subgroup of G, where $q \neq p$. Then $HQ = QH$ is a subgroup of G. Hence $L = HQ \cap N \trianglelefteq HQ$, so $Q \leq N_G(L)$. Hence $|G : N_G(L)|$ is a power of p. Therefore the functor τ is regular.

A subgroup H is said to be s-*quasinormally (respectively subnormally) embedded* [41] in a group G if every Sylow subgroup of H is also a Sylow subgroup of some s-quasinormal (respectively subnormal) subgroup of G. Note that in view of Kegel's result [263], every s-quasinormal subgroup is subnormal, so every s-quasinormally embedded subgroup is also subnormally embedded.

Example 1.9 If $\tau(G)$ be the set of all s-quasinormally embedded subgroups of G for any group G, then τ is hereditary strongly inductive by [41, Lemma 1] and τ is quasiregular (see Example 1.8 and [32, 1.2.19]). It is clear that this functor is not regular since every Sylow subgroup is s-quasinormally embedded.

Example 1.10 For any group G, let $\tau(G)$ be the set of all CAP-subgroups of G. Then τ is regular strongly inductive by Lemma 1 in [98].

A subgroup H is called a CAP^*-subgroup of G if H either covers or avoids each non-Frattini chief factor of G [287].

Example 1.11 For any group G, let $\tau(G)$ be the set of all CAP^*-subgroups of G. Then τ is Φ-regular and inductive by Lemma 2.1 in [287].

Example 1.12 Let $\tau(G)$ be the set of all $CAMP$-subgroups of G for any group G (see Definition IV, (1.1)(2) below). Then τ is strongly inductive (see Chap. 4, Lemma 1.8 and Lemma 1.9 below). Now let H be a p-subgroup of a primitive group G which is a $CAMP$-subgroup of G. Then H is subnormal in G (see Chap. 4, Lemma 1.9(a) below). Let N be a minimal normal subgroup of G. Suppose that $L = H \cap N \neq 1$. Then by Chap. 1, Lemma 5.35(6), N is a p-group, so NH is a subnormal p-subgroup of G by Chap. 1, Lemma 5.35(8). Now let M be a maximal subgroup of G such that $M_G = 1$. Then $G = N \rtimes M$ and H either covers or avoids the pair (M, G). But since $L = H \cap N \neq 1$, $H \nleq M$ and so $G = HM$. On the other hand, by Chap. 1, Lemma 2.14. $NH \cap M = 1$ since $M_G = 1$. Therefore $NH = N = H$. Hence τ is Φ-regular.

Example 1.13 Let $\tau(G)$ be the set of all s-c-permutable subgroups of G for any group G (see Chap. 2, Def. 1.4(4)). Then τ is a strongly inductive subgroup functor by Chap. 2, Lemma 1.5(3). Arguing as in Example 1.8, one can show that τ is quasiregular.

Example 1.14 For any group G, let $\tau(G)$ be the set of all completely c-permutable subgroups of G (see Chap. 2, Def. 1.4(3)). Then τ is a strongly inductive subgroup functor by Chap. 2, Lemma 1.5(3)(8), and this functor is quasiregular (see Example 1.13).

A subgroup H is called SS-quasinormal in G [289] if there is a subgroup B of G such that $HB = G$ and H permutes with all Sylow subgroups of B.

Example 1.15 Let $\tau(G)$ be the set of all SS-quasinormal subgroups of G for any group G. Then τ is a strongly inductive subgroup functor by [289, Lemma 2.1]. Arguing as in Example 1.8, one can also show that τ is regular (see [66, Lemma 7.1(6)]). This functor is also inductive and hereditary by Lemma 2.1 in [289].

A subgroup H is called S-propermutable (S-quasipermutable) in G [462] if there is a subgroup B of G such that $N_G(H)B = G$ and H permutes with all Sylow subgroups of B (with all Sylow subgroups P of B such that $(|H|, |P|) = 1$, respectively).

Example 1.16 Let $\tau(G)$ be the set of all S-propermutable (S-quasipermutable) subgroups of G for any group G. Then τ is a strongly inductive quasiregular subgroup functor by Lemma 2.3(1) and 2.3(4) in [462].

Definition 1.17 Let $\mathfrak{X}$ be a nonempty class of groups and H a subgroup of a group G. Let $\bar{G} = G/H_G$ and $\bar{H} = H/H_G$. Let τ be a subgroup functor. Then we say that that H is:

(1) *Weakly $\mathfrak{X}_\tau$-supplemented* in G if for some τ-subgroup $\bar{S}$ of $\bar{G}$ contained in $\bar{H}$ the pair $(\bar{S}, \bar{H})$ satisfies the $\mathfrak{X}$-supplement condition (I) in $\bar{G}$.
(2) *$\mathfrak{X}_\tau$-supplemented* in G if for some τ-subgroup $\bar{S}$ of $\bar{G}$ contained in $\bar{H}$ the pair $(\bar{S}, \bar{H})$ satisfies the $\mathfrak{X}$-supplement condition (I^*) in $\bar{G}$.
(3) *Strongly $\mathfrak{X}_\tau$-supplemented* in G if for some τ-subgroup S of G contained in H the pair (S, H) satisfies the $\mathfrak{X}$-supplement condition (I^{**}) in G.

Example 1.18 A subgroup H of a group G is said to be c-supplemented in G [45, 436] if G has a subgroup T such that $HT = G$ and $T \cap H \leq H_G$. Note that in this case $TH_G \cap H = H_G$. Hence every c-supplemented subgroup is $\mathfrak{X}_\tau$-supplemented and weakly $\mathfrak{X}_\tau$-supplemented, for any class $\mathfrak{X}$ of groups and for any inductive subgroup functor τ.

Example 1.19 Let H be a group of a group G. If H is c-normal in G (see [435]), then H is strongly $\mathfrak{X}_\tau$-supplemented, $\mathfrak{X}_\tau$-supplemented and weakly $\mathfrak{X}_\tau$-supplemented as well, for any class $\mathfrak{X}$ of groups and for any subgroup functor τ. Also, if H has a supplement $T \in \mathfrak{X}$ in G, then H is $\mathfrak{X}_\tau$-supplemented for any class $\mathfrak{X}$ and any subgroup functor τ, and is weakly $\mathfrak{X}_\tau$-supplemented for any hereditary saturated formation $\mathfrak{X}$ of groups and for any inductive subgroup functor τ by Chap. 1, Theorem 2.6(e).

Example 1.20 A subgroup H of a group G is said to be weakly s-permutable in G [383] if G has a subnormal subgroup T such that $HT = G$ and $T \cap H \leq H_{sG}$. It is easy to see that every weakly s-permutable subgroup is strongly $\mathfrak{X}_\tau$-supplemented, $\mathfrak{X}_\tau$-supplemented and weakly $\mathfrak{X}_\tau$-supplemented as well, for any class $\mathfrak{X}$ of groups and for the subgroup functor in Example 1.7.

Example 1.21 A subgroup H of a group G is said to be weakly c-permutable in G [146] if G has a subgroup T such that $HT = G$ and $T \cap H$ is completely c-permutable in G. It is clear that H is $\mathfrak{X}_\tau$-supplemented and weakly $\mathfrak{X}_\tau$-supplemented as well, for any class $\mathfrak{X}$ of groups and for the subgroup functor in Example 1.14.

Example 1.22 A subgroup H of a group G is said to be $\mathfrak{F}$-supplemented in G [141] if G has a subgroup T such that $HT = G$ and $(T \cap H)H_G/H_G \leq Z_{\mathfrak{F}}(G/H_G)$ Note that if $\mathfrak{F}$ is a hereditary saturated formation, then $Z_{\mathfrak{F}}(G/H_G) \cap (TH_G/H_G) \leq Z_{\mathfrak{F}}(TH_G/H_G)$ by Chap. 1, Theorem 2.7(1). Hence every $\mathfrak{F}$-supplemented subgroup is $\mathfrak{F}_\tau$-supplemented and so weakly $\mathfrak{X}_\tau$-supplemented by Chap. 1, Proposition 5.6(h), for any hereditary saturated formation $\mathfrak{F}$ and any subgroup functor τ.

Example 1.23 A subgroup H of a group G is said to be m-embedded in G [196] if G has a subnormal subgroup T such that $HT = G$ and $T \cap H \leq C \leq H$ for some $CAMP$-subgroup C of G. Every m-embedded subgroup is strongly $\mathfrak{X}_\tau$-supplemented, for any class $\mathfrak{X}$ of groups and for the subgroup functor τ in Example 1.12.

Example 1.24 A subgroup H of a group G is said to be nearly m-embedded in G [196] if G has a subgroup T such that $HT = G$ and $T \cap H \leq C \leq H$ for some $CAMP$ subgroup of C of G. Every nearly m-embedded subgroup is $\mathfrak{X}_\tau$-supplemented, for any class $\mathfrak{X}$ of groups and for the subgroup functor τ in Example 1.12.

Example 1.25 A subgroup H of a group G is said to be a CAS-subgroup of G [440] if G has a subgroup T such that $HT = G$ and $T \cap H$ is a CAP-subgroup of G. Every CAS-subgroup is $\mathfrak{X}_\tau$-supplemented, for any class $\mathfrak{X}$ of groups and for the subgroup functor τ in Example 1.10.

Example 1.26 A subgroup H of a group G is said to be weakly SS-permutable in G [239] if G has a subnormal subgroup T such that $HT = G$ and $T \cap H \leq S \leq H$ for some SS-quasinormal subgroup S of G. It is easy to see that every weakly SS-permutable subgroup is $\mathfrak{X}_\tau$-supplemented and weakly $\mathfrak{X}_\tau$-supplemented as well, for any class $\mathfrak{X}$ of groups and for the subgroup functor τ in Example 1.15.

Example 1.27 A subgroup H of a group G is said to be c-quasinormal in G [391] if G has a quasinormal subgroup T such that $HT = G$ and $T \cap H \leq S \leq H$ for some quasinormal subgroup S of G. It is clear that every c-quasinormal subgroup is $\mathfrak{X}_\tau$-supplemented and weakly $\mathfrak{X}_\tau$-supplemented as well, for any class $\mathfrak{X}$ of groups, and for the subgroup functor τ, where $\tau(G)$ is the set of all quasinormal subgroups.

Example 1.28 A subgroup H of a group G is said to be weakly s-supplemented in G [383] if G has a subgroup T such that $HT = G$ and $T \cap H \leq S \leq H$ for some s-quasinormal subgroup S of G. It is easy to see that every weakly S-supplemented subgroup is $\mathfrak{X}_\tau$-supplemented and weakly $\mathfrak{X}_\tau$-supplemented as well, for any class $\mathfrak{X}$ of groups, and for the subgroup functor τ in Example 1.8.

Example 1.29 If either $H \leq Z_{\mathfrak{X}}(G)$ or $H \in \tau(G)$, then H is $\mathfrak{X}_\tau$-supplemented and strongly $\mathfrak{X}_\tau$-supplemented, for any class $\mathfrak{X}$ of groups and for any subgroup functor τ.

Example 1.30   A subgroup H of G is said to be c^*-normal in G [439] if there is a normal subgroup K of G such that G=HK and $H \cap K$ is s-quasinormally embedded in G. It is clear that every c^*-normal subgroup is strongly $\mathfrak{X}_\tau$-supplemented, $\mathfrak{X}_\tau$-supplemented and weakly $\mathfrak{X}_\tau$-supplemented as well, for any class $\mathfrak{X}$ of groups and for the subgroup functor τ in Example 1.9.

Lemma 1.31   *Let $\mathfrak{F}$ be a nonempty hereditary solubly saturated formation and τ an inductive subgroup functor. Let G be a group, $H \leq G$ and N a normal subgroup of G.*

(1) *If $N \leq H$, then H/N is (weakly) $\mathfrak{F}_\tau$-supplemented in G/N if and only if H is $\mathfrak{F}_\tau$-supplemented (weakly $\mathfrak{F}_\tau$-supplemented, respectively) in G.*

(2) *Suppose that S is a τ-subgroup of G such that $S \leq H$ and the pair (S, H) satisfies the $\mathfrak{F}$-supplement condition (I^*) ($\mathfrak{F}$-supplement condition (I), respectively) in G. Then H is $\mathfrak{F}_\tau$-supplemented (weakly $\mathfrak{F}_\tau$-supplemented, respectively) in G.*

(3) *If H is (weakly) $\mathfrak{F}_\tau$-supplemented in G, and either $N \leq H$ or $(|H|, |N|) = 1$, then HN/N is $\mathfrak{F}_\tau$-supplemented (weakly $\mathfrak{F}_\tau$-supplemented, respectively) in G/N.*

Proof   For any subgroup V of G, we put $\bar{V} = VN/N$ and $\widehat{V} = VH_G/H_G$.

(1) First suppose that H/N is $\mathfrak{F}_\tau$-supplemented in G/N. Let f be the canonical isomorphism from $(G/N)/(H_G/N)$ onto G/H_G.

Let $\bar{T}/\bar{H}_{\bar{G}}$ be a subgroup of $\bar{G}/\bar{H}_{\bar{G}}$ such that $\bar{G}/\bar{H}_{\bar{G}} = (\bar{H}/\bar{H}_{\bar{G}})(\bar{T}/\bar{H}_{\bar{G}})$ and

$$(\bar{H}/\bar{H}_{\bar{G}}) \cap (\bar{T}/\bar{H}_{\bar{G}}) \subseteq (\bar{S}/\bar{H}_{\bar{G}})Z_{\mathfrak{F}}(\bar{T}/\bar{H}_{\bar{G}})$$

for some τ-subgroup $\bar{S}/\bar{H}_{\bar{G}}$ of $\bar{G}/\bar{H}_{\bar{G}}$ contained in $\bar{H}/\bar{H}_{\bar{G}}$.
Then

$$f((\bar{H}/\bar{H}_{\bar{G}}) \cap (\bar{T}/\bar{H}_{\bar{G}})) \subseteq f(\bar{S}/\bar{H}_{\bar{G}})f(Z_{\mathfrak{F}}(\bar{T}/\bar{H}_{\bar{G}}).$$

Note that

$$f((\bar{H}/\bar{H}_{\bar{G}}) \cap (\bar{T}/\bar{H}_{\bar{G}})) = \widehat{H} \cap \widehat{T}$$

and $\bar{H}_{\bar{G}} = (H/N)_{G/N} = H_G/N$.
On the other hand,

$$f(\bar{S}/\bar{H}_{\bar{G}}) = S/H_G = \widehat{S},$$

where $\widehat{S}$ is a τ-subgroup of $\widehat{G}$ since $\bar{S}/\bar{H}_{\bar{G}}$ is a τ-subgroup of $\bar{G}/\bar{H}_{\bar{G}}$.
Similarly, $f(Z_{\mathfrak{F}}(\bar{T}/\bar{H}_{\bar{G}})) = Z_{\mathfrak{F}}(\widehat{T})$. Therefore H is $\mathfrak{F}_\tau$-supplemented in G.
Now suppose that H is $\mathfrak{F}_\tau$-supplemented in G. Then by considering the canonical isomorphism f^{-1} from G/H_G onto $(G/N)/(H_G/N)$, one can prove analogously that H/N is $\mathfrak{F}_\tau$-supplemented in G/N. The second assertion of (1) can be proved similarly.

(2) First suppose that the pair (S, H) satisfies the $\mathfrak{F}$-supplement condition (I^*) in G. Then, by Lemma 1.2(3), the pair (S, H) satisfies the $\mathfrak{F}$-supplement condition (I^*) in G. Hence by Lemma 1.2(1), $(S/H_G, H/H_G)$ satisfies the $\mathfrak{F}$-supplement condition (I^*) in G. Thus H is $\mathfrak{F}_\tau$-supplemented in G. The second assertion of (2) can be proved similarly by using Lemma 1.4.

(3) Suppose that H is $\mathfrak{F}_\tau$-supplemented in G and let $\widehat{S}$ be a τ-subgroup of $\widehat{G}$ contained in $\widehat{H}$ such that the pair $(\widehat{S}, \widehat{H})$ satisfies the $\mathfrak{F}$-supplement condition (I^*) in $\widehat{G}$.

Then the pair $(\widehat{S}\widehat{N}/\widehat{N}, \widehat{H}\widehat{N}/\widehat{N})$ satisfies the $\mathfrak{F}$-supplement condition (I^*) in $\widehat{G}/\widehat{N}$ by Lemma 1.2(1). Let h be the canonical isomorphism from $(G/H_G)/(H_G N/H_G)$ onto $G/N H_G$. Then $h((\widehat{S}\widehat{N}/\widehat{N}) = SN/N H_G$ and $h(\widehat{H}\widehat{N}/\widehat{N}) = HN/N H_G$. Hence the pair $(SN/N H_G, HN/N H_G)$ satisfies the $\mathfrak{F}$-supplement condition (I^*) in $G/N H_G$. Note also that $SN/N H_G$ is a τ-subgroup of $G/N H_G$ since τ is inductive and so $\widehat{S}$ is a τ-subgroup of $\widehat{G}$. Hence by (2), $HN/N H_G$ is $\mathfrak{F}_\tau$-supplemented in $G/N H_G$ and so $(HN/N)/(H_G N)/N$ is $\mathfrak{F}_\tau$-supplemented in $(G/N)/(N H_G/N)$, which implies that HN/N is $\mathfrak{F}_\tau$-supplemented in G/N by (1).

The second assertion of (3) can be proved similarly.

The Supplement Conditions II, II*, and II**

Definition 1.32 Let $\mathfrak{X}$ be a nonempty class of groups and $K \leq H$ subgroups of a group G. Then we say that the pair (K, H) satisfies:

weakly $\mathfrak{X}$-supplement condition (II) in G if for some normal subgroup T of G, the subgroup HT is s-quasinormal in G and

$$H \cap T \leq K \mathrm{Int}_{\mathfrak{X}}(G).$$

$\mathfrak{X}$-supplement condition (II^*) in G if for some normal subgroup T of G, the subgroup HT is s-quasinormal in G and

$$H \cap T \leq K Z_{\mathfrak{X}}(G).$$

supplement condition (II^{**}) in G if for some s-quasinormal subgroup T of G, the subgroup HT is s-quasinormal in G and

$$H \cap T \leq K.$$

Lemma 1.33 *Let $\mathfrak{X}$ be a nonempty class of groups. Let $K \leq H \leq G$ and $N \trianglelefteq G$. Suppose that the pair (K, H) satisfies the $\mathfrak{X}$-supplement condition (II^*) (the supplement condition (II), respectively) in G.*

(1) *If either $N \leq H$ or $(|H|, |N|) = 1$, then the pair $(KN/N, HN/N)$ satisfies the $\mathfrak{X}$-supplement condition (II^*) (the supplement condition (II), respectively) in G/N.*

(2) *If $\mathfrak{X}$ is a hereditary saturated formation and $H \leq E$, then (K, H) satisfies the $\mathfrak{F}$-supplement condition (II^*) (the supplement condition (II), respectively) in E.*

(3) *If $K \leq V \leq H$, then the pair (V, H) satisfies the $\mathfrak{X}$-supplement condition (II^*) (the supplement condition (II), respectively) in G.*

Proof First suppose that the pair (K, H) satisfies the $\mathfrak{F}$-supplement condition (II^*) in G. Let T be a normal subgroup of G such that HT is s-quasinormal in G and $H \cap T \leq K Z_{\mathfrak{F}}(G)$.

(1) It is clear that TN/N is normal in G/N. On the other hand, by Chap. 1, Lemma 5.34(2), the subgroup $(HN/N)(TN/N) = HTN/N$ is s-quasinormal in G/N. Moreover, $TN \cap HN = (T \cap H)N$. Indeed, if either $N \leq T$ or $N \leq H$, it is clear. Now assume that neither $N \leq T$ nor $N \leq H$. Then $(|H|, |N|) = 1$ by hypothesis. Therefore $H \cap TN = H \cap T$, so $TN \cap HN = N(TN \cap H) = N(T \cap H)$. Hence $(HN/N) \cap (NT/N) = (NH \cap NT)/N = N(H \cap T)/N \leq NKZ_{\mathfrak{F}}(G)/N = (NK/N)(Z_{\mathfrak{F}}(G)N/N) \leq (KN/N)Z_{\mathfrak{F}}(G/N)$ by Chap. 1, Theorem 2.6(d).

(2) Let $T_0 = T \cap E$. Then T_0 is normal in E and $HT_0 = HT \cap E$ is s-quasinormal in E by Chap. 1, Lemma 5.34(4). Finally, $T_0 \cap H = T \cap H \leq KZ_{\mathfrak{F}}(G) \cap E = K(Z_{\mathfrak{F}}(G) \cap E) \leq KZ_{\mathfrak{F}}(E)$ by Chap. 1, Theorem 2.7(a).

(3) This is evident.

If the pair (K, H) satisfies the $\mathfrak{F}$-supplement condition (II) in G, then all assertions of the lemma may be proved analogously by using Chap. 1, Proposition 5.6(a–b).

Definition 1.34 Let $\mathfrak{X}$ be a nonempty class of groups and H a subgroup of a group G. Let $\bar{G} = G/H_G$ and $\bar{H} = H/H_G$. Let τ be a subgroup functor. Then we say that that H is:

(1) Weakly $\mathfrak{X}_\tau$-embedded in G if for some τ-subgroup $\bar{S}$ of $\bar{G}$ contained in $\bar{H}$ the pair $(\bar{S}, \bar{H})$ satisfies the $\mathfrak{F}$-supplement condition (II) in $\bar{G}$.

(2) $\mathfrak{X}_\tau$-embedded in G if for some τ-subgroup $\bar{S}$ of $\bar{G}$ contained in $\bar{H}$ the pair $(\bar{S}, \bar{H})$ satisfies the $\mathfrak{F}$-supplement condition (II^*) in $\bar{G}$.

(3) S_τ-embedded in G if for some τ-subgroup $\bar{S}$ of $\bar{G}$ contained in $\bar{H}$ the pair $(\bar{S}, \bar{H})$ satisfies the $\mathfrak{F}$-supplement condition (II^{**}) in $\bar{G}$.

Example 1.35 A subgroup H of a group G is said to be S-embedded in G [182] if G has a normal subgroup T such that HT is s-quasinormal in G and $T \cap H \leq S \leq H$ for some s-quasinormal subgroup S of G. In this case, the subgroup $(TH_G/H_G)(H/H_G) = TH/H_G$ is s-quasinormal in G/H_G by Chap. 1, Lemma 5.34(2) and

$$(TH_G/H_G) \cap (H/H_G) = (T \cap H)H_G/H_G \leq SH_G/H_G$$

where SH_G/H_G is s-quasinormal in G/H_G by Chap. 1, Lemma 5.34(2). Therefore every S-embedded subgroup is $\mathfrak{X}_\tau$-embedded and weakly $\mathfrak{X}_\tau$-embedded as well, for any class $\mathfrak{X}$ of groups, where τ is the subgroup functor in Example 1.8.

Example 1.36 A subgroup H of a group G is said to be $\mathfrak{X}_s$-normal in G in G [216, 242] if G has a normal subgroup T such that HT is s-quasinormal in G and $(T \cap H)H_G/H_G \leq Z_{\mathfrak{X}}(G/H_G)$. It is clear that every $\mathfrak{X}_s$-normal subgroup is $\mathfrak{X}_\tau$-embedded for every subgroup functor τ and any class $\mathfrak{X}$ of groups.

Example 1.37 A subgroup H of a group G is said to be s-embedded in G [190] if G has as s-quasinormal subgroup T such that HT is s-quasinormal in G and $T \cap H \leq S \leq H$ for some s-quasinormal subgroup S of G. Every s-embedded subgroup is S_τ-embedded , where τ is the subgroup functor in Example 1.8.

Lemma 1.38 *Let $\mathfrak{F}$ be a nonempty hereditary solubly saturated formation and τ an inductive subgroup functor. Let G be a group and N a normal subgroup of G. Suppose that H is (weakly) $\mathfrak{F}_\tau$-embedded in G.*

(1) *If $N \leq H$, then H/N is (weakly) $\mathfrak{F}_\tau$-embedded in G/N if and only if H is (weakly) $\mathfrak{F}_\tau$-embedded in G.*

(2) *Suppose that S is a τ-subgroup of G such that $S \leq H$ and the pair (S, H) a satisfies the $\mathfrak{F}$-supplement condition (II^*) (the $\mathfrak{F}$-supplement condition (II), respectively) in G. Then H is $\mathfrak{F}_\tau$-embedded (weakly $\mathfrak{F}_\tau$-embedded, respectively) in G.*

(3) *If τ is an inductive subgroup functor and either $N \leq H$ or $(|H|, |N|) = 1$, then HN/N is $\mathfrak{F}_\tau$-embedded (weakly $\mathfrak{F}_\tau$-embedded, respectively) in G/N.*

Proof See the proof of Lemma 1.31 and use Lemma 1.33.

Now we also introduce the following two special classes of subgroup functors.

Definition 1.39 Let τ be a subgroup functor. Then we say that τ is:

(1) *Hereditary* provided τ enjoys the following properties:
(A) If $H \in \tau(G)$, H is a p-group for some prime p and E is a subgroup of G containing H, then $H \in \tau(E)$.

(2) *Completely* hereditary provided τ enjoys the following properties:
(B) If $H \in \tau(G)$ and E is a subgroup of G, then $H \cap E \in \tau(E)$.

Example 1.40 Let τ be the subgroup functor in Example 1.6. Then τ is completely hereditary.

Example 1.41 Let $\tau(G)$ be the set of all subnormal subgroups of G for any group G. Then τ is a strongly inductive completely hereditary subgroup functor by Example 1.7 and Chap. 1, Lemma 5.35(2)(2).

Example 1.42 Let $\tau(G)$ be the set of all K-$\mathfrak{F}$-subnormal subgroups of G for any group G, where $\mathfrak{F}$ is a nonempty hereditary formation. Then τ is a strongly inductive and completely hereditary subgroup functor by Lemma 1.3 and Chap. 1, Lemma 5.12(1).

Example 1.43 Let τ be the subgroup functor in Example 1.8. Then τ is completely hereditary by Chap. 1, Lemma 5.34(4).

Example 1.44 Let τ be the subgroup functor in Example 1.12. Then τ is completely hereditary by Chap. 4, Lemmas 1.8 and 1.9.

Example 1.45 Let τ be the subgroup functor in Example 1.14. Then τ is completely hereditary.

Lemma 1.46 *Let $\mathfrak{F}$ be a nonempty hereditary saturated formation and τ an inductive hereditary subgroup functor. Let G be a group and $H \leq E \leq G$.*

(1) *If H is (weakly) $\mathfrak{F}_\tau$-supplemented in G, then H is (weakly) $\mathfrak{F}_\tau$-supplemented in E.*

(2) *If H is strongly $\mathfrak{F}_\tau$-supplemented in G, then H is strongly $\mathfrak{F}_\tau$-supplemented in E.*

(3) *If H is $\mathfrak{F}_\tau$-embedded in G, then H is $\mathfrak{F}_\tau$-embedded in E.*

Proof For any subgroup V of G, we put $\widehat{V} = VH_G/H_G$.

(1) By hypothesis, for some $\widehat{S} \leq \widehat{H}$, where $\widehat{S}$ is τ-subgroup of $\widehat{G}$, the pair $(\widehat{S}, \widehat{H})$ satisfies the supplement condition (I^*) in $\widehat{G}$. Hence, by Lemma 1.2(2), the pair $(\widehat{S}, \widehat{H})$ satisfies the supplement condition (I^*) in $\widehat{E}$. Hence, by Lemma 1.2(1), the pair $(\widehat{S}\widehat{H_E}/\widehat{H_E}, \widehat{H}/\widehat{H_E})$ satisfies the supplement condition (I^*) in $\widehat{E}/\widehat{H_E}$, where $\widehat{S}\widehat{H_E}/\widehat{H_E}$ is τ-subgroup of $\widehat{E}/\widehat{H_E}$. Hence SH_E/H_E is a τ-subgroup of E/H_E and the pair $(SH_E/H_E, H/H_E)$ satisfies the supplement condition (I^*) in E/H_E, so H is $\mathfrak{F}_\tau$-supplemented in E.

(2), (3) See the proof of (1).

τ-embedded Subgroups

Definition 1.47 Let τ be a subgroup functor, p a prime and H a subgroup of a group G. We say that

(I) H is *τ_p-embedded in G* and write H τ-p-emb G, if each Sylow p-subgroup of H is also a Sylow p-subgroup of some τ-subgroup of G.

(II) H is *τ-embedded in G* and write H τ-emb G, if H is τ_p-embedded in G for all primes p.

Let H is τ_p-embedded (τ-embedded) in G. Then we also say that:

(a) H is *normally p-embedded in G* (*normally embedded in G*, respectively) if τ is the subgroup functor in Example 1.6;

(b) H is *subnormally p-embedded in G* (*subnormally embedded in G*, respectively) if τ is the subgroup functor in Example 1.7;

(c) H is *s-quasinormally p-embedded in G* (*s-quasinormally embedded in G*, respectively) if τ is the subgroup functor in Example 1.8.

Lemma 1.48 *Let τ be a inductive subgroup functor and H be a τ_p-embedded subgroup of a group G, that is, H τ-p-emb G. Let $N \trianglelefteq G$. Then:*

(1) *H^x τ-p-emb G for all $x \in G$.*

(2) *H is τ_p-embedded in G if and only if each Sylow p-subgroup of H is τ_p-embedded in G.*

(3) *If τ is a complete hereditary subgroup functor and $H \leq K \leq G$, then H τ-p-emb K.*

(4) *If τ is an inductive subgroup functor, then NH/N τ-p-emb G/N.*

(5) *If τ also satisfies the property (C) in Def. 1.5. Then NH τ-p-emb G*

Proof Since H τ-p-emb G, there exists a τ-subgroup E of G such that $P \in \mathrm{Syl}_p(H) \cap \mathrm{Syl}_p(E)$. (1), (2) These assertion directly follows from the definitions.

(3) Since $E \in \tau(G)$ and τ is a complete hereditary subgroup functor, $E \cap K \in \tau(K)$. On the other hand, $P \leq E \cap K$, so $P \in \mathrm{Syl}_p(E \cap K)$. Hence H τ-p-emb K.

(4) Let P_1/N be a Sylow p-subgroup of HN/N. Then $P_1 = NP$ for some Sylow p-subgroup P of H. By the hypothesis, each Sylow p-subgroup P of H is also a Sylow p-subgroup of E. Hence $P_1/N = PN/N \in \mathrm{Syl}_p(NE/N)$. Finally, since τ is an inductive subgroup functor and $E \in \tau(G)$, we have $EN/N \in \tau(G/N)$ by Def. 1.5(B). Hence we have (4).

(5) By Chap. 2, Lemma 4.1, for any $P_2 \in \mathrm{Syl}_p(NH)$, there exist $N_p \in \mathrm{Syl}_p(N)$ and $H_p \in \mathrm{Syl}_p(H)$ such that $P_2 = N_p H_p$. Let $e \in E$ such that $(H_p)^e = P$. Then $(P_2)^e = (N_p)^e P$ is a Sylow p-subgroup of NE and so $P_2 \leq NE$. Therefore every Sylow p-subgroup of NH is a Sylow p-subgroup of NE. On the other hand, by hypothesis, $NE \in \tau(G)$. Hence NH τ-p-emb G.

For any subgroup functor τ, we write τ_{pe} (respectively τ_e) to denote the subgroup functor such that for any group G, $\tau_{pe}(G)$ is the set of all τ_p-embedded subgroups of G (respectively τ_e is the set of all τ-embedded subgroups of G).

From Lemma 1.48 we obtain the following

Lemma 1.49 *If τ is a strongly inductive completely hereditary subgroup functor and p a prime, then both functors τ_{pe} and τ_e are strongly inductive hereditary.*

Definition 1.50 τ be a subgroup functor and $H \leq G$. Then we use $H_{\tau G}$ to denote the subgroup of H generated by all those subgroups of H which belong to $\tau(G)$, and we say that $H_{\tau G}$ is the τ-*core of H in G*.

When τ is the subgroup functor in Example 1.6, then we write H_G in place of $H_{\tau G}$. If τ is the subgroup functor in Example 1.8, then $H_{\tau G}$ is written as H_{sG}.

Instead of $H_{\tau_{pe}G}$ ($H_{\tau_e G}$, respectively), we write also:

H_{peG} (H_{eG}, respectively) if τ is the subgroup functor in Example 1.6;

H_{psubeG} (H_{subeG}, respectively) if τ is the subgroup functor in Example 1.7;

H_{pseG} (H_{seG}, respectively) if τ is the subgroup functor in Example 1.8.

The following example shows that in general $H_{\tau G} \notin \tau(G)$.

Example 1.51   (see [69]). Let Ly be the Lyons simple group. Then $|Ly| = 2^8 \cdot 3^7 \cdot 5^6 \cdot 7 \cdot 11 \cdot 31 \cdot 37 \cdot 67$. Hence in view of [118] there is a group D with minimal normal subgroup N such that $C_D(N) = N \leq O_{67}(D)$, $D/N \simeq Ly$ and $N \leq \Phi(D)$. Let Q be a group of order 17. Let $G = D \wr Q = K \rtimes Q$, where K is the base group of the regular wreath product G. Then $P = \Phi(K) = N^\natural$ (we use here the terminology in [89, Chap. A]). Moreover, in view of [89, Chap. A, Proposition 18.5], P is the only minimal normal subgroup of G. It is clear also that $|P| > 67^2$.

Since P is an elementary abelian 67-group, in view of Maschke's theorem [337, (8.1.2)], $P = P_1 \times P_2 \times \ldots \times P_t$, where P_i is a minimal normal subgroup of PQ for all $i = 1, 2, \ldots, t$. Suppose that $Q \leq C_G(P_i)$ for all $i = 1, 2, \ldots, t$. Then $Q \leq C_G(P)$. Hence $PQ = P \times Q = C_G(P)$ is normal in G and consequently Q is normal in G. This contradiction shows that for some i we have $C_Q(P_i) = 1$. Hence $S := P_i \rtimes Q = Q^S$. Since Q is a Sylow 17-subgroup of G, it is s-quasinormally embedded in G. This means that $S = S_{seG}$. Suppose that P_i is an s-quasinormally embedded subgroup of G and let V be an s-quasinormal subgroup of G such that $P_i \in \mathrm{Syl}_{67}(V)$. Since 17 divides $67 + 1$ and does not divide $67 - 1$, $|P_i| = 67^2$ by Chap. 1, Lemma 3.29. By Chap. 1, Lemma 5.34(1), V is subnormal in G and V/V_G is nilpotent.

Suppose that $V_G = 1$. Then V is a subnormal nilpotent subgroup of G. It follow from Chap. 1, Lemma 5.35(6) that $V \leq F(G) = P$. Hence $V = P_i$ is s-quasinormal in G. It is clear that $O^{67}(G) = G$. By Chap. 1, Lemma 5.34(6) we deduce that P_i is normal in G. Hence $P = P_i$ and so $|P| = 67^2$. This contradiction shows that $V_G \neq 1$ and so $P \leq V_G$. But then $P \leq P_i$, a contradiction. Therefore P_i is not an s-quasinormally embedded subgroup of G. Consequently, S is not s-quasinormally embedded in G.

Lemma 1.52 *Let G be a group, $H \leq E \leq G$ and $N \trianglelefteq G$. Let τ be a subgroup functor. Then:*

(1) $\theta(H_{\tau G}) = (\theta(H))_{\tau G}$ *for any isomorphism* $\theta : G \to G^*$. *In particular,* $(H_{\tau G})^x = (H^x)_{\tau G}$ *for any* $x \in G$.
(2) $H_{\tau G} \trianglelefteq H$.
(3) *If τ is a hereditary subgroup functor, then $H_{\tau G} \leq H_{\tau E}$.*
(4) *If τ is an inductive subgroup functor, then $H_{\tau G} N/N \leq (HN/N)_{\tau(G/N)}$.*

Proof

(1) This follows from the definition of the subgroup functor.
(2) Let $x \in G$ and $i_x : G \to G$ be the inner automorphism of G. Then $i_x(\tau(G)) = \tau(i_x(G)) = \tau(G)$ and so $(H_{\tau G})^x = (H^x)_{\tau G}$. In particular, in the case when $x \in H$ we have $(H_{\tau G})^x = H_{\tau G}$. Hence $H_{\tau G} \trianglelefteq H$.
(3) Let L be any subgroup of H and $L \in \tau(G)$. In view of (A) of Definition 1.39, $L \in \tau(E)$. Hence $L \leq H_{\tau E}$ and so $H_{\tau G} \leq H_{\tau E}$.
(4) For any subgroup L of H with $L \in \tau(G)$, in view of (B) of Definition 1.5, $LN/N \in \tau(G/N)$. Clearly, $LN/N \leq HN/N$. Hence $H_{\tau G} N/N \leq (HN/N)_{\tau(G/N)}$.

3.2 Base Theorems

Lemma 2.1 *(Burnside). Let α be a p'-automorphism of a p-group P which induces the identity on $P/\Phi(P)$. The α is the identity automorphism of P.*

Lemma 2.2 *Let $\mathfrak{F}$ be a nonempty solubly saturated formation, where F is the canonical composition satellite of $\mathfrak{F}$. Let P be a normal p-subgroup of G for some prime p. If $P/\Phi(P) \leq Z_{\mathfrak{F}}(G/\Phi(P))$, then $P \leq Z_{\mathfrak{F}}(G)$.*

Proof Suppose that $P/\Phi(P) \leq Z_{\mathfrak{F}}(G/\Phi(P))$. Then $G/C_G(P/\Phi(P)) \in F(p)$ by Chap. 1, Lemma 2.26. On the other hand, $C_G(P/\Phi(P))/C_G(P)$ is a p-group by Lemma 2.1. Hence $G/C_G(P) \in \mathfrak{G}_p F(p) = F(p)$. Thus by Chap. 1, Lemma 2.26, $P \leq Z_{\mathfrak{F}}(G)$.

Lemma 2.3 *Let $\mathfrak{F}$ be a nonempty solubly saturated formation and $G = PT$, where P is a normal p-subgroup of G. If $P \cap Z_{\mathfrak{F}}(T)$ is normal in P, then $P \cap Z_{\mathfrak{F}}(T) \leq Z_{\mathfrak{F}}(G)$.*

Proof First note that since $G = PT$ and $P \cap Z_{\mathfrak{F}}(T)$ is normal in P, $P \cap Z_{\mathfrak{F}}(T)$ is normal in G.

Let F be the canonical composition satellite of $\mathfrak{F}$ and H/K a chief factor of G below $P \cap Z_{\mathfrak{F}}(T)$. Then $T/C_T(H/K) \in F(p)$ by Chap. 1, Lemma 2.26. Hence, $G/C_G(H/K) = PT/C_G(H/K) = PT/P(C_G(H/K) \cap T) \simeq T/P(C_G(H/K) \cap T) \cap T = T/(C_G(H/K) \cap T) = T/C_T(H/K) \in F(p)$. If follows that $P \cap T \leq Z_{\mathfrak{F}}(G)$ by Chap. 1, Proposition 1.15.

Lemma 2.4 *Let $L \leq V \trianglelefteq P$, where P is a Sylow p-subgroup of a group G, and N and M are different normal subgroups of G. Suppose that $|G/M : N_{G/M}((LM/M) \cap (NM/M))|$ is a power of p.*

(1) $((LM/M) \cap (NM/M))^{G/M} \leq VM/M$.
(2) *If N is a nonabelian minimal normal subgroup of G, then $L \cap N = 1$.*
(3) *If $NL \cap M = 1$, then $(L \cap N)^G \leq VM$.*
(4) *If $L \trianglelefteq G_p$ for some Sylow subgroup G_p of G and $|G : N_G(L)|$ is a power of p, then $L \trianglelefteq G$.*

Proof

(1) It is clear that $LM/M \leq VM/M \trianglelefteq PM/M$ where PM/M is a Sylow p-subgroup of G/M. On the other hand, since $|G/M : N_{G/M}((LM/M) \cap (NM/M))|$ is a power of p, we have $((LM/M) \cap (NM/M))^{G/M} = ((LM/M) \cap (NM/M))^{N_{G/M}(((LM/M) \cap (NM/M)))(PM/M)} = ((LM/M) \cap (NM/M))^{PM/M}$. Hence we have (1).
(2) Suppose that $B = L \cap N \neq 1$. Then $(LM/M) \cap (NM/M) \neq 1$ and so $N \simeq NM/M \leq ((LM/M) \cap (NM/M))^{G/M} \leq PM/M$, which implies that N is a p-group.
(3) Since $NL \cap M = 1$, $(LM/M) \cap (NM/M) = (L \cap N)M/M$. On the other hand, $((L \cap N)M/M)^{G/M} = (L \cap N)^G M/M$. Hence (3) is a corollary of Assertion (1).
(4) It is clear.

Lemma 2.5 *Let P/R be a chief factor of a group G with $|P/R| = p^n$, where p is a prime and $n > 1$. Suppose that for every normal subgroup V of G with $V < P$ we have $V \leq R$. Let H be a subgroup of P such that $R < RH < P$.*

(1) *If H is a cyclic group of prime order or order 4, and T is a supplement of H in G, then $T = G$.*
(2) *If HR/R is normal in a Sylow p-subgroup of G/R and G/H_G has a normal subgroup T/H_G such that $(H/H_G)(T/H_G)$ is s-quasinormal in G/H_G, then $P \leq T$.*

Proof (1) Assume that $T \neq G$. Then $P = H(P \cap T)$, where $P \cap T \neq P$ and $|P : P \cap T| = |G : T|$. Let $N = N_G(P \cap T)$.

Since $T \leq N$ and $P \cap T < N_P(P \cap T)$, we have that either $P \cap T$ is a normal subgroup of G or $|G : N| = 2$. The first case implies that $P \cap T \leq R$ and hence $P = RH$, a contradiction. In the second case, N is normal in G and so $N \cap P$ is a normal subgroup of G with $|P : P \cap N| = 2$. Therefore $P \cap N \leq R$ and thereby $|P/R| = 2$, a contradiction. Hence $T = G$.

(2) Assume that $P \nleq T$. Then $T_0 = T \cap P < P$, so $T_0 \leq R$ by hypothesis. Moreover, since $(T/H_G)(H/H_G) \cap P/H_G = (H/H_G)(T/H_G \cap P/H_G) = (H/H_G)(T_0/H_G)$ and $(T/H_G)(H/H_G)$ is s-quasinormal in G/H_G by hypothesis, HT_0 is s-quasinormal in G by Chap. 1, Lemma 5.34(4). Therefore the subgroup $HT_0R/R = HR/R$ is s-quasinormal in G/R. Consequently, HR/R is normal in G by Chap. 1, Lemma 5.34(6) since HR/R is normal in a Sylow p-subgroup of G/R by hypothesis. It follows that $P/R = HR/R$ and so $P = RH$, a contradiction. Thus we have (2).

Theorem 2.6 *Let $\mathfrak{F}$ be a solubly saturated formation containing all supersoluble groups and τ a Φ-quasiregular (quasiregular, respectively) inductive subgroup functor. Let P be a nonidentity normal p-subgroup of a group G for some prime p. Suppose that every maximal subgroup of P is either $\mathfrak{F}_\tau$-supplemented or $\mathfrak{F}_\tau$-embedded in G. Then $P \leq Z_{\mathfrak{F}\Phi}(G)$ ($P \leq Z_{\mathfrak{F}}(G)$, respectively).*

Proof Suppose that in this theorem is false and let (G, P) be a counterexample with $|G| + |P|$ minimal. Let $Z = Z_{\mathfrak{F}\Phi}(G)$ ($Z = Z_{\mathfrak{F}}(G)$, respectively). Let G_p be a Sylow p-subgroup of G.

(1) $P \nleq Z_{\mathfrak{U}\Phi}(G)$ ($P \nleq Z_{\mathfrak{U}}(G)$, respectively) (This follows from the hypothesis that $\mathfrak{F}$ contains all supersoluble groups and the choice of G).

(2) P *is not a minimal normal subgroup of* G.

Suppose that P is a minimal normal subgroup of G. Then $P \cap Z = 1$. Let H be a maximal subgroup of P such that H is normal in G_p. Then $H_G = 1$. First suppose that H is $\mathfrak{F}_\tau$-supplemented in G and let $S \in \tau(G)$ and T be subgroups of G such that $S \leq H$, $HT = G$ and $H \cap T \leq SZ_{\mathfrak{F}}(T)$. Suppose that $T \neq G$. Then $1 < P \cap T < P$, where $P \cap T$ is normal in G since P is abelian, which contradicts the minimality of P. Hence $T = G$, so $H = H \cap T \leq SZ$, which implies that $H = S(H \cap Z) = S$ is a τ-subgroup of G. It is clear that $P \nleq \Phi(G)$, so for some maximal subgroup M of G we have $G = P \rtimes M$. Since τ is Φ-quasiregular and HM_G/M_G is normal in G_pM_G/M_G, HM_G/M_G is normal in G/M_G by Lemma 2.4. Hence $HM_G/M_G = PM_G/M_G$, which implies that $H = P$, a contradiction. Hence H is $\mathfrak{F}_\tau$-embedded in G. Let T be a normal subgroup of G such that HT is s-quasinormal in G and $H \cap T \leq SZ$ for some subgroup $S \in \tau(G)$ contained in H. Then, by Lemma 2.5(2), $P \leq T$. Hence $H \cap T = H = S(H \cap Z) = S$. But in this case, as we have already known, is impossible. Hence we have (2).

(3) *If N is a minimal normal subgroup of G contained in P, then $P/N \leq Z_{\mathfrak{F}\Phi}(G/N)$ ($P/N \leq Z_{\mathfrak{F}}(G/N)$, respectively) and $Z \cap P = 1$.*

Indeed, by Lemma 1.31 and 1.38, the hypothesis holds for $(G/N, P/N)$. Hence $P/N \leq Z_{\mathfrak{F}\Phi}(G/N)$ ($P/N \leq Z_{\mathfrak{F}}(G/N)$, respectively) by the choice of (G, E). Hence $N \nleq Z$ by Chap. 1, Lemma 2.5.

(4) $P \leq Z_{\mathfrak{F}\Phi}(G)$.

Suppose that $P \not\leq Z_{\mathfrak{F}\Phi}(G)$. Then, in view of (4) and Chap. 1, Lemma 2.5, $\Phi(G) \cap P = 1$. Hence $P = N \times D$ for some normal subgroup D of G by Chap. 1, Proposition 2.24, where $D \neq 1$ by (2). Let R be a minimal normal subgroup of G contained in D. Then, by [89, A, 9.11], $RN/N \not\leq \Phi(G/N)$. Hence in view of (3) and the G-isomorphism $R \simeq RN/N$, R is $\mathfrak{F}$-central in G, and so $P \leq Z_{\mathfrak{F}\Phi}(G)$ by (4) and Chap. 1, Lemma 2.5. Hence we have (4).

(5) τ *is quasiregular.* (This follows from (4) and the choice of (G, E).)

(6) *If N is a minimal normal subgroup of G contained in P, then N is the unique minimal normal subgroup of G contained in P* (see the proof of (3)).

(7) $\Phi(P) \neq 1$.

Suppose that $\Phi(P) = 1$. Then P is an elementary abelian p-group. Let W be a maximal subgroup of N such that W is normal in G_p. We shall show that W is normal in G. Let B be a complement of N in P and $V = WB$. Then V is a maximal subgroup of P and, evidently, $V_G = 1$ by (6).

First suppose that V is $\mathfrak{F}_\tau$-supplemented in G and let $S \in \tau(G)$ and T be subgroups of G such that $S \leq V$, $VT = G$ and $V \cap T \leq SZ_{\mathfrak{F}}(T)$. Assume that $T = G$. Then $V = V \cap T \leq SZ$ and so $V = S(V \cap Z)$. But in view of (3), $Z \cap P = 1$. Hence $V = S$ and thereby $W = WB \cap N = V \cap N = S \cap N$. Since τ is quasiregular by (5), and W is normal in G_p, it follows that W is normal in G by Lemma 2.4. Let $T \neq G$. Then $1 \neq T \cap P < P$. Since $G = VT = PT$ and P is abelian, $T \cap P$ is normal in G. Hence $N \leq T$, which implies that $W \leq T \cap V \leq SZ_{\mathfrak{F}}(T) \cap P = S(Z_{\mathfrak{F}}(T) \cap P)$. Since P is abelian, $Z_{\mathfrak{F}}(T) \cap P$ is normal in P, so $Z_{\mathfrak{F}}(T) \cap P \leq Z \cap P = 1$ by Lemma 2.3. This implies that $W = S \cap N$ and so W is normal in G by Lemma 2.4.

Now assume that V is $\mathfrak{F}_\tau$-embedded in G. Then G has a normal subgroup T such that VT is s-quasinormal in G and $V \cap T \leq SZ$ for some τ-subgroup S of G contained in V. Assume that $N \not\leq T$. Then $T \cap P = 1$. Hence $TV \cap P = V$ and $W = V \cap N$ are s-quasinormal subgroup of G by Chap. 1, Lemma 5.34(7). Thus $O^p(G) \leq N_G(W)$ by Chap. 1, Lemma 5.34(6). Consequently, W is normal in G.

Finally, as above, we get the same conclusion in the case when $N \leq T$. Therefore W is normal in G and so $W = 1$, which implies that $|N| = p$, contrary to (3). Thus $\Phi(P) \neq 1$.

The final contradiction.

By (7), $\Phi(P) \neq 1$. Let N be a minimal normal subgroup of G contained in $\Phi(P)$. Then $P/N \leq Z_{\mathfrak{F}}(G/N)$ by (3). It follows that $P/\Phi(P) \leq Z_{\mathfrak{F}}(G/\Phi(P))$ by Chap. 1, Theorem 2.6(d). Thus $P \leq Z$ by Lemma 2.2. This contradiction completes the proof.

Let P be a nonidentity p-group. If P is not a nonabelian 2-group, we use $\Omega(P)$ to denote the subgroup $\Omega_1(P)$. Otherwise, $\Omega(P) = \Omega_2(P)$.

Lemma 2.7 (see [109, Theorem 2.4]). *Let P be a group, α a p'-automorphism of P. If $[\alpha, \Omega(P)] = 1$, then $\alpha = 1$.*

Lemma 2.8 *Let $\mathfrak{F}$ be a nonempty solubly saturated formation, P be a normal p-subgroup of a group G, where p is a prime. Let D be a characteristic subgroup of P*

*such that every nontrivial p'-automorphism of P induces a nontrivial automorphism
of D. Suppose that $D \leq Z_{\mathfrak{F}}(G)$. Then $P \leq Z_{\mathfrak{F}}(G)$.*

Proof Let F is the canonical local satellite of $\mathfrak{F}$. Let $C = C_G(P)$. Since $D \leq Z_{\mathfrak{F}}(G)$,
then $G/C_G(D) \in F(p)$ by Chap. 1, Lemma 2.26. On the other hand, we may
consider $C_G(D)/C_G(P)$ as an automorphism group of P. Then the automorphism
group stabilizes acts on every chief factor of G below P. Hence $C_G(D)/C_G(P)$ is a
p-group by Chap. 1, Lemma 2.13, which implies that $G/C_G(P) \in \mathfrak{G}_p F(p) = F(p)$.
Hence $P \leq Z_{\mathfrak{F}}(G)$ since for every chief factor H/K of G below P we have that
$G/C_G(H/K) \in F(p)$.

Lemma 2.9 (Thompson [117, Chap. 5, Theorem 3.11]). *A p-group P possesses a
characteristic subgroup C (which is called a Thompson critical subgroup of P) with
the following properties:*

  (i) *The nilpotency class of C is at most 2 and $C/Z(C)$ is elementary abelian.*
 (ii) $[P, C] \leq Z(C)$.
(iii) $C_P(C) = Z(C)$.
 (iv) *Every nontrivial p'-automorphism of P induces a nontrivial automorphism
       of C.*

Lemma 2.10   *Let C be a Thompson critical subgroup of a p-group P. Then the
group $D := \Omega(C)$ is of exponent p if p is an odd prime, or exponent 4 if P is
nonabelian 2-group. Moreover, every nontrivial p'-automorphism of P induces a
nontrivial automorphism of D.*

Proof The first assertion of the lemma may be proved as Lemma 3.9 in [117, Chap.
5]. Finally, Since $D\ char\ C\ char\ P$, D is a characteristic subgroup of P. Now by
using Lemma 2.9(iv) and Lemma 2.7, we see that every nontrivial p'-automorphism
of P induces a nontrivial automorphism of D.

Lemma 2.11   *Let $\mathfrak{F}$ be a nonempty solubly saturated formation, where F is the
canonical composition satellite of $\mathfrak{F}$. Let P be a normal p-subgroup of a group G
and $D = \Omega(C)$, where C is a Thompson critical subgroup of P. If $C \leq Z_{\mathfrak{F}}(G)$ or
$D \leq Z_{\mathfrak{F}}(G)$, then $P \leq Z_{\mathfrak{F}}(G)$.*

Proof Suppose that $C \leq Z_{\mathfrak{F}}(G)$. Then $G/C_G(C) \in F(p)$ by Chap. 1, Lemma 2.26.
On the other hand, $C_G(C)/C_G(P)$ is a p-group by Chap. 1, Lemma 2.13. Hence
$G/C_G(P) \in \mathfrak{G}_p F(p) = F(p)$. Consequently, $P \leq Z_{\mathfrak{F}}(G)$. Finally if $D \leq Z_{\mathfrak{F}}(G)$,
then $G/C_G(D) \in F(p)$ by Chap. 1, Lemma 2.26. Clearly, $C_G(C) \subseteq C_G(D)$. By
Lemma 2.7, $C_G(D)/C_G(C)$ is a p-group. On the other hand, $C_G(C)/C_G(P)$ is a
p-group by the hypothesis and Chap. 1, Lemma 2.13. So $G/C_G(P) \in \mathfrak{N}_p F(p)$.
Thus, $P \leq Z_{\mathfrak{F}}(G)$.

Theorem 2.12   *Let $\mathfrak{F}$ be a solubly saturated formation containing all supersoluble
groups and τ a quasiregular inductive subgroup functor. Let P be a nonidentity normal p-subgroup of a group G for some prime p. Suppose that every cyclic subgroup
of P of prime order or order 4 (if P is a nonabelian group) is either $\mathfrak{F}_\tau$-supplemented
or $\mathfrak{F}_\tau$-embedded in G, then $P \leq Z_{\mathfrak{F}}(G)$.*

Proof Suppose that this theorem is false and let (G, P) be a counterexample with $|G| + |P|$ minimal. Let $Z = Z_{\mathfrak{F}}(G)$.

(1) *G has a normal subgroup $R \leq P$ such that P/R is an $\mathfrak{F}$-eccentric chief factor of G, $R \leq Z$ and $V \leq R$ for any normal subgroup $V \neq P$ of G contained in P. In particular, $|P/R| > p$.*

Let P/R be a chief factor of G. Then the hypothesis holds for (G, R). Therefore $R \leq Z$ and so P/R is $\mathfrak{F}$-eccentric in G by the choice of (G, P) and Chap. 1, Lemma 2.5. It follows that $|P/R| > p$. Now let $V \neq P$ be any normal subgroup of G contained in P. Then $V \leq Z$. If $V \nleq R$, then $VR = P \leq Z$ by Chap. 1, Theorem 2.6(b). This contradiction shows that $V \leq R$.

Let L be a Thompson critical subgroup of P and $\Omega = \Omega(L)$. Let G_p be a Sylow p-subgroup of G.

(2) $\Omega = P$.

Indeed, suppose that $\Omega < P$. Then, in view of (2), $\Omega \leq Z$. Hence $P \leq Z$ by Lemma 2.11, which contradicts the choice of (G, P). Hence $\Omega = P$.

The final contradiction.

Let L/R be any minimal subgroup of $(P/R) \cap Z(G_p/R)$. Then, view of Lemma 2.4 and Claim (2), L/R is not a τ-subgroup of G/R and $L/R \ntrianglelefteq G/R$ since $|P/R| > p$. Let $x \in L \setminus R$ and $H = \langle x \rangle$. Then $L/R = HR/R$, so H is not normal in G and $H_G \leq R$.

Since $\Omega = P$ by (2), $|H|$ is ether prime or 4 by Lemma 2.10. Hence H is either $\mathfrak{F}_\tau$-supplemented or $\mathfrak{F}_\tau$-embedded in G. First we show that $H/H_G = (S/H_G)((H/H_G) \cap Z_{\mathfrak{F}}(G/H_G))$ for some τ-subgroup S/H_G of G/H_G contained in H/H_G

If H is $\mathfrak{F}_\tau$-supplemented in G and T/H_G is a subgroup of G/H_G such that $(T/H_G)(H/H_G) = G/H_G$ and $(T/H_G) \cap (H/H_G) \leq (S/H_G)Z_{\mathfrak{F}}(T/H_G)$ for some τ-subgroup S/H_G of G/H_G contained in H/H_G, then $T = G$ by Lemma 2.5(1). Therefore $H/H_G = (S/H_G)(H/H_G \cap Z_{\mathfrak{F}}(G/H_G))$.

Now assume that H is $\mathfrak{F}_\tau$-embedded in G and let T/H_G be a normal subgroup of G/H_G such that $(T/H_G)(H/H_G)$ is s-quasinormal in G/H_G and $(T/H_G) \cap (H/H_G) \leq (S/H_G)Z_{\mathfrak{F}}(G/H_G)$ for some τ-subgroup S/H_G of G/H_G contained in H/H_G. Then $P/H_G \leq T/H_G$ by Lemma 2.5(2), and so $H/H_G \leq (S/H_G)Z_{\mathfrak{F}}(G/H_G)$, which again implies that $H/H_G = (S/H_G)(H/H_G \cap Z_{\mathfrak{F}}(G/H_G))$.

Since H is cyclic, we have either $H/H_G = S/H_G$ is a τ-subgroup of G/H_G or $H/H_G \leq Z_{\mathfrak{F}}(G/H_G)$. Note that $(R/H_G)(H/H_G)/(R/H_G) = (RH/H_G)/(R/H_G) = (L/H_G)/(R/H_G)$. Hence in the former case, $(L/H_G)/(R/H_G)$ is a τ-subgroup of $(G/H_G)/(R/H_G)$ since τ is inductive, which implies that L/R is a τ-subgroup of G/R, a contradiction. Therefore we have the second case, and so by Chap. 1, Theorem 2.6(e), $(L/H_G)/(R/H_G) \leq ((P/H_G)/(R/H_G)) \cap (Z_{\mathfrak{F}}((G/H_G)/(R/H_G)))$. But then $L/R \leq (P/R) \cap Z_{\mathfrak{F}}(G/R)$. Thus P/R is $\mathfrak{F}$-central in G, contrary to (2).

The theorem is proved.

Definition 2.13 We say that a subgroup functor τ is *special* if for any subnormal subgroup H of any group G with $|H| = 4$ and $H \in \tau(G)$ the following hold:

(a) If $H = \langle x \rangle$ is cyclic, then $\langle x^2 \rangle \in \tau(G)$.
(b) If $H = A \times B$, where $|A| = 2$ and $A \in \tau(G)$, then $B \in \tau(G)$.

Example 2.14 The subgroup functor τ in Example 1.6, or 1.8 or 1.9 or 1.10 is special. In fact, for Example 1.6 and 1.10, this is evident. Now we consider the case when τ in Example 1.8. Let H be any s-quasinormal subgroup of a group G with $|H| = 4$ and $H \in \tau(G)$. Assume that $H = \langle x \rangle$ is cyclic. Then $\langle x^2 \rangle \in \tau(G)$ by Chap. 1, Lemma 5.34(5). Now assume that $H = A \times B$, where $|A| = 2$ and $A \in \tau(G)$. Let Q be a Sylow q-subgroup of G, where q is odd prime. Then $Q \leq N_G(H)$ by Chap. 1, Lemma 5.34(6) and so HQ is nilpotent, which implies that $Q \leq C_G(B)$. Hence $B \in \tau(G)$. This shows that the subgroup functors in Example 1.8 are special. Finally, let H be a subnormal subgroup of G with $|H| = 4$ such that H is a Sylow subgroup of some s-quasinormal subgroup W of G. Then in view of Chap. 1, Lemma 5.35(6), $H \leq O_2(G)$, so $H = O_2(G) \cap W$ is s-quasinormal in H by Chap. 1, Lemma 5.34(7). Hence the functor in Example 1.9 is special.

Example 2.15 A subgroup H of a group G is said to be τ-quasinormal in G if H permutes with all Sylow subgroups P of G such that $(|H|, |P|) = 1$ and $(|H|, |P^G|) \neq 1$. For any group G, let $\tau(G)$ be the set of all τ-quasinormal in G subgroups. In view of Lemma 2.1 in [301], the functor is special.

Lemma 2.16 *Let $\mathfrak{F}$ be a nonempty hereditary solubly saturated formation and τ an inductive subgroup functor. Let G be a group, $H \leq G$ and N a normal subgroup of G. Suppose that H is strongly $\mathfrak{U}_\tau$-supplemented in G. If either $N \leq H$ or $(|H|, |N|) = 1$, then HN/N is strongly $\mathfrak{U}_\tau$-supplemented in G/N.*

Proof See the proof of Lemma 1.31.
Let $1 < H < P$, where P is a p-group of order p^n. If $|H| = p^k$, where $0 < k < n$, then we call H a k-intermediate subgroup of P.

Theorem 2.17 (Guo, Skiba). *Let $\mathfrak{F}$ be a solubly hereditary saturated formation containing all supersoluble groups and τ a special quasiregular strongly inductive subgroup functor. Let P be a normal p-subgroup of G for some prime p with $|P| > p$. Suppose that there is an integer k such that every k-intermediate subgroup H of P and also, in the case when $1 = k < n - 1$ and P is a nonabelian 2-group, every cyclic subgroup H of P of order 4, either has a supplement $T \in \mathfrak{F}$ in G or is strongly $\mathfrak{U}_\tau$-supplemented in G. Then $P \leq Z_{\mathfrak{F}}(G)$.*

Proof Suppose that this theorem is false and let G be a counterexample with $|G| + |P|$ minimal. Let $Z = Z_{\mathfrak{U}}(G)$

(1) $n - 1 > k > 1$ (This follows from (2.6), (2.12) and the choice of (G, P)).
(2) *G has no maximal subgroup M with $|G : M| = p$ and $MP = G$.*
    Otherwise, by (1), the hypothesis is still true for the pair $(G, M \cap P)$. Hence $M \cap P \leq Z_{\mathfrak{F}}(G)$ by the choice of P. On the other hand, since $|P/M \cap P| = |G :$

$M| = p$, the chief factor $P/M \cap P$ of G is $\mathfrak{U}$-central in G and so $P \leq Z_{\mathfrak{F}}(G)$ since $\mathfrak{F}$ contains all supersoluble groups by hypothesis. This contradiction shows that we have (2).

(3) *If H is a subgroup of P with $|H| = p^k$, then either H has a supplement $T \in \mathfrak{F}$ in G or there is a τ-subgroup S of G such that $H = S(H \cap Z)$.*

Suppose that H has no supplement $T \in \mathfrak{F}$ in G. Then by hypothesis, H is strongly $\mathfrak{F}_\tau$-supplemented in G and so there is a K-$\mathfrak{F}$-subnormal subgroup T of G and a τ-subgroup S of G with $S \leq H$ such that $HT = G$ and $T \cap H \leq SZ_\mathfrak{X}(T)$. If $T \neq G$, then for some maximal subgroup M of G we have $|G : M| = p$ and $MP = G$, which contradicts (2). Hence $T = G$, so $H = H \cap SZ = S(H \cap Z)$.

(4) $|N| \leq p^k$ *for any minimal normal subgroup N of G contained in P.*

Assume that $p^k < |N|$. Then $N \nleq Z$. Hence $N \cap Z = 1$. Let H be any subgroup of N with $|H| = p^k$ such that H is normal in a Sylow p-subgroup of G. Then either H has a supplement $T \in \mathfrak{F}$ in G or there is a τ-subgroup S of G such that $H = S(H \cap Z)$ by (3). In the former case we have $TN = G$ and $N = H(T \cap N)$, so $T \cap N$ is a normal nonidentity subgroup of G. Hence $N \leq T$, so $G = T \in \mathfrak{F}$. Hence $P \leq Z_{\mathfrak{F}}(G)$. This contradiction shows that we have the second case. Hence $H = S$ since $H \cap Z \leq N \cap Z = 1$. Since τ is quasiregular, H is normal in G by Lemma 2.4, which contradicts the minimality of N.

(5) *If P is nonabelian 2-group, then $k > 2$.*

Suppose that $k = 2$.

(a) *If $H \leq P$ with $|H| = 4$ and H has a supplement T in G such that $T \in \mathfrak{F}$, then T is a complement of H in G, $D = T \cap P$ is normal in G and $D \leq Z_{\mathfrak{F}}(G)$.* Indeed, the choice of G implies that $G \notin \mathfrak{F}$, so $T < N_G(D)$. Hence in the case when D is not normal in G, $N_G(D)$ is a maximal subgroup of G with index 2 and $PN_G(D) = G$, which contradicts (2). Thus $D \trianglelefteq G$ and by Lemma 2.2, $D \leq Z_{\mathfrak{F}}(G)$. Finally, from (2) we deduce that T is a complement of H in G.

(b) *If $H = A \times B$, where $A \leq N$ for some minimal normal subgroup N of G contained in P and $|A| = 2 = |B|$, then H has no a supplement T in G with $T \in \mathfrak{F}$.*

Assume that this is false. Then, by (a), T is a complement of H in G, $D = T \cap P$ is normal in G, $D \leq Z_{\mathfrak{F}}(G)$. Hence $P = D \rtimes H$. It is clear that $N \nleq D$, so the minimality of N and (3) imply that $P = N \times D$ and hence $|N| = |H| = 4$. Let R be a minimal normal subgroup of G contained in D. Then $NR \leq Z_{\mathcal{F}}(G)$. Indeed, since $|N| = 4$, $|RN| > 4$ and so the hypothesis holds for (G, NR). But then $NR \leq Z_{\mathcal{F}}(G)$ by the choice of (G, P). Therefore $N \leq Z_{\mathcal{F}}(G)$, which implies that $P = N \times D \leq Z_{\mathcal{F}}(G)$. This contradiction shows that we have (b).

(c) *Every subgroup A of P with $|A| = 2$ satisfies the hypothesis of the theorem.* If $A \leq Z_{\mathfrak{F}}(G)$ or $A \in \tau(G)$ it is evident. So we may assume that neither $A \leq Z_{\mathfrak{F}}(G)$ nor $A \in \tau(G)$.

First suppose that $P \cap Z \neq 1$ and let Z_0 be a minimal normal subgroup of G contained in $P \cap Z$. Then $|Z_0| = 2$, $Z_0 \leq Z \leq Z_{\mathfrak{F}}(G)$ and $Z_0 \in \tau(G)$. Hence $A \neq Z_0$. Let $H = Z_0 \times A$. Then by (3), either H has a supplement $T \in \mathfrak{F}$ in G or $H = S(H \cap Z)$ for some τ-subgroup S of G. If we have the former case, then T is a complement of H in G, $D = T \cap P$ is normal in G and $D \leq Z_{\mathfrak{F}}(G)$ by (a). It is also clear that $Z_0 \nleq D$, so $Z_0 D \leq Z_{\mathfrak{F}}(G)$ and $P/Z_0 D$ is of order 2.

Therefore $P \leq Z_{\mathcal{F}}(G)$. This contradiction shows that we have the second case. Since $A \nleq Z$ with $A \notin \tau(G)$ and the functor τ is special, $H \notin \tau(G)$. Hence $H = S(H \cap Z)$, where $S \in \tau(G)$ and $H \cap Z$ are subgroups of order 2. But then $H = SZ_0 \in \tau(G)$ since τ is strongly inductive by hypothesis, a contradiction. Hence $P \cap Z = 1$.

Since P is a nonabelian 2-group, P has a cyclic subgroup $H = \langle x \rangle$ of order 4. Then by (3), either H has a supplement $T \in \mathfrak{F}$ in G, or $H \in \tau(G)$, or $H \leq Z \cap P$. But since $P \cap Z = 1$, the last case is impossible. Suppose we have the former case. Then $P = H(T \cap P)$, where $P/(T \cap P)$ is cyclic and $T \cap P \leq Z_{\mathfrak{F}}(G)$ by (a). Hence $P \leq Z_{\mathfrak{F}}(G)$, a contradiction. Therefore we have $H \in \tau(G)$. Since τ is special, it follows that $\langle x^2 \rangle \in \tau(G)$. Now let $L \leq Z(P) \cap N$ for some minimal normal subgroup N of G. Assume that $L \neq \langle x^2 \rangle$ and let $V = \langle x^2 \rangle \times L$. From (b) we know that V has no a supplement T in G with $T \in \mathfrak{F}$. Hence from $P \cap Z = 1$ and (3) we get $V \in \tau(G)$, so $L \in \tau(G)$ since τ is special. Now let $F = A \times L$. Then as above we get that $F \in \tau(G)$ and $A \in \tau(G)$. The case when $L = \langle x^2 \rangle$ can be considered analogously. Therefore $A \in \tau(G)$. This contradiction completes the proof of (c).

Therefore we see that the hypothesis holds for all subgroups of P of order 2 and for all subgroups of order 4, which contradicts (1). Hence we have (5).

(6) *If N is a minimal normal subgroup of G contained in P, then the hypothesis is still true for $(G/N, P/N)$.*

By (4) we have $|N| \leq p^k$. If either $p > 2$ and $|N| < p^k$ or $p = 2$ and $|N| < 2^{k-1}$, it follows from Lemma 2.16. Now let either $p > 2$ and $|N| = p^k$ or $p = 2$ and $|N| \in \{2^k, 2^{k-1}\}$.

First assume that $|N| = p^k$. Then N is noncyclic by (2) and so every subgroup of P containing N is not cyclic. Let $N \leq K \leq P$, where $|K : N| = p$. Since K is noncyclic, it has a maximal subgroup $L \neq N$. If L has a supplement T in G with $T \in \mathfrak{F}$, then $TN/N \simeq T/T \cap N \in \mathfrak{F}$ is a supplement of K/N in G/N. Otherwise, by (3), $L = S(L \cap Z)$ where $S \in \tau(G)$. Hence $K = LN = S(L \cap Z)N$, so $K/N \leq (SN/N)(Z_{\mathfrak{U}}(G/N)$ by Chap. 1, Theorem 2.6(d). Since τ is inductive, $SN/N \in \tau(G/N)$, and so K/N is strongly $\mathfrak{U}_\tau$-supplemented in G/N. Thus if P/N is abelian, the hypothesis is true for G/N.

Next suppose that $|N| = 2^k$ and P/N is a nonabelian 2-group. Then P is nonabelian and so $k > 2$ by (5). Let $N \leq K \leq V \leq P$, where $|V : N| = 4$, V/N is cyclic and $|V : K| = 2$. Since V/N is not elementary, $N \nleq \Phi(V)$. Hence for some maximal subgroup of K_1 of V we have $V = K_1 N$. Suppose that K_1 is cyclic. Then $|K_1 \cap N| = 2$ and $2 = |V : K_1| = |K_1 N : K_1| = |N : K_1 \cap N|$. This implies that $|N| = 4$. But then $k = 2$, which contradicts (5). Hence K_1 is not cyclic. Let S and R be two distinct maximal subgroup of K_1. Then $K_1 = SR$, so $V/N = K_1 N/N = (SN/N)(RN/N)$ But since V/N is cyclic, either $V/N = SN/N$ or $V/N = RN/N$. Without loss of generality, we may assume that $NS = V$. Since S is a maximal subgroup of K_1 and K_1 is a maximal subgroup of V, $|S| = |N| = 2^k$. Suppose that S have no supplement $T \in \mathfrak{F}$ in G. Then by (3), $S = S_0(S \cap Z)$ for some τ-subgroup S_0 of G. This implies that $V/N = SN/N = (S_0 N/N)((S \cap Z)N/N)$ where $S_0 N/N$ is a τ-subgroup of G/N and, by Chap. 1, Theorem 2.6(e), $(S \cap Z)N/N) \le Z_{\mathfrak{U}}(G/N)$. Thus V/N is strongly $\mathfrak{U}_\tau$-supplemented in G/N. Therefore every subgroup H/N of P/N of order 2 and every cyclic subgroup of P/N of order 4 are strongly $\mathfrak{U}_\tau$-supplemented in G/N.

Now suppose that $2^{k-1} = |N|$. If P/N is abelian, then the hypothesis holds for $(G/N, P/N)$. Suppose that P/N is nonabelian, then P is non-abelian and so $k > 2$ by (5). Hence N is not cyclic. Now, as above, one can show that every cyclic subgroup of P/N of order 4 is strongly $\mathfrak{U}_\tau$-supplemented in G/N. Therefore we have (6).

(7) $\Phi(P) \ne 1$.

Suppose that $\Phi(P) = 1$. Then P is an elementary abelian p-group. Let N be a minimal normal subgroup of G contained in P. Then the choice of G and (7) imply that $P/N \le Z_{\mathfrak{F}}(G/N)$. Suppose that there is a minimal normal subgroup $R \ne N$ of G contained in P. Since $P/R \le Z_{\mathfrak{F}}(G/R)$, by using the G-isomorphism $N \simeq NR/R$, we have $N \le Z_{\mathcal{F}}(G)$. This implies that $P \le Z_{\mathcal{F}}(G)$. This contradiction shows that N is the unique minimal normal subgroup of G contained in P and $N \nleq Z_{\mathcal{F}}(G)$. Hence $Z_{\mathcal{F}}(G) \cap P = Z \cap P = 1$.

Let N_1 be a maximal subgroup of N such that N_1 is normal in some Sylow p-subgroup of G. Denote by S a complement to N in P and by B a subgroup of S with $|N_1||B| = p^k$. Let $V = N_1 B$. If V has a supplement T in G with $T \in \mathfrak{F}$, then $T \cap P \ne 1$. But by Lemma 2.3, $T \cap P \le Z_{\mathcal{F}}(G) \cap P = 1$. This contradiction shows that $V = S(V \cap Z) = S \in \tau(G)$, which implies that $|G : N_G(N_1)| = |G : N_G(V \cap N)|$ is a power of p. Thus N_1 is normal in G by Lemma 2.4 and so $N_1 = 1$. But then $N \le Z \le Z_{\mathcal{F}}(G)$, a contradiction. Therefore $\Phi(P) \ne 1$.

The final contradiction.

By (7), (8) and the choice of G, we have $P/\Phi(P) \le Z_{\mathfrak{F}}(G/\Phi(P))$. Hence $P \le Z_{\mathcal{F}}(G)$ by Lemma 2.2. This contradiction completes the proof.

Lemma 2.18 *If $G = NT$, where T is a proper subgroup of G, $N \le Z_\infty(G)$ and N is a p-group, then $O^p(G) \ne G$.*

Proof Let $T \le M$ where M is a maximal subgroup of G. Then $G/M_G = (NM_G/M_G)(M/M_G)$ is a primitive group and $NM_G/M_G \le Z_\infty(G/M_G)$ by Chap. 1, Theorem 2.6(d). Since G/M_G is a primitive group, $F(G/M_G) = NM_G/M_G$ is the unique minimal normal subgroup of G/M_G. Hence $NM_G/M_G \le Z(G/M_G)$ and so M/M_G is normal in G/M_G. Therefore $O^p(G) \le M$ since $|G : M|$ is a power of p.

Lemma 2.19 *Let $G = NT$, where N is a minimal normal subgroup of G and T is a maximal subgroup of G.*

(1) *If $|G : T|$ divides 4, then N is abelian.*
(2) *If $N \le Z_{\mathfrak{U}}(G)$, then $|G : T|$ is a prime.*

Proof

(1) If $|G : T| = 4$, then G/T_G is isomorphic with a subgroup of the symmetric group S_4 of degree 4, which implies that $N \simeq NT_G/T_G$ is abelian. If $|G : T| = 2$, then clearly $N \leq Z(G)$.

(2) Assume that $N \leq Z_{\mathfrak{U}}(G)$. Since N is a minimal normal of G, the order of N is a prime. Thus $|G : T|$ is a prime.

Lemma 2.20 *If a subgroup H of G is s-quasinormally embedded in G and $H \leq O_p(G)$, then H is s-quasinormal in G.*

Proof If W is an s-quasinormal subgroup of G such that H is a Sylow p-subgroup of W, then $H = W \cap O_p(G)$ is s-quasinormal in G by Chap. 1, Lemma 5.34(7).

Theorem 2.21 (Guo, Skiba [203]). *Let $\mathfrak{F}$ be the formation of all p-supersoluble groups and τ an inductive subgroup functor. Suppose that either τ is regular or every subgroup in $\tau(G)$ is s-quasinormally p-embedded in G. Let E be a normal subgroup of G and P a Sylow p-subgroup of E of order $|P| = p^n$, where $n > 1$ and $(|E|, p - 1) = 1$. Suppose that every cyclic subgroup of P of prime order or order 4 (if P is a nonabelian group) is either $\mathfrak{F}_\tau$-supplemented or $\mathfrak{F}_\tau$-embedded in G. Then E is p-nilpotent.*

Proof Suppose that this theorem is false and let (G, E) be a counterexample with $|G| + |E|$ minimal. Let $Z = Z_{\mathfrak{F}}(G)$. Then:

(1) $O_{p'}(E) = 1$.

Suppose that $O_{p'}(E) \neq 1$. Since $O_{p'}(E)$ is characteristic in E, it is normal in G and the hypothesis holds for $(G/O_{p'}(E), E/O_{p'}(E))$ by Lemma 1.31(2)(2) and Lemma 1.38(3)(3). The choice of G implies that $E/O_{p'}(E)$ is p-nilpotent and consequently E is p-nilpotent, a contradiction.

(2) $O_p(E) \leq Z_\infty(E)$.

Assume $O_p(E) \neq 1$. Let H be a subgroup of $O_p(E)$. If H is $\mathfrak{F}_\tau$-supplemented in G. Then there is a subgroup T/H_G of G/H_G such that $(H/H_G)(T/H_G) = G/H_G$ and $(H/H_G) \cap (T/H_G) \leq (S/H_G)Z_{\mathfrak{F}}(T/H_G)$ for some τ-subgroup S/H_G of G/H_G contained in H/H_G. Then $(H/H_G) \cap (T/H_G) \leq (S/H_G)(Z_{\mathfrak{F}}(T/H_G) \cap (O_p(E)/H_G)) \leq (S/H_G)Z_{\mathfrak{U}}(T/H_G)$. Hence H is $\mathfrak{U}_\tau$-supplemented in G. Similarly, if H is $\mathfrak{F}_\tau$-embedded in G, then H is $\mathfrak{U}_\tau$-embedded in G. Therefore if τ is regular, then $O_p(E) \leq Z_{\mathfrak{U}}(G)$ by Theorem 2.12. On the other hand, in the case, when every subgroup in $\tau(G)$ is s-quasinormally p-embedded in G, every cyclic subgroup of P of prime order or order 4 (if P is a nonabelian group) is either $\mathfrak{F}_{\tau_0}$-supplemented or $\mathfrak{F}_{\tau_0}$-embedded in G by Lemma 2.20 for the regular inductive subgroup functor τ_0 in Example 1.8. Hence again we can get that $O_p(E) \leq Z_{\mathfrak{U}}(G)$ by Theorem 2.12. But then, since $(|E|, p - 1) = 1$, $O_p(E) \leq Z_\infty(E)$.

(3) *If V/D is a chief factor of G below E, where $D \leq O_p(E)$, then $V \leq O_p(E)$.*

Assume that this is false and let (V, D) be the pair with $|V| + |D|$ minimal such that V/D is a chief factor of G, $D \leq O_p(E)$ and $V \nleq O_p(E)$.

(a) *p divides $|V/D|$ and V is not p-nilpotent.*

Assume that V/D is a p'-group or V is p-nilpotent. Then V is p-soluble. By (2), $D \leq Z_\infty(E) \cap V \leq Z_\infty(V)$. Hence $V/C_V(D)$ is a p-group by Chap. 1, Lemma 2.26 and Chap. 1, Theorem 1.2, which implies that $D \leq Z(V)$. Hence for a Hall p'-subgroup W of V we have $V = D \times W$. Then W is characteristic in V. It follows that $W \leq O_{p'}(E)$, contrary to (1). Hence we have (a).

(b) *$p = 2$ and V/D is nonabelian. Hence $O_2(E) \leq Z_\infty(G)$.*

Since $V \nleq O_p(E)$, the choice of (V, D) and Claim (a) imply that V/D is nonabelian. But $(|E|, p - 1) = 1 = (|V|, p - 1)$, so by the Feit–Thompson's theorem we have $p = 2$. Therefore $O_2(E) \leq Z_\infty(G)$ by Theorem 2.12.

(c) *V has a 2-closed Schmidt group A such that for some cyclic subgroup H of A of order 2 or order 4 we have $H \nleq D$.*

In view of (b) and Chap. 1, Propositions 1.9 and 1.10, V has a 2-closed Schmidt group A and for the Sylow 2-subgroup A_2 of A the following hold: (i) $A_2 = A^{\mathfrak{N}}$; (ii) A_2 is of exponent 2 or exponent 4 (if A_2 is nonabelian); (iii) $\Phi(A) = Z_\infty(A)$; $A_2/\Phi(A_2)$ is a noncyclic chief factor of A. Therefore for some cyclic subgroup H of A of order 2 or order 4 we have $H \nleq \Phi(A_2)$. But $\Phi(A_2) = Z_\infty(A) \cap A_2$, so $H \nleq Z_\infty(V) \cap A \leq Z_\infty(A)$, which implies that $H \nleq D$ by (2). It is also clear that $H_G \leq D$.

Without loss of generality we may assume that $H \leq P$, so H is either $\mathfrak{F}_\tau$-supplemented or $\mathfrak{F}_\tau$-embedded in G.

For any subgroup L of G, we put $\widehat{L} = LD/D$.

(d) *For every nonidentity subgroup S/H_G of V/H_G not contained in D/H_G we have $S/H_G \notin \tau(G/H_G)$.*

Suppose that this is false. First assume that τ is regular. Then $|\widehat{G} : N_{\widehat{G}}(\widehat{H})|$ is a power of 2, so $\widehat{H} \leq O_2(\widehat{G})$ be Lemma 2.4. Hence $\widehat{V} \leq O_2(\widehat{G})$, a contradiction. Thus S/H_G is s-quasinormally p-embedded in G/H_G. Let W/H_G be an s-quasinormal subgroup of G/H_G such that S/H_G is a Sylow 2-subgroup of W/H_G. Then W/H_G is subnormal in G/H_G by Chap. Lemma 5.34(1). Let $W_0 = W \cap V$. Then S/H_G is a Sylow 2-subgroup of W_0/H_G and W_0/H_G is subnormal in V/H_G by Chap. 1, Lemma 5.35(2). Hence $(S/H_G)(D/H_G)/(D/H_G)$ is a Sylow 2-subgroup of $(W_0/H_G)(D/H_G)/(D/H_G)$ and $(W_0/H_G)(D/H_G)/(D/H_G)$ is subnormal in $(V/H_G)(D/H_G)/(D/H_G)$ by Chap. 1, Lemma 5.35(1)(8). Hence $\widehat{S}$ is a Sylow 2-subgroup of $\widehat{W_0}$ and $\widehat{W_0}$ is a subnormal subgroup of $\widehat{V}$. But since H is cyclic, it follows that some composition factor of $\widehat{V}$ has a cyclic Sylow 2-subgroup, so $\widehat{V}$ is 2-nilpotent, which implies that V is 2-nilpotent by (b). This contradiction shows that we have (d).

(e) *For any τ-subgroup S/H_G of G/H_G contained in H/H_G we have $H/H_G \nleq (S/H_G)Z_\mathfrak{F}(G/H_G)$.*

Suppose that $H/H_G \leq (S/H_G)Z_\mathfrak{F}(G/H_G)$. Then $H/H_G = (S/H_G)((H/H_G) \cap Z_\mathfrak{F}(G/H_G))$. Since H/H_G is cyclic, either $H/H_G = S/H_G$ or $H/H_G \leq Z_\mathfrak{F}(G/H_G)$. But in the former case is impossible by (d). Hence $H/H_G \leq Z_\mathfrak{F}(G/H_G)$. But then $(H/H_G)(D/H_G)(D/H_G) \leq Z_\mathfrak{F}((G/H_G)/(D/H_G))$ by Chap. 1, Theorem 2.6(d). Hence $V/D \simeq (V/H_G)/(D/H_G)$ is a 2-group, a contradiction. Therefore we have (e).

(f) $O^2(V) = V$.

Assume that $O^2(V) \neq V$. Since $O^2(V)$ is characteristic in V, it is normal in G. Hence $DO^2(V) = V$. Then in view of the G-isomorphism $V/D \simeq O^2(V)/D \cap O^2(V)$, $O^2(V)/D \cap O^2(V)$ is a nonabelian chief factor of G such that $O^2(V) \nleq O_2(V)$ and $D \cap O^2(V) \leq O_2(E)$, which contradicts the choice of (V, D). Hence we have (f).

(g) *H is not $\mathfrak{F}_\tau$-supplemented in G and so there is a normal subgroup T of G such that HT is s-quasinormal in G.*

Suppose that H is $\mathfrak{F}_\tau$-supplemented in G. Let T/H_G be a subgroup of G/H_G such that $(H/H_G)(T/H_G) = G/H_G$ and $(H/H_G) \cap (T/H_G) \leq (S/H_G)Z_{\mathfrak{F}}(T/H_G)$ for some τ-subgroup S/H_G of G/H_G contained in H/H_G.

Suppose that $T = G$. Then $(H/H_G) \leq (S/H_G)Z_{\mathfrak{F}}(G/H_G)$, which contradicts (e). Therefore $T \neq G$.

Since $HT = G$, $|G : T|$ divides 4. Let $T \leq M$ where M is a maximal subgroup of G. Then $V \nleq M$. Suppose that $|G : M| = 2$. Then from the isomorphism $G/M \simeq V/V \cap M$ we get $O^2(V) \neq V$, contrary to (f). Hence $M = T$ is a maximal subgroup of G and $|G : T| = 4$.

Note that if $TD = G$, then $V = D(T \cap V)$. Hence $O^2(V) \neq V$ by Lemma 2.18, contrary to (f). Therefore $TD \neq G$, which in view of maximality of T implies that $D \leq T$. Therefore $\widehat{G} = \widehat{H}\widehat{T} = \widehat{V}\widehat{T}$, so $\widehat{V}$ is abelian by Lemma 2.19, which contradicts (b). Therefore we have (g).

By (g) and the hypothesis, H is $\mathfrak{F}_\tau$-embedded in G. Let T/H_G be a normal subgroup of G/H_G such that $(H/H_G)(T/H_G) = HT/H_G$ is s-quasinormal in G/H_G and $(H/H_G) \cap (T/H_G) \leq (S/H_G)Z_{\mathfrak{F}}(G/H_G)$ for some τ-subgroup S/H_G of G/H_G contained in H/H_G. Then HT is s-quasinormal in G by Chap. 1, Lemma 5.34(3).

(h) $T_0 = T \cap V \neq V$.

Suppose that $T_0 = T \cap V = V$. Then $H \leq T$ and so $H/H_G \leq (S/H_G)Z_{\mathfrak{F}}(G/H_G)$, contrary to (e).

(k) $T_0 \leq D$.

If $T_0 \nleq D$ then $V = T_0D$. Hence in view of (h), $T_0/T_0 \cap D \simeq V/D$ is a nonabelian chief factor of G, which contradicts the minimality of (V, D).

Final contradiction for (3). Since HT is s-quasinormal in G, the subgroups $HT \cap V = H(T \cap V) = HT_0$ and HT_0D are s-quasinormal in G by Chap. 1, Lemma 5.34(7). Therefore, in view of (k), $HT_0D/D = \widehat{H}$ is s-quasinormal in $\widehat{G}$. Now applying Example 1.8 and Lemma 2.4, we get $\widehat{H} \leq O_2(\widehat{G} \cap \widehat{V})$. Hence $\widehat{V} \leq O_2(\widehat{G})$, so V/D is a 2-group. This contradiction completes the proof of (3).

Conclusion. In view of (3), $E = O_2(E)$. This contradiction completes the proof of the theorem.

Theorem 2.22 (Guo, Skiba [203] and Chen, Guo, Skiba [70]). *Let $\mathfrak{F}$ be the class of all p-supersoluble groups and τ any Φ-quasiregular hereditary inductive subgroup functor. Let E be a normal subgroup of a group G and P a Sylow p-subgroup of E of order $|P| = p^n$, where $n > 1$ and $(|E|, p - 1) = 1$. If every cyclic subgroup of P of* *prime order or 4 (if P is a nonabelian 2-group) is $\mathfrak{F}_\tau$-supplemented or $\mathfrak{F}_\tau$-embedded in G, then E is p-nilpotent.*

Proof Suppose that this theorem is false and let (G, E) be a counterexample with $|G| + |E|$ minimal.

Note that the hypothesis holds for every subgroup B of G containing P since τ is hereditary. Hence the choice of (G, E) implies that $E = G$. Hence every maximal subgroup of G is p-nilpotent by the choice of G. By Chap. 1, Propositions 1.10 and 1.9, $G = PQ$ is a p-closed Schmidt group, $P = G^{\mathfrak{N}}$ and Q is a cyclic Sylow q-subgroup of G for some prime q. Moreover, if $\Phi = \Phi(P)$, then P/Φ is a noncentral chief factor of G, $\Phi = P \cap Z_\infty(G)$ and P is of exponent p of exponent 4 (if P is a nonabelian 2-group) and $\Phi \leq Z_\infty(G)$. Let L/Φ be a minimal subgroup of P/Φ. Then $L = H\Phi$ and $H_G \leq \Phi$ for some cyclic subgroup H of order p or order 4.

(a) *For any τ-subgroup S/H_G of G/H_G contained in H/H_G, we have $H/H_G \nleq (S/H_G)Z_\infty(G/H_G)$.*

Indeed, if $H/H_G \leq (S/H_G)Z_\infty(G/H_G)$, then $H/H_G = (S/H_G)((H/H_G) \cap Z_\infty(G/H_G))$. Suppose that $H/H_G = S/H_G$ is a τ-subgroup of G/H_G. Then $(H/H_G)(\Phi/H_G)/(\Phi/H_G)$ is a τ-subgroup of the Schmidt group $(G/H_G)/(\Phi/H_G)$ since τ is inductive. Without loss of generality we may assume that $\Phi = 1$. Then $H = L$ is a minimal subgroup of the minimal normal subgroup P of G and Q is a maximal subgroup of G. Moreover, H is a τ-subgroup of G and so $P \leq H^G Q_G = HQ_G$ by Lemma 2.4. Hence $H = P$ is a group of order p. This contradiction shows that $H/H_G \neq S/H_G$. Hence, since H/H_G is cyclic, we have $H/H_G \leq Z_\infty(G/H_G) \cap (P/H_G)$. But since $H_G \leq \Phi \leq Z_\infty(G)$, $Z_\infty(G/H_G) = Z_\infty(G)/H_G$. Therefore $H/H_G \leq Z_\infty(G/H_G) \cap (P/H_G) = \Phi/H_G$, so $H \leq \Phi$. This contradiction show that we have (a).

(b) *H is $\mathfrak{F}_\tau$-embedded in G.*

Assume that H is not $\mathfrak{F}_\tau$-embedded in G. Then H is $\mathfrak{F}_\tau$-supplemented in G by hypothesis, so for some any τ-subgroup S/H_G of G/H_G contained in H/H_G, we have $H/H_G \leq (S/H_G)Z_\infty(G/H_G)$ by Lemma 2.5(1), contrary to (a). Hence we have (b).

Let T/H_G be a normal subgroup of G/H_G such that $(H/H_G)(T/H_G)$ is s-quasinormal in G/H_G and $(H/H_G) \cap (T/H_G) \leq (S/H_G)Z_\infty(G/H_G)$ for some τ-subgroup S/H_G of G/H_G contained in H/H_G. Then by Lemma 2.5(2), $P \leq T$. Hence $(H/H_G) = (S/H_G)((H/H_H) \cap (Z_\infty(G/H_G)))$, contrary to (a). This contradiction completes the proof.

Lemma 2.23 *Let p be a prime dividing $|G|$ such that $(|G|, p - 1) = 1$. If for some subgroup M of G we have $|G : M| = p$, then M is normal in G.*

Proof If $p = 2$, this is clear. Let $p > 2$. Then G has odd order, so is soluble by the Feit–Thompson's theorem. Without loss of generality, we may suppose that $M_G = 1$, so G is isomorphic to some subgroup of S_p. Hence M is a Hall p'-subgroup of G and so G is p-supersoluble (see [135, Theorem 1.9.4]. But then G is p-nilpotent since $(|G|, p - 1) = 1$. Thus M is normal in G.

Lemma 2.24 (see [121]). *If G is a $E_{2'}$-group, then G is a $C_{2'}$-group.*

Lemma 2.25 *Let P be a p-subgroup of G. Suppose that G is a C_π-group for some set of primes π with $p \notin \pi$. Let V_1 and V_2 be maximal subgroups of P, T_i a supplement of V_i in G. Suppose that $T_1 = N_G(H_1)$, where H_1 is a Hall π-subgroup of T_1, and that $T_1 \cap P \le V_2$. If G is not π-closed, then T_2 is not π-closed.*

Proof Suppose that T_2 is π-closed. Without loss of generality, we may assume that $T_2 = N_G(H_2)$, where H_2 is a Hall π-subgroup of T_2. Since $G = V_1 T_1 = V_2 T_2$ and $p \notin \pi$, H_1 and H_2 are Hall π-subgroups of G. Since G is a C_π-group, by hypothesis $(H_2)^x = H_1$ for some $x \in G$. Therefore, $G = V_1 T_1 = V_2 T_2 = V_2 T_2^x = V_2 N_G(H_2^x) = V_2 T_1$ by Chap. 1, Lemma 4.14 and so $P = P \cap V_2 T_1 = V_2(P \cap T_1) = V_2$, a contradiction. The lemma is proved.

Corollary 2.26 *Let P be a p-subgroup of G. Suppose that G is a C_π-group for some set of primes π with $p \notin \pi$. If every maximal subgroup of P has a π-closed supplement in G, then G is π-closed.*

Proof Suppose that this assertion is false. Let V be a maximal subgroup of P and T a π-closed supplement of V in G. Without loss of generality, we may assume that $T = N_G(H)$, where H is a Hall π-subgroup of T. It is clear that $T \cap P < P$. Let W be a maximal subgroup of P such that $T \cap P \le W$. Then, by hypothesis, W has a π-closed supplement in G, which is impossible by Lemma 2.25.

Lemma 2.27 *Let E be a normal subgroup of G and H a K-$\mathfrak{F}$-subnormal subgroup of G, where $\mathfrak{F}$ is the class of all p-supersoluble groups. Suppose that $(p-1, |E|) = 1$, $O_{p'}(E) = 1$ and $|G : H| = p^a$. Then:*

(1) *$H \cap E$ is subnormal in E.*
(2) *$Z_\mathfrak{F}(H) \cap E \le O_p(E)$, $Z_\mathfrak{F}(H) \cap E$ is normal in $H \cap E$ and $(H \cap E)/C_{H \cap E}(Z_\mathfrak{F}(H) \cap E)$ is a p-group.*

Proof

(1) By Chap. 1, Lemma 5.12(1), $H \cap E$ is K-$\mathfrak{F}$-subnormal in E. Hence there exists a chain of subgroups

$$H \cap E = H_0 < H_1 < \ldots < H_t = E$$

such that either H_{i-1} is normal in H_i or $H_i/(H_{i-1})_{H_i}$ is p-supersoluble, for all $i = 1, \ldots, t$. We shall proceed by induction on $|E|$. Clearly, $|H_{t-1} : H|$ is a power of p and $H \cap E$ is K-$\mathfrak{F}$-subnormal in H_{t-1}. Hence by induction, $H \cap E$ is subnormal in H_{t-1}.

Now let M be any maximal subgroup of E containing H_{t-1}. Then $(H_{t-1})_G \le M_G$. Hence E/M_G is p-supersoluble and so $|E : M| = p$. Then M is normal in E by Lemma 2.23. But by Chap. 1, Lemma 5.12(1), $H \cap E$ is K-$\mathfrak{F}$-subnormal in M and, clearly, $|M : H \cap E|$ is a power of p. Hence, again by induction, $H \cap E$ is subnormal in M. Consequently, $H \cap E$ subnormal in E.

(2) By (1), $H \cap E$ is subnormal in E. On the other hand, $Z_\mathfrak{F}(H) \cap E$ is normal in $H \cap E$. Hence $Z_\mathfrak{F}(H) \cap E$ is subnormal in E, so $O_{p'}(Z_\mathfrak{F}(H) \cap E) \le O_{p'}(E) = 1$,

by Chap. 1, Lemma 5.35(6). On the other hand, since $(p-1, |E|) = 1$, $Z_{\mathfrak{F}}(H) \cap E$ is p-nilpotent. Thus $Z_{\mathfrak{F}}(H) \cap E \leq O_p(E)$.

By Chap. 1, Theorem 2.7(a), $Z_{\mathfrak{F}}(H) \cap E \leq Z_{\mathfrak{F}}(H \cap E)$. Hence $(H \cap E)/C_{H \cap E}(Z_{\mathfrak{F}}(H) \cap E)$ is a p-group by Chap. 1, (1.3)(3) and Chap. 1, Lemma 2.26.

Theorem 2.28 (Guo, Skiba [203] and Chen, Guo, Skiba [70]). *Let $\mathfrak{F}$ be the class of all p-supersoluble groups and τ any Φ-regular inductive subgroup functor. Let E be a normal subgroup of a group G and P a Sylow p-subgroup of E of order $|P| = p^n$, where $n > 1$ and $(|E|, p - 1) = 1$. Suppose that at least one of the following holds:*

(I) Every maximal subgroup of P is $\mathfrak{F}_\tau$-supplemented in G and also any p-subgroup in $\tau(G)$ is subnormally embedded in G.

(II) Every maximal subgroup of P having no a p-supersoluble supplement in G is strongly $\mathfrak{F}_\tau$-supplemented in G.

(III) Every maximal subgroup of P is $\mathfrak{F}_\tau$-embedded in G.

Then E is p-nilpotent.

Proof Suppose that this theorem is false and let (G, E) be a counterexample with $|G| + |E|$ minimal.

(1) $O_{p'}(E) = 1$ (See (1) in the proof of Theorem 2.21 and use Lemma 2.16).

(2) *If $O_p(E) \neq 1$, then E is p-soluble.*

Indeed, by Lemma 1.31, 1.38, 2.16 and 1.48, and (1.48), the hypothesis holds for $(G/O_p(E), E/O_p(E))$. Hence in the case when $O_p(E) \neq 1$, $E/O_p(E)$ is p-nilpotent by the choice of (G, E), which implies the p-solubility of E.

(3) $O_p(E) \neq 1$.

Suppose that $O_p(E) = 1$. Then, $\Phi(G) = 1$ by (1). In view of (1) and Chap. 1, Lemma 5.35(6), for any subnormal subgroup L of G contained in E we have L is neither a p-group nor a p'-group. Let N be a minimal normal subgroup of G contained in E. Then N is nonabelian and so by, the Feit–Thompson's theorem, $p = 2$ divides $|N|$. Since $(|E|, p-1) = 1$, we have that $|N_2| > 2$. Note also that in view of Lemma 2.27, for every K-$\mathfrak{F}$-subnormal subgroup H of G we have $Z(H) \cap E = 1$.

(a) *For any minimal normal subgroup L of G contained in E and for any τ-subgroup S of G contained in P, we have $S \cap L = 1$.*

Assume that $S \cap L \neq 1$. Let M be a maximal subgroup of G such that $LM = G$. Since τ is Φ-regular, $|G/M_G : N_{G/M_G}((SM_G/M_G) \cap (LM_G/M_G))|$ is a power of 2. Hence L is abelian by Lemma 2.4(2) and so $L \leq O_2(E) = 1$, a contradiction. Hence we have (a).

(b) *Assertion (I) is not true.*

Assume that Assertion (I) is true. Let M be an arbitrary maximal subgroup of P. Since $O_p(E) = 1$, it is clear that $M_G = 1$. Hence there exists a subgroup T such that $G = MT$ and $M \cap T \leq SZ_{\mathfrak{F}}(T)$ for some τ-subgroup S of G contained in M.

Suppose that $S \neq 1$. Let W be a subnormal subgroup of G such that S is a Sylow 2-subgroup of W. In view of Chap. 1, Lemma 5.35(2) we may assume without loss of generality that $W \leq E$. Let L be a minimal subnormal subgroup of G contained in W. Then L neither is a 2-group nor a $2'$-group. Therefore L is a nonabelian simple group and $L_2 = S \cap L$ is a Sylow 2-subgroup of L since S is a Sylow 2-subgroup of W. It is clear that $R = L^G$ is a minimal normal subgroup of G and $S \cap R \neq 1$, contrary to (a). Therefore $S = 1$. Hence every maximal subgroup M of P has a supplement T in G such that $M \cap T \leq Z_{\mathfrak{F}}(T)$.

We now show that $V = T \cap E$ is 2-nilpotent. Let V_2 be a Sylow 2-subgroup of V containing $M \cap V$. Then $|V_2 : V \cap M| \leq |P : M| = 2$. Therefore for a Sylow 2-subgroup Q of $V Z_{\mathfrak{F}}(T)/Z_{\mathfrak{F}}(T)$ we have $|Q|$ divides 2. Hence $V Z_{\mathfrak{F}}(T)/Z_{\mathfrak{F}}(T) \simeq V/V \cap Z_{\mathfrak{F}}(T)$ is 2-nilpotent. It is well-known that the class of all 2-nilpotent groups is a hereditary saturated formation. Hence in view of Chap. 1, Theorem 2.7(a), $V \cap Z_{\mathfrak{F}}(T) \leq Z_{\mathfrak{F}}(V)$. Thus $V = T \cap E$ is 2-nilpotent by Chap. 1, Theorem 2.7(b). But $E = E \cap T M = M(T \cap E)$, so every maximal subgroup of P has a 2-nilpotent supplement T in E. It is clear that a Hall $2'$-subgroup of $T \cap E$ is also a Hall $2'$-subgroup of E. Therefore E is 2-nilpotent by Lemma 2.24 and Corollary 2.26. Hence N is a 2-group, a contradiction. Hence we have (b)

(c) *Assertion (II) is not true.*

Assume that this is false. Then for some maximal subgroup H of P having no a 2-supersoluble supplement in G, there exists a K-$\mathfrak{F}$-subnormal subgroup T of G such that $G = HT$ and $H \cap T \leq SZ_{\mathfrak{F}}(T) \leq S(Z_{\mathfrak{F}}(T) \cap H) = S$ for some τ-subgroup S of G contained in H by Lemma 2.27(2). If $N \cap O^p(E) = 1$, then $N \simeq O^p(E)N/O^p(N) \leq E/O^p(N)$ is a p-group, a contradiction. Hence $N \leq O^p(E)$. On the other hand, since $|E : T \cap E| = |G : T|$ is a p-group and $E \cap T$ is subnormal in E by Lemma 2.27, we have that $N \leq O^p(E) \leq E \cap T \leq T$ by Chap. 1, Lemma 5.35(11). Hence $N \cap H \leq T \cap H \leq S$. But since H is a maximal subgroup of P and $|N_2| > 2$, $N \cap H \neq 1$. Then $N \cap S \neq 1$, which contradicts (a). This contradiction shows that every maximal subgroup of P has a 2-supersoluble supplement in G and so E is 2-nilpotent by Lemma 2.24 and Corollary 2.26. This contradiction shows that wee have (c).

In view of (b) and (c) and the hypothesis, every maximal subgroup H of P is $\mathfrak{F}_\tau$-embedded in G. Then since $M_G \leq O_p(E) = 1$, there exists a normal subgroup T of G such that HT is s-quasinormal in G and $H \cap T \leq SZ_{\mathfrak{F}}(G)$ for some $S \in \tau(G)$. .

Note that $D = E \cap Z_{\mathfrak{F}}(G)$ is a normal 2-nilpotent subgroup of E. If $D \neq 1$, then either $O_2(D) \neq 1$ or $O_{2'}(D) \neq 1$. But $O_2(D) \leq O_2(E) = 1$ and $O_{2'}(D) \leq O_{2'}(E) = 1$ by (1). This contradiction shows that $E \cap Z_{\mathfrak{F}}(G) = 1$. By Chap. 1, Lemma 5.34(7) $1 < HT \cap N$ is s-quasinormal in G. Hence by Chap. 1, Lemma 5.34(1) and Chap. 1, Lemma 5.35(7), $HT \cap N = N$. It is clear that $N \leq T$. Hence $N \cap H \leq T \cap H \leq SZ_{\mathfrak{F}}(G)$, so $N \cap H = N \cap H \cap SZ_{\mathfrak{F}}(G) = S(H \cap Z_{\mathfrak{F}}(G)) \cap N = S \cap N \neq 1$. Now taking a maximal subgroup F of G such that $FN = G$ and arguing as above we get $N \leq O_2(G)$. This contradiction completes the proof of (3).

(4) *There is a maximal subgroup D of G such that $ND = G$, $D_G \cap E = 1$ and $E = N \rtimes M$, where $M = D \cap E$ and $N = O_p(E) = C_E(N)$ is the unique minimal normal subgroup of G and M is p-nilpotent. In particular, E is p-soluble.*

In view of (3), $O_p(E) \neq 1$. Let N be a minimal normal subgroup of G contained in $O_p(E)$. Then the hypothesis holds for $(G/N, E/N)$ by Lemma 1.31, 1.38, 2.16 and Lemma 1.48(4). Therefore E/N is p-nilpotent by the choice of (G, E), so E is p-soluble. It follows that N is the unique minimal normal subgroup of G contained in E. If $N \leq \Phi(G)$, then E is p-nilpotent by Chap. 1, (1.3)(3). Hence $N \not\leq \Phi(G)$ and so $G = N \rtimes D$ for some maximal subgroup D of G. Since $O_p(G) \leq C_G(N)$ by Chap. 1, Lemma 2.10, $O_p(G) \cap D$ is normal in G. Hence $O_p(G) \cap D \cap E$ is normal in G. Note that $E = N \rtimes (D \cap E)$, so $D_G \cap E = 1$ and

$$O_p(E) = O_p(G) \cap E = N(O_p(E) \cap D \cap E),$$

where $O_p(E) \cap D \cap E = O_p(G) \cap D \cap E$ is normal in G. Hence $O_p(E) \cap D \cap E = 1$, and so $N = O_p(E)$. Finally, since E is p-soluble and $O_{p'}(E) = 1$ by (1), we have $C_E(N) = N$ by Chap. 1, Lemma 4.17.

(5) *If H/K is a chief factor of E below N, then $|H/K| > p$*

By Proposition 4.13(c) in [89, A)], $N = N_1 \times \ldots \times N_t$, where $N_1, \ldots, N_t$ are minimal normal subgroups of E, and from the proof of this proposition we see that $|N_i| = |N_j|$ for all $i, j \in \{1, \ldots, t\}$. Hence for any chief factor H/K of E below N we have $|H/K| = |N_1|$ by Chap. 1, Lemma 2.5. Suppose that $|H/K| = p$. Since $(p - 1, |E|) = p$, $C_E(H/K) = E$. Hence $N \leq Z_\infty(E)$, which implies the p-nilpotency of E by (4). This contradiction shows that (5) holds.

(6) *If a nonidentity subgroup S of P is subnormally embedded in G, then $S \cap N \neq 1$.*

Indeed, let W be a subnormal subgroup of G such that S is a Sylow p-subgroup of W. If $S \cap N = 1$, then $W \cap N = 1$. Hence by (4) and Chap. 1, Lemma 5.35(5), $W \leq C_E(N) = N$, a contradiction. Thus (6) holds.

(7) $M = N_E(M_{p'})$, *where $M_{p'}$ is the Hall p'-subgroup of M.*

Let $J = N_E(M_{p'})$. Suppose that $M < J$. Then $J = J \cap NM = M(J \cap N)$ and therefore $J \cap N \neq 1$. Since $E = NJ$, $J \cap N$ is normal in E and $E/C_E(J \cap N)$ is a p-group. Then in view of Chap. 1, Lemma 2.26 and Chap. 1, Theorem 1.6, $J \cap N \leq Z_\infty(E)$. Hence for some minimal normal subgroup C of E contained in N we have $|C| = p$, which contradicts (5).

(8) *For every K-$\mathcal{F}$-subnormal subgroup T of G such that $TP = G = ET$ we have $Z_{\mathfrak{F}}(T) \cap E = 1$.*

Assume that $B = Z_{\mathfrak{F}}(T) \cap E \neq 1$. Then by Lemma 2.27 and Claim (4), $B \leq O_p(E) = N$, B is normal in $T \cap E$ and $(T \cap E)/C_{T \cap E}(B)$ is a p-group. Since $E = E \cap G = E \cap TP = P(E \cap T)$ and E is p-soluble by (4), some Hall p'-subgroup $E_{p'}$ of E is contained in T. Hence $1 < B \leq N_E(E_{p'})$, so $|N_E(E_{p'})| > M$. But this contradicts (7). Hence we have (8).

Final contradiction.

Let $M_p \leq D_p$, where M_p is a Sylow p-subgroup of M and D_p is a Sylow p-subgroup of D. Without loss of generality, we may suppose that $M_p \leq P$. Then

$NM_p = P$ and ND_p is a Sylow p-subgroup of G. Let $N_0 \le N$ be a normal subgroup of ND_p such that $|N : N_0| = p$. Let $W = N_0 D_p$ and $V = N_0 M_p$. Then W is maximal in ND_p and V is maximal in P such that $V_G = 1$.

(i) *For any τ-subgroup S of G contained in V, we have $S \cap N = 1$.*

Assume that this is false. Clearly $SN \cap D_G = 1$. Since τ is Φ-regular, $N \le (S \cap N)^G D_G \le W D_G$ by Lemma 2.3(3). Hence $N = N \cap N_0 D_p D_G = N_0(N \cap D_p D_G) = N_0$. This contradiction shows that we have (i).

(ii) $T_0 = T \cap E$ *is not p-nilpotent, and so V has no p-supersoluble supplement in G.*

Since $V T_0 = V(T \cap E) = VT \cap E = E$, a Hall p'-subgroup $T_{p'}$ of T_0 is a Hall p'-subgroup of E. By (4), E is p-soluble and so any two Hall p'-subgroup of E are conjugate in E. Without loss of generality we may assume that $T_{p'} \le M$. If T_0 is p-nilpotent, then $T_0 \le M$ by (7). It follows that $E = V T_0 = VM$. But since $M_p \le V$ and V is maximal in P, $VM \ne E$. This contradiction shows that we have (ii).

(iii) *Assertion (I) is not true.*

Assume that this is false. Then there exists a subgroup T such that $G = VT$ and $V \cap T \le SZ_{\mathfrak{F}}(T)$ for some τ-subgroup S of G contained in V.

Assume that $S \ne 1$. Then $S \cap N \ne 1$ by (6), contrary to (i). Hence $S = 1$. We show that in this case the subgroup $T_0 = T \cap E$ is p-nilpotent.

First note that since $S = 1$, we have

$$V \cap T_0 = V \cap T \le Z_{\mathfrak{F}}(T) \cap T_0 \le Z_{\mathfrak{F}}(T_0)$$

by Chap. 1, Theorem 2.7(a). Hence, as in the proof of (3), one can show that $T_0 = T \cap E$ is p-nilpotent. But this contradicts (ii). Hence we have (iii).

(iv) *Assertion (II) is not true.*

Assume that this false. Then in view of (ii) and (8), there exists a K-$\mathfrak{F}$-subnormal subgroup T of G such that $G = VT$ and $V \cap T \le SZ_{\mathfrak{F}}(T) \le S(Z_{\mathfrak{F}}(T) \cap E) = S$ for some τ-subgroup S of G contained in V. By Lemma 2.27, $E \cap T$ is subnormal in E. Hence $O^p(E) \le T$ by Chap. 1, Lemma 5.35(11). On the other hand, from (5) we get that $N \le O^p(E)$. Hence $1 \ne N \cap V \le T \cap V \le S$, so $S \cap N \ne 1$, which contradicts (i). Hence we have (iv).

In view of (iii) and (iv) and since $V_G = 1$, for some normal subgroup T of G and some τ-subgroup S of G contained in V, we have TV is s-quasinormal in G and $T \cap V \le SZ_{\mathfrak{F}}(G)$.

Without loss of generality, we may suppose that $T \le E$. If $T = 1$, then V is s-quasinormal in G. So, in view of Example 1.8 and Lemma 2.4 and (4), we again get that $N = N \cap N_0 D_p D_G = N_0$ as in above the proof of (i), a contradiction. Hence $T \ne 1$, so $N \le T$ by (4). Hence $1 < N \cap V \le T \cap V$. In view of (1) and (4), $V \cap Z_{\mathfrak{F}}(G) = 1$. Hence $N \cap V = N \cap V \cap SZ_{\mathfrak{F}}(G) = N \cap S(V \cap Z_{\mathfrak{F}}(G)) = N \cap S \ne 1$, which as above leads to a contradiction. The final contradiction shows that E is p-nilpotent.

The theorem is proved.

Theorem 2.29 *Let $\mathfrak{F}$ be the formation of all p-supersoluble groups and τ a regular inductive subgroup functor. Let E be normal subgroup of a group G and P a Sylow p-subgroup of E of order $|P| = p^n$, where $n > 1$ and $(|E|, p-1) = 1$. Suppose that for every k-intermediate subgroup of P, where k is some integer with $1 \le k < n$, and every cyclic subgroup of P of order 4 (if $1 = k < n - 1$ and P is a nonabelian 2-group) having no p-supersoluble supplement in G is strongly $\mathfrak{F}_\tau$-supplemented in G. Then E is p-nilpotent.*

Proof Suppose that this theorem is false and let G be a counterexample which $|G| + |E|$ minimal. Let $Z = Z_{\mathfrak{F}}(G)$. Then $Z \cap O_p(G) \le Z_\mathfrak{U}(G)$.

(1) $O_{p'}(E) = 1$ (See (1) in the proof of Theorem 2.21 and use Lemma 2.16).
(2) $1 < k < n - 1$ (This follows from Theorem 2.21, Theorem 2.28 and the choice of G).
(3) *Every k-immediate subgroup H of P either has a p-supersoluble supplement in G or has the form $H = S(H \cap Z)$ for some $S \in \tau(G)$.*

   Suppose that H has no p-supersoluble supplement in G. Then H is strongly $\mathfrak{F}_\tau$-supplemented in G. Hence there is a K-$\mathfrak{U}$-subnormal subgroup T of G such that $HT = G$ and $H \cap T \le SZ_{\mathfrak{F}}(T)$ for some subgroup $S \in \tau(G)$ contained in H. Suppose that $H \ne S(H \cap Z)$. Then $T \ne G$, so there is a maximal subgroup M of G such that $HM = G$ and $|G : M| = |E : E \cap M| = p$. Since $(p - 1, |E|) = 1$, $E \cap M$ is normal in E by Lemma 2.23 and so $E \cap M$ is normal in G. Moreover, since $k < n - 1$ by (2), the hypothesis holds for $(G, E \cap M)$. The choice of G implies that $E \cap M$ is p-nilpotent. This induces that $1 \ne O_{p'}(E \cap M) \le O_{p'}(E)$, which contradicts (1). Hence (2) holds.
(4) $D = O_p(E) \ne 1$.

   Suppose that this is false. Then $Z \cap E = 1$. Indeed, the subgroup $Z \cap E$ is p-supersoluble and normal in E. Hence in the case, where $Z \cap E \ne 1$, either $O_{p'}(Z \cap E) \ne 1$ or $O_p(Z \cap E) \ne 1$. But both these subgroups are characteristic in $Z \cap E$, so they are normal in E. In view of (1) we have $O_{p'}(Z \cap E) = 1$. Hence $1 < O_p(Z \cap E) \le O_p(E) = 1$. This contradiction shows that $Z \cap E = 1$. Let N be a minimal normal subgroup of G contained in E. Then N is nonabelian, and by (1), $p = 2$ divides $|N|$ by the Feit–Thompson's theorem. Let $N_2 = P \cap N$ be a Sylow 2-subgroup of N. Then N_2 is noncyclic by [337, (10.1.9)], in particular $|N_2| > 2$. First assume that $|N_2| \le 2^k$. Let $N_2 \le H < V \le P$, where $|H| = 2^k$ and $|V : H| = 2$. By (3) and by $Z \cap E = 1$, every k-intermediate subgroup H of P either belongs to $\tau(G)$ or has a 2-nilpotent supplement in G. If $H \in \tau(G)$, then $N_2 \le O_2(G)$ by Lemma 2.3 since τ is regular. Hence N is a 2-group, a contradiction. Hence H has a 2-nilpotent supplement T in G. It is clear that $T \ne G$, so $T \cap V < V$. Let $T \cap V \le W < V$, where $|V : W| = 2$. Since N_2 is noncyclic, $N_2 \cap W \ne 1$ and so $W \notin \tau(G)$. Hence W has a 2-nilpotent supplement in G, which in view of Lemmas 2.24 and 2.25 implies that E is 2-nilpotent. This contradiction shows that (4) holds.
(5) $|N| \le p^k$ for any minimal normal subgroup N of G contained in D (See (4) the proof of Theorem 2.17).
(6) *If $p^k = 4$, then $D \not\le Z_\infty(E)$.*

Suppose that $D \leq Z_\infty(E)$. Let V/D be a chief factor of G where $V \leq E$. If V/D is a $2'$-group, then V is 2-nilpotent. Therefore $1 < O_{2'}(V) \leq O_{2'}(E) = 1$, a contradiction. Hence 2 divides $|V/D|$ and V/D is nonabelian. By Chap. 1, Propositions 1.9 and 1.10, V has a 2-closed Schmidt subgroup A, $A^\mathfrak{N} = A_2$ is the Sylow subgroup of A and the following hold: (i) A_2 is of exponent 2 or exponent 4 (if A_2 is nonabelian); (ii) If $\Phi = \Phi(A_2)$, then A_2/Φ is a noncyclic chief factor of A; (iii) $Z_\infty(A) = \Phi(A)$. Without loss of generality, we may assume that $A_2 \leq P$. Then, by (iii), $D \cap A \leq Z_\infty(E) \cap A_2 \leq Z_\infty(A_2) \leq \Phi$ and so there exists a subgroup H of A_2 of order 4 such that $H \nleq D$.

Suppose that H has a p-supersoluble supplement T in G. Then $T \neq G$ by the choice G. If $|G : T| = 2$, then $T \trianglelefteq G$ and so $|A_2 T : T| = |G : T| = |A_2 : A_2 \cap T| = 2$. Hence $A_2 \cap T \trianglelefteq A_2$ and $A_2/(A_2 \cap T) = A_2/\Phi$ is cyclic group, a contradiction. Thus $|G : T| = 4$. If $TD = G$, then G is p-soluble and so V/D is abelian, a contradiction. Thus TD/D is a maximal subgroup of G/D such that $|G/D : TD/D|$ divides 4 and $G/D = (V/D)(DT/D)$. Hence by Lemma 2.19(1), $V/O_2(E)$ is abelian. This contradiction shows that H has no 2-supersoluble supplement in G. Therefore $H = S(H \cap Z)$, where $S \in \tau(G)$, by (3).

If $H \cap Z \nleq D$, then $(V/D) \cap (DZ/D)) \neq 1$ and so $V \leq Z$, a contradiction. Hence $H \cap Z \leq D$. Thus $HD/D = SD/D$. Then, since τ is regular, Lemma 2.3 implies that $1 < HD/D \leq O_2(G/D) \cap E/D = 1$, a contradiction. Thus (6) holds.

(7) *If G has a minimal normal subgroup N such that $N \leq D$ and E/N is not p-nilpotent, then $p^k \neq 4$.*

Let $C = C_G(N)$. Assume $p^k = 4$. Then $|N| \leq 4$ by (5), and $D \nleq Z_\infty(G)$ by (6). Suppose that $|N| = 2$. If $|D| \leq 4$, then $D \leq Z_\infty(E)$, which contradicts (5). On the other hand, if $|D| > 4$, then $D \leq Z_\mathfrak{F}(G) \cap O_p(G) \leq Z_\mathfrak{U}(G)$ by Theorem 2.17. Consequently, $D \leq Z_\infty(G)$, a contradiction. Hence $|N| = 4$ and so $|G/C| \simeq A \leq S_3$.

If $E \subseteq C$, then $N \leq Z(E)$ and so $N < D$ by (6). Hence by Theorem 2.17, $D \leq Z_\mathfrak{F}(G) \cap O_p(G) \leq Z_\mathfrak{U}(G)$ and thereby $D \leq Z_\infty(G)$, a contradiction. Thus $E \nsubseteq C$. Consequently, $E \cap C < E$.

If $P \leq C \cap E$, then the hypothesis holds for $(G, C \cap E)$. This implies that $C \cap E$ is 2-nilpotent. If $C \cap E = P$, then P is normal in G and so $C \cap E = P = D \leq Z_\infty(E)$ by Theorem 2.17, a contradiction. Hence $1 \neq O_{2'}(C \cap E) \leq O_{2'}(E)$, which contradicts (1).

Now let $P \nleq C \cap E$. Then $P \cap C$ is a maximal subgroup of P. Since $k < n-1$ by (2), the hypothesis is true for $(G, C \cap E)$, which as above leads to a contradiction. Hence we have (7).

(8) *$E = N \rtimes M$, where $N = O_p(E)$ is a minimal normal subgroup of G and M is p-nilpotent. Hence $N \nleq Z_\mathfrak{F}(E) = Z_\infty(E)$.*

By (3), $D = O_p(E) \neq 1$. Let N be a minimal normal subgroup of G contained in D. We first show that E/N is p-nilpotent. Assume that this is false. Then $p^k \neq 4$ by (7). Hence arguing as in the proof of Claim (7) in the proof of Theorem 2.17, one can show that the hypothesis holds for $(G/N, E/N)$. Hence E/N is p-nilpotent by the choice of G.

It follows that N is the unique minimal normal subgroup of G contained in $O_p(G)$. In view of Chap. 1, Corollary 2.28, $N \not\leq \Phi(G)$ and so $E = N \rtimes (B \cap E)$ for some maximal subgroup B of G. Let $M = B \cap E$, then $M \simeq E/N$ is p-nilpotent. Finally, since, clearly, $Z_{\mathfrak{F}}(E) = Z_\infty(E)$ we have $N \not\leq Z_{\mathfrak{F}}(E)$.

(9) *If H/K is a chief factor of E below N, then $|H/K| > p$ (See (5) in the proof of Theorem 2.28).*

(10) $N \leq O^p(E)$.

Suppose that $N \not\leq O^p(E)$. Then, since $E/O^p(E)$ is a p-group, from the E-isomorphism $O^p(E)N/O^p(E) \simeq N/O^p(E) \cap N \simeq N$, we deduce that some minimal normal subgroup L of E contained in N is also contained in the center $Z(E)$ of E. But then $L \leq N$ by (1) and so $|L| = |N_1| > p$, contrary to (9). Hence we have (10).

(11) *P has a maximal subgroup V such that V neither contains N nor has a p-nilpotent supplement in E.*

In view of (8), every maximal subgroup of P containing N has a p-nilpotent supplement M in E. If every maximal subgroup of P not containing N also has a p-nilpotent supplement M in E, then E is p-nilpotent by Lemma 2.24 and Corollary 2.26. Hence we have (11).

Final contradiction.

Let $W = V \cap N$. Then W is a maximal subgroup of N and $|N : W| = p$. In view of (5), for some subgroup L of V we have $|WL| = p^k$. Let $H = WL$. Then H has no p-supersoluble supplement in G since $H \leq V$. Hence there is a K-$\mathfrak{U}$-subnormal subgroup T of G such that $HT = G$ and $H \cap T = S(H \cap Z) \cap T \leq S(Z \cap H) = S \in \tau(G)$ by (2) and since $H \cap Z \leq E \cap Z = 1$ by (8). By Chap. 1, Lemma 5.12, $T \cap E$ is K-$\mathfrak{U}$-subnormal in E. Hence $T \cap E$ is subnormal in E by Lemma 2.27. Then by (10) and Chap. 1, Lemma 5.35(11), $N \leq O^p(E) \leq T \cap E$. Thus $W = V \cap N = H \cap N = S \cap N \neq 1$. Since τ is regular, $|G : N_G(W)|$ is a power of p and $W = V \cap N$ is normal in P. Since $|EN_G(W) : N_G(W)| = |E : E \cap N_G(W)| = |E : N_E(W)|$ is a p'-number, W is normal in E. Hence $|N_i| = p$, contrary to (9). The final contradiction completes the proof.

Conditions Under Which a Group Belongs to a Solubly Saturated Formation In conclusion of this section, we give general conditions under which a group belongs to (solubly) saturated formation.

Lemma 2.30 *Let $\mathfrak{F}$ be a solubly saturated formation containing all supersoluble groups and G a group with a normal subgroup E such that $G/E \in \mathfrak{F}$. If E is cyclic, then $G \in \mathfrak{F}$.*

Proof Without loss of generality, we may assume that E is a minimal normal subgroup of G. Then $|E| = p$ for some prime p. We know that $\mathfrak{F} = CLF(f)$ for some formation function (see Chap. 1, Theorem 1.6). It is clear that the group $H = E \rtimes (G/C_G(E))$ is supersoluble. Hence $H \in \mathfrak{F}$ and so $G/C_G(E) \in f(p)$ by Chap. 1, Proposition 1.5. Let $C^p/E = C^p(G/E)$. Then $C^p(G) = C^p \cap C_G(E)$ by the definition of $C^p(G)$ and Chap. 1, Lemma 2.5. But since $G/E \in \mathfrak{F}$, $G/C^p(G) \simeq (G/E)/C^p(G/E) \in f(p)$ and so $G \in \mathfrak{F}$.

Theorem 2.31 (Guo, Skiba [203] and Chen, Guo, Skiba [70]). *Let $\mathfrak{F}$ be a solubly saturated formation containing all supersoluble groups and $X \leq E$ be normal subgroups of a group G such that $G/E \in \mathfrak{F}$ and $X = F^*(E)$ or $X = E$. Let τ be a Φ-regular hereditary or regular inductive subgroup functor. Suppose that for every noncyclic Sylow subgroup P of X at least one of the following holds:*

(I) Every maximal subgroup of P is $\mathfrak{U}_\tau$-supplemented in G and also any subgroup in $\tau(G)$ is p-subnormally embedded in G.

(II) Every maximal subgroup of P is $\mathfrak{U}_\tau$-embedded in G.

Then $G \in \mathfrak{F}$. Moreover, in the case when τ is regular, then $E \leq Z_\mathfrak{U}(G)$.

Proof Suppose that this corollary is false and let (G, E) be a counterexample with $|G| + |E|$ minimal.

Let p be the smallest prime dividing $|X|$ and P a Sylow p-subgroup of X. First we show that $X \leq Z_\mathfrak{U}(G)$. If $X = P$, this directly follows from Theorem 2.6. Assume that $X \neq P$. Note that in the case when P is cyclic, X is p-nilpotent by Chap. 1, Lemma 3.39. On the other hand, if P is not cyclic, then X is p-nilpotent by Theorem 2.28. Let V be the Hall p'-subgroup of X. Then V is nonidentity characteristic in X and so it is normal in G. It is clear that the hypothesis holds for (G, V), so $V \leq Z_\mathfrak{U}(G)$ by the choice of (G, E). On the other hand, by Lemmas 1.31, 1.38 and 1.48, the hypothesis holds for $(G/V, X/V)$. Hence again the choice of (G, E) implies that $X/V \leq Z_\mathfrak{U}(G/V)$ and so $X \leq Z_\mathfrak{U}(G)$ by Chap. 1, Lemma 2.5. Now using Chap. 1, Theorem 2.6 we get $E \leq Z_\mathfrak{U}(G)$ and so $G \in \mathfrak{F}$ by Lemma 2.30.

Theorem 2.32 (Guo, Skiba [203] and Chen, Guo, Skiba [70]). *Let $\mathfrak{F}$ be a solubly saturated formation containing all supersoluble groups and $X \leq E$ be normal subgroups of a group G such that $G/E \in \mathfrak{F}$ and $X = F^*(E)$ or $X = E$. Let τ be an inductive subgroup functor and either τ is regular, or every subgroup in $\tau(G)$ is s-quasinormally p-embedded in G, or τ is hereditary and quasiregular. Suppose that for every noncyclic Sylow subgroup P of X every cyclic subgroup of P of prime order or order 4 (if P is a nonabelian group) is either $\mathfrak{U}_\tau$-supplemented or $\mathfrak{U}_\tau$-embedded in G. Then $G \in \mathfrak{F}$ and $E \leq Z_\mathfrak{U}(G)$.*

Proof See the proof of Theorem 2.31 and use Theorems 2.12, 2.21 and 2.22.

Theorem 2.33 (Guo, Skiba). *Let $\mathfrak{F}$ be a solubly saturated formation containing all supersoluble groups and $X \leq E$ be normal subgroups of a group G such that $G/E \in \mathfrak{F}$ and $X = F^*(E)$ or $X = E$. Let τ a regular special strongly inductive subgroup functor. Suppose that for every noncyclic Sylow subgroup P of X every k-intermediate subgroup of P and also, in the case when $1 = k < n - 1$ and P is a nonabelian 2-group, every cyclic subgroup of P of order 4 having no a p-supersoluble supplement in G is strongly $\mathfrak{U}_\tau$-supplemented in G. Then $G \in \mathfrak{F}$ and $E \leq Z_\mathfrak{U}(G)$.*

Proof See Theorem 2.31 and use Theorems 2.17 and 2.29.

Lemma 2.34 *Let P be a normal p-subgroup of G with $|P| > p$ and $P \cap \Phi(G) = 1$. Suppose that either τ is Φ-quasinormal and every maximal subgroup of P is either*

$\mathfrak{U}_\tau$-supplemented or $\mathfrak{U}_\tau$-embedded in G, or τ is quasiregular and every k-immediate subgroup of P is strongly $\mathfrak{U}_\tau$-supplemented. Then some maximal subgroup of P is normal in G.

Proof Let G_p be a Sylow p-subgroup of G and $Z = Z_\mathfrak{U}(G)$. Since $P \cap \Phi(G) = 1$, $P = N_1 \times N_2 \times \ldots \times N_t$, where N_i is a minimal normal subgroup of G, for all $i = 1, 2, \ldots t$ by Chap. 1, Proposition 2.24. We may assume, clearly, that $|N_i| \neq p$ for all $i = 1, 2, \ldots, t$. Then $P \cap Z = 1$.

Arguing similarly as in the proof of Theorem 2.17 (see Claim (4)), one can show that $|N_i| \leq p^k$ for all $i = 1, 2, \ldots t$. Hence $t > 1$.

If for some i we have $|N_i| < p^k$, then by Lemmas 1.31, 1.38 and 2.16 the hypothesis holds for G/N_i and so by induction some maximal subgroup M/N_i of P/N_i is normal in G. Then a maximal subgroup M of P is normal in G. Hence we may assume that $|N_i| = p^k$ for all $i = 1, 2, \ldots t$. Therefore we now only need to consider the case when every k-immediate subgroup of P is strongly $\mathfrak{U}_\tau$-supplemented. Let $H = A \times L$ where A is a maximal subgroup of N_1 and L is a minimal subgroup of N_2 such that L is normal in G_p. Then H is strongly $\mathfrak{U}_\tau$-supplemented in G.

Let T be a K-$\mathfrak{U}$-subnormal subgroup of G such that $HT = G$ and $H \cap T \leq SZ_\mathfrak{U}(T)$ for some τ-subgroup S of G contained in H. First assume that $T \neq G$. Then for some maximal subgroup M of G we have $T \leq M$ and $|G : M| = p$. Hence $M \cap P$ is normal in G. But then $P/M \cap P$ is a chief factor of G with order p. Hence by Chap. 1, Lemma 2.5, $|N_i| = p^k = p$. This contradiction shows that $T = G$, and so $H = S$. But then $H \cap N_2 = L$ is normal in G by Lemma 2.4 and so $|N_2| = p$. This contradiction completes the proof of the lemma.

Lemma 2.35 *Let E be a normal subgroup of G. Let τ be a Φ-regular inductive subgroup functor. If either every maximal subgroup of every noncyclic Sylow subgroup of E is $\mathfrak{U}_\tau$-embedded in G or every maximal subgroup of every noncyclic Sylow subgroup of E is $\mathfrak{U}_\tau$-supplemented in G and every τ-subgroup of G is subnormally embedded in G, then E is supersoluble.*

Proof Suppose that this lemma is false and let (G, E) be a counterexample with $|G| + |E|$ minimal. Let P be a Sylow p-subgroup of E where p is the smallest prime dividing $|E|$. By Theorem 2.29 and Chap. 1, Lemma 3.39, E is p-nilpotent. Let V be the Hall p'-subgroup of E. Then V is normal in G and the hypothesis holds for (G, V). Hence V is supersoluble by the choice of (G, E). Thus a Sylow q-subgroup Q of V, where q is the largest prime dividing $|V|$, is normal and so characteristic in V. It follows that Q is normal in G and the hypothesis holds for $(G/Q, E/Q)$ by Lemmas 1.31 and 1.38. The choice of (G, E) implies that E/Q is supersoluble. On the other hand, by Theorem 2.6, $Q \leq Z_{\mathfrak{U}\Phi}(G)$, so E is supersoluble by Chap. 1, Theorem 2.27.

Theorem 2.36 (Guo, Skiba [203]). *Let $\mathcal{F}$ be a saturated formation containing all supersoluble groups and G be a group with normal subgroups $X \leq E$ such that $G/E \in \mathcal{F}$. Let τ be any Φ-regular hereditary inductive subgroup functor and every τ-subgroup of G be subnormally embedded in G. Suppose that every maximal*

subgroup of every noncyclic Sylow subgroup of X is $\mathfrak{U}_\tau$-supplemented in G. If $X = E$
or $X = F^*(E)$, then $G \in \mathfrak{F}$.

Proof Assume that this theorem is false and let (G, E) be a counterexample with
$|G| + |E|$ minimal. Let $F = F(E)$ and $F^* = F^*(E)$. Let p be prime divisor of $|F^*|$
and P the Sylow p-subgroup of F^*. Then:

(1) X is supersoluble (This follows from Lemma 2.35).
(2) $X = F^* \neq E$.
   Indeed, suppose that $X = E$. Then E is q-nilpotent, where q be smallest prime
   divisor of $|E|$ by (1). Let V be the Hall q'-subgroup of X. If $V = 1$, then
   $E \leq Z_{\mathfrak{U}}(G)$ by Theorem 2.6, so $G \in \mathfrak{F}$ by Lemma 2.30. But this contradicts
   the choice of (G, E). Hence $V \neq 1$. Since V is characteristic in X, it is normal
   in G. Moreover, the hypothesis holds for $(G/V, X/V)$ by Lemma 1.31. Hence
   $G/V \in \mathfrak{F}$ by the choice of (G, E). Now we see that the hypothesis holds also
   for (G, V) and so $G \in \mathfrak{F}$ again by the choice of (G, E). This contradiction
   shows that we have (2).
(3) $F^* = F$ and $C_G(F) = C_G(F^*) \leq F$.
   Since $X = F^*$ is soluble by (1) and (2), $F^* = F$ by Chap. 1, Proposition
   2.9(8). We have also $C_G(F) = C_G(F^*) \leq F$ by Chap. 1, Proposition 2.9(5).
(4) *Every proper normal subgroup W of G with $F \leq W \leq E$ is supersoluble.*
   By Chap. 1, Proposition 2.9(1)(3), $F^*(E) = F^*(F^*(E)) \leq F^*(W) \leq F^*(E)$.
   It follows that $F^*(W) = F^*(E) = F^*$. Thus the hypothesis is still true for
   (W, W) by Lemma 1.46(1). Note that $W/W = 1 \in \mathfrak{U}$. The minimal choice of
   G implies that W is supersoluble.
(5) *If $E \neq G$, E is supersoluble* (It follows directly from (4)).
(6) *If L is a minimal normal subgroup of G and $L \leq P$, then $|L| > p$.*
   Assume that $|L| = p$. Let $C_0 = C_E(L)$. Then the hypothesis is true for
   $(G/L, C_0/L)$. Indeed, clearly, $G/C_0 = G/(E \cap C_G(L))$ is supersoluble. Besides, since $L \leq Z(C_0)$ and evidently $F = F^* \leq C_0$ and $L \leq Z(F)$, we have
   $F^*(C_0/L) = F^*(C_0)/L = F^*/L$ by Chap. 1, Proposition 2.9(1). Hence the
   hypothesis is still true for G/L. This implies that $G/L \in \mathcal{F}$ and thereby $G \in \mathcal{F}$
   by Lemma 2.30, a contradiction.
(7) $\Phi(G) \cap P \neq 1$ and $F^*(E/L) \neq F^*/L$ *for every minimal normal subgroup L*
   of G contained in $\Phi(G) \cap P$.
   Suppose that $\Phi(G) \cap P = 1$. Then P is the direct product of some minimal
   normal subgroups of G by Chap. 1, Proposition 2.24. Hence by Lemma 2.34,
   P has a maximal subgroup M which is normal in G. Now by Chap. 1, Lemma
   2.5, G has a minimal normal subgroup L with order p contained in P, which
   contradicts (6). Thus $\Phi(G) \cap P \neq 1$. Let $L \leq \Phi(G) \cap P$ and L be a minimal
   normal subgroup of G. Assume that $F^*(E/L) = F^*/L$. Then the hypothesis
   is true for G/L and so $G/L \in \mathcal{F}$ by the choice of G. But then $G \in \mathcal{F}$ since
   $L \leq \Phi(G)$. This contradiction shows that $F^*(E/L) \neq F^*/L$.
(8) *E is not soluble and $E = G$.*
   Assume that E is soluble. Let L be a minimal normal subgroup of G contained in
   $\Phi(G) \cap P$. By [89, A, 9.3(c)], $F/L = F(E/L)$. On the other hand, $F^*(E/L) =$

$F(E/L)$ by Chap. 1, Proposition 2.9(8). Hence $F^*(E/L) = F(E/L) = F/L = F^*/L$ by (3), which contradicts (7). Therefore E is not soluble and so $E = G$ by (5).

(9) G *has a unique maximal normal subgroup M containing F, M is supersoluble and G/M is a nonabelian simple group* (This directly follows from (4) and (8)).

(10) G/F *is a nonabelian simple group and G/L is a quasinilpotent group if L is a minimal normal subgroup of G contained in $\Phi(G) \cap P$.*
Let L be a minimal normal subgroup of G contained in $\Phi(G) \cap P$. Then by (7), $F^*(E/L) \neq F^*/L$. Thus $F/L = F^*/L$ is a proper subgroup of $F^*(G/L)$ by Chap. 1, Proposition 2.9(2). By Chap. 1, Proposition 2.9(4), $F^*(G/L) = F(G/L)E(G/L)$, where $E(G/L)$ is the layer of G/L. By (9), every chief series of G has only one nonabelian factor. But since $E(G/L)/Z(E(G/L))$ is a direct product of simple nonabelian groups, we see that $F^*(G/L) = G/L$ is a quasinilpotent group. Since $F(G/L) \cap E(G/L) = Z(E(G/L))$ by Chap. 1, Proposition 2.9(4), $G/F \simeq (G/L)/(F/L)$ is a simple nonabelian group.

(11) $F^* = P$.
Assume that $P \neq F$ and let Q be a Sylow q-subgroup of F, where $q \neq p$. By (10) and Chap. 1, Corollary 3.12(2), $Q \leq Z_\infty(G)$. Hence by Chap. 1, Proposition 2.9(6), $F^*(G/Q) = F^*/Q$ and so the hypothesis is still true for $(G/Q, G/Q)$. Hence G/Q is supersoluble by the choice of G. It follows that G is soluble, which contradicts (8).

(12) p *is the largest prime dividing $|G|$ and every Sylow q-subgroup Q of G where $q \neq p$ is abelian.*
Let $D = PQ$. Then $D < G$ by (8). By Lemma 2.35, $D = PQ$ is supersoluble. Since $O_q(D) \leq C_G(P)$, we have $C_G(P) \leq P$ by (3) and (11). Hence $O_q(D) = 1$. Consequently, $p > q$ and $F(D) = P$. Hence p is the largest prime dividing $|G|$ and $D/P \simeq Q$ is abelian.
Final contradiction.
By (8) and the Feit–Thompson's theorem, 2 divides $||G|$. By (12), a Sylow 2-subgroup of G/P is abelian. Hence by [250, XI, Theorem 13.7], G/P is isomorphic to one of the following: (a) $PSL(2, 2^f)$; (b) $PSL(2, q)$, where 8 divides $q - 3$ or $q - 5$; (c) The Janko group J_1; (d) A Ree group. It is not difficult to show that in any case G/P has a nonabelian supersoluble subgroup V/P such that p does not divide $|V/P|$. Hence in view of (3) and (11), we have $C_V(P) \leq P$ and so $P = F(V)$. On the other hand, V is supersoluble by Lemma 2.35. Thus V/P is abelian, a contradiction. The theorem is proved.

Theorem 2.37 (see Guo, Skiba [203] and [70]). *Let $\mathcal{F}$ be a saturated formation containing all supersoluble groups and G a group with normal subgroups $X \leq E$ such that $G/E \in \mathcal{F}$. Let τ be any Φ-regular hereditary inductive subgroup functor. Suppose that every maximal subgroup of every noncyclic Sylow subgroup of X is $\mathfrak{U}_\tau$-embedded in G. If $X = E$ or $X = F^*(E)$, then $G \in \mathfrak{F}$ and $E \leq Z_{\mathfrak{U}}(G)$.*

Proof See the proof of Theorem 2.36 and use Lemmas 1.46(3) and 1.38 instead of Lemmas 1.46(1) and (1.32), respectively.

3.3 Groups with Some Special Supplement Conditions for Subgroups

The general theory in previous two section covers a large number of the results in a long list papers (see Sect. 3.5 below). Nevertheless, there are some useful applications supplement conditions which are not covered by the general theory. Some of them we consider in this section.

SE-**supplemented subgroups.** A subgroup H of a group G is said to be SE-*supplemented* in G [208] if there is a subgroup T of G such that $HT = G$ and $H \cap T \leq H_{seG}$ (see Definition 1.50).

Lemma 3.1 *Let G be a group, $H \leq E \leq G$ and N a normal subgroup of G.*

(1) $H_{seG} \leq H_{seE}$.
(2) $H_{seG}N/N \leq (HN/N)_{se(G/N)}$.
(3) *If either $N \leq H$ or $(|N|, |H|) = 1$, then $H_{seG}N/N = (HN/N)_{se(G/N)}$.*

Proof (1), (2) These follow from Example 1.43 and Lemma 1.52(3)(4).
   (3) In view of (2) we have only to prove that $(HN/N)_{se(G/N)} \leq H_{seG}N/N$.
   Let V/N be an s-quasinormally embedded subgroup of G/N such that $V/N \leq HN/N$. Then $V = V \cap HN = N(V \cap H)$. We now show that $U := V \cap H$ is s-quasinormally embedded in G. Let p be any prime dividing $|U|$ and $P \in \mathrm{Syl}_p(U)$. Let W/N be an s-quasinormal subgroup of G/N such that $PN/N \in \mathrm{Syl}_p(W/N)$. Then W is s-quasinormal in G by Chap. 1, Lemma 5.34(3), and $PN/N = W_pN/N$ for some Sylow p-subgroup W_p of W. Then $W_p \in \mathrm{Syl}_p(W_pN)$. First suppose that $N \leq H$. Then $U = V$ and $P \in \mathrm{Syl}_p(PN)$. Therefore P is a Sylow p-subgroup of W.
   Now assume that $(|N|, |H|) = 1$. Then $P \in \mathrm{Syl}_p(V)$. Hence $PN/N \in \mathrm{Syl}_p(V/N)$. Consequently, $PN = W_pN$. It is clear that P and $W_p \in \mathrm{Syl}_p(PN)$. Hence $P = (W_p)^n$ for some $n \in N$. Since W is s-quasinormal in G, W^n is s-quasinormal in G. Moreover, $P = (W_p)^n \in \mathrm{Syl}_p(W^n)$. Thus, U is s-quasinormally embedded in G. It follows that $U \leq H_{seG}$. Therefore $V/N = UN/N \leq H_{seG}N/N$ and so $(HN/N)_{se(G/N)} \leq H_{seG}N/N$.

Lemma 3.2 *Let H be an SE-supplemented subgroup of G and N a normal subgroup of G.*

(1) *If $H \leq K \leq G$, then H is SE-supplemented in K.*
(2) *If $N \leq H$, then H/N is SE-supplemented in G/N.*
(3) *If $(|N|, |H|) = 1$, then HN/N is SE-supplemented in G/N.*

Proof By using Lemma 3.1, the proof is similar to the proof of Lemmas 1.31 and 1.46.

Theorem 3.3 (Guo, Skiba, Yang [208]). *Let E be a normal subgroup of a group G, $p_1 < p_2 < \ldots < p_n$ the set of all prime divisors of $|E|$ and $\pi_i = \{p_1, p_2, \ldots, p_i\}$. Suppose that for each $p \in \pi_i$, the maximal subgroups of any Sylow p-subgroup of*

E are SE-supplemented in G. Then E has a normal Hall π_i'-subgroup $E_{\pi_i'}$ and each chief factor of G between E and $E_{\pi_i'}$ is cyclic.

Proof Assume that this theorem is false and let (G, E) be a counterexample for which $|G| + |E|$ is minimal. Let $p = p_1$ be the smallest prime dividing $|E|$ and P a Sylow p-subgroup of E. Let $Z = Z_{\mathfrak{U}}(G)$. We proceed the proof via the following steps.

(1) *E is p-nilpotent.*

We may consider, without loss of generality, that $i = 1$. Assume that E is not p-nilpotent. Then:

(a) $E = G$.

Indeed, if $E < G$, then $|E| + |E| < |G| + |E|$. Hence the hypothesis is true for (E, E) by Lemma 3.2(1). The choice of (G, E) implies that E is p-nilpotent, a contradiction.

(b) $O_{p'}(G) = 1$.

Let $D = O_{p'}(G)$. By Lemma 3.2(3), the hypothesis is true for $(G/D, ED/D)$. Hence, if $D \neq 1$, then G/D is p-nilpotent by the choice of (G, E). It follows that G is p-nilpotent, a contradiction.

(c) *If $P \leq V < G$, then V is p-nilpotent.*

In fact, by Lemma 3.2(1), the hypothesis holds for V. Hence V is p-nilpotent by the choice of G.

(d) $O_{p'}(L) = 1$ *for all s-quasinormal subgroups L of G.*

By Chap. 1, Lemma 5.34(1), L is subnormal in G. It follows that $O_{p'}(L)$ is subnormal in G. Hence $O_{p'}(L) \leq O_{p'}(G) = 1$ by Chap. 1, Lemma 5.35(6).

(e) *If N is an abelian minimal normal subgroup of G, then G/N is p-nilpotent.*

In view of (b), N is a p-group and so $N \leq P$. Thus the hypothesis is true for $(G/N, E/N)$ by Lemma 3.2(2). The choice of (G, E) implies that G/N is p-nilpotent.

(f) *G is p-soluble.*

In view of (e), we need only to show that G has an abelian minimal normal subgroup. Suppose that this is false. Then $p = 2$ by the Feit-Thompson's theorem. By Chap. 1, Lemma 5.34(1) and Chap. 1, Lemma 5.35(6), we see that every nonidentity subgroup of P is not s-quasinormal in G. Hence for every nonidentity s-quasinormally embedded in G subgroup $L \leq V$, where V is a maximal subgroup of P, and for any s-quasinormal subgroup W of G such that $L \in \mathrm{Syl}_2(W)$ we have $L \neq W$. Moreover, $W_G \neq 1$. Indeed, if $W_G = 1$, then W is nilpotent by Chap. 1, Lemma 5.34(1). Hence $O_{2'}(W) \neq 1$, which contradicts (d). Note also that for any minimal normal subgroup N of G we have $NP = G$ (otherwise, N is 2-nilpotent by (c), a contradiction). It follows that N is the unique minimal normal subgroup of G. Therefore $N \leq W$ (since $W_G \neq 1$) and consequently $N \cap P = N \cap L$.

Now we show that $V_{seG} \neq 1$ for any maximal subgroup V of P. In fact, suppose that $V_{seG} = 1$ and let T be a subgroup of G such that $VT = G$ and $V \cap T \leq V_{seG} = 1$. Then T is a complement of V in G. This induces that T is 2-nilpotent since the order of a Sylow 2-subgroup of T is equal to 2. We may, therefore, assume that $T = N_G(H_1)$ for some Hall $2'$-subgroup H_1 of G. It is clear that $H_1 \leq N$. By Lemma 2.24, any two Hall $2'$-subgroups of N are conjugate in N. By the Frattini argument, $G = NT$. Then $P = (P \cap N)(P \cap T^x)$ for some $x \in G$ by Chap. 1, Lemma 4.14. Let $T_1 = T^x = N_G(H_1{}^x)$. It is clear that $P \cap T_1 \neq P$. Hence we can choose a maximal subgroup V_1 in P containing $P \cap T_1$. By the hypothesis, there exists a subgroup T_2 such that $G = V_1 T_2$, where $V_1 \cap T_2 \leq (V_1)_{seG}$. If $(V_1)_{seG} = 1$, then as above, we have that T_2 is 2-nilpotent and so G is 2-nilpotent by Lemma 2.24 and Corollary 2.26. This contradiction shows that $(V_1)_{seG} \neq 1$. Let L be any nonidentity s-quasinormally embedded subgroup of G contained in V_1 and W be an s-quasinormal subgroup of G such that $L \in \mathrm{Syl}_2(W)$. Then $L \cap N = P \cap N$, which implies $P = (P \cap N)(P \cap T_1) = (L \cap N)(P \cap T_1) \leq V_1$, a contradiction. Therefore for every maximal subgroup V of P we have $(V_1)_{seG} \neq 1$. But then from above, we know that $N \cap P \leq V$. Hence $N \cap P \leq \Phi(P)$ and so N is 2-nilpotent by [248, Chap. IV, Theorem 4.7], a contradiction. Hence, (f) holds.

The final contradiction for (1).

Let N be any minimal normal subgroup of G. Then in view of (b) and (f), N is a p-group and so G/N is p-nilpotent by (e). This implies that N is the unique minimal normal subgroup of G and $N \nleq \Phi(G)$. Hence G is a primitive group and thereby $N = C_G(N) = F(G)$ by Chap. 1, Proposition 1.4. Let M be a maximal subgroup of G such that $G = N \rtimes M$. Let $M_p \in \mathrm{Syl}_p(M)$ and V be a maximal subgroup of P such that $M_p \leq V$. Then $VM \neq G$ and so $VM^x \neq G$ for all $x \in G$ by Chap. 1, Lemma 4.14. Since V is SE-supplemented in G, there is a subgroup T of G such that $VT = G$ and $V \cap T \leq V_{seG}$. Suppose that $V_{seG} = 1$. Then T is a complement of V in G. It follows that $|T_p| = p$, where $T_p \in \mathrm{Syl}_p(T)$. Then T is p-nilpotent since p is the smallest prime dividing $|G|$. Hence $T_{p'} \trianglelefteq T$, where $T_{p'}$ is a Hall p'-subgroup of T. Since G is p-soluble, any two Hall p'-subgroups of G are conjugate. Therefore there is an element $x \in G$ such that $T_{p'} \leq M^x$. If $T_p \leq M^x$, then $T \leq M^x$ and $G = VT = VM^x$, a contradiction. Hence $T_p \nleq M^x$. But $G/N \simeq M^x \leq N_G(T_{p'})$ (since G/N is p-nilpotent) and $T_p \leq N_G(T_{p'})$. Therefore $G = \langle M^x, T_p \rangle = N_G(T_{p'})$, which contradicts (b). Hence $V_{seG} \neq 1$. Let $L \neq 1$ be an s-quasinormally embedded subgroup of G such that $L \leq V$ and W an s-quasinormal subgroup of G such that $L \in \mathrm{Syl}_p(W)$. Suppose that $L = W$. Then by Chap. 1, Lemma 5.34(6), $N \leq L^G \leq V$, a contradiction. Hence $L \neq W$. Then in view of (b) and Chap. 1, Lemma 5.34(1), we have $W_G \neq 1$. This implies that $N \leq L \leq V$ and so $V = VN = P$. This final contradiction shows that (1) holds.

(2) $E = P$

Let $E_{p'} \neq 1$ be the Hall p'-subgroup of E.

Suppose that $E \neq P$. Then $E_{p'} \neq 1$. Hence every chief factor of $G/E_{p'}$ below $E/E_{p'}$ is cyclic by the choice of (G, E). On the other hand, the minimality of (G, E) implies that $E_{p'}$ has a normal Hall π_i'-subgroup V and each chief factor of G between $E_{p'}$ and V is cyclic. Hence V is a normal Hall π_i'-subgroup of

E and each chief factor of G between E and V is cyclic, which contradicts the choice of (G, E). Hence $E = P$.

(3) *If N is a minimal normal subgroup of G contained in P, then $P/N \leq Z_{\mathfrak{U}}(G/N)$, N is the only minimal normal subgroup of G contained in P and $|N| > p$.*

Indeed, by Lemma 3.2(2), the hypothesis holds on G/N for any minimal normal subgroup N of G contained in P. Hence $P/N \leq Z_{\mathfrak{U}}(G/N)$ by the choice of $(G, E) = (G, P)$. If $|N| = p$, then $P \leq Z_{\mathfrak{U}}(G)$, a contradiction. If G has two minimal normal subgroups R and N contained in P, then $NR/R \leq P/R$ and from the G-isomorphism $RN/N \simeq N$ we have $|N| = p$, a contradiction. Hence, (3) holds.

(4) $\Phi(P) \neq 1$.

Suppose that $\Phi(P) = 1$. Then P is an elementary abelian p-group. Let N_1 be any maximal subgroup of N. We show that N_1 is s-quasinormal in G. Let B be a complement of N in P and $V = N_1 B$. Then V is a maximal subgroup of P. Hence V is SE-supplemented in G. Let T be a subgroup of G such that $G = TV$ and $T \cap V \leq V_{seG}$. If $T = G$, then $V = V_{seG}$ is s-quasinormal in G by Lemma 2.20. Hence $V \cap N = V_{seG} \cap N = N_1 B \cap N = N_1 (B \cap N) = N_1$ is s-quasinormal in G by Chap. 1, Lemma 5.34(7). Now assume that $T \neq G$. Then $1 \neq T \cap P < P$. Since $G = VT = PT$ and P is abelian, $T \cap P$ is normal in G. Hence $N \leq T \cap P \leq T$ and consequently $N_1 \leq N \cap T \cap V \leq N \cap V_{seG} \leq N$. Clearly, $N \nleq V$. Hence $N_1 = N \cap V_{seG}$. By Lemma 2.20 and since the subgroup generated by all s-quasinormal in G subgroup of V is also s-quasinormal in G (see Chap. 1, Lemma 5.34(7)), we see that V_{seG} is s-quasinormal in G. Thus by Chap. 1, Lemma 5.34(7), N_1 is s-quasinormal in G. This shows that every maximal subgroup of N is s-quasinormal in G. Hence some maximal subgroup of N is normal in G by Example 1.8 and Lemma 2.34. This contradiction shows that $\Phi(P) \neq 1$.

The final contradiction.

By (4), $\Phi(P) \neq 1$. Let N be a minimal normal subgroup of G contained in $\Phi(P)$. Then the hypothesis is still true for G/N. Hence $P/N \leq Z_{\mathfrak{U}}(G/N)$ by the choice of (G, E). This means that $P/\Phi(P) \leq Z_{\mathfrak{U}}(G/\Phi(P))$. Then by Lemma 2.2, we obtain that $P \leq Z$. This final contradiction completes the proof.

Theorem 3.4 (Guo, Skiba, Yang [208]). *Let $\mathcal{F}$ be a solubly saturated formation containing all supersoluble groups and G a group with normal subgroups $X \leq E$ such that $G/E \in \mathcal{F}$. Suppose that the maximal subgroups of every Sylow subgroup of X are SE-supplemented in G. If either $X = E$ or $X = F^*(E)$, then $G \in \mathfrak{F}$ and $E \leq Z_{\mathfrak{U}}(G)$.*

Proof See the proof of Theorem 2.31 and use Theorem 3.3.

S_p-**embedded Subgroups** Let $H \leq G$, H_{sG} be the subgroup of G generated by all those subgroups of H which are s-quasinormal in G and H^{sG} be the intersection of all s-quasinormal subgroups of G containing H. Then we say that H_{sG} is the S-core of H in G and H^{sG} is the S-closure of H in G.

The s-quasinormal subgroups of a group G form a sublattice of the lattice $L(G)$ of all subgroups of G (Kegel, see Chap. 1, Lemma 5.34(7)). This important result shows that for any subgroup H of a group G both subgroups H_G and H^{sG} are s-quasinormal in G. Based on this fact, we give some new applications of s-quasinormal subgroups in the theory of soluble groups. Our main tools here are the following two concepts.

Definition 3.5 Let H be a subgroup of a group G and p a prime. We say that:

(1) H is S_p-*embedded* in G if G has an s-quasinormal subgroup T such that $|T \cap H|_p = |T \cap H_{sG}|_p$ and $HT = H^{sG}$.
(2) H is s-*embedded* in G if H is S_p-embedded in G for all primes p.

Lemma 3.6 *Let G be a group and $H \leq K \leq G$. Then the following statements hold.*

(1) H_{sG} *is an s-quasinormal subgroup of G and $H_G \leq H_{sG}$.*
(2) $H_{sG} \leq H_{sK}$.
(3) *Suppose that H is normal in G. Then $(K/H)_{s(G/H)} = K_{sG}/H$.*
(4) *If H is either a Sylow subgroup of G or a maximal subgroup of G, then $H_{sG} = H_G$.*

Proof The assertions (1)–(3) follow from Chap. 1, Lemma 5.34(2)(3)(4)(7).

(4) By Chap. 1, Lemma 5.34(1)(7), H_{sG} is subnormal in G and so in the case when H is a Sylow subgroup of G, $H_{sG} = H_G$ by Chap. 1, Lemma 5.35(6). Now assume that H is a maximal subgroup of G. If $D = H_G \neq 1$, then, by induction, $(H/D)_{s(G/D)} = (H/D)_{(G/D)} = D/D$ and hence $H_{sG} = D$. We may, therefore, assume that $D = 1$. In this case, G is a primitive group. Let N be a minimal normal subgroup of G and $C = C_G(N)$. Then from Chap. 1, Proposition 1.4 we know that either N is the only minimal normal subgroup of G and $C \leq N$ or G has precisely two minimal normal subgroups N and R, $N \simeq R$ is nonabelian, $R = C$ and $N \cap H = 1 = R \cap H$. Let L be a minimal subnormal subgroup of G contained in H. If $L \leq N$, then $L^G = L^{NH} = L^H \leq D = 1$, a contradiction. Hence $L \nleq N$ and analogously $L \nleq R$. Hence $L \cap N = 1 = L \cap R$. But by Chap. 1, Lemma 5.35(5), $NL = N \times L$ and so $L \leq C = R$, a contradiction. Hence $H_{sG} = 1 = D$.

Lemma 3.7 *Let G be a group and $H \leq K \leq G$.*

(1) H^{sG} *is an s-quasinormal subgroup of G and $H^{sG} \leq H^G$.*
(2) $H^{sK} \leq H^{sG}$.
(3) *If H is normal in G, then $(K/H)^{s(G/H)} = K^{sG}/H$.*
(4) *If H is either a Hall subgroup of G or a maximal subgroup of G, then $H^{sG} = H^G$.*

Proof Assertions (1–3) follow from I, Chap. 1, Lemma 5.34.

(4) Suppose that H is a Hall π-subgroup of G and $x \in G$. Then H^x is also a Hall π-subgroup of G and so by Assertion (1) and Chap. 1, Lemma 5.35(4), $H^x \cap H^{sG}$ is a Hall π-subgroup of H^{sG}. But since $H \leq H^{sG}$, $H^x \leq H^{sG}$. It follows that $H^G \leq H^{sG} \leq H^G$. Now assume that H is a maximal subgroup of G. By (1) and

Chap. 1, Lemma 5.35(1), H^{sG} is subnormal in G. Thus, either $H = H^G = H^{sG}$ or $H^G = H^{sG} = G$.

Lemma 3.8 *Let H be a normal subgroup of G and $H \le K \le G$.*

(1) *If H is p-soluble and K/H is S_p-embedded in G/H, then K is S_p-embedded in G.*

(2) *If K is S_p-embedded in G, then K/H is S_p-embedded in G/H.*

(3) *If L is an S_p-embedded subgroup of G and $L \le K$, then L is S_p-embedded in K.*

(4) *The subgroup HE/H is S_p-embedded in G/H, for every S_p-embedded in G subgroup E satisfying $(|H|, |E|) = 1$.*

Proof (1) We prove that K is S_p-embedded in G by induction on $|G|$. Let L be a minimal normal subgroup of G such that $L \le H$. Then, obviously, $(K/L)/(H/L)$ is S_p-embedded in $(G/L)/(H/L)$. If $L \ne H$, then by induction, K/L is S_p-embedded in G/L. We may, therefore, assume that H is a minimal normal subgroup of G. Let T/H be an s-quasinormal subgroup of G/H such that $KT/H = (K/H)(T/H) = (K/H)^{s(G/H)}$ and $|(T/H) \cap (K/H)|_p = |(T/H) \cap (K/H)_{s(G/H)}|_p$. By Chap. 1, Lemma 5.34(3), T is s-quasinormal in G. By Lemma 3.7(3), $(K/H)^{s(G/H)} = K^{sG}/H$. Hence $K^{sG} = KT$. Since H is p-soluble, H is either a p-group or a p'-group. Then, since $|(T/H) \cap (K/H)|_p = |(T/H) \cap (K/H)_{s(G/H)}|_p$, we obtain $|T \cap K|_p = |T \cap K_{sG}|_p$. Hence K is S_p-embedded in G.

(2) Assume that $KT = K^{sG}$ and $|T \cap K|_p = |T \cap K_{sG}|_p$, for some s-quasinormal subgroup T of G. Then HT/H is an s-quasinormal in G/H and $(HT/H)(K/H) = KT/H = K^{sG}/H = (K/H)^{sG}$. Clearly, $H \le K_{sG}$. Hence $H \cap T \cap K = H \cap T = H \cap T \cap K_{sG}$. This implies that $|(TH/H) \cap (K/H)|_p = |H(T \cap K)/H|_p = |H(T \cap K_{sG})/H|_p = |(TH/H) \cap (K/H)_{s(G/H)}|_p$. Thus K/H is S_p-embedded in G/H.

(3) Let T be an s-quasinormal subgroup of G such that $LT = L^{sG}$ and $|T \cap L|_p = |T \cap L_{sG}|_p$. Let $T_0 = T \cap L^{sK}$. Then $T_0 = K \cap T \cap L^{sK}$ and $T_0 \cap L_{sK} = T \cap L_{sK}$. By Chap. 1, Lemma 5.34(4), $K \cap T$ is s-quasinormal in K and so by Chap. 1, Lemma 5.34(7), T_0 is s-quasinormal in K. Moreover, by Lemma 3.7(2), $L^{sK} \le L^{sG}$ and so $L^{sK} = L^{sK} \cap L^{sG} = L^{sK} \cap LT = L(L^{sK} \cap T) = LT_0$. Finally, we show that $|T_0 \cap L|_p = |T_0 \cap L_{sK}|_p$. In fact, we only need to prove that $|P_1| \le |P_2|$, for some Sylow p-subgroups P_1 of $T_0 \cap L$ and some Sylow p-subgroup P_2 of $T_0 \cap L_{sK}$. Since $H \cap T_0 \le L \cap T$, we have $P_1 \le P_3$, for some Sylow p-subgroup P_3 of $L \cap T$. On the other hand, by Lemma 3.6(2), $L_{sG} \le L_{sK}$. Hence $T \cap L_{sG} \le T \cap L_{sK} = T_0 \cap L_{sK}$. It follows that $|P_1| \le |P_3| = |T \cap L|_p = |T \cap L_{sG}|_p \le |P_2|$. Hence L is S_p-embedded in K.

(4) By (2), we only need to prove that HE is S_p-embedded in G. Assume that E is S_p-embedded in G and let T be an s-quasinormal subgroup of G such that $ET = E^{sG}$ and $|T \cap E|_p = |T \cap E_{sG}|_p$. Let $T_0 = HT$. Then, obviously, T_0 is an s-quasinormal subgroup of G and $HET_0 = HE^{sG} = (HE)^{sG}$. Next we show that $|T_0 \cap HE|_p = |T_0 \cap (HE)_{sG}|_p$.

Since $(|E|, |H|) = 1$, E is a Hall π-subgroup of EH and H is a Hall π'-subgroup of EH, for some set π of primes. If p divides $|H|$, then E is p'-group. Hence $|T_0 \cap HE|_p = |H| = |T_0 \cap HE_{sG}|_p = |T_0 \cap (HE)_{sG}|_p$. Now we assume that p divide $|E|$. In this case, H is a p'-group. Let $D = T \cap HE$. By Chap. 1, Theorem 3.35(2) and Chap. 1, Lemma 5.34(1), D is subnormal in HE and so $D = (D \cap H)(D \cap E) \leq H(T \cap E)$. It follows that $T_0 \cap HE = H(T \cap HE) = HD \leq H(T \cap E)$ and so $|T_0 \cap HE|_p = |T \cap E|_p = |T \cap E_{sG}|_p \leq |HT \cap (HE)_{sG}|_p \leq |T_0 \cap HE|_p$. Therefore $|T_0 \cap HE|_p = |T_0 \cap (HE)_{sG}|_p$. This shows that HE is S_p-embedded in G.

Lemma 3.9 *Let $G = P \rtimes Q$, where P is the Sylow 2-subgroup of G, and $|Q| = q$ where $q \neq 2$ is a prime. Suppose that all maximal subgroups of P or all cyclic subgroups H of P with prime order and order 4 (if P is nonabelian and $H \not\subseteq Z_\infty(G)$) are s-embedded in G. Then G is nilpotent.*

Proof Suppose that this lemma is false and let G be a counterexample of minimal order. Then $|P| > 2$. We first prove that some maximal subgroup of P is not s-embedded in G. Indeed, suppose that every maximal subgroup of P is s-embedded in G. Then by Lemma 3.8(2), the hypothesis is still true for G/L for any minimal normal subgroup L of G and so G/L is nilpotent. It follows that L is the only minimal normal subgroup of G and $L \not\subseteq \Phi(G)$. But then $L = C_G(L)$ and hence $L = P$ is abelian by Chap. 1, Proposition 1.4. It follows from Chap. 1, Lemma 5.34(6) that every S-quasinormal subgroup T of P is normal in G. This shows that every maximal subgroup of P is $\mathfrak{F}_\tau$-embedded in G, where $\mathfrak{F}$ is the class of all p-supersoluble groups and τ is the subgroup functor in Example 1.6. Hence G is p-nilpotent by Theorem 2.29, a contradiction. Hence some maximal subgroup of P is not S-embedded in G and thereby by hypothesis every cyclic subgroup H of P with prime order and order 4 (if P is nonabelian and $H \not\subseteq Z_\infty(G)$) is s-embedded in G. By Lemma 3.8(3), the hypothesis holds for any subgroup of G. The choice of G implies that G is a Schmidt group. Then by Chap. 1, Proposition 1.9 we have $O^2(G) = G$, so by Chap. 1, Lemma 5.34(6), every s-quasinormal subgroup of G contained in P is normal in G. This shows that every cyclic subgroup H of P with prime order or 4 (if P is nonabelian and $H \not\subseteq Z_\infty(G)$) is $\mathfrak{F}_\tau$-embedded in G, where $\mathfrak{F}$ is the class of all p-supersoluble groups and τ is the subgroup functor in Example 1.6. Hence G is 2-nilpotent by Theorem 2.21 and consequently G is nilpotent, a contradiction.

Lemma 3.10 *Suppose that for every noncyclic Sylow subgroup Q of G, either all maximal subgroups of Q or all cyclic subgroup H of Q with prime order and order 4 (if Q is a nonabelian 2-group and $H \not\subseteq Z_\infty(G)$) are s-embedded in G. Then G is soluble.*

Proof Suppose that this lemma is false and let G be a counterexample of minimal order. Let P be a Sylow p-subgroup of G, where p is the smallest prime dividing $|G|$. Then $p = 2$ by the Feit–Thompson's theorem.

(1) *The hypothesis holds for every Hall subgroup of G and for every quotient G/X, where X is a Hall normal subgroup of G* (This follows directly from Lemma 3.8).

(2) *G is not 2-nilpotent* (Otherwise, G is soluble by the Feit–Thompson's theorem).

(3) *P is not cyclic* (This follows from (2) and Chap. 1, Lemma 3.39).

(4) *Every maximal subgroup of P is S-embedded in G.*

Suppose that some maximal subgroup of P is not S-embedded in G. Then by hypothesis every cyclic subgroup H of P of order 2 and order 4 (if P is a nonabelian 2-group and $H \nleq Z_\infty(G)$) is S-embedded in G. Assume that $D = O^2(G) \neq G$. Then by Lemma 3.8(3), the hypothesis still holds for D and so D is soluble by the choice of G. But then G is soluble, which contradicts the choice of G. Hence $D = G$. It follows from Chap. 1, Lemma 5.34(6) that every s-quasinormal subgroup of P is normal in G. Then as in the proof of Lemma 3.9, we get that G is p-nilpotent. Let W be the Hall p'-subgroup of G. Then by (1) and the choice of G, W is soluble, so G is soluble, a contradiction. Hence we have (4).

(5) *Every nonidentity normal subgroup of G is nonsoluble.*

Indeed, if G has a nonidentity soluble normal subgroup R and N is a minimal normal subgroup of G contained in R, then the hypothesis holds for G/N by (4) and Lemma 3.8(2), so G/R is soluble and so G is soluble, a contradiction.

Final contradiction.

Suppose that for some maximal subgroup V of P we have $V_{sG} \neq 1$. By Chap. 1, Lemma 5.34(1)(7) and Chap. 1, Lemma 5.35(6), we have $V_{sG} \leq D = O_2(G)$. By Lemma 3.8(2)(4), the hypothesis holds for G/D. Hence by the choice of G, G/D is soluble, which implies the solubility of G. Therefore $V_{sG} = 1$ for all maximal subgroups V of P. By (3), $P = V_1 V_2$ for some maximal subgroups V_1 and V_2 of P. Since V_i is s-embedded in G, G has an s-quasinormal subgroup T_i such that $D_i = V_i^{sG} = V_i T_i$ and $T_i \cap V_i \leq (V_i)_{sG} = 1$. By Chap. 1, Lemma 5.34(1), T_i is subnormal in G. Hence by Chap. 1, Lemma 5.35(4), $P \cap T_i$ is a Sylow 2-subgroup of T. But since $T_i \cap V_i = 1$, $|T_i \cap P| \leq 2$. This implies that T_i is soluble and hence by Chap. 1, Lemma 5.35(7), T_i is contained in some soluble normal subgroup R of G, contrary to (5). This contradiction completes the proof.

Theorem 3.11 (Guo, Skiba [190]). *A group G is soluble if for every noncyclic Sylow subgroup P of $F^*(G)$ every maximal subgroup of P or every cyclic subgroup H of P with prime order and order 4 (if P is a nonabelian 2-group and $H \nleq Z_\infty(G)$) is s-embedded in G.*

Proof Suppose that this theorem is false and let G be a counterexample of minimal order. Put $F = F(G)$ and $F^* = F^*(G)$. Let p be the largest prime divisor of $|F|$ and P the Sylow p-subgroup of F. We proceed the proof via the following steps:

(1) $F^* = F$ and $C_G(F) = C_G(F^*) \leq F$.

By Lemma 3.8(3), the hypothesis is still true for F^*. Hence F^* is soluble by Lemma 3.10. It follows by Chap. 1, Proposition 2.9(8) that $F^* = F$. Finally, by Chap. 1, Proposition 2.9(5), $C_G(F) = C_G(F^*) \leq F$.

(2) *Every proper normal subgroup X of G containing F is soluble.*
By Chap. 1, Proposition 2.9(1), $F^*(X) \leq F^* = F \leq X$. Since $F^* = F$ is a normal nilpotent subgroup of X, $F^* \leq F^*(X)$. Hence $F^*(X) = F^*$. This implies that the hypothesis still holds for X. Thus X is soluble by the choice of G.

(3) $FO^p(G) = G$.
Indeed, if $FO^p(G) \neq G$, then $FO^p(G)$ is soluble by (2) and thereby G is soluble, which contradicts the choice of G.

(4) *If H is a normal subgroup of P and H is s-quasinormal in G, then H is normal in G* (This follows directly from (3) and Chap. 1, Lemma 5.34(6)).

(5) $p \neq 2$.
Assume that $p = 2$. Since p is the largest prime divisor of $|F|$, $F^* = F$ is a 2-group by (1). Let Q be a subgroup of G of prime order q, where $q \neq 2$, and let $X = FQ$. By Lemmas (3.8)(3) and (3.9), X is nilpotent and so Q is normal in X. It follows that $Q \leq C_G(F)$. But by (1), $C_G(F) = C_G(F^*) \leq F$, a contradiction. Hence we have (5).

(6) *Every maximal subgroup of P is S-embedded in G.*
Suppose that some maximal subgroup of P is not S-embedded in G. Then by hypothesis every cyclic subgroup H of P with prime order and order 4 (if P is a nonabelian 2-group and $H \not\subseteq Z_\infty(G)$) is S-embedded in G. Assume that $O^p(G) \neq G$. By Chap. 1, Proposition 2.9(1), $F^*(O^p(G)) \leq F^* = F$. It is clear that for any Sylow q-subgroup Q of F where $q \neq p$ we have $Q \leq O^p(G)$ and $O_p(O^p(G)) \leq O_p(G) \leq F$. Hence by Lemma 3.8(3), the hypothesis is still true for $O^p(G)$. Then by the choice of G, we have that $O^p(G)$ is soluble. It follows that G is soluble. This contradiction shows that $O^p(G) = G$ and hence by Chap. 1, Lemma 5.34(6) every s-quasinormal subgroup of P is normal in G. Therefore, every cyclic subgroup H of P with prime order and order 4 (if P is a nonabelian 2-group and $H \not\subseteq Z_\infty(G)$) is $\mathfrak{U}_\tau$-embedded in G, where τ is the subgroup functor in Example 1.6. It follows from Theorem 2.12 that $P \leq Z_\mathfrak{U}(G)$. Let H/K be a chief factor of G where $H \leq P$ and $C = C_G(H/K)$. Since $|H/K| = p$, G/C is abelian p'-group. Since $F^* = F \leq C$, by Chap. 1, Proposition 2.9(1) $F^*(C) = F^*$. This means that the hypothesis is still true for C. If $C \neq G$, then C is soluble by the choice of G and hence G is soluble, a contradiction. Hence $C_G(H/K) = G$ for any chief factor H/K of G where $H \leq P$. Let

$$1 = D_0 \leq D_1 \leq \ldots \leq D_t = P \qquad (*)$$

be a chief series of G below P and $C_0 = C_G(P)$. Then $G/C_0 \simeq E \leq \mathrm{Aut}(P)$ and G/C_0 stabilizes the series $(*)$. Hence by Chap. 1, Lemma 2.13, G/C_0 is a p-group. But since $O^p(G) = G$, $C_0 = G$. It follows that $C_G(P) = G$. Hence by Chap. 1, Proposition 2.6(6), $F^*(G/P) = F^*/P$. Then by Lemma 3.8(4), the hypothesis still holds for G/P. It follows that G/P is soluble by the choice of G, which implies the solubility of G. This contradiction completes the proof of (6).

(7) *If L is a minimal subgroup of P, then L is not normal in G.*

Assume that L is normal in G. Then $F \le C = C_G(L)$ and G/C is cyclic. Suppose that $C \ne G$. Then by (2), C is soluble and hence does G, which contradicts the choice of G. Hence $C = G$. In this case, $F^*(G/L) = F^*/L$ by Chap. 1, Proposition 2.9(6). Then in view of (6) and Lemma 3.8, we see that the hypothesis still holds for G/L and so G/L is soluble by the choice of G. But then G is soluble, a contradiction.

(8) $\Phi(G) \cap P \ne 1$ *and if L is a minimal normal subgroup of G contained in $\Phi(G) \cap P$, then $F^*(G/L) \ne F^*/L$.*

Suppose that $\Phi(G) \cap P = 1$. Then P is the direct product of some minimal normal subgroups of G by Chap. 1, Proposition 2.24. Hence P is abelian and so every s-quasinormal subgroup of P is normal in P. Therefore by (4) every maximal subgroup of P is $\mathfrak{U}_\tau$-embedded in G, where τ is the subgroup functor in Example 1.6. Hence $P \le Z_{\mathfrak{U}}(G)$ by Example 1.6 and Theorem 2.6, which contradicts (7). Thus $\Phi(G) \cap P \ne 1$. Let L be some minimal normal subgroup of G contained in $\Phi(G) \cap P$. If $F^*(G/L) = F^*/L$, then by (6) the hypothesis is true for G/L. Hence G/L is soluble by the choice of G, which implies the solubility of G. This contradiction shows that $F^*(G/L) \ne F^*/L$.

(9) *G has a unique maximal normal subgroup M containing F, M is soluble and G/M is a nonabelian simple group* (This directly follows from (2) and the choice of G).

(10) *G/F is a nonabelian simple group and if L is a minimal normal subgroup of G contained in $\Phi(G) \cap P$, then G/L is a quasinilpotent group.*

Let L be a minimal normal subgroup of G contained in $\Phi(G) \cap P$. Then by (8), $F^*(G/L) \ne F^*/L$. Thus $F/L = F^*/L$ is a proper subgroup of $F^*(G/L)$ by Chap. 1, Proposition 2.9(2). Now by Chap. 1, Proposition 2.9(4), $F^*(G/L) = F(G/L)E(G/L)$, where $E(G/L)$ is the layer of G/L. By (9), every chief series of G has a unique nonabelian factor. But since $E(G/L)/Z(E(G/L))$ is the direct product of some simple nonabelian groups (see [250, p. 128]), $F^*(G/L) = G/L$ is a quasinilpotent group. Since $F(G/L) \cap E(G/L) = Z(E(G/L))$ by Lemma Chap. 1, Proposition 2.9(4), $G/F \simeq (G/L)/(F/L)$ is a simple nonabelian group.

(11) *$F = P$.*

Assume that $P \ne F$ and let Q be a Sylow q-subgroup of F where $q \ne p$. Then from the G-isomorphism $Q \simeq QP/P$, by (10) and [250, X, 13.7], we have $Q \le Z_\infty(G)$. Hence by Chap. 1, Proposition 2.9, $F^*(G/Q) = F^*/Q$ and so by Lemma 3.8 the hypothesis is still true for G/Q. The choice of G implies that G/Q is soluble. It follows that G is soluble, a contradiction.

(12) *$\Phi(P) = 1$.*

Assume that $\Phi(P) \ne 1$ and let L be a minimal normal subgroup of G contained in $\Phi(P)$. Then by (10), G/L is quasinilpotent and hence G is quasinilpotent by Chap. 1, Proposition 2.9(2). But then $F = F^* = G$, a contradiction. Thus (12) holds.

(13) If H is a subgroup of P and H is s-quasinormal in G, then H is normal in G. (This follows directly from (4) and (12)).

Final contradiction.

By (8), $\Phi(G) \cap P \neq 1$. Let N be a minimal normal subgroup of G contained in $\Phi(G) \cap P$. By (12), for some maximal subgroup V of P, we have $P = NV$. First assume that V is s-quasinormal in G. Then by (13), V is normal G. But by (7), $|N| > p$ and so $N \neq N \cap V \neq 1$, which contradicts the minimality of N. Therefore V is not s-quasinormal in G. By (6), there exists an s-quasinormal subgroup T of G such that $P = V^{sG} = VT$ and $T \cap V \leq V_{sG}$. By (13), both subgroups T and V_{sG} are normal in G. Hence $V_{sG} = V_G$.

Suppose that $N \leq T$. Then $V \cap N \leq V_G \cap T = V \cap T$. It is clear that $1 \neq V \cap N \subseteq V \cap T = V_G \cap T \trianglelefteq G$. It follows that $V \cap N = N$, a contradiction. Thus $N \nsubseteq T$ and consequently $N \cap T = 1$. Let L be a minimal normal subgroup of G contained in T. Then by (10) and [250, X, 13.7], $LN/N \leq Z_\infty(G/N)$ and thereby $L \leq Z(G)$, which contradicts (7). The theorem is proved.

Lemma 3.12   [450, Chap. 4, Theorem 1.6]. *Let p be an odd prime number and* **F** *field of characteristic p. Let G be a completely reducible soluble linear group of degree n over* **F**. *Suppose that a Sylow p-subgroup of G has order $p^{\lambda(n)}$. Then $\lambda(n) \leq n - 1$ with equality only if $n = 1$ or $n = 2$ and $p = 3$.*

Lemma 3.13 . *Suppose that every maximal subgroup E of G with $(|G : E|, p) = 1$ is normal in G. Let P be a Sylow p-subgroup of G. Then G is p-closed and G/P is nilpotent.*

Proof   Suppose that this lemma is false and let G be a counterexample of minimal order. Obviously, the hypothesis is true for any factor group of G. Hence G has a unique minimal normal subgroup, L say, and G/L is p-closed with nilpotent factor $(G/L)/(PL/L)$. If L is a p-group, then G is p-closed with nilpotent factor G/P, which contradicts the choice of G. Hence L is not a p-group. Since the class of all p-closed groups G is a saturated formation (see Chap. 1, (1.3)(3)), we see that $L \nsubseteq \Phi(G)$. Let M be a maximal subgroup of G such that $ML = G$. Suppose that p divides $|G : M|$. Then p divides $|L|$ and by Fratinni argument, for some maximal subgroup E of G we have $EL = G$ and p does not divide $|G : E|$. Hence E is normal in G by hypothesis, which implies $|E| = 1$. Consequently, $G = L$. This contradiction completes the proof.

Theorem 3.14   (Guo, Skiba [199]) *A soluble group G is p-supersoluble if and only if every 2-maximal subgroup E of G with $O_{p'}(G) \leq E$ and $|G : E|$ is not a power of p, both has a supplement in E^{sG} with cyclic Sylow p-subgroups and is S_p-embedded in G.*

Proof   First suppose that G is p-supersoluble and let E be any 2-maximal subgroup of G such that $O_{p'}(G) \leq E$ and $|G : E|$ is not a power of p. We show that E has a supplement T in E^{sG} with cyclic Sylow p-subgroups and E is S_p-embedded in G. If $E = E^{sG}$, then it is evident. We may, therefore, assume that $E \neq E^{sG}$. Suppose that $O_{p'}(G) \neq 1$. Then, by induction, $E/O_{p'}(G)$ is S_p-embedded in $G/O_{p'}(G)$ and

$E/O_{p'}(G)$ has a supplement $T/O_{p'}(G)$ in $(E/O_{p'}(G))^{s(G/O_{p'}(G))}$ with cyclic Sylow p-subgroups. By Lemma 3.8(1), E is S_p-embedded in G. On the other hand, since $(E/O_{p'}(G))^{s(G/O_{p'}(G))} = E^{sG}/O_{p'}(G)$ by Lemma 3.7(3), T is a supplement of E in E^{sG} and clearly the Sylow p-subgroups of T is cyclic.

Now suppose that $O_{p'}(G) = 1$. Then by Chap. 2, Lemma 4.10, G is supersoluble, $P = O_p(G) = F(G)$ is a Sylow p-subgroup of G and G/P is abelian. Let M be a maximal subgroup of G such that E is maximal in M. Since G is p-supersoluble and $|G : E|$ is not a power of p, one of $|M : E|$ and $|G : M|$ is a p'-number (see [135, Theorem 1.9.4]). Hence $|G : E| = pn$, where $(p, n) = 1$. It follows that $P \cap E$ is a maximal subgroup of P with $|P : P \cap E| = p$ and so $\Phi(P) \leq E$. Since $\Phi(P)$ is a characteristic subgroup of P, it is normal in G. Hence $\Phi(P) \leq E_{sG}$. If $\Phi(P) \neq 1$, then as above we can show that E is S_p-embedded in G. Besides, PE is normal in G by Chap. 2, Lemma 4.10(iii) and $|PE : E| = |P : E \cap P| = p$. Since $E \subseteq E^{sG} \subseteq PE$, $E^{sG} = PE$. Hence E has a cyclic supplement $\langle x \rangle$ in E^{sG}, where $x \in PE$ and $x \notin E$.

Finally, assume that $\Phi(P) = 1$. Then P is an elementary abelian p-group and $P = P_1 \times P_2 \times \ldots \times P_t$, where P_i is a minimal normal subgroup of G, for all $i = 1, 2, \ldots, t$. It is clear that for some i, $P_i \not\leq E \cap P$. Hence $P_i(E \cap P) = P$. Since G is p-supersoluble, $|P_i| = p$. It follows that $P_i \cap E \cap P = 1$ and $EP_i = E$. Since $E \neq E^{sG}$, E is not s-quasinormal in G. But since PE is normal in G, by Chap. 2, Lemma 4.10(iii), $PE = E^{sG} = P_i(E \cap P)E = P_iE$. Since $P_i \cap E \trianglelefteq E$, $P_i \cap E \subseteq E_{sG}$. Hence $|P_i \cap E|_p = |P_i \cap E_{sG}|_p$. This implies that E has a supplement P_i in E^{sG}, which is a cyclic Sylow p-subgroup and E is S_p-embedded in G.

Conversely, assume that G is soluble and every 2-maximal subgroup E of G with $O_{p'}(G) \leq E$ and $|G : E|$ is not a power of p has a supplement in E^{sG} with cyclic Sylow p-subgroups and E is S_p-embedded in G. We show that G is p-supersoluble. Assume that this is false and let G be a counterexample of minimal order. Then

(1) *G has a unique minimal normal subgroup L, G/L is p-supersoluble, p divides $|L|$ and $L \not\leq \Phi(G)$.*

Let L be a minimal normal subgroup of G and E/L a 2-maximal subgroup of G/L such that $O_{p'}(G/L) \leq E/L$ and $|G/L : E/L|$ is not a power of p. Since $O_{p'}(G)L/L \leq O_{p'}(G/L)$, $O_{p'}(G) \leq E$. Besides, $|G : E|$ is not a power of p. Hence by hypothesis, E is S_p-embedded in G and there is a subgroup T such that $ET = E^{sG}$ and a Sylow p-subgroup of T is cyclic. Obviously, $(EL)^{sG} = E^{sG}L$. Hence $(E/L)(TL/L) = E^{sG}L/L = (E/L)^{s(G/L)}$ by Lemma 3.7(3) and clearly TL/L has a cyclic Sylow p-subgroup. Besides, E/L is S_p-embedded in G/L by Lemma 3.8(2). This shows that the hypothesis still holds for G/L. The minimal choice of G implies that G/L is p-supersoluble. It is well known that the class of all p-supersoluble groups is a saturated formation. Hence we see that (1) holds.

(2) *$G = L \rtimes M$ for some maximal subgroup M of G, $L = C_G(L) = F(G) = O_p(G)$ and $p \neq |L|$.*

By (1), there exists a maximal subgroup M of G such that $G = LM$. Since G is soluble, L is either a p'-group or a p-group. In the former case, G is clearly

p-supersoluble, a contradiction. Hence L is a abelian p-group. It follows that $L = C_G(L) = F(G) = O_p(G)$ and $|L| > p$ since G is not p-supersoluble.

(3) *L is not a Sylow p-subgroup of G.*

Assume that L is a Sylow p-subgroup of G and let E be a maximal subgroup of M. Then $|G : E| = |L||M : E| \neq p^a$ and $O_p(G) = 1 \leq E$. Hence by hypothesis E is S_p-embedded in G and E has a supplement X in E^{sG} with cyclic Sylow p-subgroups. Suppose that $E = 1$ and let V be a maximal subgroup of L. Then by hypothesis, V is S_p-embedded in G. Let T be an s-quasinormal subgroup of G such that $V^{sG} = VT$ and $|T \cap V|_p = |T \cap V_{sG}|_p$. Since L is a Sylow p-subgroup of G, the subgroups V_{sG}, V^{sG} and T are normal in G by Chap. 1, Lemma 5.34(6). This implies that $T = L$ and $V = 1$. But then $|L| = p$, which contradicts (2). Therefore $E \neq 1$. Let q be prime dividing $|M : E|$ and Q be a Sylow q-subgroup of M. Clearly, Q is a Sylow q-subgroup of G and $A = E^{sG}Q = QE^{sG}$ is a subgroup of G. Since E is maximal in M and $Q \nleq E$, we have $\langle E, Q \rangle = M$. Hence $M \leq A$ and so $LA = G$. Since L is a minimal normal subgroup of G, either $L \cap A = L$ or $L \cap A = 1$. In the former case, $G = A = E^{sG}Q$ and so L is a Sylow p-subgroup of any supplement of E in E^{sG}. Therefore L is cyclic and hence $|L| = p$, which contradicts (2). Thus $L \cap A = 1$. Obviously, $E^{sG} \leq M$. Since $M_G = 1$, $(E^{sG})_G = 1$. Hence E^{sG} is nilpotent by Chap. 1, Lemma 5.34(1). Then by Chap. 1, Lemma 5.34(1) and Chap. 1, Lemma 5.35(7), $E^{sG} \leq L$, a contradiction. This shows that L is not a Sylow p-subgroup of G.

(4) *M has a nonnormal maximal subgroup E such that $(|M : E|, p) = 1$.*

Suppose that every maximal subgroup E of M with $(|M : E|, p) = 1$ is normal in M. Then by Lemma 3.13, M is p-closed. Besides, by (3), p divides $|M|$. But by (2), we have $O_p(G/L) = O_p(G/C_G(L)) = 1$ by Chap. 1, Lemma 2.15. Hence L is a Sylow p-subgroup of G, which contradicts (3). Hence, (4) holds.

(5) *$E^{sG} = G$.*

Indeed, suppose that $D = E^{sG} \neq G$. Since D is subnormal in G by Chap. 1, Lemma 5.34(7) and Lemma 3.7(1), $M \nleq D$. Hence $E = D \cap M$ is a subnormal subgroup of M by Chap. 1, Lemma 5.35(2). But since E is maximal in M, E is normal in M, which contradicts (4).

(6) *If T is a supplement of E in G, then $O_{p'}(T) = 1$.*

Since $G = ET = MT$, $M = M \cap ET = E(M \cap T)$ and $|L| = |G : M| = |T : M \cap T|$. It follows that $O_{p'}(T) \leq M$. Hence $(O_{p'}(T))^G = O_{p'}(T)^{TM} = O_{p'}(T)^M \leq M_G = 1$. Consequently, $O_{p'}(T) = 1$.

(7) *$p = 3$ and $|L| = 9$.*

Let T be a supplement of E in $E^{sG} = G$ with cyclic Sylow p-subgroups. Then $G = ET = MT$. Let T_p be a Sylow p-subgroup of T. Suppose that $p = 2$. Then by Chap. 1, Lemma 3.39, T is 2-nilpotent. But by (6), $T_{2'} = 1$ and so T is a 2-group. It follows that 2 divides $|M : E|$, which contradicts (4). Therefore p must be an odd number. Suppose that either $p \neq 3$ or $p = 3$ and $|L| > 3^2$. Let $|L| = p^a$ and p^b be the order of a Sylow p-subgroup of M. Then by Lemma 3.12, $b < a - 1$. Since T_p is cyclic, $|T_p \cap L| \leq p$. It follows from $G = MT = L \rtimes M$

that $|L| \leq |T_p|$. Hence $p^{a+b} < p^{2a-1} \leq |LT_p| \leq p^{a+b}$. This contradiction shows that $p = 3$ and $a = 2$.

(8) *The order of a Sylow 2-subgroup P of T is 2.*

Let T be a supplement of E in G. Since T is 3-soluble and its Sylow 3-subgroups are cyclic, T is 3-supersoluble. But by (6), $O_{3'}(T) = 1$. Hence T is supersoluble by Chap. 2, Lemma 4.10. Then, for a Sylow 3-subgroup P of T, we have $P = F(T)$. Thus $C_T(P) \leq P$ by [135, Theorem 1.8.18]. It follows from [117, Chap. 5, Lemma 4.1(iii)] that $|T/P| = 2$. Therefore, (8) holds.

Final contradiction.

In view of (7), M is isomorphic with some subgroup of $GL(2, 3)$. Hence $|M| \leq 48$. It follows that $|M : E| = 2^m$ for some $m > 1$ (since E is not normal in M). But since $G = ET$, we see that $|M : E|||P| = 2$, which is impossible. The final contradiction completes the proof.

Corollary 3.15 *Let G be a soluble group and p be a prime with $(p - 1, |G|) = 1$. Then G is p-nilpotent if and only if every 2-maximal subgroup E of G with $O_{p'}(G) \leq E$ and $|G : E|$ is not a power of p, both has a supplement in E^{sG} with cyclic Sylow p-subgroups and is S_p-embedded in G.*

Lemma 3.16 (Guo, Skiba [199]). *Suppose that G is not a p-group. Then the following are equivalent.*

(1) G *is p-soluble.*

(2) *Every maximal subgroup of G is S_p-embedded in G.*

(3) G *has two maximal p-soluble S_p-embedded subgroups M_1 and M_2, whose indices $|G : M_1|$ and $|G : M_2|$ are coprime.*

(4) *For every maximal subgroup M of G, either $|G : M|$ is a power of p or M is S_p-embedded in G.*

Proof (1) $\Rightarrow$ (2). Let M be a maximal subgroup of G and H/K a chief factor of G such that $HM = G$ and $K \leq M$. If $M_G \neq 1$, M/M_G is S_p-embedded in G/M_G by induction and consequently M is S_p-embedded in G by Lemma 3.8(1). Suppose that $M_G = 1$. Then $K = 1$. If H is a p'-group, then $|H \cap M|_p = 1 = |H \cap M_{sG}|_p$. Hence M is S_p-embedded in G. If H is an abelian p-group, then $H \cap M = H \cap M_{sG} = K = 1$. Then M is also S_p-embedded in G.

(3) $\Rightarrow$ (1). Assume that this is false and let G be a counterexample of minimal order. Let N be a minimal normal subgroup of G. If $N \leq M_1 \cap M_2$, then M_1/N and M_2/N are p-soluble maximal subgroups of G/N whose indices $|G/N : M_1/N| = |G : M_1|$ and $|G/N : M_2/N| = |G : M_2|$ are coprime and M_1/N and M_2/N are S_p-embedded in G by Lemma 3.8(2). This shows that the hypothesis holds for G/N. Therefore G/N is p-soluble by the choice of G. On the other hand, if $N \not\leq M_1 \cap M_2$, for example $N \not\subseteq M_1$, then $G/N \simeq M_1/M_1 \cap N$ is p-soluble. Therefore, N is the only minimal normal subgroup N of G, $N \neq \Phi(G)$ and N is a nonabelian group with p divides $|N|$. Then, clearly, $G = NM_1 = NM_2$ and $(M_1)_G = 1 = (M_2)_G$.

Let N_p be a Sylow p-subgroup of N and P a Sylow p-subgroup of G containing N_p. Since $|G : M_1|$ and $|G : M_2|$ are coprime. Without loss of generality, we may assume that P is contained in at least one of the subgroups M_1 and M_2, for example, $P \leq M_1$. Since M_1 is S_p-embedded in G, $M^{sG} = MT$ and $|M \cap T|_p = |T \cap M_{sG}|_p$, for some s-quasinormal subgroup T of G. By Chap. 1, Lemma 5.34(1), M^{sG}, T and M_{sG} are subnormal in G. Then since $(M_1)_G = 1$ and by Lemma 3.6(4), we have that $(M_1)_{sG} = 1$ and $(M_1)^{sG} = G$. This shows that T is a complement of M_1 in G. Since $P \leq M_1$, p does not divide $|T|$. Let $N = N_1 \times N_2 \times \ldots \times N_t$, where $N_1, N_2, \ldots, N_t$ are isomorphic simple nonabelian groups. Let L be a minimal subnormal subgroup of G contained in T. Since, obviously, $C_G(N) = 1$, $L \nleq C_G(N)$ and so $L \leq N$ by Chap. 1, Lemma 5.35(5). Hence $L = N_i$, for some i. It follows that p divides $|L|$ and therefore p divides $|T|$, a contradiction.

(1) $\Rightarrow$ (3). Since G is p-soluble and G is not a p-group, then there are two maximal subgroups M_1 and M_2 of G such that $|G : M_1| = p^a$ for some $a \in \mathbb{N}$ and p does not divide $|G : M_2|$ by [135, Theorem 1.7.13]. Then $(|G : M_1|, |G : M_2|) = 1$. By (2), we see that M_1 and M_2 are S_p-embedded in G. Thus (3) holds.

(4) $\Rightarrow$ (1). Let L be a minimal normal subgroup of G. Clearly, the hypothesis is true for G/L. By induction, G/L is p-soluble. We may, therefore, assume that L is nonabelian, p divides $|L|$ and L is the only minimal normal subroup of G. Thus $C_G(L) = 1$. By the Frattini argument, for any Sylow p-subgroup P of L, there is a maximal subgroup M of G such that $LM = G$ and $N_G(P) \leq M$. It is clear that $M_G = 1$ and p does not divides $|G : M|$. By hypothesis, G has an s-quasinormal subgroup T such that $M^{sG} = M^G = G = MT$ and $|T \cap M|_p = |T \cap M_{sG}|_p$. But by Lemma 3.6(4), $M_{sG} = M_G = 1$. This implies that $T \cap M$ is a p'-group. Let X be a minimal subnormal subgroup of G contained in T. Since $C_G(L) = 1$, $X \leq L$ by Chap. 1, Lemma 5.35(5) and so p divides $|X|$. It follows that p divides $|T|$. Since T is subnormal in G by Chap. 1, Lemma 5.34(4), $|T \cap M|_p \neq 1$. This contradiction shows that G is p-soluble. The theorem is proved.

Theorem 3.17 (Guo, Skiba [199]). *A finite group G is p-soluble if for every 2-maximal subgroup E of G such that $O_{p'}(G) \leq E$ and $|G : E|$ is not a power of p, G has an s-quasinormal subgroup T such that $E^{sG} = ET$ and $|E \cap T|_p = |E_{sG} \cap T|_p$.*

Proof Assume that this theorem is false and let G be a counterexample of minimal order. Then p divides $|G|$. We proceed the proof via the following steps.

(1) *G is not simple.*

Suppose that G is a simple nonabelian group. Then $O_{p'}(G) = 1$. Let M be a maximal subgroup of G containing a Sylow p-subgroup P of G and E any maximal subgroup of M. Then $|G : E|$ is not a power of p. Hence by hypothesis, E is S_p-embedded in G. Let T be an s-quasinormal subgroup of G such that $TE = E^{sG}$ and $|T \cap E|_p = |T \cap E_{sG}|_p$. By Lemmas 3.6(1) and 3.7(1) and Chap. 1, Lemma 5.34(1), E^{sG}, E_{sG} and T are all subnormal subgroups of G. Since G is a simple group, we have that $T = G$ and $E_{sG} = 1$. It follows from $|T \cap E|_p = |T \cap E_{sG}|_p = 1$ that $|E|_p = 1$. If $M = P$, then $|M| = p$ and so G is soluble by [248, IV, Theorem 7.4]. Otherwise, we may assume that $P \leq E$, which implies $P = 1$. This contradiction shows that (1) holds.

(2) *G has a unique minimal normal subgroup L, G/L is p-soluble, p divides |L| and L $\not\leq \Phi(G)$.*

Let L be a minimal normal subgroup of G. Then by Lemma 3.8(2), the hypothesis still holds for G/L. The minimal choice of G implies that G/L is p-soluble. Hence (2) holds.

(3) *G is p-soluble.*

By (1), $L \neq G$. Let M be any maximal subgroup of G containing L. Suppose that $|G : M| = p^a$. Then for every maximal subgroup E of M with $|M : E|$ is not a power of p, we have that $|G : E|$ is also not a power of p. Hence E is S_p-embedded in G by (2) and hypothesis. It follows from Lemma 3.8(3) that E is S_p-embedded in M. Hence M is p-soluble by Lemma 3.16. Consequently, L is p-soluble and thereby G is p-soluble. The contradiction completes the proof.

From Theorem 3.14 and 3.17, we directly get

Corollary 3.18 (Guo, Skiba [190]). *Suppose that for every 2-maximal subgroup E of G such that $|G : E|$ is not a power prime, G has an s-quasinormal cyclic subgroup T satisfying $E^{sG} = ET$ and $E \cap T = E_{sG} \cap T$. Then G is supersoluble.*

3.4 *G*-covering Systems of Subgroups

For any nonidentity subgroup H of G, let Σ_H denote a set of subgroups of G which contains at least one supplement in G of each maximal subgroup of H.

Let $\mathcal{F}$ be a class of groups. We call a set Σ of subgroups of G a *G-covering subgroup system* for the class $\mathfrak{F}$ if $G \in \mathfrak{F}$ whenever $\Sigma \subseteq \mathfrak{F}$.

If P is a Sylow p-subgroup of G and p is odd, then the set $\{N_G(J(P)), C_G(Z(G))\}$ (Thompson) and the one-element set $\{N_G(Z(J(P)))\}$ (Glauberman–Thompson) are G-covering subgroup systems for the class of all p-nilpotent groups. Now let Σ_i be any set of subgroups $H_1, H_2, \ldots, H_i$ of G whose indices $|G : H_1|, |G : H_2|, \ldots, |G : H_i|$ are pairwise coprime. Then Σ_3 is a G-covering subgroup system for the classes of all soluble (Wielandt), all nilpotent (Kegel) and all abelian (Doerk) groups. It is not difficult to find examples which show that Σ_3 is not a G-covering subgroup system for the classes of all supersoluble and all metanilpotent groups. However, Σ_4 is a G-covering subgroup system for both these classes (Doerk, Kramer). According to [53], the set Σ of all normalizers of all Sylow subgroups of G is a G-covering subgroup system for the class of all nilpotent groups and, by Fedri and Serens [102], Σ is not a G-covering subgroup system for the class of all supersoluble groups. In this section, we develop some new aspect of the theory G-covering subgroup systems.

Recall that a class $\mathfrak{X}$ of groups is said to be a homomorph if $G/N \in \mathfrak{X}$ whenever $G \in \mathfrak{X}$.

Lemma 4.1 *Let $\mathfrak{F}$ be a homomorph, $1 < E \leq G$ and N be a normal subgroup of G. Suppose that every maximal subgroup of E has a supplement T in G such that $T \in \mathfrak{F}$. If $E \not\leq N$, then every maximal subgroup of EN/N has a supplement $\bar{W}$ in G/N such that $\bar{W} \in \mathfrak{F}$.*

Proof Let V/N be a maximal subgroup of EN/N. Then $V = N(V \cap E)$, where $V \cap E \leq W$ for some maximal subgroup W of E. By hypothesis, $WT = G$ for some subgroup $T \in \mathfrak{F}$ of G. Since $V \leq NW \leq NE$ and V is maximal in NE, either $V = NW$ or $NW = NE$. In the former case, $(V/N)(TN/N) = (NW/N)(NT/N) = G/N$ and $TN/N \simeq T/T \cap N \in \mathfrak{F}$ since $\mathfrak{F}$ is a homomorph. The second case is impossible, since otherwise we have $E = E \cap NW = W(E \cap N) \leq W(E \cap V) = W$, which contradicts the choice of W. The lemma is proved.

Lemma 4.2 *Suppose that $G = AB$, where A and B are subgroups of G such that $1 \neq A \cap B \neq A$. Then A is not an abelian minimal normal subgroup of G.*

Proof Otherwise, $A \cap B$ is normal in G, which contradicts the minimality of A.

Lemma 4.3 (see [366, Theorem 8.3]) *Let $\mathcal{F}$ be a nonempty totally local formation (see [135, Definition 4.5.6]). Then there is a formation function f such that $\mathfrak{F} = LF(f)$ and $f(r)$ is a totally local formation contained in $\mathfrak{F}$ for all $r \in \pi(\mathfrak{F})$, and $f(r) = \varnothing$ for all primes $r \notin \pi(\mathfrak{F})$.*

Lemma 4.4 *Let P be a p-subgroup of G. Let $\mathcal{F}$ be a class of groups. Suppose that every maximal subgroup of P has a supplement $T \in \mathfrak{F}$ in G. Then every maximal subgroup of P^x has a supplement $U \in \mathfrak{F}$ in G, for all $x \in G$.*

Proof If V is a maximal subgroup of P^x, then $V^{x^{-1}}$ is a maximal subgroup of P and so $V^{x^{-1}}$ has a supplement $T \in \mathfrak{F}$ in G. But then $U = T^x \in \mathfrak{F}$ is a supplement of V in G by Chap. 1, Lemma 4.14.

For a set $\mathfrak{X}$ of groups, we write form$\mathfrak{X}$ to denote the intersection of all formations containing $\mathfrak{X}$.

Lemma 4.5 (see [135, Theorem 3.1.15]). *Let $\mathcal{F}$ be a nonempty saturated formation. Then $\mathfrak{F} = LF(f)$, where $f(p) = \text{form}(G/O_{p'p}(G) \mid G \in \mathfrak{F})$ for all $p \in \pi(\mathfrak{F})$, and $f(p) = \varnothing$ for all primes $p \notin \pi(\mathfrak{F})$.*

Lemma 4.6 (see [135, Lemma 3.1.1]). *Let $\mathfrak{F} = LF(f)$ be a nonempty saturated formation, where f satisfies that $f(r) \subseteq \mathfrak{F}$ for all primes r. If $G/O_p(G) \in f(p)$, then $G \in \mathfrak{F}$.*

Theorem 4.7 (Guo, Huang, Skiba [149, Theorem A]). *Let S be a nonidentity p-subgroup of G, where $p \notin \pi$. Then Σ_S is a G-covering subgroup system for a formation $\mathcal{F}$ if at least one of the following conditions holds:*

(i) *$\mathcal{F}$ is the class of all π-closed groups and any two Hall π-subgroups of G are conjugate.*

(ii) *$\mathcal{F}$ is the class of all r-decomposable groups, r is a prime.*

(iii) *$\mathcal{F}$ is a totally local formation contained in the class of all φ-dispersive groups.*

Proof Let G_p be a Sylow p-subgroup of G containing S.

(i) Let V be a maximal subgroup of S and T a π-closed supplement of V in G. Without loss of generality we may assume that $T = N_G(H)$, where H is a Hall π-subgroup of T. If $T \cap S = S$, then $G = T$ is π-closed. Assume that $T \cap S < S$.

Let W be a maximal subgroup of S such that $T \cap S \leq W$. Then, by hypothesis, W has a π-closed supplement in G, which implies that G is π-closed by Lemma 2.25.

(ii) Suppose that the theorem in this case is false and let G be a counterexample of minimal order. Then:

(1) *G/N is r-decomposable, for every minimal normal subgroup N of G.*
    Assume that $S \leq N$ and T is a r-decomposable supplement of a maximal subgroup V of S in G. Then $G/N = VT/N = NT/N \simeq T/N \cap T$ and so G/N is r-decomposable. Now assume that $S \nleq N$. Then SN/N is a nonidentity p-subgroup of G/N and every maximal subgroup of SN/N has a r-decomposable supplement in G/N by Lemma 4.1. This shows that the hypothesis is true for G/N. The minimal choice of G implies that G/N is r-decomposable.

(2) $O_r(G) \neq 1$.
    Let V be a maximal subgroup of S and T a r-decomposable supplement of V in G. Then $T_p = G_p \cap T$ is a Sylow p-subgroup of T. Suppose that $p = r$. Then T_p is normal in T and $(T_p)^G = (T_p)^{TG_p} = (T_p)^{G_p} \leq G_p$. Hence $O_p(G) = O_r(G) \neq 1$. Finally, if $r \neq p$, then G is r-closed by (i).

(3) $G = N \rtimes M$, *where M is a r-decomposable maximal subgroup of G and $N = C_G(N) = O_r(G) = F(G)$ is a minimal normal subgroup of G.*
    Let N be a minimal normal subgroup of G. It is well known that the class of all r-decomposable groups is a saturated formation (see for example, [359, Chap. I, Sect. 2]). Hence in view of (1), N is the only minimal normal subgroup of G and $N \nleq \Phi(G)$. Moreover, in view of (2), N is a r-group. Hence $G = N \rtimes M$, where $M \simeq G/N$ is a r-decomposable maximal subgroup of G and $N = C_G(N) = O_r(G) = F(G)$ by Chap. 1, Proposition 1.4.

Final contradiction for the case (ii). Let V be a maximal subgroup of S and T a r-decomposable supplement of V in G. Since G/N is r-decomposable by (1) and $N = C_G(N) = O_r(G)$ by (3), G/N is a r'-group. Hence $N = P$ is an abelian Sylow r-subgroup of G. Since G is not r-decomposable, $1 < T \cap P < P$, which contradicts (4.2). This final contradiction shows that Σ_S is a G-covering subgroup system for the class of all r-decomposable groups.

(iii) Suppose that the theorem in this case is false and let G be a counterexample of minimal order. Let $p_1 \varphi p_2 \varphi \ldots \varphi p_t$, where $\{p_1, p_2, \ldots, p_t\} = \pi(G)$ and $r = p_1$. Let N be a minimal normal subgroup of G. Then:

(a) $O_r(G) \neq 1$.
    If $r \neq p$, then G is r-closed by (i). Suppose that $r = p$. Let V be a maximal subgroup of S and $T \in \mathfrak{F}$ be a supplement of V in G. Then $TG_r = G$ and so $1 \neq T \cap G_r = T_r$ is a Sylow r-subgroup of T. This shows that T_r is normal in T. Therefore $(T_r)^G = (T_r)^{TG_r} = (T_r)^{G_r} \leq G_r$. Thus (a) holds.

(b) $G = N \rtimes M$, *where $M \in \mathfrak{F}$ is a maximal subgroup of G and $N = C_G(N) = O_r(G) = F(G)$ is the Sylow r-subgroup of G.*
    It is easy to similarly show as above that $G/R \in \mathfrak{F}$ for any minimal normal subgroup R of G. Hence G has a unique minimal normal subgroup N and $N = G^{\mathfrak{F}} \nleq \Phi(G)$. Hence, in view of (a), $N \leq O_r(G)$. This implies that $N = C_G(N) =$

$F(G) = O_r(G)$ by Chap. 1, Proposition 1.4. Since every group in $\mathfrak{F}$ is a φ-dispersive group, G/N is a φ-dispersive group and so its Sylow r-subgroup is normal. But $O_r(G/N) = O_r(G/C_G(N)) = 1$ (see Chap. 1, Lemma 2.15), so N is the Sylow r-subgroup of G.

Final contradiction for the case (iii).

By Lemma 4.3, $\mathfrak{F} = LF(f)$, where $f(r)$ is a totally local formation contained in $\mathfrak{F}$ for all $r \in \pi(\mathfrak{F})$, and $f(r) = \varnothing$ for all primes $r \notin \pi(\mathfrak{F})$. In view of (b) and Lemma 4.4, we may, without loss of generality, assume that $S \leq M$. Let V be a maximal subgroup of S and $T \in \mathfrak{F}$ be a supplement of V in G. Then $N \leq T$ and hence $O_{r'}(T) = 1$ by (b). On the other hand, since $T \in \mathfrak{F}$, $T/O_{r',r}(T) \in f(r)$ by Lemma 4.5. Therefore, the supplement $E = T \cap M \simeq T/O_r(T)$ of V in M belongs to $f(r)$. But since $|M| < |G|$, $M \in f(r)$ by Lemma 4.3 and the choice of G. Hence $G \in \mathfrak{F}$ by (b) and Lemma 4.6, a contradiction.

The theorem is proved.

As a immediate corollaries of Theorem 4.7, we have following:

Corollary 4.8 *Let S be a nonidentity p-subgroup of G. Then Σ_S is a G-covering subgroup system for any formation $\mathcal{F}$ in the following list:*

(i) *$\mathcal{F}$ is the class of all nilpotent π-groups.*
(ii) *$\mathcal{F}$ is the class of all φ-dispersive groups.*
(iii) *$\mathcal{F}$ is the class of all φ-dispersive groups G of p-length $l_p(G) \leq t$ ($t \in \mathbb{N}$).*
(iv) *$\mathcal{F}$ is the class of all φ-dispersive groups G contained in the class $\mathfrak{N}^t$ ($t \in \mathbb{N}$).*

Theorem 4.9 (Guo, Huang, Skiba [149, Theorem B]). *Let P and Q be a nonidentity p-subgroup and a q-subgroup of G, respectively. Then $\Sigma = \Sigma_P \cup \Sigma_Q$ is a G-covering subgroup system for any formation contained in $\mathfrak{N}$ and for any saturated formation $\mathfrak{F}$ contained in the class of all p-soluble groups A with p-length $l_p(A) \leq 1$ such that $\mathfrak{F} = \mathfrak{G}_{p'}\mathfrak{F}$.*

Proof First suppose that $\mathcal{F}$ is a formation contained in $\mathfrak{N}$. Suppose that the theorem in this case is false and let G be a counterexample. In view of Theorem 4.7(i), G is nilpotent since the class of all nilpotent groups is the intersection of all classes of p-closed groups, for all $p \in \mathbb{P}$. Let R be any Sylow r-subgroup of G, where $r \neq p$. Let V be a maximal subgroup of P and $VT = G$ for some $T \in \mathfrak{F}$. Then $R \leq T$ and so $R \in \mathfrak{F}$ since T is nilpotent. Similarly, we can show that the Sylow p-subgroup of G belongs to $\mathcal{F}$ by considering $p \neq q$. But then $G \in \mathfrak{F}$, a contradiction.

Now assume that $\mathfrak{F}$ is a saturated formation contained in the class of all p-soluble groups A with p-length $l_p(A) \leq 1$ and $\mathfrak{F} = \mathfrak{G}_{p'}\mathfrak{F}$. Suppose that in this case the theorem is false and let G be a counterexample of minimal order. Then:

(1) $G/N \in \mathfrak{F}$ *for any minimal normal subgroup N of G* (see (1) in the proof (ii) of Theorem 4.7 and the proof of Lemma 4.1).
(2) $O_{p'}(G) = 1$.
   Indeed, if $O_{p'}(G) \neq 1$ and N is a minimal normal subgroup of G contained in $O_{p'}(G)$, then $G/N \in \mathfrak{F}$ by (1) and so $G \in \mathfrak{F}$ since $\mathfrak{F} = \mathfrak{G}_{p'}\mathfrak{F}$ by hypothesis, a contradiction. Hence we have (2).

(3) *G has a unique minimal normal subgroup, N say, $N \nleq \Phi(G)$ and p divides $|N|$.*
   By (2), p divides $|N|$. Moreover, since the class $\mathfrak{F}$ is a saturated formation, $N \nleq \Phi(G)$ and N is the only minimal normal subgroup of G.

(4) *N is a p-group.*
   In view of (2) and (3), we only need to prove that G has a nonidentity p-soluble normal subgroup.
   If for every maximal subgroup V of Q and for every supplement $T \in \mathfrak{F}$ of V in G we have $O_{p'}(T) = 1$, then T is p-closed since $\mathfrak{F}$ is contained in the class of all p-soluble groups A with p-length $l_p(A) \leq 1$. Hence G is p-closed by Theorem 4.7(i).
   Now assume that some maximal subgroup V of Q and for some supplement $T \in \mathfrak{F}$ of V in G we have $O_{p'}(T) \neq 1$. Let r be a prime dividing $|O_{p'}(T)|$ and R a Sylow r-subgroup of $O_{p'}(T)$. By Frattini argument, $T = O_{p'}(T)N_T(R)$. Hence, for some Sylow p-subgroup G_p of G we have $G_p \leq N_T(R)$. Let W be a maximal subgroup of P and $M \in \mathfrak{F}$ be a supplement of W in G. Then for any $x \in G$, we have $G = WM = G_p M = G_p M^x$ by Chap. 1, Lemma 4.14. Therefore we may assume that $R \leq M$. Then $R^G = R^{G_p M} = R^M \leq M$. Therefore R^G is p-soluble. Hence (4) holds.

(5) $N = C_G(N) = F(G)$. *(This follows directly from (3), (4))*

(6) *P is not contained in N.*
   Suppose that $P \leq N$. Let V be a maximal subgroup of P and $T \in \mathfrak{F}$ be a supplement of V in G. Then $1 \neq T \cap N \neq N$, which contradicts Lemma 4.2.

Final contradiction. Let V be a maximal subgroup of Q and $T \in \mathfrak{F}$ be a supplement of V in G. Then $N \leq T$ and hence $O_{p'}(T) = 1$ by (5). But then T is p-closed. Therefore G is p-closed by Theorem 4.7(i), which contradicts (6). The theorem is proved.

From Theorem 4.9, we get

Corollary 4.10 *Let P and Q be nonidentity p-subgroup and q-subgroup of G, respectively. Then $\Sigma = \Sigma_P \cup \Sigma_Q$ is a G-covering subgroup system for any formation $\mathcal{F}$ in the following list:*

  (i) *$\mathcal{F}$ is the class of all supersoluble groups.*
 (ii) *$\mathcal{F}$ is the class of all groups G with $G' \leq F(G)$.*
(iii) *$\mathcal{F}$ is the class of all metanilpotent groups.*
 (iv) *$\mathcal{F}$ is the class of all p-supersoluble groups.*
  (v) *$\mathcal{F}$ is the class of all p-nilpotent groups.*
 (vi) *$\mathcal{F}$ is the class of all nilpotent groups G of class $c(G) \leq t$ $(t \in \mathbb{N})$.*

Lemma 4.11 (Wielandt [447]). *If G has three soluble subgroups A_1, A_2, and A_3 whose indices $|G : A_1|$, $|G : A_2|$, $|G : A_3|$ are pairwise coprime, then G is itself soluble.*

Theorem 4.12 (Guo, Huang, Skiba [149, Theorem C]). *Let $p_1, \ldots, p_t$ be some set of pairwise distinct prime divisors of $|G|$ and S_i a nonidentity p_i-subgroup of G. Let*

$\Sigma_i = \Sigma_{S_i}$, $i = 1, \ldots, t$. *Then the set* $\Sigma = \Sigma_1 \cup \ldots \cup \Sigma_t$ *is a G-covering subgroup system for the class* $\mathfrak{N}^t$.

Proof Suppose that this Theorem is false and let G be a counterexample with minimal $|G| + t$. Then in view of Theorem 4.7(ii), $t > 1$. Note that the class $\mathfrak{N}^t$ is a saturated formation (see Chap. 1, (1.3)(8)). We now proceed the proof by the following steps.

(1) $G/N \in \mathfrak{N}^t$ *for any minimal normal subgroup N of G* (see (1) in the proof of (ii) of Theorem 4.7 and the proof of Lemma 4.1).

(2) G *is soluble.*

In view of (1), we only need to prove that G has a nonidentity soluble normal subgroup. In view of Lemma 4.11, we only need to consider the case when $t = 2$, that is, each maximal subgroup of S_1 and each maximal subgroup of S_2 has a metanilpotent supplement in G.

First assume that some maximal subgroup V of S_1 has a metanilpotent supplement T in G such that $O_r(T) \neq 1$, for some prime $r \neq p_2$. Let W be a maximal subgroup of S_2 and H a metanilpotent supplement of W in G. It is clear that a Sylow r-subgroup of H is a Sylow r-subgroup of G. Hence, $O_r(T) \leq H^x$, for some $x \in G$. Since $(|G : T|, |G : H^x|) = 1$, $G = TH^x$. Hence $(O_r(T))^G = (O_r(T))^{T H^x} = (O_r(T))^{H^x} \leq H^x$ is a normal soluble subgroup of G. Now we assume that, for any maximal subgroup V of S_1 and for any metanilpotent supplement T of V in G, we have $O_r(T) = 1$ for all primes $r \neq p_2$. Then $O_{p_2}(T)$ is a Sylow p_2-subgroup of T. Thus G is p_2-closed by Theorem 4.7(i). Hence we have (2).

(3) $G = N \rtimes M$, *where* $M \in \mathfrak{N}^t$ *is a maximal subgroup of G and* $N = C_G(N) = O_r(G) = F(G)$ *is a minimal normal subgroup of G for some prime r.*

Since $\mathcal{F}$ is a saturated formation, then in view of (1), N is the only minimal normal subgroup of G and $N \nleq \Phi(G)$. Moreover, in view of (2), N is a r-group for some prime r and so $G = N \rtimes M$ for some maximal subgroup $M \simeq G/N \in \mathfrak{N}^t$ of G and $N = C_G(N) = O_r(G) = F(G)$ by Chap. 1, Prop. 1.4.

Final contradiction. In view of (3) and Lemma 4.11, we may, without loss of the generality, assume that $S_2, S_3, \ldots, S_t \leq M$ and that $r \neq p_i$ for all $1 < i \leq t$. Let $1 < i \leq t$, V be a maximal subgroup of S_i and $T \in \mathfrak{N}^t$ a supplement of V in G. Then $V(T \cap M) = VT \cap M = M$. Hence $T \cap M$ is a supplement of V in M.

We now show that $U = T \cap M \in \mathfrak{N}^{t-1}$. It is clear that $N \leq T$. Hence $O_{r'}(T) = 1$ by (3) and so $F(T) = O_r(T)$. It follows that $T/O_r(T) \in \mathfrak{N}^{t-1}$, which implies $U O_r(T)/O_r(T) \simeq U/U \cap O_r(T) \in \mathfrak{N}^{t-1}$. Therefore $U/O_r(U) \in \mathfrak{N}^{t-1}$. On the other hand, $M \in \mathfrak{N}^t$ by (3), which implies $U F(M)/F(M) \simeq U/U \cap F(M) = U/(T \cap F(M)) \in \mathfrak{N}^{t-1}$. Since $G = N \rtimes M$ and $N = C_G(N)$ by (3), $O_r(M) = 1$ by Chap. 1, Lemma 2.15. Hence $F(M)$ is a r'-group. Therefore $T \cap F(M) \cap O_r(U) = 1$. It follows that $U \simeq U/(T \cap F(M)) \cap O_r(U) \in \mathfrak{N}^{t-1}$. This shows that each maximal subgroup of every subgroup $S_2, S_3, \ldots, S_t$ has a supplement $U \in \mathfrak{N}^{t-1}$ in M. Therefore $M \in \mathfrak{N}^{t-1}$ by the choice of (G, t). This induces that $G \in \mathfrak{N}^t$. The final contradiction completes the proof.

Theorems 4.7, 4.9, and 4.12 cover some main results in [173, 174, 195]. In particular, we have

Corollary 4.13 [173]. *Let Σ be a set of subgroups which contains at least one supplement of each maximal subgroup of every Sylow subgroup of G. Then Σ is a G-covering subgroup system for the classes of all nilpotent and all supersoluble groups.*

Corollary 4.14 [195]. *Let P and Q be nonconjugate Sylow subgroups of G. Then $\Sigma_P \cup \Sigma_Q$ is a G-covering subgroup system for the classes of all metanilpotent, all supersoluble, all p-supersoluble, all p-nilpotent, and all abelian groups.*

Corollary 4.15 [173]. *Let Σ be a set of subgroups which contains at least one supplement of each maximal subgroup of every Sylow subgroup of G. If G is p-soluble, then Σ is a G-covering subgroup system for the classes of all p-nilpotent and all p-supersoluble groups.*

Corollary 4.16 [195]. *Let $p_1, \dots, p_t$ be some set of pairwise distinct prime divisors of $|G|$. Let P_i be a Sylow p_i-subgroup of G and $\Sigma_i = \Sigma_{P_i}$, $i = 1, \dots, t$. Let r be a prime. Then:*

(1) Σ_1 *is a G-covering subgroup system for the classes of all r-decomposable and all nilpotent groups.*
(2) *The set $\Sigma_1 \cup \dots \cup \Sigma_t$ is a G-covering subgroup system for the class $\mathfrak{N}^t$.*

3.5 Additional Information and Some Problems

I. The theorems in Sects. 3.1 and 3.2 unify and generalize main results of many papers. But in fact, the list of such results it too long, so in this section we consider only a few examples for applications of the theorems. By analogy with it, the readers can easily obtain another corollaries of the results in Sects. 3.1 and 3.2.

II. In view of Examples 1.6, 1.8, 2.14 and 2.15, among corollaries of Theorem 2.33 we meet main results in [37, 255, 293, 368, 383, 398].

III. In view of Example 1.6, Theorems 2.31 and 2.32 cover Theorems 3.3 and 3.9 in [329], Theorem 2 in [39], Theorem 4.1 in [45], Theorems 1.1 and 1.2 in [441], Main Theorem in [15], Theorem 3 in [58], Theorem 1 in [395], Theorems 2.7 and 2.10 in [220], Theorems 4.1 and 4.2 in [435].

IV. In view of Example 1.8, Theorems 2.31 and 2.32 cover a lot of know results, in particular, Theorems 3.1, 3.4, and 3.6 in [311], Theorem A in [384], Theorems A and B in [213], Theorems 3.1 and 3.6 in [270], Theorems A, B, D, and E in [190], Theorems A–D in [182], Theorems 3.1 and 3.4 in [348], Theorem 3.1 in [10], Theorems 1.3 and 1.4 in [12], Theorems 4.1 and 4.2 in [19], Theorem 2 in [395].

V. In view of Example 1.10, Theorem 2.31 covers Theorems C and D in [98], Theorem 4.7 in [440], Theorems 4.1 and 4.3 in [438].

VI. In view of Examples 1.14 and 1.21, Theorem 2.32 covers Theorem 1.3 in [146]. Similarly, Theorem 2.32 covers Lemma 3.2 and Theorem 3.3 in [240].

VII. In view the of Example 1.22 and 1.30, Theorems 2.31 and 2.32 cover Theorems 1.2 and 1.4 in [4], Theorems 3.1 and 3.2 in [141], Theorem 3.3 in [210], Theorems 3.1–3.3 in [242], Theorems 3.1 and 3.2 in [216] and Theorem 3.4 and 3.12 in [314].

VIII. In view of Example 1.12, Theorems 2.36 and 2.32 cover Theorems 5.1 and 5.2 in [196].

IX. In view of Example 1.9 and 1.30, Theorem 2.32 covers Theorems 1.2 in [296], Theorems 3.1, 3.2 in [470], Theorems 3.12 and 3.13 in [309]

X. In view of Example 1.15, Theorems 2.31 and 2.32 cover Theorems 3.1, 3.4–3.7 in [288], Theorem 1.5 in [289].

XI. In addition to the above mentioned application, many results in [3, 16, 41, 65, 224, 277, 284, 301, 302, 439, 455, 467, 476] can also be developed and unified by our Theorems 2.31–2.33, 2.36, 2.37.

XII. We now not know the answer to the following questions:

Problem 5.1 Is the subgroup functor in Example 1.14 regular?

Problem 5.2 Is the subgroup functors in Example 1.16 hereditary?

XIII. Since all known G-covering subgroup systems of groups are related some concrete formations of groups, in this connection, L.A. Shemetkov asked at Gomel Algebraic seminar the following **question:** *Is it true that the theory of the G-covering subgroup systems in* [195], [173] *can be extended to an arbitrary soluble saturated formation?* Theorems 4.8 and 4.9 give a partial solution to this still open problem.

XIV. The following example shows that in general Σ_P, where P is a Sylow p-subgroup of G, is not a G-covering subgroup system for the class of all supersoluble groups.

Example 5.3   Suppose that q divides $p - 1$ and $q > 2$. Let $Q = \langle x, y, z \mid x^q = y^q = z^q = 1, [x, z] = [y, z] = 1,$ and $[x, y] = z \rangle$. Then $|Z(Q)| = q$ and Q is of exponent q (see [117, p. 203]). Hence there is a simple $\mathbb{F}_p[Q]$-module P which is faithful for Q. Let $G = P \rtimes Q$. Let V be a maximal subgroup of Q. Then V has a complement L in Q, so $T = PL$ is a complement of V in G. Since q divides $p - 1$, T is supersoluble by [450, Chap. 1, Theorem 1.9]. Thus each maximal subgroup of Q has a supersoluble complement in G. But since $F(G) = P$, $G/F(G)$ is not abelian, so G is not supersoluble by [450, Chap. 1, Theorem 1.9]. Therefore Σ_Q is not a G-covering subgroup system for the class of all supersoluble groups.

XV. Let π be a set of primes. Recall that a group G is said to be π-separable if it has a chief series each factor of which is either a π-group or a π'-group.

Let Σ be a set of subgroups which contains at least one supplement to G of each maximal subgroup of every Sylow subgroup of G.

Problem 5.4 Is Σ a G-covering subgroup system for the classes of all π-separable and all p-soluble groups?

Problem 5.5 Is Σ a G-covering subgroup system for any soluble saturated formation?

XVI. The example in Sect. 5 in [195] shows that Σ_S is not necessarily a G-covering subgroup system for the formation of all supersoluble groups.

XVII. Some other applications of the theory of subgroups functors see the books [34, 259, 345, 381] and the recent papers [26, 27, 258, 413, 409].

XVIII. A subgroup H of a group G is called *modular* [343] if H is a modular element (in the sense of Kurosh [343, p. 43]) of the lattice of all subgroups of G.

Example 5.5 Let $\tau(G)$ be the set of all modular subgroups of G for any group G. Then τ is hereditary by [343, p. 201]. From Theorem 5.2.5 in [343] it is easy to see that τ is regular.

In view of Example 5.5, Theorems 2.31 and 2.32 cover Theorems 1.3 in [412], Theorem 1.2 in [407].

IXX. A. S. Kondrat'ev proved in [268] that if the normalizer of any Sylow subgroup of a group G is of odd index, then G is 2-nilpotent. N. Chigura proved in [72] that a group G is p-nilpotent if $p \neq 3$ and $(|G : N_G(G_r)|, p) = 1$ for every $r \in \pi(G)$. J. Zhang [464] proved that if the index of the normalizer of any Sylow subgroup of a group G is primary, then G is soluble. In [128] W. Guo proved that the index of the normalizer of every Sylow subgroup of a group G is an odd number or a prime power number if and only if G is soluble and $G = HK$, where H and K are Hall subgroups of G, H is 2-nilpotent, K is nilpotent and is normal in some Hall $2'$-subgroup of G. He also proved with K.P. Shum in [171] that if in a group G, the normalizer of any Sylow 2, 3-subgroup is of prime power index, then G is soluble. As a develop of above results, A. S. Kondratev and W. Guo proved the following

Theorem 5.6 (Kondratev, Guo [269]). *If the normalizer of a Sylow 3-subgroup of a finite group G is of odd index in G, then the nonabelian composition factors of G are isomorphic to one of the following groups: $L_2(q)$ for $q \equiv \pm1(mod\ 12)$; $L_n(q)$ for $n \in \{3,4,5\}$ and $q \equiv -1(mod\ 12)$; $U_n(q)$ for $n \in \{3,4,5\}$ and $q \equiv 1(mod\ 12)$; $PSp_4(q)$ for $q \equiv \pm1(mod\ 12)$; $Sz(q)$; M_{11}.*

Since all primary groups are nilpotent, we see that if the index of the normalizer of any Sylow subgroup of a group G is primary, then all these normalizers has a nilpotent Hall supplement in G, but not the converse. Hence it is interesting to determine the structures of groups in which the normalizer of any Sylow subgroup has a nilpotent Hall supplement. In [280], the following theorem was proved.

Theorem (Li, Guo, Huang [280]). If the normalizer of any Sylow subgroup of a group G has a nilpotent Hall supplement in G, then G is soluble.

Chapter 4
Groups With Given Maximal Chains of Subgroups

4.1 Σ-Embedded Subgroups

Σ-Embedded and Quasipermutable Subgroups Let A be a subgroup of a group G, $K \leq H \leq G$. Then we say: (i) A *covers the pair* (K, H) if $AH = AK$; (ii) A *avoids* (K, H) if $A \cap H = A \cap K$. Note that if A is a quasinormal subgroup of G, then for every maximal pair (K, H) of G (that is, K is a maximal subgroup of H), A either covers or avoids (K, H). Indeed, by Chap. 1, Lemma 5.35(2), $A \cap H$ is quasinormal in H, so in the case when $A \cap H \nleq K$, $(A \cap H)K = H$ in view of the maximality of the pair (K, H). This observation leads us to the following generalization of the quasinormality.

Definition 1.1 Let A be a subgroup of G and $\Sigma = \{G_0 \leq G_1 \leq \ldots \leq G_n\}$ some subgroup series of G. Then we say that:

(1) A *is* Σ-*embedded* in G if A either covers or avoids every maximal pair (K, H) such that $G_{i-1} \leq K < H \leq G_i$, for some i;
(2) A *is quasipermutable in* G *or a CAMP-subgroup of* G if A is $\{1 \leq G\}$-embedded in G.

The following examples show that a Σ-embedded subgroup (even a quasipermutable subgroup) of G is not necessarily quasinormal.

Example 1.2 Let p and q be primes, where q divides $p - 1$. Let $A = \langle a \rangle$ be a cyclic group with order p^2 and B a group with $|B| = q$. Put $G = A \wr B = K \rtimes B$, where $K = A_1 \times A_2 \times \ldots \times A_q$ is the base subgroup of the regular wreath product G and $L = \langle a^p \rangle^\natural$ (here we use the terminology in [89, A, p. 63]). Then $G/L \simeq \langle a^p \rangle \wr B$, $L \leq \Phi(G)$ and G is supersoluble. Let R be a subgroup of A_1 with $|R| = p$. Assume that R is quasinormal in G. Then $B \leq N_G(R)$ by Chap. 1, Lemma 5.34(6), and so R is normal in G, a contradiction. Hence R is not quasinormal in G. However, obviously, R is quasipermutable in G (see Theorem 1.24 below).

© Springer-Verlag Berlin Heidelberg 2015
W. Guo, *Structure Theory for Canonical Classes of Finite Groups*,
DOI 10.1007/978-3-662-45747-4_4

Example 1.3 Let G be a soluble group and $\Sigma = \{1 = G_0 \leq G_1 \leq \ldots \leq G_n = G\}$ some chief series of G. Then every subgroup of G is Σ-embedded in G.

Example 1.4 Every subgroup of any nilpotent group G is quasipermutable in G.

Example 1.5 Let G be a metanilpotent group and $\Sigma = \{1 \leq F(G) \leq G\}$. Then every subgroup of G is Σ-embedded and some Sylow subgroup of G is not quasipermutable in G (see below Lemma 1.9).

General Properties of Σ-Embedded Subgroups Note that for any subgroups M, K, and H of G, where $K \leq H$, $M \cap H = M \cap K$ is equivalent to $M \cap H \leq K$ and $MH = MK$ is equivalent to $H \leq K(M \cap H)$.

Lemma 1.6 *Let $M \leq G$, N be a normal subgroup of G and (K, H) a maximal pair of G such that M either covers or avoids (K, H). Then*

(1) If N avoids (K, H), then (KN, HN) is a maximal pair of G and $|HN : KN| = |H : K|$.

(2) If M covers (avoids) (KN, HN), then MN covers (avoids, respectively) (K, H).

(3) If $H \leq V$ and M covers (avoids) (K, H), then $M \cap V$ covers (avoids, respectively) (K, H).

(4) If M covers (avoids) (K, H) and $N \leq K$, then MN covers (avoids, respectively) (K, H).

(5) If $A \leq B \leq C \leq D$ and M covers (avoids) (A, D), then M covers (avoids, respectively) (B, D).

Proof

(1) Suppose that $H \cap N \leq K$. Let T be a subgroup of G such that $KN \leq T \leq HN$. Then $T = N(T \cap H)$ and $K \leq T \cap H \leq H$. Hence either $T \cap H = K$ or $T \cap H = H$. In the former case, we have $T = (T \cap H)N = KN$. If $T \cap H = H$, then $T = HN$. Hence (KN, HN) is a maximal pair of G. Finally, since $H \cap N = K \cap N$, $|HN : KN| = |H : K|$.

(2) Since $K \leq KN \cap H \leq H$, obviously, N either covers or avoids (K, H). If N covers (K, H), then, clearly, NM covers (K, H). Suppose that N avoids (K, H). Then by (1), (KN, HN) is a maximal pair of G. If M covers (KN, HN), then $H \leq HN \leq NK(M \cap NH)$. Hence $H = K(H \cap N(M \cap NH)) \leq K(H \cap NM) \leq H$, which induces that NM covers (K, H). Suppose that M avoids (KN, HN). Then $KN \cap M = HN \cap M$. Hence $HN \cap MN = (HN \cap M)N = (KN \cap M)N \leq KN$ and $MN \cap H \leq KN \cap H = K(N \cap H) = K(N \cap K) = K$. Thus MN avoids H/K.

(3) Since $M \cap V \cap H = M \cap H$ and M either covers or avoids (K, H), obviously, either $H = K((M \cap V) \cap H)$ or $M \cap V \cap H \leq K$.

(4) If $M \cap H \leq K$, then $NM \cap H = N(M \cap H) \leq NK = K$. If $MK = MH$, then $MNK = MNH$.

(5) It is evident. The lemma is proved.

Let $V \leq G$ and N be a normal subgroup of G. If $\Sigma = \{G_0 \leq G_1 \leq \ldots \leq G_n\}$ is some subgroup series of G, then we use $\Sigma N/N$ to denote the series

$$G_0 N/N \leq G_1 N/N \leq \ldots \leq G_n N/N$$

and $\Sigma \cap V$ to denote the series

$$G_0 \cap V \leq G_1 \cap V \leq \ldots \leq G_n \cap V.$$

Note that if p a prime, A is a subgroup of a p-soluble group G and $\Sigma = \{G_0 \leq G_1 \leq \ldots \leq G_n\}$ is some chief series of G, then A either covers or avoids every maximal pair (K, H) such that $G_{i-1} \leq K < H \leq G_i$ and p divides $|G_i : G_{i-1}|$, for some i. On the other hand, if (K, H) is a maximal pair such that $G_{i-1} \leq K < H \leq G_i$ and p does not divide $|G_i : G_{i-1}|$, then G may have a subgroup A such that A neither covers nor avoids (K, H). For example, let $G = C_7 \wr A_5$, where C_7 is a group of order 7. Let $H = A_4$ and K be a Sylow 3-subgroup of H. Let A be a subgroup of H of order 2. Then A neither covers nor avoids the maximal pair (K, H) and A either covers or avoids every maximal pair (L, T) where T is a 7-subgroup of G. This is a motivation for the following generalization of the Σ-embedded subgroups.

Definition 1.7 Let A be a subgroup of G, p a prime and $\Sigma = \{G_0 \leq G_1 \leq \ldots \leq G_n\}$ some subgroup series of G. Then we say that A is Σ_p-embedded in G if A either covers or avoids every maximal pair (K, H) such that $G_{i-1} \leq K < H \leq G_i$, for some i, and p divides $|G_i : G_{i-1}|$.

Let p be a prime. A subgroup E of G is called semi-p-cover-avoiding [101] if G has a chief series $\Sigma = \{G_0 \leq G_1 \leq \ldots \leq G_n\}$ such that E either covers or avoids each factor G_i/G_{i-1} with p divides $|G_i/G_{i-1}|$. It is clear that every semi-p-cover-avoiding subgroup is Σ_p-embedded in G for some chief series Σ of G. The example before Definition 1.7 shows that the inverse is not true in general.

Lemma 1.8 *Let $M \leq G$, N, and R be normal subgroups of G.*

(1) If $E \leq V$ and M is $\{E \leq G\}$-embedded in G, then $M \cap V$ is $\{E \leq V\}$-embedded in V.

(2) If $R \leq N$ and M is $\{R \leq G\}$-embedded in G, then NM is $\{R \leq G\}$-embedded in G and NM/N is $\{1 \leq G/N\}$-embedded in G/N.

(3) If M is Σ_p-embedded in G for some composition series Σ of G and $N \leq M$, then M/N is $(\Sigma N/N)_p$-embedded in G/N.

(4) If M is Σ-embedded in G for some composition series Σ of G and $N \leq M$, then M/N is $\Sigma N/N$-embedded in G/N.

(5) If M is Σ_p-embedded in G for some composition series Σ of G and $M \leq V \leq G$, then M is $(\Sigma \cap V)_p$-embedded in V.

(6) If M is Σ-embedded in G for some subnormal series Σ of G and $M \leq V \leq G$, then M is $(\Sigma \cap V)$-embedded in V.

Proof

(1) It follows from Lemma 1.6(3).

(2) Let (K, H) be a maximal pair such that $R \leq K < H \leq G$. If $NH = NK$, then NM covers (K, H). If $N \cap H \leq K$, then by Lemma 1.6(1), (NK, NH) is a maximal pair and $R \leq NK$. Hence M either covers or avoids (NK, NH). It follows from Lemma 1.6(2) that NM either covers or avoids (K, H). Hence NM is $\{R \leq G\}$-embedded in G. Consequently, NM/N is $\{1 \leq G/N\}$-embedded in G/N.

(3) Let $\Sigma = \{G_0 < G_1 < \ldots < G_n\}$ be a composition series of G such that M is Σ_p-embedded in G. Then

$$\Sigma N/N = \{G_0 N/N \leq G_1 N/N \leq \ldots \leq G_n N/N\}.$$

Let (K, H) be a maximal pair such that $NG_{i-1} \leq K < H \leq NG_i$, where p divides $|NG_i/NG_{i-1}|$. Then $K = N(K \cap G_i)$, $H = N(H \cap G_i)$ and $|H \cap G_i : K \cap G_i| = (|H \cap G_i||N|/|G_i \cap N|) : (|K \cap G_i||N|/|G_i \cap N|) = (|H \cap G_i||N|/|G_i \cap H \cap N|) : (|K \cap G_i||N|/|G_i \cap K \cap N|) = |H : K|$. Besides, clearly, $1 \neq NG_i/NG_{i-1} \simeq G_i/G_{i-1}$. Hence p divides $|G_i/G_{i-1}|$.

We show that $(K \cap G_i, H \cap G_i)$ is a maximal pair. Indeed, since $K \neq H$, N does not cover $(K \cap G_i, H \cap G_i)$. Hence, there exists a maximal pair (L, T) such that $K \cap G_i \leq L < T \leq H \cap G_i$ and N does not cover (L, T). Then by Lemma 1.6(1), (NL, NK) is a maximal pair and $|TN : LN| = |T : L|$. However $K = N(K \cap G_i) \leq NL < NT \leq N(H \cap G_i) = H$. Hence, $(K \cap G_i, H \cap G_i)$ is a maximal pair such that $G_{i-1} \leq K \cap G_i < H \cap G_i \leq G_i$. By hypothesis, M either covers or avoids $(K \cap G_i, H \cap G_i)$. If M covers $(K \cap G_i, H \cap G_i)$, then MN covers the pair $(N(K \cap G_i), N(H \cap G_i)) = (K, H)$. Assume that M avoids $(K \cap G_i, H \cap G_i)$. Since $N \leq M$, $M \cap H = M \cap N(H \cap G_i) = N(M \cap H \cap G_i) = N(M \cap K \cap G_i) \leq N(K \cap G_i) = K$. Hence M avoids (K, H). It follows that M/N is $(\Sigma N/N)_p$-embedded in G/N.

(4) See the proof of (3)

(5) Let $\Sigma = \{1 = G_0 < G_1 < \ldots < G_n\}$. Then

$$\Sigma \cap V = \{G_0 \cap V \leq G_1 \cap V \leq \ldots \leq G_n \cap V\}.$$

Let (K, H) be a maximal pair such that $G_{i-1} \cap V \leq K < H \leq G_i \cap V$ and p divides $|G_i \cap V/G_{i-1} \cap V|$. Then $KG_{i-1} \neq HG_{i-1}$ (Otherwise, $H = H \cap KG_{i-1} = K(H \cap G_{i-1}) \leq K(V \cap G_{i-1}) \leq K$, which is impossible) and so G_{i-1} avoids (K, H). Hence by Lemma 1.6(1), $(G_{i-1}K, G_{i-1}H)$ is a maximal pair and $G_{i-1} \leq G_{i-1}K < G_{i-1}H \leq G_i$. Since $(G_i \cap V)/(G_{i-1} \cap V) \simeq (G_i \cap V)G_{i-1}/G_{i-1}$, p divides $|G_i/G_{i-1}|$. Hence by hypothesis, M either covers or avoids $(G_{i-1}K, G_{i-1}H)$. Suppose that $M \cap G_{i-1}K = M \cap G_{i-1}H$. Then $M \cap K = M \cap V \cap G_{i-1}K = M \cap V \cap G_{i-1}H = M \cap H$. Hence M avoids (K, H). On the other hand, if M covers $(G_{i-1}K, G_{i-1}H)$, then $H = V \cap G_{i-1}H \leq V \cap MG_{i-1}K = M(G_{i-1}K \cap V) = MK$. Hence M covers (K, H). This means that M is $(\Sigma \cap V)_p$-embedded in V.

(6) See the proof of (5).

Lemma 1.9 *Let $N \leq M$ and E be subgroups of a group G. Suppose that N is normal in G and E is quasipermutable in G. Then*

(1) E is subnormal in G.
(2) EN/N is quasipermutable in G/N.
(3) $E \cap M$ is quasipermutable in M.

Proof

(1) Suppose that it is false and let G be a counterexample of minimal order. Then $E \leq M$, for some maximal subgroup M of G. If $EM^x = G$ for some $x \in G$, then $MM^x = G$, which contradicts Chap. 1, Lemma 4.14. Hence $E \leq M_G$. Obviously, E is also quasipermutable in M_G. Hence E is subnormal in M_G by the choice of G. It follows that E is subnormal in G, a contradiction.

(2) This follows from Lemma 1.8(2).

(3) This follows from Lemma 1.6(3).

Lemma 1.10 *If every Sylow subgroup of G is $\{H \leq G\}$-embedded in G, then H is subnormal in G.*

Proof Let M be a maximal subgroup of G such that $H \leq M$. Let p be a prime dividing $|G : M|$ and P a Sylow p-subgroup of G. Then $P \nsubseteq M$ and so $PM = G$. Clearly, for every Sylow q-subgroup Q of G where $q \neq p$, we have that $QM \neq G$, which implies $Q \leq M_G$. Hence G/M_G is a p-group and so M is normal in G. For every Sylow subgroup M_p of M, there is a Sylow subgroup P of G such that $M_p = P \cap M$. Hence by Lemma 1.8(1), M_p is $\{H \leq M\}$-embedded in M. By induction, H is subnormal in M. Consequently, H is subnormal in G.

Lemma 1.11 *Suppose that G is factorized $G = MN$, where N is a minimal normal subgroup of G and M is a subgroup of G. If E is a subnormal subgroup of G such that $E \leq M$, then $E \leq M_G$.*

Proof By Chap. 1, Lemma 5.35(5), we have $N \leq N_G(E)$. Then, $E^G = E^{NM} = E^M \leq M$. Hence $E \leq E^G \leq M_G$.

Characterizations of some classes of groups in terms of Σ-embedded and quasipermutable subgroups.

Theorem 1.12 (Guo, Skiba [196, Theorem 3.2]). *Let p be a prime. Then the following assertions are equivalent:*

(a) *G is p-soluble.*

(b) *Every subgroup of G is Σ_p-embedded in G, for all composition series Σ of G.*

(c) *Every maximal subgroup of G is Σ_p-embedded in G, for some composition series Σ of G.*

(d) *Every 2-maximal subgroup of G is Σ_p-embedded in G, for some composition series Σ of G.*

(e) *Some Sylow p-subgroup of G is Σ_p-embedded in G, for some composition series Σ of G.*

(f) *Either G is a group of prime power order or G has two p-soluble maximal subgroups M_1 and M_2 such that $(|G : M_1|, |G : M_2|) = r^a q^b$ for some primes r,*

q and some $a, b \in \{0\} \cup \mathbb{N}$ and M_i is Σ_{ip}-embedded in G for some composition series Σ_i of G, $i = 1, 2$.

(g) G is nilpotent or every nonsupersoluble Schmidt subgroup of G is Σ_p-embedded in G for some composition series Σ of G.

Proof (a) $\Rightarrow$ (b) It is clear.

(c), (d) $\Rightarrow$ (a) Suppose that every 2-maximal (every maximal) subgroup M of G is Σ_p-embedded in G for some composition series $\Sigma = \Sigma(M)$ of G. We shall show that G is p-soluble.

Suppose that this is false and let G be a counterexample with minimal order. Then G/N is p-soluble for any minimal normal subgroup N of G. Indeed, if N is a maximal subgroup of G, it is clear. Otherwise, by Lemma 1.8(3) the hypothesis is still true for G/N. Therefore G/N is p-soluble by the choice of G. It follows that N is the only minimal normal subgroup of G, N is non-abelian and p divides $|N|$.

Let $N = N_1 \times N_2 \times \ldots \times N_t$ be a direct product of isomorphic simple groups. We now prove that G has a maximal subgroup M such that p does not divide $|G : M|$, $NM = G$, and $N_i \neq M \cap N_i \neq 1$ for all $i = 1, 2, \ldots, t$.

Let $N_p \leq P$, where N_p and P are Sylow p-subgroups of N and G, respectively. Then $P \leq N_G(N_p)$ and G has a maximal subgroup M such that $N_G(N_p) \leq M$ and so $P \leq M$. Let P_i be a Sylow p-subgroup of N_i. Then for some $x \in G$, we have $P_i \leq P^x$. By Frattini's argument, $G = NM$. Hence $x = mn$ where $n \in N$ and $m \in M$. Then $P_i \leq (P^m)^n$, where $P^m \leq M$. It follows that $(P_i)^{n^{-1}} \leq M$. Since N_i is normal in N, $(P_i)^{n^{-1}} \leq N_i$. Thus $M \cap N_i \neq 1$, for all $i = 1, 2, \ldots, t$. If for some i we have $N_i \leq M$, then by Lemma 1.11, $N_i \leq M_G = 1$, a contradiction. Therefore $M \cap N_i \neq N_i$, for all i.

Let $E_i = M \cap N_i$ and A_i be a maximal subgroup of M such that $E_i \leq A_i$ (in case (c)) or $A_i = M$ (in case (c)). Then $E_i = A_i \cap N_i$. By hypothesis, A_i is $(\Sigma_i)_p$-embedded in G for some composition series Σ_i of G. Note that if L is a minimal subnormal subgroup of G, then by Chap. 1, Lemma 5.35(5), $N \leq N_G(L)$. Hence $L \leq N$. Without loss of generality, we may assume that A_i is $\{1 \leq N_i\}$-embedded in G. By Lemma 1.8(1), $E_i = A_i \cap N_i$ is $\{1 \leq N_i\}$-embedded in N_i. Hence by Lemma 1.9(1), E_i is subnormal in N_i, which contradicts $N_i \neq M \cap N_i \neq 1$. Hence (c) $\Rightarrow$ (a) and (d) $\Rightarrow$ (a) hold.

(e) $\Rightarrow$ (a) Suppose that it is false and let G be a counterexample with minimal order. Then p divides $|G|$. Let P be a Sylow p-subgroup of G and

$$\Gamma_P = \{1 = G_0 < N = G_1 < \ldots < G_n = G\}$$

be a composition series of G such that P is $(\Gamma_P)_p$-embedded in G. Since G is not p-soluble, there is an index i such that G_i/G_{i-1} is a simple non-abelian group and $p||G_i/G_{i-1}|$. Hence $G_i \neq G_{i-1}(G_i \cap P) \not\subseteq G_{i-1}$. Then P is $\{G_{i-1} \leq G_i\}$-embedded in G. By Lemma 1.8(1), $G_i \cap P$ is $\{G_{i-1} \leq G_i\}$-embedded in G_i. Hence by Lemma 1.8(4), $G_{i-1}(G_i \cap P)/G_{i-1}$ is $\{1 \leq G_i/G_{i-1}\}$-embedded in G_i/G_{i-1}. Then by Lemma 1.9(1), $G_{i-1}(G_i \cap P)/G_{i-1}$ is subnormal in G_i/G_{i-1}. This implies that G_i/G_{i-1} is not simple, a contradiction. So (e) $\Rightarrow$ (a).

(a) $\Rightarrow$ (f) If G is a p-soluble group and G is not a group of prime power order, then G has a maximal subgroup M_1 such that $|G : M_1| = p^a$ for some $a \in \mathbb{N}$ and a maximal subgroup M_2 such that p does not divide $|G : M_2|$. Hence $(|G : M_1|, |G : M_2|) = 1$. Now by (b), we see that (f) holds.

(f) $\Rightarrow$ (a) Suppose that G has two p-soluble maximal subgroups M_1 and M_2 such that $(|G : M_1|, |G : M_2|) = r^a q^b$ for some primes r, q and some $a, b \in \{0\} \cup \mathbb{N}$ and M_i is Σ_{ip}-embedded in G for some composition series Σ_i of G. We prove that G is p-soluble. Assume that this is false and let G be a counterexample of minimal order.

Let N be a minimal normal subgroup of G and $\pi = \{p_1, p_2, \ldots, p_t\}$ be the set of all primes dividing $|N|$. First we claim that N is the only minimal normal subgroup of G, G/N is p-soluble and $t > 2$. Indeed, if $N \leq M_1 \cap M_2$, then the hypothesis is still true for G/N by Lemma 1.8(3). Hence G/N is p-soluble by the choice of G. Assume that $N \not\leq M_1 \cap M_2$, for example, $N \not\leq M_1$. Then $G/N \simeq M_1/M_1 \cap N$ is also p-soluble since M_1 is p-soluble. This implies that N is the only minimal normal subgroup of G and N is non-abelian by the choice of G. Hence $t > 2$ and every minimal subnormal subgroup of G is contained in N. Since M_1 and M_2 are p-soluble, $M_1N = G$ and $M_2N = G$.

Let $p_i \in \pi$. Suppose that any Sylow p_i-subgroup of G is neither contained in M_1 nor in M_2. Then p_1 divides $(|G : M_1|, |G : M_2|)$. Since $(|G : M_1|, |G : M_2|) = r^a q^b$ for some primes r and q and $t > 2$, there exists some i and a Sylow p_i-subgroup Q of G such that either $Q \leq M_1$ or $Q \leq M_2$. Without loss of generality, we may assume that $i = 1$ and $Q \leq M_1$. Let L be a minimal subnormal subgroup of G such that M_1 is $\{1 \leq L\}$-embedded in G. Then, similar to the proof of (c) $\Rightarrow$ (a), we can obtain that $1 \neq M_1 \cap L \neq L$ and which leads to a contradiction. Therefore (f) $\Rightarrow$ (a) holds.

(a) $\Rightarrow$ (g) It follows from (b).

(g) $\Rightarrow$ (a). Suppose that this is false and let G be a counterexample of minimal order. Then p divides $|G|$. By Lemma 1.8(6), the hypothesis is still true for every subgroup of G. Hence all subgroups of G are p-soluble by the choice of G. Besides, obviously, G is not q-nilpotent for the smallest prime q dividing $|G|$. Hence by Chap. 1, Proposition 1.10, G has a q-closed Schmidt subgroup H. It is easy to see that H is not supersoluble and $H \neq G$. If G is a simple group, then $\{1 \leq G\}$ is a unique composition series of G. By Lemma 1.9(1), H is subnormal in G, a contradiction. Therefore G is not simple. Let R be a normal subgroup of G such that G/R is a simple non-abelian group and p divides $|G/R|$. Then every proper subnormal subgroup L of G is contained in R (otherwise $G = LR$ is p-soluble since it is the product of the p-soluble subgroups L and R and R is normal, which contradicts the choice of G). Besides, $R = \Phi(G)$ since all maximal subgroups of G are p-soluble. Thus, by hypothesis, the nonsupersoluble Schmidt subgroup H of G either covers or avoids every maximal pair of the form (M, G). Let $H \leq M$, where M is a maximal subgroup of G. Then for every $x \in G$ we have $HM^x \neq G$. Hence $H \leq M_G$. Obviously, $\Phi(G) \subseteq M_G$. If $M_G \not\subseteq \Phi(G)$, then there exists a maximal subgroup S of G such that $G = M_G S$. This implies that G is p-soluble, a contradiction. Thus $H \leq M_G = \Phi(G) \leq F(G)$, a contradiction.

The theorem is proved.

Corollary 1.13 [100, Theorem 2.1]. *Let p be a prime divisor of $|G|$. Then the following assertions are equivalent:*

(1) G is p-soluble.
(2) Every maximal subgroup of G is semi p-cover-avoiding in G.
(3) Some Sylow p-subgroup of G is semi p-cover-avoiding in G.

Corollary 1.14 [100, Theorem 2.2]. *The following assertions are equivalent:*

(1) G is soluble.
(2) Every maximal subgroup of G is semi cover-avoiding in G.
(3) Every Sylow subgroup of G is semi cover-avoiding in G.

Corollary 1.15 *Let G be a group and p a prime. Then the following are equivalent:*

(a) G is soluble.
(b) Every subgroup of G is Σ-embedded in G, for every composition series Σ of G.
(c) Every maximal subgroup M of G is Σ-embedded in G, for some composition series Σ of G.
(d) Every 2-maximal subgroup of G is Σ-embedded in G, for some composition series Σ of G.
(e) Every Sylow subgroup P of G is Σ-embedded in G, for some composition series Σ of G.
(f) Either G is a group of prime power order or G has two p-soluble maximal subgroups M_1 and M_2 such that $(|G : M_1|, |G : M_2|) = r^a q^b$ for some primes r, q and some $a, b \in \{0\} \cup \mathbb{N}$ and M_i (i = 1, 2) is Σ_i-embedded in G for some composition series Σ_i of G.
(g) Either G is nilpotent or every nonsupersoluble Schmidt subgroup H of G is Σ-embedded in G for some composition series Σ of G.

Corollary 1.16 [224, Theorem 3.4]. *If every 2-maximal subgroup of G is a CAP-subgroup of G, then G is soluble.*

Corollary 1.17 [226, Theorem 3.4, Corollary 3.7]. *If every 2-maximal subgroup of G is a semi-cover-avoiding subgroup of G, then G is soluble.*

Theorem 1.18 (Guo, Skiba [196]). *A group G is soluble and its Fitting length $l(G) \le n$ if and only if for some subgroup series $\Sigma = \{1 \le G_1 \le G_2 \le \ldots \le G_n = G\}$ of G, every Sylow subgroup of G is Σ-embedded in G.*

Proof First suppose that G is soluble and its Fitting length $l(G) \le n$. Then $F_{l(G)}(G) = G$, where $F_1(G) = F(G)$ and $F_i(G)/F_{i-1}(G) = F(G/F_{i-1}(G))$ for all $i > 1$. Then, obviously, for every maximal pair (K, H) between $F_i(G)$ and $F_{i+1}(G)$, we have $p = |H : K|$. Hence every subgroup of G is Σ-embedded in G, where $\Sigma = 1 \le F_1(G) \le F_2(G) \le \ldots \le F_{l(G)}(G)$.

Now suppose that there is a subgroup series $\Sigma = \{1 \le G_1 \le G_2 \le \ldots \le G_n = G\}$ such that every Sylow subgroup of G is Σ-embedded in G. We prove by induction on n that G is soluble and its Fitting length $l(G) \le n$. If $n = 1$, then by Lemma 1.9 every Sylow subgroup of G is subnormal and so G is nilpotent by Chap. 1, Lemma 5.35(6). Now suppose that $n > 1$. By Lemma 1.10, G_{n-1} is subnormal in

G. For every Sylow subgroup P of G_{n-1}, there is a Sylow subgroup G_p of G such that $P = G_p \cap G_{n-1}$. Hence by hypothesis and Lemma 1.8(1), P is Σ_1-embedded in G_{n-1}, where $\Sigma_1 = \{1 \le G_1 \le G_2 \le \ldots \le G_{n-1}\}$. Therefore the hypothesis is true for (Σ_1, G_{n-1}). By induction, $l(G_{n-1}) \le n - 1$. Then $G_{n-1} \le F_{n-1}(G)$. If $P/F_{n-1}(G)$ is a Sylow subgroup of $G/F_{n-1}(G)$, then for some Sylow subgroup G_p of G we have $P/F_{n-1}(G) = F_{n-1}(G)G_p/F_{n-1}(G)$. Hence by Lemma 1.8(2), every Sylow subgroup of $G/F_{n-1}(G)$ is $\{1 \le G/F_{n-1}(G)\}$-embedded in $G/F_{n-1}(G)$. It follows that $G/F_{n-1}(G)$ is nilpotent. This completes the proof.

Corollary 1.19 *A group G is soluble if and only if every Sylow subgroup of G is Σ-embedded in G, for some $\Sigma = \{1 \le G_1 \le G_2 \le \ldots \le G_n = G\}$.*

Corollary 1.20 *The following assertions are equivalent:*

(1) G is metanilpotent.
(2) For some $\Sigma = \{1 \le H \le G\}$, where H is a subgroup of G, every Sylow subgroup of G is Σ-embedded in G.

Theorem 1.21 (Guo, Skiba [196, Theorem 6.1]). *Let p be a prime. Then the following assertions are equivalent:*

(1) G is p-supersoluble.
(2) G is p-soluble and every subnormal subgroup of G is $\{1 \le G\}_p$-embedded in G.
(3) $G = AB$, where A is p-nilpotent, B is p-supersoluble and every subgroup of G is $\{A \cap B \le G\}$-embedded in G.
(4) $G = AB$, where A is p-supersoluble and B has a maximal series all members of which are quasipermutable in G.

Proof (1) $\Rightarrow$ (2) Since G is p-supersoluble, for every maximal pair (K, H), where K is maximal in H and $p||H : K|$, we have $|H : K| = p$. Let V be any subnormal subgroup of G. By Chap. 1, Lemma 5.35(2), $V \cap H$ is subnormal in H. Hence, without loss of generality, we may assume that $H = G$. If $V \subseteq K$, then $V \cap K = V \cap H$, that is, V avoids (K, H). Suppose that $V \nsubseteq K$. If $K_G \ne 1$, then by induction, $(K_G V/K_G)(K/K_G) = G/K_G$. Hence $VK = G$, that is, V covers (K, H). Hence we may assume that $K_G = 1$. In this case, G is primitive and if V_0 is a minimal subnormal subgroup of G contained in V, then $V_0 \nsubseteq K$ by Lemma 1.11. We may, therefore, assume that $V = V_0$ is a simple group. Then by considering the permutation representation of G on the right coset of K, we can see that $G \simeq G/K_G$ is isomorphic to some subgroup of the symmetric group S_p of degree p. It follows that the order of every Sylow p-subgroups of G is p. Hence K is a Hall p'-subgroup of G. Let $E = V^G$ be the normal closure of V in G. Then $EK = G$ and so p divides $|V|$. Hence $G = KV$. Therefore (2) holds.

(1) $\Rightarrow$ (3) Since G is p-supersoluble, $G/O_{p'p}(G)$ is abelian. Hence every subgroup of G is $\{O_{p', p}(G) \le G\}$-embedded in G. We may, therefore, take $A = O_{p'p}(G)$ and $G = B$.

The implication (1) $\Rightarrow$ (4) is evident. It is enough to take $A = G$ and $B = 1$.

(2) $\Rightarrow$ (1) Suppose that this is false and let G be a counterexample with minimal order. It is clear that the hypothesis is still true for every quotient of G by Lemma 1.8. Hence by the choice of G, G has a unique minimal normal subgroup N, G/N is p-supersoluble and $N \nsubseteq \Phi(G)$. Obviously, $O_{p'}(G) = 1$. It follows that $G = N \rtimes M$ for some maximal subgroup M of G and $N = F(G) = C_G(N) = O_p(G)$ is not cyclic. Let L be a minimal subgroup of N. Then $L \neq N$ and by hypothesis L covers (M, G). Hence $ML = G$. This induces that $|N| = |G : M| \leq |L| < |N|$, a contradiction. Hence G is p-supersoluble.

(3) $\Rightarrow$ (1) Suppose that this is false and let G be a counterexample of minimal order. Then $A \neq G \neq B$. Clearly, a $\{A \cap B \leq G\}$-embedded subgroup is a $\{A \leq G\}$-embedded subgroup and a $\{B \leq G\}$-embedded subgroup as well. By Lemma 1.10, A and B are subnormal in G. Hence G is p-soluble and by Chap. 2, Lemma 4.10, the p-length $l_p(G) = 1$. Let N be a minimal normal subgroup of G. Then the hypothesis is still true for G/N by Lemma 1.8(4). Hence G/N is p-supersoluble by the choice of G. This implies that $O_{p'}(G) = 1$ and N is the only minimal normal subgroup of G. Since G is p-soluble and $l_p(G) = 1$, we have that $G = N \rtimes E$ for some maximal subgroup E of G and $N = C_G(N) = O_p(G)$ is the Sylow p-subgroup of G. Since A subnormal in G, $O_{p'}(A)$ is subnormal in G. Hence $O_{p'}(A) \leq O_{p'}(G) = 1$ by Chap. 1, Lemma 5.35(6). This means that A is a p-group, which implies $A \leq N$. Similarly, we have that $B \leq N$. It follows from $G = AB$ that $G = N$. This contradiction shows that (1) holds.

(4) $\Rightarrow$ (1) Suppose that this implication is false and let G be a counterexample of minimal order. By Lemma 1.9, we see that B is a subnormal soluble subgroup of G. Hence $B \leq R(G)$ by Chap. 1, Lemma 5.35(7), where $R(G)$ is the soluble radical of G, which implies that $G = AB = AR(G)$ is p-soluble.

Let $1 = B_0 \leq B_1 \leq \dots \leq B_t = B$, where B_{i-1} is a maximal subgroup of B_i and B_{i-1} is $\{1 \leq G\}$-embedded in G for all $i = 1, 2, \dots, t$. Let N be a minimal normal subgroup of G and N avoid (B_{i-1}, B_i). Then by Lemma 1.6(1), $(B_{i-1}N/N, B_iN/N)$ is a maximal pair. Hence by Lemma 1.8(2) the hypothesis holds for $G/N = (AN/N)(BN/N)$. This shows that G/N is p-supersoluble. Thus N is the only minimal normal subgroup of G, $N \nsubseteq \Phi(G)$ and $O_{p'}(G) = 1$. Therefore $G = N \rtimes M$ for some maximal subgroup M of G and $N = O_p(G) = C_G(N)$. By Lemma 1.9, B_1 is subnormal in G. Hence $B_1 \leq N$ by Chap. 1, Lemma 5.35(6). Since B_1 is $\{1 \leq G\}$-embedded in G, $G = MB_1$. Hence $|N|$ is prime, which implies the p-supersolubility of G. This contradiction completes the proof.

Corollary 1.22 *A group G is supersoluble if and only if $G = AB$, where A is supersoluble and every subgroup of B is quasipermutable in G.*

Corollary 1.23 *A group G is supersoluble if and only if $G = AB$, where A is nilpotent, B is supersoluble and every subgroup of G is $\{A \cap B \leq G\}$-embedded in G.*

From Theorem 1.21 we also obtain

Theorem 1.24 *A soluble group G is supersoluble if and only if every subnormal subgroup of G is quasipermutable in G.*

4.2   Groups with Permutability for 2-Maximal and 3-Maximal Subgroups

The Structure of the Group in Which Every 3-Maximal Subgroup Permutes with All Maximal Subgroups Recall that a group G is said to be *primary* if $|G| = p^a$ for some prime p.

Lemma 2.1 . *Let E be a maximal subgroup of a group G. Suppose that E commutes with all 3-maximal subgroups in G. Then the following assertions hold:*

1) *every 2-maximal subgroup in E is contained in E_G; in particular, $|E : E_G| \in \{1, p, pq\}$, where p and q are primes (not necessarily distinct);*
2) *every maximal subgroup of E which differs from E_G is a cyclic primary group.*

Proof

(1) Suppose that $E \neq E_G$ and E/E_G is not a group of prime order. Let V be an arbitrary 2-maximal subgroup of E. Then V^x is a 3-maximal subgroup of G for any $x \in G$, and so $EV^x = V^x E$ is a subgroup of G. If $EV^x = G$, then $G = EE^x$, which contradicts Chap. 1, Lemma 4.14. Then, since E is maximal, we have that $EV^x = E$, which implies $V \leq E_G$. Therefore, every maximal subgroup of E/E_G is of prime order. Consequently, $|E/E_G| = pq$ for some prime p and q(not necessarily distinct).

(2) Let T_1 be a maximal subgroup of E which differs from E_G, and let T_2 be a maximal subgroup of T_1. Then T_2 is a 3-maximal subgroup of G. Therefore, by (1), it follows that $T_2 \leq E_G$. Since

$$T_2 \leq T_1 \cap E_G \leq T_1,$$

we see that either $T_1 = T_1 \cap E_G$ or $T_2 = T_1 \cap E_G$. It is clear that the first case is impossible. Thus, $T_2 = T_1 \cap E_G$ is a unique maximal subgroup of T_1. This implies that T_1 is a cyclic primary group.

Lemma 2.2 *Let G be a p-solvable group. If all maximal subgroups of G whose index is a power of p are normal in G, then G is p-nilpotent.*

Proof See the proof of [248, Theorem 9.3 on p. 717].

Lemma 2.3 *Let $X = F(G)$ be the Fitting subgroup of G. If every maximal subgroup of G is X-permutable with all 3-maximal subgroups in G, then G is solvable, and its p-length $l_p(G)$ does not exceed 2 for any prime p.*

Proof Let $\mathfrak{F}$ be the class of all solvable groups A with $l_p(A) \leq 2$ for any prime p. Suppose that $G \notin \mathfrak{F}$ and let G be a counterexample of minimal order.

It is well known that $\mathfrak{F}$ is a saturated Fitting formation (see [359, I, Theorems 4.3, 5.5, and 5.6]). Let H be a minimal normal subgroup of G. Then $G/H \in \mathfrak{F}$. Indeed, if $H = G$, then this is obvious. Otherwise, by Chap. 2, Lemma 1.5, the condition of the theorems is satisfied for G/H, and so $G/H \in \mathfrak{F}$ by the choice of G. Hence

H is a unique minimal normal subgroup of G and $H \nsubseteq \Phi(G)$. Let M be a maximal subgroup of G such that $H \nsubseteq M$. Then $M_G = 1$ and $G = HM$.

We first prove that G is solvable. Suppose that $X = 1$. Then by Lemma 2.1, for any maximal subgroup E of G with $E_G = 1$, there are primes p and q (not necessarily different) such that $|E| \in \{p, pq\}$. If $H = G$ is a simple group, then all maximal subgroups of G are supersolvable, and so G is solvable (see [248, V, Theorem 26.3]). This contradiction shows that $H \neq G$. Since $G = HM$ and $|M| \in \{p, pq\}$, H is either maximal or 2-maximal in G. Let V be an arbitrary 3-maximal subgroup of G contained in H. Then MV^x is a subgroup of G for any $x \in G$, which yields $V^x \leq M$ because $|M| \in \{p, pq\}$. Thus $V^G \leq M$ and so $V = 1$. It follows that either H is a group of prime order or all maximal subgroups of H are of prime order, which implies the solvability of the subgroup H. This contradiction shows that $X \neq 1$. Thus from above we see that G is solvable. It follows that $G = H \rtimes M$. Since $C_G(H) \cap M$ is a normal subgroup of G and $M_G = 1$,

$$C_G(H) = C_G(H) \cap HM = H(C_G(H) \cap M) = H.$$

Since the Fitting subgroup is contained in the centralizer of any principal factor of a given group [248, VI, Theorem 5.4], it follows that

$$H = C_G(H) = O_q(G) = F(G)$$

for some prime q.

Let p be a prime such that $l_p(G) > 2$. Then $q = p$ and the subgroup M is not nilpotent. Hence M has a nonnormal maximal subgroup, E say. Let V be a maximal subgroup of E. Then HE is a maximal subgroup of G and V is a 3-maximal subgroup of G. Therefore, $V \leq (HE)_G$ by Lemma 2.1. It can readily be seen that HV is maximal subgroup of HE. Then since

$$HV \leq (HE)_G \leq HE,$$

we have $HV = (HE)_G$ (because E is a nonnormal subgroup of M). Hence $V = HV \cap M$ is normal in M. This means that every maximal subgroup of E is normal in M, and so E is nilpotent. Assume that M has a nonnormal maximal subgroup E such that E contains some Sylow p-subgroup P of M. Suppose that $E = P$. Let L be a minimal normal subgroup of M. Since M is solvable and $H = O_p(G)$, L is a q-group for some prime $q \neq p$. So $M = LP$ because $P = E$ is maximal in M. In this case, $l_p(G) \leq 2$, which contradicts the choice of G. Hence $P \neq E$, and therefore P is normal in M. In this case, $l_p(G) = 1$. Therefore, the index of any nonnormal maximal subgroup of M is a power of p. By Lemma 2.2, M is q-nilpotent for any prime $q \neq p$. It follows from [248, I, Lemma 2.13,b)] that M is p-closed, which implies that $l_p(G) = 1$. This contradiction completes the proof.

Lemma 2.4 *Let G be a nonnilpotent primitive group. Then every 3-maximal subgroup of G commutes with all maximal subgroups of G if and only if G is a group of one of the following types:*

1) $G = N \rtimes M$, where N is a group of prime order p and M is a group of prime order $q \neq p$;

2) $G = N \rtimes M$, where N is a minimal normal subgroup of G of order p^2 and M is a group of prime order $q \neq p$;

3) $G = N \rtimes M$, where N is a group of prime order p and M is a cyclic group of order qr for some primes $q \neq p$ and $r \neq p$ (not necessarily distinct).

Proof Let M be a maximal subgroup of G with $M_G = 1$. Suppose that every 3-maximal subgroup of G commutes with any maximal subgroup of G. Then G is solvable by Lemma 2.3, and so

$$G = N \rtimes M, \text{ where } N = F(G) = O_p(G) = C_G(N)$$

is the unique minimal normal subgroup of G for some prime p. Suppose that M is not a group of prime order. Then by Lemma 2.1, $|M| = rq$ for some primes q and r (not necessarily distinct). In this case, N is a 2-maximal subgroup of G. Suppose that $|N| > p$. Let N_1 be a maximal subgroup of N. Then N_1 is a 3-maximal subgroup of G. By the hypothesis, the subgroup N_1 commutes with M, and thereby $M < N_1 M < G$, which contradicts the maximality of M. Thus $|N| = p$ and

$$M \simeq G/N = G/C_G(N)$$

is a cyclic group. Then $q \neq p \neq r$ (see [359, I, Lemma 3.9]). In this case, G is a group of type (3). If $|M|$ is a prime, then G is a group of one of the types (1) and (2). Finally, note that, if G is a group of type (1), then the 3-maximal subgroups are absent in G and, if G is a group of one of the types (2) and (3), then the identity subgroup is the unique 3-maximal subgroup of G.

In any nilpotent group, all its maximal subgroups are normal, and therefore any 3-maximal subgroup commutes with any maximal subgroup. In the nonnilpotent case, the following theorem holds.

Theorem 2.5 (Guo, Legchekova, Skiba [151, Theorem 3.1]). *Let G be a non-nilpotent group. Then every 3-maximal subgroup of G commutes with all maximal subgroups of G if and only if one of the following three possibilities holds:*

1) $|G| = p^\alpha q^\beta r^\gamma$, where p, q, and r are primes and $\alpha + \beta + \gamma \leq 3$,

2) G is isomorphic to $SL(2,3)$,

3) G is a supersolvable group of one of the following types:

   a) $G = P \rtimes M$, where P is a group of prime order p and M is a group with cyclic Sylow subgroups such that $|M| = rq^\alpha$, where $\alpha > 1$, r and q are distinct primes, $q \neq p$, the subgroup M_G is q-closed, and $|M : M_G| = q$;

   b) $G = P \rtimes Q$, where P is a group of prime order p, Q is a cyclic q-group with $|Q| > q^2$ ($q \neq p$), and $|Q : Q_G| = q^2$;

   c) $G = P \rtimes Q$, where P is a group of prime order p, Q is a q-group with $|Q| > q^2$ ($q \neq p$), $|Q : Q_G| = q$, and all maximal subgroups of Q distinct from Q_G are cyclic.

Proof Necessity. Assume that every 3-maximal subgroup of G is permutable with all maximal subgroups of G and that G is not a group with $|G| = p^\alpha q^\beta r^\gamma$, where p, q, and r are primes and $\alpha + \beta + \gamma \leq 3$.

Since G is a nonnilpotent group, some maximal subgroup M of G is not normal. Hence $G/M_G = N/M_G \rtimes (M/M_G)$ is a nonnilpotent primitive group satisfying the assumption of Lemma 2.4, and therefore the quotient G/M_G has the structure described in Lemma 2.4. In particular, $M_G \neq 1$ according to our original assumption concerning G.

Consider the following formally possible cases.

I. Let $|M/M_G| = qr$, where q and r are primes (not necessarily different).

It follows from Lemma 2.4 that $|N/M_G| = p$, where p is a prime, $q \neq p \neq r$, and M/M_G is a cyclic group. Since $|M/M_G| = qr$, M_G is not a maximal subgroup of M. Then by Lemma 2.1(2), every maximal subgroup of M is a cyclic primary group. Hence M is a supersolvable group by [248, IV, Theorem 2.11]. Suppose that $q \neq r$. Then $|M| = qr$, and so $M_G = 1$. The contradiction shows that $q = r$, and therefore M is a Sylow q-subgroup of G. Since G/M_G is supersolvable by Lemma 2.4(3) and M_G is a cyclic group, G is supersolvable. Assume that q is the largest prime divisor of $|G|$. Then M is normal in G, which contradicts the choice of the subgroup M. Therefore, p is the greatest prime divisor of $|G|$ and the Sylow p-subgroup P of G is normal in G. This implies that $G = P \rtimes M$, where $P \simeq N/M_G$ is a group of order p, M is a q-group with $|M| > q^2$ and $|M : M_G| = q^2$. Suppose that M is not a cyclic group. Since every 2-maximal subgroup of M is contained in M_G, we see that $M_G \neq 1$ is the unique 2-maximal subgroup of M. Therefore, $|M_G| = 2$ by [248, III, Theorem 8.2 and 8.4], and M is isomorphic to the quaternion group of order 8. In this case, $M_G \leq Z(M)$. However, since M/M_G is a cyclic group, this means that M is an abelian group. The contradiction shows that M is a cyclic group. Thus G is a group of type (2).

II. Let $|M/M_G| = q$. Then either $|N/M_G| = p^2$ or $|N/M_G| = p$ $(p \neq q)$ by Lemma 2.4.

Let P be a Sylow p-subgroup in G.

1. Assume that $|N/M_G| = p^2$.

By Lemma 2.1(2), every maximal subgroup of M which differs from M_G is a cyclic primary group.

Assume that M is a q-group. Then $|P| = p^2$. Let P_1 be a maximal subgroup of P. Suppose that $M_G \neq 1$. Then, it is clear that $P_1 M_G$ is a 2-maximal subgroup of G. Let T be a maximal subgroup of $P_1 M_G$ such that $P_1 \leq T$. Then T is a 3-maximal subgroup of G. Therefore $TM = P_1 M$ is a subgroup of G. This implies that $M < P_1 M < G$. This contradiction shows that $M_G = 1$ and $G = P \rtimes M$, where P is a minimal normal subgroup in G of order p^2 and $|M| = q$. Thus, $|G| = p^2 q$, which contradicts our assumption.

Now suppose that $|\pi(M)| = 2$. Let $\pi(M) = \{q, r\}$. Assume that R is a Sylow r-subgroup of M, Q is a Sylow q-subgroup of M and M_1 is a maximal subgroup of M such that Q is contained in M_1. Then $M_1 = Q$. Since Q is a cyclic Sylow q-subgroup of M, it follows that M is a q-supersolvable group.

Assume that $|Q| > q^2$. Let Q_1 be a Sylow q-subgroup of M_G and Q_2 be a maximal subgroup of Q_1. Then $Q_2 \neq 1$. Suppose that $O_{q'}(M) = 1$. Then M is supersolvable and $q > r$. This means that Q is normal in M, and so $|R| = r$. It is also clear that Q_1 is a characteristic subgroup of M_G. Thus Q_1 is normal in G. Since Q is a cyclic group, Q_2 is also a normal subgroup of G.

Assume that $r = p$. Let P_1 be a maximal subgroup of P such that $R \leq P_1$. Then $T = Q_2 P_1$ is a 3-maximal subgroup of G, and hence T and M are permutable. In this case,

$$M < TM = P_1 Q_1 M = P_1 M < G,$$

which contradicts the maximality of M. Therefore $r \neq p$. Now suppose that P_1 is a maximal subgroup of P. Then $T = P_1 Q_1$ is a 3-maximal subgroup of G and

$$M < TM = P_1 Q_1 M = P_1 M < G.$$

This contradiction shows that $O_{q'}(M) \neq 1$. Hence $O_r(M) \neq 1$. It is easy to see that $R = O_r(M)$. Thus $M = R \rtimes Q$. Since R is a Sylow r-subgroup of M_G, we have R char M_G. It follows that $R \lhd G$. Since M_G/R is a cyclic group, RQ_2 is a normal subgroup of G. Suppose that $r = p$. Let P_1 be a maximal subgroup of P such that $R \leq P_1$. Then

$$M < (P_1 R Q_2)M = P_1 M < G,$$

which contradicts the maximality of M. One can similarly prove that the case $r \neq p$ is also impossible. This shows that $|Q| \leq q^2$. It can readily be seen that the case $|Q| = q^2$ also leads to a contradiction.

Now let $|Q| = q$. Suppose that $r = p$. Then $M_G \leq P$, and so $N = PM_G = P$ is a normal Sylow p-subgroup of G. Hence $G = NM = PM = P \rtimes Q$. Let P_1 be a maximal subgroup of P such that M_G is a maximal subgroup of P_1. Let P_2 be a maximal subgroup of P_1. Then, clearly, P_2 is a 3-maximal subgroup of G. By hypothesis, $P_2^x M = M P_2^x$ for any $x \in G$. Hence $P_2 \leq M_G$. This implies that M_G is a unique maximal subgroup of P_1. Consequently, P_1 is a cyclic group. Let $T \neq P_1$ be a maximal subgroup of P. Since

$$p = |P : T| = |T P_1 : T| = |P_1 : T \cap P_1|,$$

M_G is a maximal subgroup of T. Let $P_2 \neq M_G$ be a maximal subgroup of T. Then P_2 is a 3-maximal subgroup of G. Hence P_2 permutable with M, and thereby

$$P_2 M = P_2 M_G Q = TQ < G.$$

This yields that $P_2 = M_G$. Hence, M_G is the unique 2-maximal subgroup of P. By [248, III, Theorem 8.4], we have that either P is a cyclic group or it is isomorphic to the quaternion group of order 8. In the first case, we have P_1 char $P \lhd G$, which implies $P_1 \lhd G$. Hence $M < P_1 Q < G$, a contradiction. In the other case, $G = P \rtimes Q$, where Q is a group of order 3, and therefore $G \simeq SL(2,3)$. The case $r \neq p$ is impossible. Indeed, if P_1 is a maximal subgroup of P and E is a maximal subgroup of $P_1 M_G$ containing P_1, then $M < EM < G$, which contradicts the maximality of M.

2. Finally, let us consider the case that $|G : E_G| = pq$ for any nonnormal maximal subgroup E of G, where p and q are distinct primes. In this case, for any maximal subgroup E of G, the quotient group G/E_G is supersolvable, and therefore G is a supersolvable group. Let

$$|M/M_G| = q \text{ and } |N/M_G| = p$$

for a maximal subgroup M such that

$$G/M_G = N/M_G \rtimes (M/M_G).$$

Let $M_1 \neq M_G$ be a maximal subgroup of M. Then one can prove as above that M_1 is a cyclic primary group, and so $|M| = q^\alpha r$ for some $\alpha \in N$ and for some prime r. Thus, $|G| = pq^\alpha r$, where $\alpha \geq 1$. Let P be a Sylow p-subgroup of G, Q and R be a Sylow q-subgroup and a Sylow r-subgroup of M, respectively. It is clear that Q is a nonnormal subgroup of G.

Suppose first that p, q, and r are distinct primes. Then $|P| = p$ and $|R| = r$. We claim that G is a group of type (1). Assume first that P is a normal subgroup of G. Let Q_1 be a maximal subgroup of Q. Then Q_1 is a 3-maximal subgroup of G and $G = P \rtimes M$. We prove that M_G is a q-closed group. Let T be a subgroup of G with $|G : T| = r$. Since, by assumption, $Q_1^x T = T Q_1^x$ for any $x \in G$, it follows that $Q_1 \leq T_G$. Hence $|T : T_G| \leq q$. Since Q is cyclic, Q_1 is normal in M. Hence M_G is a q-closed group. It follows that G is a group of type (1). Now assume that P is nonnormal in G. Then r is the largest prime divisor of the order of G, and therefore R is normal in G. Let T be a maximal subgroup of G such that $P \leq T$ and $|G : T| = r$. Suppose that T is normal in G. If $P \lhd T$, then P char $T \lhd G$, and so P is normal in G. This contradicts the structure of G/M_G. Hence T is a nonnormal subgroup of G. Then, by our assumption on G, the quotient G/T_G is a group of type (1) in Lemma 2.2. As above, this enables us to prove that G is a group of type (1). If $r = p$, then it can be proved in a similar way that G is a group of type (1).

Now suppose that $r = q$. Then M is a Sylow q-subgroup of G. Since G is supersolvable, it follows that P is normal in G, and so G is a group of type (3).

Sufficiency Let T be a maximal subgroup of G and K be a 3-maximal subgroup of G. We claim that K and T commute. This is obvious both if $|G| = p^\alpha q^\beta r^\gamma$, where p, q, and r are primes and $\alpha + \beta + \gamma \leq 3$, and if G is isomorphic to $SL(2, 3)$. Now assume that G is a supersolvable group of one of the types a)-c). Then $|G : T|$ is a prime.

Assume that G is a group of type a). Suppose that $r \neq p$ (the case $r = p$ can be treated in a similar way). Let Q and R be the Sylow q-subgroup and the Sylow r-subgroup of M, respectively. Then $|R| = r$ and, since the group G is supersolvable, it follows that at least one of the subgroups R or Q is normal in M. Suppose that R is normal in M. Then R (which is characteristic in M_G) is normal in G. If $|G : T| = q$, then $P \leq T$. Hence

$$TM = G \text{ and } T = T \cap PM = P(T \cap M).$$

Since $|M : M_G| = q$ with M_G is q-closed and Q is cyclic, every maximal subgroup Q_1 in Q is normal in G. Since

$$q = |G : T| = |M : T \cap M|,$$

we see that $T \cap M = RQ_1$ is a normal subgroup of G, and hence the subgroup T is normal in G. Therefore, $TK = KT$. Suppose that $|G : T| = p$. Then $T = RQ^x = M^x$ for some $x \in G$. If

$$|G : K| = rq^2 \text{ or } |G : K| = q^3,$$

then $G = TK = KT$. Let $|G : K| = pqr$. Then K is a q-group. Hence $K \leq Q^y$ and $|Q^y : K| = q$ for some $y \in G$. In this case,

$$K = M_G \cap Q^y \leq M^x = T,$$

which yields $TK = T = KT$. Similarly, if $|G : K| = pq^2$, then $K \leq M_G$. Hence $TK = T = KT$. The case $|G : T| = r$ can be treated in a similar way. Now suppose that the subgroup Q is normal in M. Note that every maximal subgroup Q_1 of Q is normal in G. Indeed, since the subgroup M_G is q-closed and it is clear that Q_1 is a Sylow q-subgroup of M_G, we have that Q_1 is characteristic in M_G, and so it is normal in G. Consequently, $RQ_1 = M_G$ is the unique subgroup of G whose index is equal to q. Then the group M is nilpotent, and therefore R is normal in M. We arrive at the case studied above. The cases in which G is a group of one of the types b) and c) can be treated in a similar way.

This completes the proof of the theorem.

Corollary 2.6 *If every* 3-*maximal subgroup of a group G commutes with all maximal subgroups of G and $|\pi(G)| > 3$, then G is nilpotent.*

We note that all classes of groups belonging to the types described in Theorem 2.5 are nonempty.

Let p, r, and q be primes, and q divide both the number $p - 1$ and $r - 1$. Let P, R, and Q be groups of order p, r, and q, respectively. Then both the groups $Aut(P)$ and $Aut(R)$ have subgroups of order q, and therefore there exists a non-abelian group $M = R \rtimes Q$, where $Q = \langle a \rangle$ is a cyclic group of order q^2 and $C_Q(R) = \langle a^q \rangle$, and a supersolvable nonnilpotent group

$$G = P \rtimes M, \text{ where } C_M(P) = R\langle a^q \rangle = R \times \langle a^q \rangle = M_G.$$

It is clear that G is a group of type a) in Theorem 2.5. A group of type b) can be constructed in a similar way.

Let

$$Q = \langle x, y | x^9 = y^3 = 1, x^y = x^4 \rangle,$$

and let Z_7 be a group of order 7. Then $\Omega_1(Q) = \Omega$ is an abelian group of order 9. Since the quotient group Q/Ω is isomorphic to a subgroup Z_3 of order 3 in the automorphism group $Aut(Z_7)$, we can construct the group $G = Z_7 \rtimes Q$. It is clear that $\langle y \rangle$ is a 3-maximal subgroup of G and $\langle y \rangle$ is a nonnormal subgroup of G. It can readily be proved that all maximal subgroups of Q distinct from Ω are cyclic, and therefore G is a group of type c). This example shows that the class of groups in which every 3-maximal subgroup commutes with all maximal subgroups is broader than the class of groups all of whose 3-maximal subgroups are normal.

The structure of the nonnilpotent group in which every 3-maximal subgroup permutes with all 2-maximal subgroups.

Lemma 2.7 *Let G be a group and $X = F(G)$ the Fitting subgroup of G. If every 2-maximal subgroup of G X-permutes with every maximal subgroup of G, then G is a metanilpotent group.*

Proof Assume that the assertion is false and let G be a counterexample of minimal order.

Let N be a minimal normal subgroup of G. By Chap. 2, Lemma 1.5, the hypothesis still holds on G/N. Hence, by the choice of G, we have that G/N is metanilpotent. Since the class of all metanilpotent groups is a saturated formation (see [359, p. 36]), N is the only minimal normal subgroup of G and $\Phi(G) = 1$.

We now prove that G is soluble. In fact, let M be a maximal subgroup of G and M_1 a maximal subgroup of M. Then, by the hypothesis, $M_1^x M = M M_1^x$ for every $x \in G$. But by Chap. 1, Lemma 4.14, $M_1^x M \neq G$ and so $M_1^x \leq M$. It follows that $M_1 \leq M_G$. If $|M_1| = 1$, then M is cyclic and so G is soluble (see [248, IV, Theorem 7.4]). If every 2-maximal subgroup M_1 of G is nontrivial, then $M_G \neq 1$ for all maximal subgroup M of G. It follows that $N \leq \Phi(G) = 1$, a contradiction. Hence G is soluble.

Since N is the only minimal normal subgroup of G, $N = O_p(G) = F(G) = X = C_G(N)$ for some prime p and $G = N \rtimes M$. If M is nilpotent, then G is metanilpotent. Hence, we may assume that M is a nonnilpotent maximal subgroup of G with $O_p(M) = 1$ and $M_G = 1$. Then, obviously, M has a nonnormal maximal subgroup M_1. Let $T = NM_1$. Then T is a maximal subgroup of G. Since $M_1 = M_1(M \cap N) = M \cap M_1 N = M \cap T$, T is not normal in G. Since M_1 is a 2-maximal subgroup of G, M_1^x is also a 2-maximal subgroup of G for all $x \in G$. Hence, by our hypothesis, M_1^x is X-permutable with T. Since $X = N \leq T$, $M_1^x T = T M_1^x$. Then, by Chap. 1, Lemma 4.14, we have $M_1^x \leq T$ for all $x \in G$ and hence $M_1 \leq T_G$. It follows that $T = NM_1 \leq T_G$ and $T \trianglelefteq G$. This contradiction completes the proof.

Lemma 2.8 *Let G be a group. Then every maximal subgroup of G permutes with every 2-maximal subgroup of G if and only if either G is a nilpotent group or G is a supersoluble group of order $|G| = pq^\beta$ such that a Sylow q-subgroup $Q =\langle x \rangle$ of G is cyclic and $Q_G =\langle x^q \rangle$.*

Proof The sufficiency part is obvious. We only need to prove the necessity part. Firstly by Lemma 2.7, G is soluble.

Assume that G is a nonnilpotent group. Then G has a nonnormal maximal subgroup M. Let $|G : M| = p^\alpha$ and M_1 be an arbitrary maximal subgroup of M. Then $M_1 \leq M_G$ (see the proof of Lemma 2.7). Thus $M_1 = M_G \trianglelefteq G$. This shows that M has a unique maximal subgroup, and consequently, M is a cyclic group of prime power order. Let $|M| = q^\beta$. Now, we prove that $\alpha = 1$. Assume $\alpha \neq 1$ and let P be a Sylow p-subgroup of G and P_1 a maximal subgroup of P. Then $P_1 M_1$ is a 2-maximal subgroup of G. By the hypothesis, $P_1 M_1 M = P_1 M$ is a subgroup of G, which contradicts the maximality of M. Hence $|G| = pq^\beta$. Since $|G/M_1| = pq$, every maximal subgroup of G/M_1 has a prime index. Hence G/M_1 is a supersoluble group. It follows that G is a supersoluble group since M_1 is cyclic. This completes the proof.

Corollary 2.9   *Let G be a group. If every 2-maximal subgroup of G is permutable with every maximal subgroup of G and $|\pi(G)| > 2$, then G is a nilpotent group.*

Corollary 2.10   (Poljakov [327, Theorem 2, Corollary 2.6]). *Let G be a group. If every 2-maximal subgroup of G permutes with every maximal subgroup of G, then G is supersoluble.*

Now we describe the structure of the nonnilpotent groups in which every 3-maximal subgroup permutes with every 2-maximal subgroup. First, we prove the following proposition.

Proposition 2.11   *Let G be a group and $X = F(G)$ the Fitting subgroup of G. If very 3-maximal subgroup of G X-permutes with every 2-maximal subgroup of G, then G is soluble and $l_p(G) \leq 1$ for all primes p.*

Proof   Let $\mathfrak{F}$ be the class of asoluble groups G such that $l_p(G) \leq 1$ for all primes p. It is well known that $\mathfrak{F}$ is a saturated Fitting formation (see, for example, [359, Theorems 5.4, 5.6]). We prove that $G \in \mathfrak{F}$ by induction on $|G|$. Firstly, we prove that G is soluble. Indeed, if $X = 1$, then every 2-maximal subgroup of G permutes with every 3-maximal subgroup of G and so by Lemma 2.8 every maximal subgroup of G is supersoluble. In this case, G is soluble by the well known Huppert's Theorem [248, VI, Theorem 9.6]. Now assume that $X \neq 1$. Then, by Chap. 2, Lemma 1.5, we know that the hypothesis still holds for G/X. Hence, by induction, G/X is soluble and consequently, G is a soluble group. Now, let H be a minimal normal subgroup of G. By induction, $G/H \in \mathfrak{F}$. So H is the only minimal normal subgroup of G and $H \nsubseteq \Phi(G)$ since $\mathfrak{F}$ is a saturated formation. This means that $H = O_p(H) = X = C_G(H)$ for some prime p and $G = H \rtimes M$ for some maximal subgroup M of G. Since $G/H \in \mathfrak{F}$ and H is a p-group, $l_q(G) \leq 1$ for any prime $q \neq p$. In order to prove that $G \in \mathfrak{F}$, we need only to prove that $l_p(G) \leq 1$.

Assume that $|G : T| = q \neq p$ for some normal maximal subgroup T of G. Then $H \leq T$ and $H \subseteq F(T)$. On the other hand, since $F(T)$ char $T \trianglelefteq G$, $F(T) \trianglelefteq G$ and so $F(T) \leq X = H$. This implies that $X = F(T)$. Now, by the hypothesis, every 2-maximal subgroup of T is $F(T)$–permutable with every maximal subgroup of T. Hence, by Lemma 2.8, $l_p(T) \leq 1$ and thereby $l_p(G) \leq 1$.

Now assume that $|G : T| = p$ for all normal maximal subgroups T of G. Let M_2/H be a 2-maximal subgroup of G/H and M_1/H a maximal subgroup of G/H. Then, $M_2 = M_2 \cap HM = H(M_2 \cap M)$ and $M_1 = H(M_1 \cap M)$. Let T be a maximal subgroup of G such that M_2 is a maximal subgroup of T. We prove that $M_2 \cap M$ is a maximal subgroup of $T \cap M$ and $T \cap M$ is maximal in M. Indeed, since $H(M_2 \cap M) = M_2 \neq T = H(T \cap M)$, $M_2 \cap M$ is a proper subgroup of $T \cap M$.

Assume that G has a subgroup D such that $M_2 \cap M < D < T \cap M$. Then, since $H(M_2 \cap M)$ is maximal in $H(T \cap M)$, $H(M_2 \cap M) = HD$ or $HD = H(T \cap M)$. If $H(M_2 \cap M) = HD$, then $D = D \cap (M_2 \cap M)H = (M_2 \cap M)(D \cap H) = M_2 \cap M < D$, a contradiction. Analogously, $HD \neq H(T \cap M)$. Thus $M_2 \cap M$ is maximal in $T \cap M$. By using similar arguments, we can prove that $M_1 \cap M$ and $T \cap M$ are maximal in M. Hence $M_2 \cap M$ is a 3-maximal subgroup of G and $M_1 \cap M$ is a 2-maximal subgroup of G. Hence, by the hypothesis, there exists an element $h \in H$ such that $(M_2 \cap M)(M_1 \cap M)^h = (M_1 \cap M)^h (M_2 \cap M)$. It follows that $(M_2 \cap M)(M_1 \cap M)H = (M_1 \cap M)(M_2 \cap M)H$ and so $(M_2/H)(M_1/H) = (M_1/H)(M_2/H)$. This shows that every 2-maximal subgroup of G/H is permutable with every maximal subgroup of G/H. If G/H is nilpotent, then, clearly, $l_p(G) \leq 1$ since H is a p-group. Therefore, by Lemma 2.8, G/H is a supersoluble group of order $|G/H| = pq^\alpha$ and G/H has a normal maximal subgroup T/H such that $|G/H : T/H| = q$. This induces that G has a normal maximal subgroup T such that $|G : T| = q \neq p$. This contradiction completes the proof.

Lemma 2.12 *Let G be a Schmidt group. Then every 2-maximal subgroup of G permutes with all 3-maximal subgroups of G if and only if G is one of the groups of the following types:*

(1) G is a Schmidt group with abelian Sylow subgroups;
(2) $G = P \rtimes Q$ is a Schmidt group, where P is a quaternion group of order 8 and Q is a group of order 3.

Proof We only need to prove the necessity part as the sufficiency part is straightforward. Assume that the conclusion is false and let G be a counterexample of minimal order.

Since G is a Schmidt group, by Chap. 1, Proposition 1.9, $G = P \rtimes Q$, where P is a Sylow p-subgroup of G and Q is a cyclic q-subgroup of G.

We first prove that $|Q|$ is a prime. Assume that it is false. Then Q has a proper subgroup Z_q such that $|Z_q| = q$. By Chap. 1, Proposition 1.9, $Z_q \subseteq Z(G)$. This means that Z_q is a normal subgroup of G and $G/Z_q = (PZ_q/Z_q) \rtimes (Q/Z_q)$ is a Schmidt group. Clearly, every 2-maximal subgroup of G/Z_q permutes with every 3-maximal subgroup of G/Z_q. Since $|Z_q| \neq 1$, $|G/Z_q| < |G|$. Hence, by the choice of G, either G/Z_q is a Schmidt group with abelian Sylow subgroups or $G/Z_q = (PZ_q/Z_q) \rtimes (Q/Z_q)$, where PZ_q/Z_q is a quaternion group of order 8 and Q/Z_q is a group of order 3. Suppose that the first case holds. Then $PZ_q/Z_q \cong P$ is an abelian subgroup and so G is a Schmidt group with abelian Sylow subgroups, a contradiction. Thus $G/Z_q = (PZ_q/Z_q) \rtimes (Q/Z_q)$ where PZ_q/Z_q is a quaternion group of order 8 and Q/Z_q is a group of order 3. It follows that $G = P \rtimes Q$ where $P \cong PZ_q/Z_q$ is a quaternion group of order 8 and Q is a cyclic group of order 3^2. Let M be a maximal subgroup of G such that $Q \leq M$. Then, clearly, $|G : M| = 4$. Hence Q is a maximal subgroup of M and consequently, Q is a 2-maximal subgroup of G. Let H be a subgroup of G such that $|G : H| = 2 \cdot 3^2$. Then, obviously, H is a 3-maximal subgroup of G. By the hypotheses, Q are permutable with H and so QH is a subgroup of G with $|G : QH| = 2$. Then, Q char $HQ \trianglelefteq Q$ and hence Q a normal subgroup of G. This contradiction shows that Q is a group of order q.

If P is abelian, then the conclusion has held. So we may assume that P is a non-abelian group. Let T_1 be a maximal subgroup of P and T_2 be a maximal subgroup of T_1. Then, T_2 is a 3-maximal subgroup of G. Since P is a non-abelian group, T_2 is a non-identity subgroup.

By Chap. 1, Proposition 1.9, $P' \times Q$ is a maximal subgroup of G. Let M be a maximal subgroup of $P' \times Q$ such that $Q \le M$. Then

$$M = M \cap P'Q = Q(M \cap P')$$

and M is a 2-maximal subgroup of G. By the hypotheses, M and T_2 are permutable. If $MT_2 = G$, then, by Chap. 1, Proposition 1.9, we have

$$G = T_2 M = T_2(M \cap P')Q \le T_2 P'Q = T_2\Phi(P)Q.$$

Hence $G = T_2\Phi(P)Q$. Since $T_2\Phi(P) \le T_1$, $|G| = |T_2\Phi(P)Q|$ divides $|T_1||Q|$. But $|T_1||Q| < |G|$, a contradiction. Therefore MT_2 is a proper subgroup of G and hence MT_2 is nilpotent. Then

$$T_2 \le C_P(Q) = P' = \Phi(P).$$

This implies that every 2-maximal subgroup T_2 of P is contained in $P' = \Phi(P)$. Hence either P is a cyclic or P' is the unique 2-maximal subgroup of P. The first case is impossible since P is non-abelian. As $P/\Phi(P)$ is an elementary abelian group, by [248, III, Theorem 8.2], P is a quaternion group of order 8. Finally, since Q is a group of prime order, $Q \simeq G/P$ is isomorphic to some subgroup of the automorphism group $Aut P$ and consequently $q = 3$. This contradiction completes the proof.

The following theorem give the structure of the nonnilpotent groups in which every 3-maximal subgroup permutes with every 2-maximal subgroup.

Theorem 2.13 (Guo, Legchekova, Skiba [150]). *Let G be a nonnilpotent group. Then every 3-maximal subgroup of G permutes with all 2-maximal subgroups of G if and only if either $|G| = p^\alpha q^\beta r^\gamma$ where p, q, r are primes (two of them are probably the same) such that $\alpha + \beta + \gamma \le 3$; or G is one of the Schmidt groups of the following types:*

(1) G is a Schmidt group with abelian Sylow subgroups;
(2) $G = P \rtimes Q$ is a Schmidt group where P is a quaternion group of order 8 and Q is a group of order 3;
    or G is one of the supersoluble groups of the following types:
(3) $G = P \rtimes Q$ and $|Q : C_Q(P)| = q$ is prime, where P is a group of prime order $p \ne q$, Q is a noncyclic group of order $|Q| = q^\alpha$ where $\alpha > 2$ and all maximal subgroups of Q different from $C_Q(P)$ are cyclic;
(4) $G = P \rtimes Q$, where P is a group of order p^2 (p is a prime), all maximal subgroups of P are normal in G, $Q = \langle a \rangle$ is cyclic q-group ($q \ne p$) of order $|Q| > q$ and $C_Q(P) = \langle a^q \rangle$;
(5) $G = P \rtimes Q$, where $P = \langle a \rangle$ is a cyclic group of order p^3 (p is a prime), Q is a group of prime order $q \ne p$ and $C_G(a^{p^2}) = P$;

(6) $G = (P \times Q) \rtimes R$, where P is a group of prime order p, Q is a group of prime order $q \neq p$, R is a cyclic group of prime power order $|R| = r^\alpha$ where $\alpha > 1$; moreover, R is nonnormal in G, but every maximal subgroup of R is normal in G.

Proof The sufficiency part can be directly examined. We only need to prove the necessity part.

Suppose that every 3-maximal subgroup of G permutes with all 2-maximal subgroups of G. Then by Lemma 2.8, every maximal subgroup of G is either a nilpotent group or a supersoluble Schmidt group. If every maximal subgroup of G is nilpotent, then G is a Schmidt group. Hence, by Lemma 2.12, G is either a Schmidt group of type (1) or a Schmidt group of type (2). Now, we assume that G has a nonnilpotent maximal subgroup. By Proposition 2.11, G is soluble and so each of its maximal subgroups has a prime power index. It follows from Corollary 2.9 that $\pi(G) \leq 3$, that is, $|G| = p^\alpha q^\beta r^\gamma$ for some primes p, q, r. If $\alpha + \beta + \gamma > 3$, then we consider the following possible cases.

I) $\pi(G) = 2$.

Assume that G has no normal Sylow subgroup. Since G is soluble, there exists a normal subgroup N of G such that $|G : N|$ is a prime, say p. We claim that N is a nonnilpotent group. Indeed, if N is nilpotent, then a Sylow q-subgroup K of N is normal in N. Since K char $N \lhd G$, K is a normal subgroup in G. However $|G : N| = p$, so K is a normal Sylow q-subgroup of G. This contradiction shows that N is a nonnilpotent group. Let P be a Sylow p-subgroup of G and Q be a Sylow q-subgroup of G where $q \neq p$. Since $Q \leq N$ and Q is nonnormal in G, Q is nonnormal in N. Hence Lemma 2.8, $N = T \rtimes Q$ where T is a group of order p, Q is a cyclic q-group and $p > q$. If $N_G(Q)$ is a nilpotent group, then $Q \leq Z(N_G(Q))$ and by the Burnside Theorem (cf. [337, Theorem 10.1.8]), G is q-nilpotent. It follows that $P \lhd G$, a contradiction. Thus, $N_G(Q)$ is a nonnilpotent group and clearly it is a maximal subgroup of G. Since Q is a normal subgroup in $N_G(Q)$, by Lemma 2.8, $N_G(Q) = Q \rtimes R$, where R is a cyclic p-group and $p < q$. This contradiction shows that G has a normal Sylow subgroup, P say. Let $C = C_Q(P)$. Then clearly, $|Q : C| > 1$.

Let H be a nonnilpotent maximal subgroup of G. Then, H is a Schmidt group. Since $|G : H|$ is a prime power, $H \supseteq P$ or $H \supseteq Q$. It follows that $|G : H| = q$ (since $P \lhd G$) or $|G : H| = p^\alpha$.

First assume that G has a Schmidt subgroup H such that $|G : H| = q$. Then $H = P \rtimes (Q \cap H)$ and P is a group of order p by Chap. 2, Lemma 4.10.

Let $|Q : C| = q$. Suppose that $Q =< a >$ is a cyclic group. Then, obviously, $G = P \rtimes Q$ is a Schmidt group and so H is nilpotent, which contradicts our assumption. Thus, Q is a noncyclic group. Let Q_1 be a maximal subgroup of Q different from C. If PQ_1 is a nilpotent group, then $G = PQ_1C$ is nilpotent, a contradiction. Thus, PQ_1 is a Schmidt group and so Q_1 is cyclic. In this case, G is a group of type (3).

Let $|Q : C| > q$. Then $|Q : C| = q^2$. Thus, C is a 2-maximal subgroup of Q. Let Q_1 be a maximal subgroup of Q. If PQ_1 is nilpotent, then $Q_1 \subseteq C$ and consequently $|Q : C| = q$. This contradiction shows that PQ_1 is a Schmidt group. Hence Q_1 is a cyclic group. This means that all maximal subgroups of Q are cyclic. Thus C is the unique 2-maximal subgroup of Q. Besides, $G/C = (PC/C) \rtimes (Q/C)$, where Q/C is an elementary abelian subgroup of order q^2 and $C_{G/C}(PC/C) = PC/C$. Thus Q/C is an irreducible abelian automorphism group of PD/D. It follows that Q/C is a cyclic group and consequently, Q is cyclic, a contradiction.

Now assume that the index of every Schmidt subgroup H of G in G is a power of p.

Since H is a Schmidt subgroup of G, $|P| \geq p^2$. Obviously, $Q^x \leq H$ for some $x \in G$. Without loss of generality, we may assume that $Q \leq H$. Since $P \lhd G$, $H \cap P \lhd H$.

Since H is a maximal subgroup of G, by the hypothesis, every 2-maximal subgroup of H permutes with every maximal subgroup of H. Hence, by Lemma 2.8, $H \cap P$ is a group of prime order, Q is a cyclic maximal subgroup of H and all proper subgroups of Q are normal in G.

Assume that $|P| > p^3$. Let Q_1 be a maximal subgroup of Q. Then PQ_1 is a nilpotent maximal subgroup of G. Hence Q_1 is normal in PQ_1. But Q_1 is normal in Q, so Q_1 is a normal subgroup of G. Since $|P| > p^3$, P has a 2-maximal subgroup T such that $H \cap P < T$. This means that TQ_1 is a 2-maximal subgroup of PQ_1 and so TQ_1 is a 3-maximal subgroup of G. Then, by the hypothesis, Q permutes with TQ_1. Hence

$$H = (H \cap P)Q < TQ_1Q = TQ < G,$$

which contradicts the maximality of H. Therefore $|P| \leq p^3$.

Assume that $|P| = p^3$. Suppose that $|Q| \geq q^2$. Let P_1 be a maximal subgroup of P and Q_1 be a 2-maximal subgroup of Q. By using the same argument as above, we see that Q_1 is a normal subgroup of G. Hence $T_3 = P_1Q_1$ is a subgroup of G. By using the above arguments, we also see that $H < T_3Q < G$ which contradicts the maximality H. Hence $|Q| = q$. It is not difficult to show that $H \cap P$ is a normal subgroup of G.

Suppose that P has a subgroup Z_p of order p which is different from $H \cap P$. Since $|Q| = q$, P is a maximal subgroup of G. Therefore, Z_p is a 3-maximal subgroup of G. By the hypothesis, Z_p and Q are permutable. So we have that

$$H < Z_pQ(H \cap P) < G.$$

This contradiction shows that P has a unique subgroup of order p. Then, in view of [248, III, Theorem 8.2], P is either a quaternion group of order 8 or a cyclic group of order p^3. In the former case, a subgroup Z_2 of order 2 of P is contained in $Z(G)$, and so $H = Z_2Q$ is not a Schmidt group, a contradiction. In the latter case, $P = \langle a \rangle$ is a cyclic subgroup of order p^3. By [117, V, Theorem 2.4], we know that $P = C_G(a^{p^2})$. Hence G is a group of type (5).

Now let $|P| = p^2$. Clearly $|Q| > q$ since $\alpha + \beta + \gamma > 3$. Let Q_1 be a maximal subgroup of Q and Q_2 be a maximal subgroup of Q_1. Since $|Q| > q$, Q_1 is non-identity group. Since PQ_1 is nilpotent, $Q_1 = C$. But $Q = \langle a \rangle$ is a cyclic subgroup, so $C = Q_1 = \langle a^q \rangle$.

Let Z_p be a subgroup of order p of P contained in H. It is clear that $Z_p Q_2$ is a 3-maximal subgroup of G and Q is a 2-maximal subgroup of G. Hence by the hypothesis, $Z_p Q_2$ and Q are permutable. Since P is an abelian subgroup, Z_p is a normal subgroup in P. However, since $P \lhd G$,

$$P \cap (Z_p Q_2 Q) = P \cap Z_p Q = Z_p$$

is normal in $Z_p Q$ and consequently, Z_p is normal in G. In this case, G is a group of type (4).

II) $\pi(G) = 3$.

Let P be a Sylow p-subgroup of G, Q a Sylow q-subgroup of G and R a Sylow r-subgroup of G respectively, where p, q, r are distinct prime divisors of $|G|$. Let M be a normal maximal subgroup of G and let, for example, $|G : M| = q$. By Lemma 2.8, M is either a nilpotent group or nonnilpotent group of order $|M| = pr^\alpha$ such that a Sylow r-subgroup R is cyclic and $R_G = \langle a^r \rangle$.

1) Assume that M is a nilpotent group.

Let M_p be a Sylow p-subgroup of M, M_r a Sylow r-subgroup of M and M_q a Sylow q-subgroup of M respectively. Then $M = M_p \times M_r \times M_q$ and M_p and M_r are normal subgroups in G. Since $|G : M| = q$, M_p and M_r are Sylow subgroups of G and so $M_p = P$ and $M_r = R$. By Sylow theorem, $M_q \le Q^x$ for some $x \in G$. Without loss of generality, we may let $M_q \le Q$. Thus, $M_q = M \cap Q$. It is clear that Q is nonnormal in G.

If PQ and RQ are nilpotent, then $Q \lhd PQ$ and $Q \lhd RQ$. It follows that Q is normal in G. Thus, one of PQ and RQ is nonnilpotent. Since every maximal subgroup of G is either a nilpotent group or a Schmidt group, we may consider the following possible cases.

a) PQ and RQ are all Schmidt groups.

Since P char $M \lhd G$, $P \lhd PQ$. Hence, by Lemma 2.8, P is a group of order p, Q is cyclic and Q is nonnormal in PQ but the maximal subgroup Q_1 of Q is normal in PQ. Analogously, R is a group of order r, Q_1 is normal in RQ. Thus, Q_1 is a normal subgroup of G. In this case, G is a group of type (6).

b) One of PQ and RQ is a nilpotent group, another is a Schmidt group.

Without loss of generality, assume PQ is a Schmidt group and RQ is a nilpotent group. Then PQ is a maximal subgroup of G. By Lemma 2.8, P is a group of order p, Q is cyclic and Q is nonnormal in PQ but the maximal subgroup Q_1 of Q is normal in PQ. Since RQ is nilpotent, Q_1 is also normal in RQ. This implies that $Q_1 \unlhd G$. Let R_1 is a maximal subgroup of R. Obviously $PQ \subseteq C_G(R)$ and PQR_1 is a maximal subgroup of G. However, PQ is maximal in G, this shows that $R_1 = 1$. In this case, G is a group of type (6).

2) Assume that M is a nonnilpotent group.

Then by Lemma 2.8, $M = P \rtimes R$, where P is a group of order p and R is a cyclic nonnormal subgroup of M and the maximal subgroup R_1 of R is normal in M. It is easy to see that R_1 is a characteristic subgroup of M. Hence R_1 is normal in G. Since $|G : M| = q$, Q is a group of prime order q.

Now we prove that Q is a normal subgroup in G. Indeed, since $|G : R_1 Q| = rp$, $S = R_1 Q$ is a 2-maximal subgroup of G. Since $\alpha + \beta + \gamma > 3$, we have $|R| > r$. Let

R_2 be a 2-maximal subgroup of R. Then $(R_2Q)^x = R_2Q^x$ is a 3-maximal subgroup of G for every $x \in G$. By the hypothesis, $S(R_2Q^x) = SQ^x$ is a subgroup of G. Because

$$|SQ^x| = \frac{|S||Q^x|}{|S \cap Q^x|} = \frac{|R_1||Q||Q^x|}{|S \cap Q^x|}$$

divides $|G| = pqr^\alpha$, we have $S \cap Q^x = Q^x$. Hence $Q^x \leq S$ for all $x \in G$ and consequently $Q \leq S_G$. Since R_1 is a normal subgroup of G, $R_1 \leq S_G$ and so $S = QR_1 = S_G$, that is, S is a normal subgroup of G. Because S is a 2-maximal subgroup of G, S is nilpotent. Thus, Q char $S \vartriangleleft G$. It follows that Q is normal in G. In this case, G is also a group of type (6). This completes the proof.

Corollary 2.14 *Let G be a group. If every 2-maximal subgroup of G permutes with all 3-maximal subgroups of G and $|\pi(G)| > 3$, then G is a nilpotent group.*

Now we show that all classes groups of types (1)-(6) in Theorem 2.13 are nonempty.

The existence of the groups of type (1) is obvious. The following examples shows that the groups in Theorem 2.13 of types (2–6) exist.

Example 2.15 It is well known that the automorphism group $Aut\, Q_8$ of a quaternion group Q_8 of order 8 has an element α of order 3. Let $G = Q_8 \rtimes <\alpha>$. Then, clearly, G is a group of type (2).

Example 2.16 Let $Q = <x, y \mid x^9 = y^3 = 1, x^y = x^{-1}>$ and Z_7 be a group of order 7. In view of [117, V, 4.3], $\Omega_1(Q) = \Omega$ is an abelian group of order 9. Since Q/Ω is isomorphic to the subgroup Z_3 of order 3 of the automorphism group $Aut(Z_7)$, we may form $G = Z_7 \rtimes Q$. Then, clearly, $<y>$ is a 3-maximal subgroup of G and $<y>$ is not-normal in G. It is not difficult to check that all maximal subgroups of Q different from Ω are cyclic. Hence, G is a group of type (3).

Example 2.17 Let P and Q be the groups of order 5 and order 3 respectively, and let $V = P \times Q$. Let α be an automorphism of P of order 2 and R a cyclic group of order 4. Let $(xy)^\alpha = x^\alpha y$ for all $x \in P$ and $y \in Q$, then α is an automorphism of V. Thus we may form the group $G = V \rtimes R$, where $C_R(P)$ is a maximal subgroup of R. Then, G is a group of type (6).

Example 2.18 As in Example 2.17, we may construct the group $A = P \rtimes Z$ where P is a group of order 5, Z is a cyclic group of order 4 and $C_A(P)$ is a maximal subgroup of P. Let $A_i = A$, where $i = 1, 2$, and $\varphi : A \to A/P$ be the canonical epimorphism from A to A/P. Put $G = A_1A_2 = \{(a_1, a_2) \mid a_1^\varphi = a_2^\varphi\}$ be the direct product of the groups A_1 and A_2 with the join factor group A/P (see [248, I, Theorem 9.11]). Then, it is not difficult to check that G is a group of type (4).

Example 2.19 Let P be a cyclic group of order p^3, where p is odd prime. Then $|Aut\, P| = p^2(p - 1)$. Now let q be a prime divisor of $p - 1$ and Z a group of order q of $Aut\, P$. Put $G = P \rtimes Z$ and $P_1 = <a^{p^2}>$ be a subgroup of P of order p. Then $P_1 = \Omega_1(P)$. If $C_Z(P_1) = Z$, then $Z = 1$ by [117, V, Theorem 2.4]. This contradiction shows that $C_Z(P_1) = 1$, and consequently, $C_G(a^{p^2}) = P$. Thus G is a group of type (5).

The structure of nonnilpotent groups in which every two 3-maximal subgroups are permutable.

Lemma 2.20 [273, Theorem 5.1.9]. *Take a p-group P and a maximal abelian subgroup A of P. Suppose that* $|P'| = p$. *Then*

$$|P : A| = |A/Z(P)| \text{ and } |P/Z(P)| = |A/Z(P)|^2.$$

The next lemma describes the nonnilpotent groups in which every two 2-maximal subgroups are permutable.

Lemma 2.21 *Let G be a nonnilpotent group. Then the following conditions are equivalent:*

(1) every two 2-maximal subgroups of G are permutable;
(2) G is a Schmidt group with abelian Sylow subgroups.

Proof (1) $\Rightarrow$ (2). Suppose that G is a nonnilpotent group and every two 2-maximal subgroups of G are permutable. This implies that all maximal subgroups of G are nilpotent. Thus, G is a Schmidt group. We now verity that G is a group with abelian Sylow subgroups. Chap. 1, Proposition 1.9(1) yields $G = P \rtimes Q$, where P is a Sylow p-subgroup of G and Q is a cyclic q-subgroup of G.

Suppose that $P' \neq 1$. Take a maximal subgroup E of $P'Q$ whose index is equal to p. Then Chap. 1, Proposition 1.9(5) implies that E is a 2-maximal subgroup of G, and $Q \leq E$. The hypotheses of the lemma imply that EE^x is a subgroup of G for all $x \in G$. According to [248, VI, Lemma 4.7] there exists $y \in EE^x$ such that $Q^y = QQ^x$. Hence, $Q = Q^x$ is a normal subgroup of G. This contradiction shows that $P' = 1$, which implies that P is abelian.

(2) $\Rightarrow$ (1). Assume that $G = P \rtimes \langle a \rangle$ is a Schmidt group with abelian Sylow subgroups, where P and $\langle a \rangle$ are Sylow p- and q-subgroups of G respectively. Chap. 1, Proposition 1.9(5) implies that G has precisely two classes of maximal subgroups, whose representatives are $\langle a \rangle$ and $P\langle a^q \rangle$. The groups of the form $\langle a^q \rangle$, $P\langle a^{q^2} \rangle$ and $P_1\langle a^q \rangle$, where P_1 is some maximal subgroup of P, are representatives of three classes of 2-maximal subgroups of G. By Chap. 1, Proposition 1.9(2), $\langle a^q \rangle$ is contained in $Z(G)$. Hence $\langle a^q \rangle$ and $\langle a^{q^2} \rangle$ are normal subgroups of G. This implies that the 2-maximal subgroups $\langle a^q \rangle$ and $P\langle a^{q^2} \rangle$ are normal in G. Since every two 2-maximal subgroups of G of the form $P_1\langle a^q \rangle$ and $P_2\langle a^q \rangle$ are permutable, where P_1 and P_2 are maximal subgroups of P, it follows that every two 2-maximal subgroups of G are permutable. This completes the proof.

Corollary 2.22 *If every two 2-maximal subgroups of a group G are permutable and* $|\pi(G)| > 2$ *then G is a nilpotent group.*

Corollary 2.23 *Given a nonnilpotent group G, every 2-maximal subgroup is normal if and only if G is a Schmidt group whose normal Sylow subgroups have prime order.*

Corollary 2.24 [245, Corollary 2.6]. *If every second maximal subgroup of G is normal in G, then G is a supersoluble group.*

Lemma 2.25 [305]. *Suppose that $P = H \rtimes C_p$, where H is an elementary p-group, $|C_p| = p$, and every two 2-maximal subgroups of P are permutable. Then P is an abelian group.*

The next lemma can be proved by induction on n.

Lemma 2.26 *Take an n-maximal subgroup E of a q-nilpotent group G. Then $|G : E| = q^\alpha s$, where $\alpha \leq n$ and $(s, q) = 1$.*

Lemma 2.27 *Take a subgroup T of G and a cyclic q-subgroup $Q = \langle a \rangle$ of G. If $T \cap Q^x = 1$ for every $x \in G$ and G is a q-nilpotent group, then $|Q|$ divides $|G : T|$.*

Proof We can assume that $T \neq G$. By induction on $|G : T|$, we show that $|Q|$ divides $|G : T|$. Suppose that $T \leq M$, where M is a maximal subgroup of G.

Suppose that $\langle a \rangle^x \leq M$ for some $x \in G$. Since $(\langle a \rangle^x)^m \cap T = 1$ for all $m \in M$, by induction $|\langle a \rangle^x| = |\langle a \rangle|$ divides $|M : T|$. It follows that $|Q| = |\langle a \rangle|$ divides $|G : T|$.

We may, therefore, assume that $\langle a \rangle^x \nleq M$ for all $x \in G$. Then M is a normal subgroup of G. Since $(\langle a \rangle \cap M)^m \cap T = 1$ for all $m \in M$, by induction $|\langle a \rangle \cap M|$ divides $|M : T|$. This implies that $|\langle a \rangle \cap M|$ divides $|G : T|$. Since

$$G/M = \langle a \rangle M/M \simeq \langle a \rangle / M \cap \langle a \rangle,$$

it follows that $|\langle a \rangle : M \cap \langle a \rangle|$ divides $|G : T|$. Consequently, $|Q| = |\langle a \rangle : M \cap \langle a \rangle| |\langle a \rangle \cap M|$ divides $|G : T|$. The proof is completed.

Following [49], we denote the intersection of all 2-maximal subgroups of G by $\Phi^2(G)$.

Let p, q, and r be distinct primes. In the following theorems, P, Q and R denote some Sylow p-, q-, and r-subgroups of G respectively.

Theorem 2.28 (Guo, Lutsenko and Skiba [155]). *Given a nonnilpotent group G, every two 3-maximal subgroups of G are permutable if and only if G is a group of one of the following types:*

I. *G is a Schmidt group of one of the forms:*
    (a) G is a group with abelian Sylow subgroups;
    (b) $G = P \rtimes Q$, where P is isomorphic to either the group $M_3(p)$ (see [117, p.190]) or the quaternion group of order 8;
    (c) $G = P \rtimes Q$, where $|P| > p^3$, $|\Phi(P)| = p$ and $\Phi(P) = \Phi^2(P)$.
II. *G is a biprimary group which is not a Schmidt group and is of one of the following forms:*
    (1) $G = P \rtimes Q$, where P is a minimal normal subgroup of G, while Q is a cyclic group, and $P \rtimes \Phi(Q)$ is a Schmidt group;
    (2) $G = (P \rtimes Q_1) \times C_q$, where P is a minimal normal subgroup of G, $|C_q| = q$ and $P Q_1$ is a Schmidt group;
    (3) $G = P \rtimes Q$, where P is a minimal normal subgroup of G, $Q = \langle a \rangle \times \langle b \rangle$, $|a| = |b| = q$, while $P\langle a \rangle$ and $P\langle b \rangle$ are Schmidt groups;
    (4) $G = P \rtimes Q$, where $|P| = p$, $p > 2$ and Q is isomorphic to the quaternion group of order 8;

(5) $G = (P \rtimes Q_1)C_q$, where P is a minimal normal subgroup of G, $Q_1 = \langle a \rangle$, $C_q = \langle b \rangle$, $|Q_1 C_q| = q^\beta$, $|a| = q^{\beta-1}$ ($\beta \geq 3$), PQ_1 is a Schmidt group, $a^b = a^{1+q^{\beta-2}}$ and $[P, C_1] = 1$ for every subgroup C_1 isomorphic to C_q;

(6) $G = P \rtimes Q$, where $\Phi(P)$ is a minimal normal subgroup of G, both $\Phi(P)Q$ and $G/\Phi(P)$ are Schmidt groups, the maximal subgroup of Q coincides with $Z(G)$, and every two 2-maximal subgroups of P are permutable;

(7) G is a subdirect product of two isomorphic distinct Schmidt groups with abelian Sylow subgroups;

(8) $G = (P_1 \times C_p) \rtimes Q$, where P_1 is a minimal normal p-subgroup of G, $|C_p| = p$, $P_1 Q$ is a Schmidt group, a maximal subgroup of Q is contained in $Z(G)$, and $[C_p, Q] = 1$;

(9) $G = (P_1 \rtimes Q) \rtimes C_p$, where P_1 is a minimal normal p-subgroup of G, $|Q| = q$, $|C_p| = p$, $N_G(Q) = Q \rtimes C_p$ and $P_1 C_p$ is an abelian group;

III. the order of G has precisely three prime divisors p, q, and r, and G is a group of one of the following forms:

(i) $G = (P \rtimes Q)R$, where P and R are minimal normal subgroups of G, while Q is a cyclic group, and $F(G) = PR\Phi(Q)$;

(ii) $G = R \rtimes (P \times Q)$, where $|P| = p$, $|Q| = q$ and $R = F(G)$ is a minimal normal subgroup of G.

Proof Necessity. Assume first that G is a nonsoluble group. Since by assumption every two 3-maximal subgroups of G are permutable, every 2-maximal subgroup of G is nilpotent; thus, according to [400] and [253] G is isomorphic to either A_5 or $SL(2, 5)$.

Suppose that $G \simeq A_5$. The group A_5 has three classes of maximal subgroups, which are isomorphic to A_4, D_6 and D_{10} respectively. Consider the maximal subgroup A_4 of A_5. Since A_4 lacks subgroups of order 6, the maximal subgroups of A_4 are its Sylow 2- and 3-subgroups. Consequently, the subgroups of order 2 are 2-maximal subgroups of A_4. By assumption, every two subgroups of order 2 in A_5 are permutable, and so every two Sylow 2-subgroups of A_5 are permutable. If K is some Sylow 2-subgroup of A_5, then KK^x is a subgroup of A_5 for all $x \in A_5$. Since A_5 is a simple group, $K \neq K^x$ for some $x \in A_5$. However, A_5 lacks subgroups of order $|KK^x|$; a contradiction. Since $A_5 \simeq SL(2, 5)/Z(SL(2, 5))$, some pairs of nonpermutable 3-maximal subgroups exist in $SL(2, 5)$.

Therefore, G is a soluble group in which every two 3-maximal subgroups are permutable. It is clear that every proper subgroup H of G is a nilpotent or Schmidt group. Furthermore, if H is a Schmidt group, then H is maximal in G. Since in a soluble group the index of every maximal subgroup is a prime power and the order of every Schmidt group has two distinct prime divisors, it follows that $\pi(G) \leq 3$.

I. Assume first that $G = P \rtimes Q$ is a Schmidt group.

If P is an abelian group, then G is a group of type I(a).

Assume now that P is a nonabelian group. Then by Chap. 1, Proposition 1.9, $\Phi(P) = P' = Z(P)$.

Suppose that $|\Phi(P)| > p$. Take a 2-maximal subgroup E of $\Phi(P)Q$ whose index is equal to p^2. Then E is a 3-maximal subgroup of G by Chap. 1, Proposition 1.9(5), and $Q \leq E$. The hypotheses of the theorem imply that EE^x is a subgroup of G for all

$x \in G$. By [248, VI, Lemma 4.7], there exists $y \in EE^x$ such that $Q^y = QQ^x$, and so $Q = Q^x$ is a normal subgroup of G. This contradiction shows that $|\Phi(P)| = p$.

In order to verify that every two 2-maximal subgroups of P are permutable, take arbitrary 2-maximal subgroups P_1 and P_2 of P and a maximal subgroup Q_1 of Q. Since Q_1 is a normal subgroup of G by Chap. 1, Proposition 1.9(2), $P_1 Q_1$ and $P_2 Q_1$ are 3-maximal subgroups of G. By assumption we have $(P_1 Q_1)(P_2 Q_1) = (P_2 Q_1)(P_1 Q_1)$. Hence, $L = Q_1(P_1 P_2)$ is a subgroup of G. According to [248, VI, Lemma 4.7], the group L includes a Sylow p-subgroup L_p such that $L_p = P_1 P_2$. This implies that P_1 and P_2 are permutable.

Assume that there exists a 2-maximal subgroup T in P with $\Phi(P) \nsubseteq T$. Since $P/\Phi(P)$ is an abelian group and $T\Phi(P)/\Phi(P) \le P/\Phi(P)$, it follows that $T \simeq T\Phi(P)/\Phi(P)$ is also an abelian group. Thus $T \times \Phi(P)$ is an abelian maximal subgroup of P. Then, Lemma 2.20 yields $|P| = p^3$. It follows from [117, V, Teorem 5.1] that P is isomorphic to one of the following groups: $M_3(p)$, $M(p)$, D or Q, where D is a dihedral group, Q is the quaternion group of order 8, $M_3(p) = \langle a, b \mid a^{p^2} = b^p = 1, a^b = a^{1+p} \rangle$ and $M(p) = \langle x, y, z \mid x^p = y^p = z^p = 1, [x, z] = [y, z] = 1, [x, y] = z \rangle$ (see [117, pp. 190–191, 203]).

If P is isomorphic to $M(p)$, then $P = \Omega_1(P) = \{g \in G \mid g^p = 1\}$. However, every subgroup of P of order p is a 2-maximal subgroup. Since in P every two 2-maximal subgroups are permutable, it follows that P is a non-abelian group; a contradiction. If P is isomorphic to a dihedral group, then by [117, V, Theorem 4.3], we see that $P = \Omega_1(P)$, which is impossible as we showed above. Consequently, P is isomorphic to either $M_3(p)$ or the quaternion group of order 8.

Therefore, G is a group of type I(b).

Suppose now that $\Phi(P)$ lies in every 2-maximal subgroup of P. This implies that $\Phi(P) = \Phi^2(P)$. If $|P| = p^3$, then G is a group of type I(b). However, if $|P| > p^3$, then G is a group of type I(c).

II. Assume now that G is not a Schmidt group and $\pi(G) = \{p, q\}$, where $p \ne q$.

Suppose first that G includes a normal Sylow subgroup P, that is, $G = P \rtimes Q$. Assume that G includes a pair of Schmidt subgroups of the form $A = P \rtimes Q_1$ and $B = P_1 \rtimes Q$ (with $P_1 < P$ and $Q_1 < Q$). Then, it is clear that A and B are maximal subgroups of G. Consequently, every two 2-maximal subgroups of A are permutable. Hence P is an abelian group by Lemma 2.21. Then by Chap. 1, Proposition 1.9(6), P is a minimal normal subgroup of A, and consequently of G itself. On the other hand, every two 2-maximal subgroups of B are also permutable. Thus we can show similarly that P_1 is a minimal normal subgroup of G. This contradicts the fact that P is a minimal normal subgroup of G. Therefore, all Schmidt subgroups of G include either a Sylow p-subgroup of G or a Sylow q-subgroup of G.

If all Schmidt subgroups of G include a Sylow p-subgroup of G, then $G = P \rtimes Q$, where P is a minimal normal subgroup of G.

Suppose that Q is an abelian group. If Q is cyclic then G is a group of type II(1).

Assume now that Q is a noncyclic group. Take a Schmidt subgroup H of G. Then $|G : H| = q$ and $H = P \rtimes Q_1$, where Q_1 is a cyclic maximal subgroup of Q. By the fundamental theorem of finite abelian groups, $Q = Q_1 \times C_q$, where $|C_q| = q$. Take a maximal subgroup Q_2 of Q_1. Then $P Q_2 C_q$ is a maximal subgroup of G. It is clear that this subgroup is not a Schmidt group; consequently, it is a nilpotent group. This implies that $[P, C_q] = 1$; thus, G is a group of type II(2).

If we assume that the maximal subgroup Q_2 of Q_1 is trivial, then PC_q is a maximal subgroup of G. If PC_q is a nilpotent group, then G is a group of type II(2). However, if PC_q is a Schmidt group, then G is a group of type II(3).

Suppose now that Q is a non-abelian group with $|Q| = q^\beta$ (where $\beta \in \mathbb{N}$). If $q = 2$ and $\beta = 3$, then [117, V, Theorem 4.4] implies that Q is isomorphic to a quaternion or dihedral group. In the latter case $Q = \langle a \rangle \rtimes \langle b \rangle$, where $|a| = 2^2, |b| = 2$, and $a^b = a^{-1}$. Then Q has exactly three maximal subgroups: $\langle a \rangle, \langle a^2 \rangle \langle b \rangle$, and $\langle a^2 \rangle \langle ab \rangle$. Hence, the subgroups of the form $\langle a^2 \rangle, \langle b \rangle$, and $\langle ab \rangle$ are 2-maximal subgroups of Q. Since P is a minimal normal subgroup of G, it follows that Q is a maximal subgroup of G. Then by assumption $\langle ab \rangle \langle b \rangle = \langle b \rangle \langle ab \rangle$, and so $ab = ba$. Therefore, $a^b = a = a^{-1}$, and so $a^2 = 1$; a contradiction. Consequently, Q is isomorphic to the quaternion group of order 8. It follows from $q = 2$ and Chap. 1, Proposition 1.9(9) that $|P| = p$, and so G is a group of type II(4).

If now $q = 2$ and $\beta > 3$ or q is an odd prime, then [117, V, Theorem 4.4] implies that Q is isomorphic to one of the groups $M_\beta(q), D_\beta, Q_\beta$, and S_β (see [117, pp. 190–191]). If Q is isomorphic to one of the groups D_β, Q_β, or S_β, then [117, V, Theorem 4.3] shows that $Q/Z(Q)$ is isomorphic to $D_{\beta-1}$. In this case $Q/Z(Q)$ includes exactly two noncyclic maximal subgroups. Consequently, Q includes at least two noncyclic 2-maximal subgroups; thus, Q includes at least two noncyclic maximal subgroups. This means that G includes some maximal subgroups M_1 and M_2 which include P and are not Schmidt groups. Then M_1 and M_2 are nilpotent normal subgroups of G; thus, $G = M_1 M_2$ is a nilpotent group; a contradiction. Thus, Q cannot be isomorphic to any of the groups D_β, Q_β or S_β.

Therefore, Q is isomorphic to

$$M_\beta(q) = \langle a, b \mid a^{q^{\beta-1}} = b^q = 1, a^b = a^{1+q^{\beta-2}} \rangle$$

(see [117, p. 190]). In this case, G is of the form $G = (P \rtimes Q_1) \rtimes C_q$, where P is a minimal normal subgroup of G, $Q_1 = \langle a \rangle$, $|a| = q^{\beta-1}$, $C_q = \langle b \rangle$, $|b| = q$, PQ_1 is a Schmidt group, and $a^b = a^{1+q^{\beta-2}}$. Consider now the maximal subgroup PQ_2C_q of G, where Q_2 is a maximal subgroup of Q_1. It is clear that $P \times Q_2C_q$ is a nilpotent group, and so $[P, C_q] = 1$.

Take the set $\mathcal{C}$ of all subgroups of G isomorphic to C_q. Assume that $[P, C_1] \neq 1$ for some subgroup C_1 in $\mathcal{C}$. This implies that PC_1 is a Schmidt group. Then PC_1 is a maximal subgroup of G, and so $|Q| = q^2$, which is impossible. Hence, G is a group of type II(5).

Assume now that every Schmidt subgroup of G contains some Sylow q-subgroup of G. This means that $Q = \langle a \rangle$ is a cyclic group and $P\langle a^q \rangle$ is a maximal subgroup of G of index q. Since $P\langle a^q \rangle$ is not a Schmidt group, it is a nilpotent group. Consequently, $P\langle a^q \rangle = P \times \langle a^q \rangle = F(G)$; hence, $\langle a^q \rangle$ is contained in $Z(G)$. Then either every maximal subgroup of G which includes a Sylow q-subgroup of G is a Schmidt group or G includes a nilpotent maximal subgroup which includes a Sylow q-subgroup of G.

Assume that the first case holds. Take an arbitrary maximal subgroup M of G of the form $P_1 Q^x$, where $P_1 < P$. Since M is a Schmidt group, Lemma 2.21 and Chap. 1, Proposition 1.9 imply that P_1 is a minimal normal subgroup of M.

Suppose first that P is a non-abelian group. Then $\Phi(P) \neq 1$. Suppose that $P_1 \neq \Phi(P)$. Then $\Phi(P)P_1 Q$ is a subgroup of G. Since $P_1 Q$ is a maximal subgroup of G, $\Phi(P)P_1 Q = P_1 Q$ or $\Phi(P)P_1 Q = G$. If $\Phi(P)P_1 Q = P_1 Q$, then $\Phi(P) \leq P_1$. Since P_1 is a minimal normal subgroup of $P_1 Q$, it follows that $\Phi(P) = P_1$, which contradicts our assumption. Hence $\Phi(P)P_1 Q = G$. Then $\Phi(P)P_1 = P$, and so $P_1 = P$, which is impossible. Therefore, $P_1 = \Phi(P)$. Since $P_1 = \Phi(P)$ is a minimal normal subgroup of M, we have that $P_1 = \Phi(P) = P' = Z(P)$ is a minimal normal subgroup of G.

Since $\Phi(P) \subseteq \Phi(G)$ and G is not a nilpotent group, $G/\Phi(P)$ is a Schmidt group with abelian Sylow subgroups. Since every maximal subgroup of G which includes a Sylow q-subgroup of G is a Schmidt group with abelian Sylow subgroups, a maximal subgroup of Q coincides with $Z(G)$. As in case I(b), we can show that every two 2-maximal subgroups of P are permutable. Hence, G is a group of type II(6).

Assume that P is a non-abelian group. Suppose that $\Phi(P) \neq 1$. In this case, $P_1 = \Phi(P)$ is a minimal normal subgroup of G and $G/\Phi(P)$ is a Schmidt group with abelian Sylow subgroups. Consequently, G is a group of type II(6).

Suppose that $\Phi(P) = 1$. In this case, P is an elementary p-group. Consider the maximal subgroup $M = P_1 Q^x$ of G. Since M is a Schmidt group and P is an abelian group, P_1 is a minimal normal subgroup of M by Chap. 1, Proposition 1.9 and consequently of G itself.

Consider a maximal subgroup T of G with $G = P_1 \rtimes T$. Since $T = P_2 \rtimes Q$ is a Schmidt group (in the case under consideration) and P is an abelian group, Chap. 1, Proposition 1.9 implies that P_2 is a minimal normal subgroup of T, and consequently of G itself. Since $P_1 \cap P_2 = 1$ and both P_1 and P_2 are minimal normal subgroups of G, it follows that G is a subdirect product of two isomorphic distinct Schmidt groups. Therefore, G is a group of type II(7).

Assume that G includes a nilpotent maximal subgroup M which includes a subgroup Q. Then M is of the form $P_1 \times Q$, where $P_1 < P$. As in case I(b), we can show that $|P_1| = p$.

Take a Schmidt subgroup H of G which includes Q. Then H is a maximal subgroup of G; hence, $HM = G$. Take a Sylow p-subgroup H_p of H. Observe that P_1 is not included into H_p since $P_1 Q$ is a maximal subgroup of G. Then $H_p \cap P_1 = 1$ since $|P_1| = p$; thus, $H \cap M = Q$. This yields

$$|G : H| = \frac{|M||H|}{|M \cap H||H|} = \frac{|M|}{|M \cap H|} = \frac{|M|}{|Q|} = \frac{|P_1 Q|}{|Q|} = |P_1| = p;$$

hence, $|P : H_p| = p$. Therefore, H_p is a maximal subgroup of P. Since H is a Schmidt group, the subgroup H_p is normal in H, and so H_p is a normal subgroup of G. Lemma 2.21 and Chap. 1, Proposition 1.9 imply that H_p is a minimal normal subgroup of H, and consequently of G itself.

Since P_1Q and H_pQ are maximal subgroups of G, it follows that

$$\Phi(G) \leq P_1Q \cap H_pQ = (P_1 \cap H_pQ)Q.$$

Verify that $P_1 \cap H_pQ = 1$. If $P_1 \cap H_pQ \neq 1$ then $P_1 < H_pQ$ since $|P_1| = p$. Consequently, $P_1Q < H_pQ$, which is impossible since P_1Q is a maximal subgroup of G. Thus $P_1 \cap H_pQ = 1$, and so $\Phi(G) \leq Q$. Since $P' \subseteq \Phi(P) \subseteq \Phi(G)$, it follows that $P' = 1$. Therefore, P is an abelian group. Since the maximal subgroup P_1Q of G is a nilpotent group, $[P_1, Q] = 1$.

This shows that G is a group of type II(8).

Assume that G lacks normal Sylow subgroups. Take a normal subgroup H of G of index p. Then Q is a subgroup of H. Suppose that Q is a normal subgroup of H. Then Q is a normal subgroup of G since H is a normal subgroup of G, which contradicts the case under consideration. Hence $H = P_1 \rtimes Q$ is a Schmidt subgroup of G, and so Q is a cyclic group. Then by Lemma 2.21 and Chap. 1, Proposition 1.9, P_1 is a minimal normal subgroup of H, and thereby of G itself.

Assume that $N_G(Q)$ is a nilpotent subgroup of G. Then $Q \leq Z(N_G(Q))$ since Q is an abelian group. It follows from [230, Theorem 14.3.1] that in this case G possesses a normal q-supplement, which contradicts the case under consideration. Consequently, $N_G(Q) = Q \rtimes \langle b \rangle$ is a Schmidt subgroup of G. Since Q is a cyclic group, $|Q| = q$.

Since H and $N_G(Q)$ are maximal in G, we have $G = HN_G(Q)$. Hence, $P_1\langle b \rangle$ is a Sylow p-subgroup of G. By the choice of H we have $|G : H| = p$; thereby $P_1 \cap \langle b \rangle = \langle b^p \rangle$. Since $Q \rtimes \langle b \rangle$ is a Schmidt group, $Q\langle b^p \rangle$ is a nilpotent group, and so $\langle b^p \rangle \leq C_{P_1}(Q)$. On the other hand, since $P_1 \rtimes Q$ is a Schmidt group with abelian Sylow subgroups; $C_{P_1}(Q) = 1$ by Chap. 1, Proposition 1.9. Thus, $\langle b^p \rangle = 1$, so that $|\langle b \rangle| = p$.

Therefore, $P_1\langle b \rangle = P_1 \rtimes \langle b \rangle$ is a maximal subgroup of G and P_1 is an elementary group. By hypotheses, every two 2-maximal subgroups of $P_1\langle b \rangle$ are permutable, and Lemma 2.25 implies that $P_1\langle b \rangle$ is an abelian group.

Consequently, G is a group of type II(9).

III. Finally, consider the case $\pi(G) = \{p, q, r\}$, where p, q, and r are distinct prime divisors of $|G|$.

Denote by M some normal subgroup of G with $|G : M| = q$. Then M is a nilpotent or Schmidt group.

Assume that M is a nilpotent group. Then $G = (P \times R) \rtimes Q$ and $M = P \times R \times Q_1$, where Q_1 is some maximal subgroup of Q. The subgroups PQ and RQ cannot both be nilpotent; thus, either both PQ and RQ are Schmidt groups, or one of them, for instance RQ, is a nilpotent group and the other is a Schmidt group.

Assume that we have the first case. Then PQ and RQ are maximal subgroups of G. By assumption, every two 2-maximal subgroups of PQ are permutable, as well as every two 2-maximal subgroups of RQ do. Lemma 2.21 and Chap. 1, Proposition 1.9 imply that P and R are minimal normal subgroups of G. Moreover, by Chap. 1, Proposition 1.9, $Q = \langle a \rangle$ is a cyclic group and $\langle a^q \rangle$ is a subgroup of $Z(\langle PQ, RQ \rangle) = Z(G)$. Assume now that $PQ = P\langle a \rangle$ is a Schmidt group and RQ is a nilpotent group. Then PQ is a maximal subgroup of G; thus, $G = PQ \times R$, where $|R| = r$. By assumption, every two 2-maximal subgroups of PQ are permutable. Hence by

Lemma 2.21 and Chap. 1, Proposition 1.9, P is a minimal normal subgroup of G. Since $\langle a^q \rangle$ is a characteristic subgroup of Q and Q is a normal subgroup of RQ, $\langle a^q \rangle$ is a normal subgroup of RQ. Chap. 1, Proposition 1.9 shows that $\langle a^q \rangle$ is a normal subgroup of PQ, and so of G itself. Therefore, G is a group of type III(i).

Assume that M is a Schmidt group and G is not a group of type III(i). Without loss of generality, we can assume that $M = R \rtimes P$, where $P = \langle b \rangle$ is a cyclic group. Then $G = M \rtimes Q = (R \rtimes P) \rtimes Q$, where Q is a group of prime order q, and Q is not a normal subgroup of G. Indeed, if Q is a normal subgroup of G, then $G = M \times Q$ is a group of type III(i).

Since $M = R \rtimes P$ is a Schmidt group in which every two 2-maximal subgroups are permutable, R is a minimal normal subgroup of M by Lemma 2.21 and Chap. 1, Proposition 1.9 and consequently of G as well. Assume that RQ is a nilpotent group. If PQ is also a nilpotent group, then Q is a normal subgroup of G, which contradicts the case under consideration. Hence, $PQ = P \rtimes Q$ is a Schmidt group. Since $|P| = p$ because P is cyclic, $C_G(R) = RQ$, and so Q is a normal subgroup of G, which again contradicts the case under study.

Therefore, $RQ = R \rtimes Q$ is a Schmidt group. Assume that $PQ = P \rtimes Q$ is also a Schmidt group. Since P is then a cyclic group, $|P| = p$. However, then $p - 1 = q\alpha$ for some nonnegative integer α by Chap. 1, Proposition 1.9(9). Similarly, since RQ and RP are Schmidt groups, $r^n - 1 = q\beta$ and $r^n - 1 = p\gamma$ for some nonnegative integers n, β, and γ. Hence, $p = q\beta\gamma^{-1} = 1 + q\alpha$, which is impossible. Therefore, PQ is a nilpotent group. Consequently, $G = R \rtimes (P \times Q)$. Furthermore, the maximality of RQ in G implies that $P = \langle b \rangle$ is a group of prime order p. This yields $R = F(G)$.

In this case G is a group of type III(ii).

Sufficiency Assume that G is a group of type II(2). Then G has precisely two classes of maximal nilpotent subgroups, whose representatives are some subgroups $Q_1 C_q$, $P Q_2 C_q$ (where Q_2 is a maximal subgroup of Q_1), and one maximal Schmidt subgroup PQ_1. In this case, the representatives of 3-maximal subgroups of G are the subgroups Q_2, $P Q_3$, $P_1 Q_2$, $Q_3 C_q$, $P Q_4 C_q$, $P_1 Q_3 C_q$, and $P_2 Q_2 C_q$, where Q_3 is a maximal subgroup of Q_2, Q_4 is a maximal subgroup of Q_3, P_1 is a maximal subgroup of P, and P_2 is a 2-maximal subgroup of P. By hypotheses, Q_2, $P Q_3$, $Q_3 C_q$, and $P Q_4 C_q$ are normal subgroups of G. Since P is an abelian group, all its subgroups are normal. Hence, every two 3-maximal subgroups of G are permutable.

Assume that G is a group of type II(5). We show that every two 3-maximal subgroups T and E of G are permutable.

Consider firstly the case $T \cap Q_1^x \neq 1$ and $E \cap Q_1^x \neq 1$ for some $x \in G$. Then Q_1 has a proper subgroup Z such that $Z \leq T \cap E$ and $|Z| = q$. Since $P Q_1$ is a Schmidt group, $Z \leq C_G(P)$. Hence Z is normal in G. Then $G/Z = (PZ/Z) \rtimes (Q_1 C_q/Z)$, where $|Q_1 C_q/Z| = q^\alpha, \alpha \geq 2$ for odd q and $\alpha > 2$ for $q = 2$. By [117, V, Theorem 4.3], we have $Z = Q'$, and so $Q_1 C_q/Z$ is an abelian group of type $(q^{\alpha-1}, q)$. According to case II(2), $(T/Z)(E/Z) = (E/Z)(T/Z)$ and thereby $TE = ET$.

Suppose now that at least one of the subgroups T or E, for instance T, enjoys the property that $T \cap Q_1^x = 1$ for all $x \in G$. Then, Lemma 2.27 implies that $|Q_1|$ divides $|G : T|$. Since G is a q-nilpotent group, by Lemma 2.26, either $|Q_1| = q^3$ or $|Q_1| = q^2$.

Verify that if $|\langle x \rangle| = q^2$ and $\langle x \rangle \subseteq Q$ then $\langle x \rangle$ is not contained in any 3-maximal subgroup T of G. Suppose that it is not true, that is, that $\langle x \rangle \subseteq T$. Take a maximal subgroup M of G such that T is a 2-maximal subgroup of M. If p divides $|G : M|$, then $G = P \rtimes M$, where $|M| = |Q| = q^3$, and so $\langle x \rangle$ is not contained in T. If q divides $|G : M|$, then $|G : M| = q$ since G is q-nilpotent. Then $M = P \rtimes \langle x \rangle$ since $\langle x \rangle$ is contained in M. Hence M is a Schmidt group. Since $\langle x \rangle \subseteq T \subseteq M$, we have that $T = T_p\langle x \rangle$ is a 2-maximal subgroup of M, where T_p is a Sylow p-subgroup of T. However, the only 2-maximal subgroups in the Schmidt group M are of the form P, $\langle x \rangle^q$, and $P_1\langle x \rangle^q$, where P_1 is some maximal subgroup of P. The resulting contradiction shows that $\langle x \rangle$ is not contained in T for any 3-maximal subgroup T of G.

We can show similarly that if $|\langle x \rangle| = q^3$ and $\langle x \rangle \subseteq Q$, then $\langle x \rangle$ is not contained in any 3-maximal subgroup T of G.

Take $T = T_p T_q$ and $E = E_p E_q$, where T_p, E_p, T_q, and E_q are Sylow p- and q-subgroups of T and E respectively. Since T_q and E_q do not include Q_1, as we showed above, it follows that $E_q \leq Q_2 \rtimes Z_1$, where Q_2 is a maximal subgroup of Q_1 and $Z_1 \simeq C_q$. Similarly, $T_q \leq Q_2 \rtimes Z_2$, where $Z_2 \simeq C_q$. It is clear that every subgroup of Q_1 is normal in G. Moreover, [117, V, Theorem 4.3] implies that $\Omega_1(Q)$ is an abelian group. Since $Z_1 \leq \Omega_1(Q)$ and $Z_2 \leq \Omega_1(Q)$, it follows that $Z_1 Z_2 = Z_2 Z_1$. Then since P is an abelian group,

$$TE = T_p T_q E_p E_q = T_p E_p T_q E_q = E_p E_q T_p T_q = ET.$$

Consequently, every two 3-maximal subgroups of G are permutable.

Similarly we can verify all remaining cases. The proof of the theorem is complete.

Corollary 2.29 *If every two 3-maximal subgroups of a group G are permutable and $|\pi(G)| > 3$ then G is a nilpotent group.*

Observe, for instance, that in the Schmidt group $G = P \rtimes Q$, where P is isomorphic to the quaternion group of order 8 and $|Q| = 3^2$, every two 3-maximal subgroups are permutable since G is a group of type I(b) in Theorem 2.28. However, not every 3-maximal subgroup of this group is normal in it. In this regard, it is a natural question to describe the nonnilpotent groups in which all 3-maximal subgroups are normal. Using Theorem 2.28, we may prove the following theorem which solves this problem.

Theorem 2.30 (Guo, Lutsenko and Skiba [155]). *Given a nonnilpotent group G, every 3-maximal subgroup of G is normal in G if and only if or $|G| = p^\alpha q^\beta r^\gamma$, where p, q, and r are primes and $\alpha + \beta + \gamma \leq 3$, or G is isomorphic to $SL(2,3)$, or G is a supersoluble group of one of the following types:*

(1) *$G = P \rtimes Q$, where $|P| = p$, $|Q| = q^\beta$ ($\beta \geq 3$); the group Q is either abelian or isomorphic to the quaternion group of order 8 or $M_\beta(q)$ ($\beta > 4$); every element of Q of order smaller than $q^{\beta-1}$ belongs to $C_G(P)$;*

(2) *$G = P \rtimes Q$, where P is a cyclic group of order p^2, both $\Phi(P)Q$ and $G/\Phi(P)$ are Schmidt groups, and the maximal subgroup of Q coincides with $Z(G)$;*

(3) *$G = (P_1 \times P_2) \rtimes Q$, where $|P_1| = |P_2| = p$, $P_1 Q$ is a Schmidt group, and $P_2 Q$ is either a nilpotent group or a Schmidt group;*

(4) $G = (P \rtimes Q)R$, where P and R are minimal normal subgroups of G, $|P| = p$, $|R| = r$, Q is a cyclic group, and $F(G) = PR\Phi(Q)$.

Corollary 2.31 [245] *Assume that every third maximal subgroup of G is normal in G. Then the commutator group G' of G is a nilpotent group, and the order of every principal quotient of G is not divisible by p^3 for any prime p.*

The classes of groups which are described in Theorem 2.28 and consequently in Theorem 2.30 are pairwise disjoint. It is easy to construct examples to show that all classes of groups in these theorems are nonempty.

Example 2.32   Take the noncyclic group P of order 4, the cyclic group Q of order 9, and the maximal subgroup V of Q. There exists a homomorphism f from Q into $Aut(P)$ with $Kerf = V$. Hence, we can consider the group $G = P \rtimes Q$, where $C_Q(P) = V$. It is clear that G is a Schmidt group of type I(a) of Theorem 2.28.

Example 2.33   The group $SL(2, 3)$ is a Schmidt group of type I(b) of Theorem 2.28. Indeed, the automorphism group $Aut(Q_8)$ of the quaternion group Q_8 of order 8 contains an element α of order 3. Take the group $G = Q_8 \rtimes <\alpha>$ isomorphic to $SL(2, 3)$. Verify that G is a Schmidt group. Take the unique subgroup Z of order 2 in Q_8. Then $Z \lhd G$, and so $C_G(Z) = G$. It is clear that Q_8/Z is a principal quotient of G and $Z \leq \Phi(G)$. It is also clear that $Z\langle\alpha\rangle$ is a maximal subgroup of G and every maximal subgroup M of G with $|G : M| = 2^\alpha$ is conjugate to $Z\langle\alpha\rangle$. Therefore, G is a Schmidt group.

Example 2.34   Take the noncyclic group P of order 4 and the cyclic group Q of order 3^3. Then according to Example 2.32 the group $G = P \rtimes Q_1$, where Q_1 is a maximal subgroup of Q, is a Schmidt group. Therefore, $G = P \rtimes Q$ is a group of type II(1) of Theorem 2.28.

Example 2.35   The direct product of the Schmidt group $G = P \rtimes Q$, where P is the noncyclic group of order 4 and Q is the cyclic group of order 9, and the group Z_3 of order 3 is a group of type II(2) of Theorem 2.28.

Example 2.36   Take the group Z_3 of order 3 and the quaternion group Q_8 of order 8. Then $Aut(Z_3)$ includes a subgroup of order 2; thus, there is a homomorphism f from Q_8 into $Aut(Z_3)$ with $Kerf = V$, where V is the subgroup of Q_8 with $|Q_8 : V| = 2$. So, we can consider $G = Z_3 \rtimes Q_8$, which is a group of type II(4) of Theorem 2.28.

Example 2.37   Take $Q = \langle x, y \mid x^9 = y^3 = 1, x^y = x^4 \rangle$ and the order 7 group Z_7. Then $\Omega_1(Q) = \Omega$ is an abelian group of order 9. Since Q/Ω is isomorphic to the order 3 subgroup Z_3 of $Aut(Z_7)$, we can construct $G = Z_7 \rtimes Q$. It is clear that $\langle y \rangle$ is a 3-maximal subgroup of G and $\langle y \rangle$ is not normal in G. It is easy to verify that all maximal subgroups of Q distinct from Ω are cyclic, and so G is a group of type II(5) of Theorem 2.28.

Example 2.38   Take now the cyclic group P of order 3^2. Then [117, V, Lemma 4.1] implies that $|Aut(P)| = 3 \cdot 2$. Take the Sylow 2-subgroup Q of $Aut(P)$. We can consider the group $G = P \rtimes Q$, where $C_Q(P) = 1$. If $C_Q(\Phi(P)) = Q$, then [117, V, Theorem 2.4] implies that $Q = 1$; a contradiction. Consequently, $\Phi(P)Q$ is a maximal Schmidt subgroup of G, and every maximal subgroup M of G with $|G : M| = 3$ is conjugate to $\Phi(P)Q$. It is easy to observe that in G every 3-maximal subgroup is normal. Therefore, G is an example of a group of type II(6) of Theorem 2.28.

Example 2.39 Take $M = R \rtimes Q$, where R is the group of order 5 and Q is the Sylow 2-subgroup of $Aut(R)$. Take the order 7 group P. Then $Aut(P)$ includes an order 2 subgroup; thus, there exists a homomorphism f from M into $Aut(P)$ with $Ker f = V$, where V is the subgroup of M with $|M : V| = 2$. Consequently, we can consider the group $G = P \rtimes M$, where $C_M(P) = V$. It is clear that G is a group of type III(i) of Theorem 2.28.

4.3   Finite Groups of Spencer Height ≤ 3

If $M_n < M_{n-1} < \ldots < M_1 < M_o = G$, where M_i is a maximal subgroup of M_{i-1}, $i = 1, 2, \ldots, n$, then we say that the chain $M_n < M_{n-1} < \ldots < M_1$ is a *maximal chain of G of length n* and M_n is an *n-maximal subgroup of G*. If a subgroup H of G is n-maximal but not i-maximal for all $i < n$, then H is said to be a *strictly* [11] n-maximal subgroup of G.

The relationship between n-maximal subgroups (where $n > 1$) of a group G and the structure of G was studied by many authors. One of the first results in this direction were obtained by Huppert [245]. He proved that if all 2-maximal subgroups of a group G are normal, then G is supersoluble; and that if all 3-maximal subgroups of G are normal, then G is soluble of rank $r(G) \leq 2$ (see [248, p. 685]). Janko [254] described the nonsoluble groups in which all 4-maximal subgroups are normal. Later, these results were developed in various directions. For instance, Asaad [11] improved the abovementioned results of Huppert by considering only strictly n-maximal subgroups for $n = 2, 3$. In [310], Mann described the structure of the group G in which each n-maximal subgroup is subnormal in the case when $k \leq n-1$, where k is the number of distinct prime divisors of the order of G. He proved that G is nilpotent if G is soluble and $k > n$. Deskins [84] and Spencer [393] analyzed the group G in which every n-maximal chain contains at least one subnormal in G subgroup.

Our main goal in this section is to describe the groups of Spencer hight 3 in the following sense.

Definition   (see [393]). A group G is said to be a group of *Spencer height $h(G) = n$* if every maximal chain of G of length n contains a proper subnormal entry and there exists at least one maximal chain of G of length $n - 1$ that contains no proper subnormal entry.

Note that since a subnormal maximal subgroup is normal, $h(G) = 1$ if and only if G is nilpotent. From Theorem 5 in [393] $h(G) = 2$ if and only if G is a Schmidt group with abelian Sylow subgroup. Every group G of Spencer height 3 is soluble

([393, Theorem 6]. If G is a non-nilpotent group with $h(G) = 3$, then $|\pi(G)| \leq 3$, and if, in addition, $|\pi(G)| = 2$, then one of Sylow subgroups of G is normal in G (These follow from Theorems 2 and 3 in [393]).

The following lemma is well known (see, for example, [310, Lemma 4]).

Lemma 3.1 *Let M be a maximal subgroup of G. If L is subnormal in G and $L \subseteq M$, then $L \subseteq M_G$.*

Lemma 3.2 [393].

(1) If H is a nonnormal maximal subgroup of G, then $h(H) \leq h(G) - 1$.
(2) If H is a normal subgroup of G, then $h(G/H) \leq h(G)$.

Lemma 3.3 [393, Theorem 3]. *If G is soluble and $h(G) - |\pi(G)| \leq 1$, then G is a φ-dispersive group for some ordering φ of the prime divisors of $|G|$.*

Lemma 3.4 [393, Theorem 4]. *Let G be soluble groups with $2 \leq h(G) = |\pi(G)|$. If there exist at least two nonisomorphic nonnormal Sylow subgroups of G, then all non-normal Sylow subgroups of G are of prime order.*

Theorem 3.5 (Guo, Andreeva, Skiba [145]) *G is a group of Spencer height 3 if and only if $|G| = p^\alpha q^\beta r^\gamma$, where $p \neq q$ and $3 \leq \alpha + \beta + \gamma$, and G is one of the following types:*

I. $G = P \rtimes Q$ and G satisfies at least one of the following:
   (1) G is a Schmidt group, where $|\Phi(P)|$ divides p.
   (2) P is a minimal normal subgroup of G and either $|Q : Q_G| = q^2$ and all maximal subgroups of Q are cyclic or Q is not cyclic, $|Q : Q_G| = q$ and any maximal subgroup of Q different from Q_G is cyclic.
   (3) $G = G^\mathfrak{N} \rtimes M$, where $G^\mathfrak{N}$ is a minimal normal subgroup of G, $M = M_p \times Q$ is a representative of the unique conjugate class of non-normal maximal subgroups of G, $|M_p| = p$ and $Q = \langle a \rangle$ is cyclic and $|Q : Q_G| = q$.
   (4) $\Phi(P) = 1$ and G is the subdirect product of a non-normal maximal subgroup A and a subgroup B, where $A = A_p \rtimes Q$ is a Schmidt group with abelian Sylow subgroups and A_p is a minimal normal subgroup of G, $B = B_p \rtimes Q$ and either B is nilpotent, $|B_p| = p$ and $B_p \leq Z(G)$ or $B \simeq A$.
   (5) $\Phi(P) > 1$, $A = \Phi(P) \rtimes Q$ is a representative of the unique conjugate class of non-normal maximal subgroups of G, A is a Schmidt group with abelian Sylow subgroups and $|Q : Q_G| = q$.
II. $G = P \rtimes (R \rtimes Q)$ and G has only three classes of maximal subgroups, the representatives of which are a Hall r'-subgroup A, a non-normal Hall p'-subgroup L and a normal subgroup M of G such that $|G : M| = q$. Moreover, the following statements hold:
   (a) L is either a Schmidt group with abelian Sylow subgroups or a nilpotent group with $|R| = r$.
   (b) $A = P \rtimes Q$ is a Schmidt group with abelian Sylow subgroups and $|Q : Q_G| = q$. Moreover, if A is normal in G, then L is nilpotent.

(c) P is a minimal normal subgroup of G, and either R is a minimal normal
  subgroup of G or $|R| = r$ and $|Q| = q$.

Proof We only need to prove the necessity part as the sufficiency part is
straightforward.

Let $h(G) = 3$, then G is not nilpotent by [393, Theorem 5]. By [393, Theorem 2
and 3], $h(G) - |\pi(G)| \leq 1$. Hence by Lemma 3.3, G has a normal Sylow subgroup.
Without loss of generality, we may suppose that P is normal in G. We now proceed
the proof via the following steps.

(1) *Any non-normal maximal subgroup of G is either a nilpotent group or a
Schmidt group with abelian Sylow subgroups.*

Let A be a non-normal maximal subgroup of G. Then A is not subnormal in
G. Hence, by Lemma 3.2, $h(A) \leq 2$. Therefore A is either a nilpotent group or a
Schmidt group with abelian Sylow subgroups by [393, Theorem 5].

(2) *G is soluble and $|G| = p^\alpha q^\beta r^\gamma$, where $3 \leq \alpha + \beta + \gamma$.*

By [393, Theorem 6], G is soluble. Hence $|\pi(G)| \leq 3$ by Chap. 1, Proposition
1.9 and (1). Now let $1 = G_0 < G_1 < \ldots < G_n = G$ be a composition series of
G. Then 1 is the unique n-maximal subgroup of G. Hence $3 \leq n = \alpha + \beta + \gamma$.
Moreover, if $3 = \alpha + \beta + \gamma$, then 1 is the unique 3-maximal subgroup of G and
$|G| = pqr$ or $|G| = p^2 q$ or $|G| = pq^2$. Hence G is a group of one of the type in
Theorem.

In the following, we suppose that $3 < \alpha + \beta + \gamma$.

(3) *If $E < T < M < G$ be a maximal chain of G, then $E \leq M_G$. In particular,
$|M : M_G| \in \{1, s, st\}$ for some primes s and t.*

Suppose that M is non-normal in G. Then M is not subnormal in G. Hence,
by hypothesis, either E or T is subnormal in G. Hence $E \leq M_G$ by Lemma 3.1.
Therefore every maximal subgroup of M/M_G has prime order and so $|M/M_G| = st$
for some primes s and t.

(4) *Let $E < T < M < G$ be a maximal chain of G. If T and M both are not
subnormal in G, then T is a cyclic group of prime power order and $E \leq O_s(G)$,
where s divides $|T|$.*

By hypothesis, every maximal subgroup of T is subnormal in G. Hence in view
of Chap. 1, Lemma 5.35(8), T is a cyclic of prime power order. Moreover, by
hypothesis, E is subnormal in G, so $E \leq O_s(G)$, where s divides $|T|$, by Chap. 1,
Lemma 5.35(6).

(5) *If M is a non-normal maximal subgroup of G and all maximal subgroup of M
are subnormal in G, then M is a cyclic group of prime power order and $|M : M_G|$
is a prime (see (4)).*

(6) *Suppose that all non-normal maximal subgroups of G are nilpotent. Then G
is a group of type I(1)-(3) if $G = PQ$ or G is of type II if $|\pi(G)| = 3$.*

It is easy to see that a maximal subgroup M of G is non-normal in G if and
only if $G^{\mathfrak{N}} \not\subseteq M$, where $\mathfrak{N}$ is the class of all nilpotent groups. Hence by Chap. 1,
Proposition 1.8, the following statements hold:

1) $G^{\mathfrak{N}}$ is an s-group, for some prime s.
2) $G^{\mathfrak{N}}/\Phi(G^{\mathfrak{N}})$ is a non-central chief factor of G.
3) All non-normal maximal subgroups of G are conjugate in G.

Without loss of generality, we may assume that $s = p$. Then if M is any non-normal maximal subgroup of G, then M is nilpotent and $|G : M|$ is a power of p. We may, therefore, assume that $Q \leq M$ and $R \leq M$. Note also that since M is nilpotent, every subgroup of M_G is subnormal in G.

First suppose that $G = PQ$. Then Q is not normal in G. By Chap. 1, Lemma 5.35(6), Q is not contained in M_G. Hence by (3), either $Q = M$ or Q is maximal in M. Assume that $Q = M$. Then P is a minimal normal subgroup of G of order p^n and $|Q : Q_G|$ is either q or q^2 by (3). Let $|Q : Q_G| = q$. If Q is cyclic, then G is a Schmidt group, that is, G is a group of type I(1). On the other hand, if Q is non-cyclic, then every maximal subgroup Q_1 of Q different from Q_G is not subnormal in G and so Q_1 is cyclic by (4). Hence, in this case, G is a group of type I(2). Now let $|Q : Q_G| = q^2$. Then every maximal subgroup Q_1 of Q is not subnormal and thereby Q_1 is cyclic. Hence in this case G is again a group of type I(2). Finally, suppose that Q is a maximal subgroup of M. Then $M = M_p \times Q$, where $M_p = P \cap M$ is a group of order p. Moreover, Q is cyclic and $Q_1 = O_q(G)$, where Q_1 is the maximal subgroup of Q by (4). Hence in this case G is a group of type I(3).

Now consider the case when $|\pi(G)| = 3$. Since M is not subnormal in G, at least one of the subgroups Q and R is not subnormal in G by Chap. 1, Lemma 5.35(6). Without loss of generality, we may assume that Q is not subnormal in G. Then Q is maximal in M, so $|R| = r$ and P is a minimal normal subgroup of G. Finally, if R is not normal in G, then $|Q| = q$ by Lemma 3.4 and so G is a group of type II.

(7) *Suppose that a maximal subgroup A of G is either a Schmidt group or nilpotent, and $|G : A| = p^a$ for some $a \in \mathbb{N}$. Then A_p is a minimal normal subgroup of G.*

Since A is maximal in G and $A_p < N_P(A_p)$, $N_G(A_p) = G$. Note that A_p is a minimal normal subgroup of A by Chap. 1, Proposition 1.9(9) since A_p is abelian by (1). Hence A_p is a minimal normal subgroup of G.

(8) *Suppose that $G = P \rtimes Q$ and at least one of the non-normal maximal subgroups of G is non-nilpotent. Then G is a group of type I(4) if $\Phi(P) = 1$, or G is of type I(5) if $\Phi(P) \neq 1$.*

Suppose that a non-normal maximal subgroup A of G is nonnilpotent. Then by (1), A is a Schmidt group with abelian Sylow subgroups. Hence $A = A_s \rtimes A_t$ is a Schmidt group with abelian Sylow subgroups A_s and A_t, for some primes s and t. Since P is normal in G, $s = p$ and $t = q$. On the other hand, since A is non-normal in G, $|G : A| = p^a$ for some $a \in \mathbb{N}$. Hence A contains some Sylow q-subgroup of G. Without loss of generality, we may assume $A_q = Q$. Therefore, $A = A_p \rtimes Q$. Note that A_p is a minimal normal subgroup of G by (7). Moreover, since Q is not subnormal in G, the maximal subgroup Q_1 of Q is subnormal in G and so $Q_1 \leq O_q(G)$ by (4), which implies $|Q : Q_G| = q$.

Suppose that $\Phi(P) = 1$. Then, by Maschkes's theorem, there is a normal non-identity p-subgroup P_1 of G such that $P = P_1 \times A_p$. Hence there is a maximal subgroup B of G such that $|G : B| = p^c$ for some $c \in \mathbb{N}$ and $B \neq A^x$ for all $x \in G$. Without loss of generality, we may assume that $B = B_p \rtimes Q$, where $B_p = P \cap B$. We now prove that B is non-normal in G. Assume that it is false. Then $A_p \cap B$ is a normal subgroup of G. This means that $A_p \cap B = 1$ since A_p is a minimal normal subgroup of G. Therefore, $A \cap B = A_p Q \cap B = Q(A_p \cap B) = Q$ is a normal subgroup of A. This contradiction shows that B is non-normal in G. Hence, by (1), B is either nilpotent or a Schmidt group. Since B is maximal in G, B_p is a minimal normal subgroup of G by (7). Since A is maximal in G and $B_p \not\leq A$, $G = B_p \rtimes A$. Hence $P = B_p \times A_p$. On the other hand, since $A_p \not\leq B$, $G = A_p \rtimes B = B_p \rtimes A$. Therefore,

$$G/A_p \simeq A_p B/A_p \simeq B/A_p \cap B \simeq B$$

and

$$G/B_p \simeq B_p A/B_p \simeq A/A \cap B_p \simeq A.$$

This implies that G is the subdirect product of the subgroups A and B. Suppose that B is nilpotent. Since Q is not subnormal in G, then all subgroups of B containing Q are not subnormal in G, which implies $|B_p| = p$. Since P is abelian and $G = BP$, $B_p \leq Z(G)$. Now suppose that B is a Schmidt group. Then $B \simeq A$ by Chap. 1, Proposition 1.9 since $Q \leq A \cap B$. Therefore, G is a group of type I(4).

Now suppose that $\Phi(P) \neq 1$. Since P is normal in G and $\Phi(P)$ is characteristic in P, $\Phi(P)$ is normal in G and $\Phi(P) \leq \Phi(G)$. Hence $\Phi(P) \leq A_p$. Since A_p is a minimal normal subgroup of G, $G = PA$ and $A = \Phi(P) \rtimes Q$. Now we show that every non-normal maximal subgroup of G is conjugate with A. Suppose that G has a non-normal maximal subgroup $M \neq A^x$ for all $x \in G$. Then $|G : M| = p^c$ for some $c \in \mathbb{N}$ and, by (1), M is either nilpotent or a Schmidt group with abelian Sylow subgroups. Without loss of generality, we may assume that $Q \leq M$ and so $Q_G \leq M$. First suppose that M be a Schmidt group with abelian Sylow subgroups. Then $M = M_p \rtimes Q$. Moreover, by (7), M_p is a minimal normal subgroup of G. Recall that A_p is also a minimal normal subgroup of G. Hence $M_p = A_p = \Phi(P)$. Thus, $A_G = \Phi(P)Q_G = M_G$ and so, by [89, Chap. A, Theorem 16.1], $M = A^x$ for some $x \in G$. This contradiction shows that M is nilpotent. Since $1 \neq \Phi(P) \leq M$, Q is contained in some maximal subgroup L of M, where L is not subnormal in G. Since Q is not subnormal, $Q = L$ since $h(G) = 3$. Hence $|M_p| = p$, $M_G = \Phi(P)Q_G = A_G$ and so M and A are conjugate in G by [89, Chap. A, Theorem 16.1]. This contradiction shows that all non-normal maximal subgroups of G are conjugate of A. Hence G is a group of type I(5).

(9) *If* $|\pi(G)| = 3$ *and G has a non-nilpotent non-normal maximal subgroup, then G is a group of type II.*

Suppose that $\pi(G) = \{p, q, r\}$. In view of Lemma 3.3, we may assume, without loss of generality, that P and PR are normal in G and Q is not normal in G. Let A be a non-normal non-nilpotent maximal subgroup of G. Then by (1), $A = A_s \rtimes A_t$ is a Schmidt group with abelian Sylow subgroups A_s and A_t for some primes s and t. First we assume that $s = r$ and $t = q$. Since G is soluble by (2) and A is maximal in G, $|G : A| = |P|$. We may assume that $A_s = R$ and $A_q = Q$. By using the same arguments as above, we can prove that $|Q : Q_G| = q$. Since, clearly, $G = PA$, P is a minimal normal subgroup of G. Note that every maximal subgroup B of G with $|G : B| = p^a$ for some $a \in \mathbb{N}$ is conjugate of A. Indeed, since P is a minimal normal subgroup of G, B is a Hall p'-subgroup of G. Hence A an B are conjugate in G by [89, Chap. I, Theorem 3.3]. This shows that all maximal subgroups of G

with p-power index are conjugate. If R is also not normal in G, then $|Q| = q$ and $|R| = p$ by Lemma 3.4.

We now prove that if L is any maximal subgroup of G such that $|G : L| = r^b$ for some $b \in \mathbb{N}$, then L is a non-normal Hall r'-subgroup of G. Indeed, clearly, $Q^x \leq L$ for some $x \in G$. Then $L \cap A^x = Q^x$. Hence L is non-normal in G. From above we know that either R is a minimal normal subgroup of G or $|R| = r$. Therefore, $L \cap R = 1$ and so L is a Hall r'-subgroup of G.

Now we show that every maximal subgroup M of G with $|G : M| = q^c$ for some $c \in \mathbb{N}$ is normal in G and M is the only normal maximal subgroup of G. In fact, since PR is normal in G, G has a normal maximal subgroup T such that $|G : T| = q$. Suppose that M is not normal in G. Then by (1), M is either a nilpotent group or a Schmidt group with abelian Sylow subgroups. First assume that M is a Schmidt group with abelian Sylow subgroups. Then R is not normal in G and so $|Q| = q$ by Lemma 3.4. Hence $M = T$, a contradiction. Thus M is nilpotent. But then, since R is normal in A, R is normal in $AM = G$ and therefore M is normal in G. This shows that every maximal subgroup M with $|G : M| = q^c$ is normal in G and $M = PQ_G R = T$.

Finally, let L be a Hall r'-subgroup of G. From above we have known that L is a non-normal maximal subgroup of G. Without loss of generality, we may assume that $L = PQ$. First suppose that L is nilpotent. Since Q is not subnormal in G, all subgroups of L containing Q are not subnormal in G, which implies $|P| = p$. If R is not a normal subgroup of G, then $|R| = r$ and $|Q| = q$ by Lemma 3.4. Hence $|G| = pqr$, which contradicts the assumption that $3 < \alpha + \beta + \gamma$. Therefore, if L is nilpotent, then $|P| = p$ and R is a normal subgroup of G of order r. If $L = P \rtimes Q$ is a Schmidt group with abelian Sylow subgroups, then P is a minimal normal subgroup of G by Chap. 1, Proposition 1.9. In view of all above proof, we see that G is a group of type II. Analogously it may be shown that G is a group of type II in case when $s = p$ and $t = q$. Theorem is proved.

It is easy to see that all groups with types in Theorem always exist.

4.4   On $\hat{\theta}$-Pairs for Maximal Subgroups of a Finite Group

Recall that for a maximal subgroup M of G, a subgroup C of G is said to be a completion for M [83] if C is not contained in M, while every proper subgroup of C that is normal in G is contained in M. In [473], Zhao introduced the concept of θ-completions for maximal subgroups: for a maximal subgroup M of G, a subgroup C of G is said to be a θ-completion for M if C is not contained in M while M_G, the core of M in G, is contained in C and no non-trivial proper subgroup of C/M_G is normal in G/M_G. Besides, Mukherjee and Bhattacharya [321] gave the concept of θ-pairs for maximal subgroups, which has a close relationship with the concepts of completions and θ-completions: a pair of subgroups (C, D) of G is called a θ-pair for a maximal subgroup M of G if the following hold: (i) $D \trianglelefteq G$, $D \leq C$; (ii)

$\langle M, C \rangle = G$, $\langle M, D \rangle = M$; (iii) no nontrivial proper subgroup of C/D is normal in G/D.

We denote the set of completions, θ-completions, and θ-pairs for a maximal subgroup M of G by $I(M)$, $\theta I(M)$, and $\theta(M)$, respectively. It is clear that $I(M)$, $\theta I(M)$, and $\theta(M)$ can be partially ordered by set-theoretic inclusion. The maximal elements of $I(M)$ ($\theta I(M)$, and $\theta(M)$) with respect to this partial ordering are called maximal completions (maximal θ-completions, and maximal θ-pairs, respectively). Also, an element C of $I(M)$ ($\theta I(M)$) or an element (C, D) of $\theta(M)$ is called *normal* if C is a normal subgroup of G.

Obviously, if $C \in I(M)$ ($C \in \theta I(M)$), then $(C, (C \cap M)_G)$ ((C, M_G), respectively) is a θ-pair for M. But the converse does not hold in general (cf. Example in [25]). Moreover, a normal completion (a normal θ-completion) C is maximal in $I(M)$ ($\theta I(M)$, respectively). However, a normal θ-pair (C, D) is not necessarily maximal in $\theta(M)$. For example, let $G = Z_6 = Z_2 \times Z_3$ and $M = Z_3$. Clearly $(Z_2, 1)$ is a normal θ-pair for M. But since (G, M) is a θ-pair for M, $(Z_2, 1)$ is not maximal in $\theta(M)$.

There has been a lot of interest in using the notions of completions, θ-completions, and θ-pairs, to investigate the structure of finite groups (see, for example, [25, 33, 48, 86, 93, 94, 218, 285, 291, 321, 469, 471, 472, 473, 474, 475]). For the purpose of weakening or dispensing the maximality imposed on completions and θ-completions, the concepts of s-completions and s-θ-completions (strong θ-completions) for maximal subgroups were first defined by Li et al. in [292] and [90].

Recall that for a maximal subgroup M of G, a completion (θ-completion) C for M is called an *s-completion* (*s-θ-completion*, respectively) if either $C = G$ or there exists a subgroup E of G which is not a completion (θ-completion, respectively) for M such that E contains C as a maximal subgroup. Actually, a maximal completion (a maximal θ-completion) is an s-completion (s-θ-completion, respectively), but the converse is not true in general (cf. Example 1.2 in [292]).

As a continuation of the abovementioned ideas, we introduce the concept of $\hat{\theta}$-pairs for maximal subgroups as follows:

Definition 4.1 (Chen, Guo [64]). For a maximal subgroup M of G, a θ-pair (C, D) for M is called a $\hat{\theta}$-*pair* if either $C = G$ or there exists a subgroup E of G such that C is a maximal subgroup of E and $(EM_G)_G$ is not contained in M.

In fact, many interesting results concerning s-completions, s-θ-completions, and maximal θ-pairs may be nontrivially generalized based on the concept of $\hat{\theta}$-pairs (see below Lemma 4.3). The aim of this section is to exhibit some new characterizations for a finite group to be solvable, supersolvable, nilpotent, etc.

For the sake of convenient argument, we introduce the notion of $\hat{\theta}$-completions for maximal subgroups.

Definition 4.2 (Chen, Guo [64]). For a maximal subgroup M of G, a θ-completion C for M is called a $\hat{\theta}$-*completion* if either $C = G$ or there exists a subgroup E of G such that C is a maximal subgroup of E and E_G is not contained in M.

It is clear that if C is a $\hat{\theta}$-completion for a maximal subgroup M of G, then (C, M_G) is a $\hat{\theta}$-pair for M.

Lemma 4.3 *Suppose that M is a maximal subgroup of G.*

(1) If C is an s-completion for M, then $(C, (C \cap M)_G)$ is a $\hat{\theta}$-pair for M.
(2) If C is an s-θ-completion for M, then C is a $\hat{\theta}$-completion for M and (C, M_G) is a $\hat{\theta}$-pair for M.
(3) If (C, D) is a maximal θ-pair for M, then (C, D) is a $\hat{\theta}$-pair for M.
(4) If C is a normal θ-completion for M, then C is a $\hat{\theta}$-completion for M.
(5) If (C, D) is a normal θ-pair for M, then (C, D) is a $\hat{\theta}$-pair for M.

Proof

(1) Obviously, $(C, (C \cap M)_G)$ is a θ-pair for M. If $C = G$, then it is trivial. Now assume that $C < G$. Then there exists a subgroup E such that C is maximal in E and E is not a completion for M. Since E is not a completion, we may take a normal subgroup K of G such that $1 < K < E$ and $K \not\leq M$. Then $(EM_G)_G \not\leq M$. Therefore, $(C, (C \cap M)_G)$ is a $\hat{\theta}$-pair for M.
(2) can be similarly discussed as (1).
(3) and (4) are evident.
(5) If $C = G$, then it is trivial. Now suppose that $C < G$. Let E be an arbitrary subgroup of G such that C is maximal in E. As $C \leq (EM_G)_G$ and $C \not\leq M$, we have that $(EM_G)_G \not\leq M$. Hence (C, D) is a $\hat{\theta}$-pair for M.

Remark

(1) For any maximal subgroup M of G, the $\hat{\theta}$-completions and the $\hat{\theta}$-pairs for M always exist. For example, let A/M_G be a minimal normal subgroup of G/M_G, then it is clear that A is a normal θ-completion for M, and so it is a $\hat{\theta}$-completion for M.
(2) For a maximal subgroup M of G, a $\hat{\theta}$-completion for M is not necessarily an s-θ-completion. For example, let $G = A_5$ be the alternating group of degree 5, $M = A_4$, and $C \in Syl_5(G)$. Then $M_G = 1$ and C is a θ-completion for M. Since C is maximal in G and G is also a θ-completion for M, C is a $\hat{\theta}$-completion, but not an s-θ-completion for M.
(3) For a maximal subgroup M of G, a $\hat{\theta}$-pair for M is not necessarily a maximal θ-pair. For example, let $G = S_4$ be the symmetric group of degree 4, $M = A_4$, and $C = S_3$. Obviously, $(C \cap M)_G = 1$ and $(C, 1)$ is a θ-pair for M. Since C is maximal in G and (G, M) is also a θ-pair for M, we see that $(C, 1)$ is a $\hat{\theta}$-pair, but not a maximal θ-pair for M.

Lemma 4.4 *Suppose that M is a maximal subgroup of G, $N \trianglelefteq G$ and $N \leq M$. Then C is a $\hat{\theta}$-completion for M if and only if C/N is a $\hat{\theta}$-completion for M/N.*

Proof It is obvious that C is a θ-completion for M if and only if C/N is a θ-completion for M/N. Suppose that C is a $\hat{\theta}$-completion for M. If $C = G$, then it is trivial. We may, therefore, assume that there exists a subgroup E of G such that C is maximal in E and $E_G \not\leq M$. Then C/N is maximal in E/N, and $(E/N)_{(G/N)} = E_G/N \not\leq M/N$. Hence C/N is a $\hat{\theta}$-completion for M/N. With the similar argument as above, the converse also holds.

We say that a class of groups $\mathfrak{X}$ is *section-closed* if whenever G is an $\mathfrak{X}$-group (that is, $G \in \mathfrak{X}$), every section A/B of G is also an $\mathfrak{X}$-group.

Lemma 4.5 *Let M be a maximal subgroup of G and $\mathfrak{X}$ a class of groups which is section-closed. Then the following statements are equivalent:*
(a) There exists a $\hat{\theta}$-completion C for M such that C/M_G is an $\mathfrak{X}$-group;
(b) There exists a $\hat{\theta}$-pair (C, D) for M such that C/D is an $\mathfrak{X}$-group.

Proof It is trivial that the statement (a) implies (b). Now we prove that (b) implies (a). If $C = G$, then it is obvious. We may, therefore, assume that there exists a subgroup E of G such that C is maximal in E and $(EM_G)_G \not\leq M$. Since clearly, $D = (C \cap M)_G \leq C \cap M_G$ and C/D is an $\mathfrak{X}$-group, $CM_G/M_G \cong C/C \cap M_G$ is also an $\mathfrak{X}$-group. Suppose that CM_G is a θ-completion for M, then either $(CM_G)_G \leq M$ or $CM_G \trianglelefteq G$. If $CM_G = EM_G$, then since $(CM_G)_G = (EM_G)_G \not\leq M$, we have that $CM_G \trianglelefteq G$. Hence CM_G is a normal θ-completion for M, and consequently, CM_G is a $\hat{\theta}$-completion for M by Lemma 4.3(4). If $CM_G < EM_G$, then it is easy to see that CM_G is maximal in EM_G. As $(EM_G)_G \not\leq M$, CM_G is a $\hat{\theta}$-completion for M by Definition 4.2. Now consider that CM_G is not a θ-completion for M. Then there exists a minimal normal subgroup L/M_G of G/M_G, which is properly contained in CM_G/M_G. It is clear that L is a $\hat{\theta}$-completion for M such that L/M_G is an $\mathfrak{X}$-group. The lemma is thus proved.

With similar argument as the proof of Lemma 4.5, we can easily obtain the next lemma.

Lemma 4.6 *Let M be a maximal subgroup of G and $\mathfrak{X}$ a class of groups which is section-closed. Then the following statements are equivalent:*
(a) There exists a $\hat{\theta}$-completion C for M such that $C^g \not\leq M$ for every $g \in G$ and C/M_G is an $\mathfrak{X}$-group;
(b) There exists a $\hat{\theta}$-pair (C, D) for M such that $C^g \not\leq M$ for every $g \in G$ and C/D is an $\mathfrak{X}$-group.

Recall that a maximal subgroup M of G is said to be *c-maximal* if M has a composite index in G.

Lemma 4.7 *Let $N \trianglelefteq G$ and p be the largest prime divisor of $|N|$. If $P \in Syl_p(N)$, then either $P \trianglelefteq G$ or any maximal subgroup of G containing $N_G(P)$ is c-maximal.*

Proof Suppose that $P \ntrianglelefteq G$. Let M be an arbitrary maximal subgroup of G containing $N_G(P)$. Then by Frattini's argument, $G = N_G(P)N = MN$. If $|G : M| = |N : M \cap N| = r$ for some prime r, then $N/(M \cap N)_N$ is isomorphic to some subgroup of S_r, where S_r denotes the symmetric group of degree r. Hence r is the largest prime divisor of $|N : (M \cap N)_N|$. Since $P \leq M \cap N$, we have that $r < p$, and so $P \leq (M \cap N)_N$. By Frattini's argument again, $N = N_N(P)(M \cap N)_N \leq M \cap N$. It follows that $N \leq M$, a contradiction. Thus M is c-maximal.

Lemma 4.8 [330, Corollary 1]. *Let G be a finite group containing a nilpotent maximal subgroup S, and P denote the Sylow 2-subgroup of S. If any subgroup H of P such that $C_P(H) \leq H$ is a normal subgroup of P, then G is solvable.*

Lemma 4.9 *Let M be a solvable maximal subgroup of G. If $N \trianglelefteq G$ such that $G = MN$ and $M \cap N = 1$, then N is solvable.*

Proof We may assume that $M > 1$ and $M_G = 1$. Since M is solvable, M has a nontrivial normal p-group P for some prime p. By [273, Lemma 8.1.3], there exists a P-invariant Sylow p-subgroup H of N. If $H > 1$, then $N_H(P) > 1$. Since $M_G = 1$, we have that $N_G(P) = M$, and so $M \cap N \geq N_G(P) \cap H = N_H(P) > 1$, a contradiction. This shows that N is a p'-group. Since $C_N(P) \leq N_N(P) = M \cap N = 1$, N has a unique P-invariant Sylow q-subgroup Q for any prime divisor q of $|N|$ by [273, Lemma 8.2.3]. Note that for any $m \in M$ and for any $x \in P$, $(Q^m)^x = (Q^{mxm^{-1}})^m = Q^m$. This implies that Q^m is also P-invariant. Consequently, $Q^m = Q$, and so Q is M-invariant. Therefore, $G = MQ$, and thereby $N = Q$ is solvable.

Lemma 4.10 [117, Chap. 8, Theorem 3.1]. *Let P be a Sylow p-subgroup of G and p an odd prime. If $N_G(Z(J(P)))$ has a normal p-complement, then so also does G.*

Lemma 4.11 [42, Lemma 3.1]. *Let $\mathfrak{F}$ be a saturated formation of characteristic π and H a subnormal subgroup of G containing $O_\pi(\Phi(G))$ such that $H/O_\pi(\Phi(G))$ belongs to $\mathfrak{F}$. Then H belongs to $\mathfrak{F}$.*

Lemma 4.12 [52, Theorem 3]. *Let $L(G)$ denote the intersection of c-maximal subgroups M of a group G (if no such M exists, then $L(G) = G$). Then $L(G)$ is supersolvable.*

Lemma 4.13 [229, Theorem 12.5.1]. *The groups G of order p^n which contain a cyclic subgroup of index p are of the following types:*

(1) $G = \langle a \mid a^{p^n} = 1, n \geq 1 \rangle$;
(2) $G = \langle a, b \mid a^{p^{n-1}} = 1, b^p = 1, bab^{-1} = a, n \geq 2 \rangle$;
(3) $G = \langle a, b \mid a^{p^{n-1}} = 1, b^p = 1, bab^{-1} = a^{1+p^{n-2}}, p \text{ is odd}, n \geq 3 \rangle$;
(4) $G = \langle a, b \mid a^{2^{n-1}} = 1, b^2 = a^{2^{n-2}}, bab^{-1} = a^{-1}, n \geq 3 \rangle$;
(5) $G = \langle a, b \mid a^{2^{n-1}} = 1, b^2 = 1, bab^{-1} = a^{-1}, n \geq 3 \rangle$;
(6) $G = \langle a, b \mid a^{2^{n-1}} = 1, b^2 = 1, bab^{-1} = a^{1+2^{n-2}}, n \geq 4 \rangle$;
(7) $G = \langle a, b \mid a^{2^{n-1}} = 1, b^2 = 1, bab^{-1} = a^{-1+2^{n-2}}, n \geq 4 \rangle$.

Let $\mathfrak{F} = LF(f)$ be a saturated formation, where f is a formation function. Then we give the following definition of f-central that generalizes the classical notion.

Definition 4.14 (Chen, Guo [64]). Let A and B be subgroups of a group G such that $B \trianglelefteq G$ and $B \leq A$. We say that A/B is f-central in G if $(G/B)^{f(p)} \leq C_{G/B}(A/B)$ for each prime p dividing $|A/B|$. Here $(G/B)^{f(p)}$ denotes the $f(p)$-residual of G/B.

Theorem 4.15 (Chen, Guo [64]). *Let G be a group and $\mathfrak{X} = \langle X \mid X$ is a nilpotent group, and for the Sylow 2-subgroup P of X, every subgroup H of P with $C_P(H) \leq$*

H is normal in P). For every given normal subgroup S of G, we denote by $\Phi_{\mathfrak{X},S}(G)$
the intersection of all c-maximal subgroups M of G such that

    (i) $S \nleq M$;
    (ii) M fails to have a $\hat{\theta}$-*pair* (C, D) *with* C/D *is an* $\mathfrak{X}$-*group* ($\Phi_{\mathfrak{X},S}(G) = G$ *if no such M exists).*
Then $\Phi_{\mathfrak{X},S}(G) \cap S$ *is the maximal normal solvable subgroup of G contained in S.*

Proof First, we claim that the class of groups $\mathfrak{X}$ is section-closed. Suppose that X is an $\mathfrak{X}$-group, that is, X is nilpotent, and for $P \in Syl_2(X)$, every subgroup H of P with $C_P(H) \leq H$ is normal in P. Let A/B be a section of the group X. Then $(P \cap A)B/B \in Syl_2(A/B)$. Let K/B be a subgroup of $(P \cap A)B/B$ such that $C_{(P\cap A)B/B}(K/B) \leq K/B$. Then, clearly, $K = (P \cap A \cap K)B$ and $C_{(P\cap A)B}(K/B) \leq K$. It follows that $C_{P\cap A}(P \cap A \cap K) \leq P \cap A \cap K$. Since

$$C_P(C_P(P \cap A \cap K)(P \cap A \cap K)) \leq C_P(P \cap A \cap K)(P \cap A \cap K),$$

we have

$$C_P(P \cap A \cap K)(P \cap A \cap K) \trianglelefteq P.$$

This implies that $P \cap A \cap K \trianglelefteq P \cap A$, and so

$$K/B = (P \cap A \cap K)B/B \trianglelefteq (P \cap A)B/B.$$

This shows that A/B is an $\mathfrak{X}$-group. Hence the claim holds.

It is clear that $\Phi_{\mathfrak{X},S}(G)$ is a normal subgroup of G. Now let N be an arbitrary normal solvable subgroup of G contained in S. If G has no c-maximal subgroup, then $G = \Phi_{\mathfrak{X},S}(G)$, and G is solvable by [248, Chap. VI, Satz 9.4]. In this case, the theorem holds trivially. We may, therefore, assume that G contains at least a c-maximal subgroup M. If $N \nleq M$, then $S \nleq M$. Now take a minimal normal subgroup L/M_G of G/M_G contained in NM_G/M_G. It is easy to see that (L, M_G) is a normal θ-pair for M. Hence (L, M_G) is a $\hat{\theta}$-pair for M by Lemma 4.3(5). Clearly, L/M_G is elementary abelian, and so L/M_G is an $\mathfrak{X}$-group. This shows that $N \leq \Phi_{\mathfrak{X},S}(G)$.

By Lemma 4.5, a maximal subgroup M of G has no $\hat{\theta}$-pair (C, D) with C/D is an $\mathfrak{X}$-group if and only if M has no $\hat{\theta}$-completion C with C/M_G is an $\mathfrak{X}$-group. (i) Let T be a c-maximal subgroup of G which has no $\hat{\theta}$-completion C such that $S \nleq T$ and C/T_G is an $\mathfrak{X}$-group. Then $N \leq T$ and T/N is a c-maximal subgroup of G/N. Suppose that T/N has a $\hat{\theta}$-completion A/N such that $(A/N)/(T_G/N)$ is an $\mathfrak{X}$-group. By Lemma 4.4, T has a $\hat{\theta}$-completion A, and evidently, A/T_G is an $\mathfrak{X}$-group. This contradiction shows that $\Phi_{\mathfrak{X},S/N}(G/N) \leq \Phi_{\mathfrak{X},S}(G)/N$. (ii) Conversely, let T/N be a c-maximal subgroup of G/N which has no $\hat{\theta}$-completion C/N such that $S/N \nleq T/N$ and $(C/N)/(T_G/N)$ is an $\mathfrak{X}$-group. With the similar discussion as (i), we can see that T is also a c-maximal subgroup of G which has no $\hat{\theta}$-completion C such that $S \nleq T$ and C/T_G is an $\mathfrak{X}$-group. This implies that $\Phi_{\mathfrak{X},S}(G)/N \leq \Phi_{\mathfrak{X},S/N}(G/N)$. Therefore, $\Phi_{\mathfrak{X},S/N}(G/N) = \Phi_{\mathfrak{X},S}(G)/N$.

If $N \neq 1$, then $(G/N, S/N)$ satisfies the hypothesis of the theorem by Lemma 4.4. Hence by induction, $(\Phi_{\mathfrak{X},S}(G) \cap S)/N = \Phi_{\mathfrak{X},S/N}(G/N) \cap S/N$ is the maximal normal solvable subgroup of G/N contained in S/N. It follows that $\Phi_{\mathfrak{X},S}(G) \cap S$ is the maximal normal solvable subgroup of G contained in S. So we may assume that S contains no nontrivial normal solvable subgroup of G. In this case, we shall show that $\Phi_{\mathfrak{X},S}(G) \cap S = 1$.

Suppose that $\Phi_{\mathfrak{X},S}(G) \cap S > 1$, and let L be a minimal normal subgroup of G contained in $\Phi_{\mathfrak{X},S}(G) \cap S$. Then L is nonsolvable. Let p be the largest prime divisor of $|L|$ and $P \in Syl_p(L)$. Since $P \ntrianglelefteq G$, there exists a c-maximal subgroup M of G containing $N_G(P)$ by Lemma 4.7. By Frattini's argument,

$$G = N_G(P)L = ML.$$

Hence $\Phi_{\mathfrak{X},S}(G) \cap S \nleq M$. This means that M has a $\hat{\theta}$-pair (C, D) such that C/D is an $\mathfrak{X}$-group. It follows from Lemma 4.5 that M has a $\hat{\theta}$-completion C such that C/M_G is an $\mathfrak{X}$-group. If $C = G$, then G/M_G is nilpotent, and consequently, $L \cong LM_G/M_G$ is solvable, which is absurd. We may, therefore, assume that there exists a subgroup E such that C is maximal in E and $E_G \nleq M$. Note that G/M_G is primitive and LM_G/M_G is a nonsolvable minimal normal subgroup of G/M_G. By [89, Chap. A, Theorem 15.2], either G/M_G has a unique non-abelian minimal normal subgroup LM_G/M_G or G/M_G has exactly two non-abelian minimal normal subgroups LM_G/M_G and L^*/M_G. In the former case, since $E_G/M_G > 1$, $LM_G/M_G \leq E/M_G$. If $LM_G/M_G \leq C/M_G$, then LM_G/M_G is nilpotent, which is impossible. Hence $LM_G/M_G \nleq C/M_G$. Then, since C/M_G is maximal in E/M_G, we have that $E/M_G = (C/M_G)(LM_G/M_G)$. As C/M_G is an $\mathfrak{X}$-group, by Lemma 4.8, E/M_G is solvable, and so is LM_G/M_G, a contradiction. In the latter case, either $LM_G/M_G \leq E/M_G$ or $L^*/M_G \leq E/M_G$. With the similar argument as above, at least one of LM_G/M_G and L^*/M_G is solvable. The final contradiction completes the proof.

Corollary 4.16 *A group G is solvable if and only if for any c-maximal subgroup M of G, there exists a $\hat{\theta}$-pair (C, D) such that C/D is nilpotent in which the nilpotent class of the Sylow 2-subgroup is at most 2.*

Proof For the sufficiency part, let $P/D \in Syl_2(C/D)$ with nilpotent class at most 2. Then $(P/D)/Z(P/D)$ is abelian. Therefore, for any subgroup H/D of P/D with $C_{P/D}(H/D) \leq H/D$, we have that $Z(P/D) \leq H/D$, and so $H/D \trianglelefteq P/D$. Take $S = G$. Then G is solvable by Theorem 4.15. Conversely, assume that G is solvable. Let A/M_G be a minimal normal subgroup of G/M_G. Then clearly, (A, M_G) is a normal θ-pair for M, and so it is a $\hat{\theta}$-pair for M by Lemma 4.3. Since G is solvable, A/M_G is elementary abelian. This shows that the necessity part also holds. $\quad\blacksquare$

Remark Note that [86, Theorem 1], [25, Proposition 1], and [473, Theorem 3.1] can be directly generalized by Corollary 4.16. Furthermore, with similar technology and a slight improvement, [292, Theorem 3.1], [91, Theorem 3.1], and [475, Theorem 3.1] can also be extended by using the notion of $\hat{\theta}$-pairs.

Theorem 4.17 (Chen, Guo [64]). *Let G be a group and $\mathfrak{X} = \langle X \mid X$ is a group such that there exists at most one odd prime q such that X is not q-closed but X is*

$\{2, q\}$-*closed). For any given normal subgroup S of G, we denote by $\widetilde{\Phi}_{\mathfrak{X},S}(G)$ the intersection of all maximal subgroups M of G such that*

(i) $S \nleq M$;

(ii) M fails to have a $\hat{\theta}$-pair (C, D) such that $C^g \nleq M$ for every $g \in G$ and C/D is an $\mathfrak{X}$-group ($\widetilde{\Phi}_{\mathfrak{X},S}(G) = G$ if no such M exists).

Then $\widetilde{\Phi}_{\mathfrak{X},S}(G) \cap S$ is the maximal normal solvable subgroup of G contained in S.

Proof Obviously, the class of groups $\mathfrak{X}$ is section-closed and $\widetilde{\Phi}_{\mathfrak{X},S}(G)$ is a normal subgroup of G. Let N be an arbitrary normal solvable subgroup of G contained in S and M a maximal subgroup of G. If $N \nleq M$, then $S \nleq M$ and there exists a $\hat{\theta}$-pair (L, M_G) for M such that L/M_G is a minimal normal solvable subgroup of G/M_G by Lemma 4.3(5). Hence L/M_G is an $\mathfrak{X}$-group, which implies that $N \le \widetilde{\Phi}_{\mathfrak{X},S}(G)$. By using Lemmas 4.4 and 4.6, analogously as the proof of Theorem 4.15, we have that $\widetilde{\Phi}_{\mathfrak{X},S/N}(G/N) = \widetilde{\Phi}_{\mathfrak{X},S}(G)/N$. If $N \ne 1$, then the theorem holds by induction. We may, therefore, assume that S contains no nontrivial normal solvable subgroup of G. We shall show that $\widetilde{\Phi}_{\mathfrak{X},S}(G) \cap S = 1$.

Suppose that $\widetilde{\Phi}_{\mathfrak{X},S}(G) \cap S > 1$, and let L be a minimal normal subgroup of G contained in $\widetilde{\Phi}_{\mathfrak{X},S}(G) \cap S$. Evidently L is nonsolvable. If $L \le \Phi(G)$, then L is solvable, a contradiction. Thus $L \nleq \Phi(G)$, and thereby we may choose a maximal subgroup M of G such that $G = ML$. It is easy to see that for any nontrivial normal subgroup K of G,

$$\widetilde{\Phi}_{\mathfrak{X},S}(G)K/K \le \widetilde{\Phi}_{\mathfrak{X},SK/K}(G/K).$$

Using Lemma 4.4, the theorem holds for $(G/K, SK/K)$ by induction. Hence $\widetilde{\Phi}_{\mathfrak{X},SK/K}(G/K) \cap SK/K$ is solvable. As $(\widetilde{\Phi}_{\mathfrak{X},S}(G) \cap S)K/K \le (\widetilde{\Phi}_{\mathfrak{X},S}(G)K \cap SK)/K \le \widetilde{\Phi}_{\mathfrak{X},SK/K}(G/K) \cap SK/K$, $(\widetilde{\Phi}_{\mathfrak{X},S}(G) \cap S)K/K$ is also solvable. It follows that $(\widetilde{\Phi}_{\mathfrak{X},S}(G) \cap S)/L$ is solvable. If $L \cap K = 1$, then $\widetilde{\Phi}_{\mathfrak{X},S}(G) \cap S$ is solvable, and so $\widetilde{\Phi}_{\mathfrak{X},S}(G) \cap S = 1$. We may, therefore, assume that L is contained in any nontrivial normal subgroup of G. This shows that L is the unique minimal normal subgroup of G, and so $M_G = 1$.

Since $\widetilde{\Phi}_{\mathfrak{X},S}(G) \nleq M$, by hypothesis and Lemma 4.6, M has a $\hat{\theta}$-completion C such that $C^g \nleq M$ for every $g \in G$ and there exists at most one odd prime q such that C is not q-closed but C is $\{2, q\}$-closed. If $C = G$, then G is 2-closed, and so G is solvable by Feit–Thompson Theorem, which is impossible. Hence G has a subgroup E such that C is maximal in E and $E_G \nleq M$. This implies that $L \le E$. If $L \le C$, then L is 2-closed. Therefore, L is solvable, a contradiction. It follows that $E = CL$. Note that C is solvable. If $C \cap L = 1$, then L is solvable by Lemma 4.9, a contradiction again. Thus $C \cap L > 1$.

Suppose that $C \cap L$ has a nontrivial Sylow p-subgroup P for a prime $p \ne q$ (where the prime p may be 2). Then since $C \cap L$ is p-closed, $P \trianglelefteq C \cap L$, and consequently $P \trianglelefteq C$. As C is maximal in E, $N_E(P) = E$ or C. If $P \trianglelefteq E$, then $P \trianglelefteq L$, which contradicts the fact that L is nonsolvable. Therefore, $N_E(P) = C$, and so $N_L(P) = C \cap L$. If $P \notin Syl_p(L)$, then L has a Sylow p-subgroup P^* properly containing P. It follows that $P < N_{P^*}(P) = C \cap P^* = P$, a contradiction. Hence $P \in Syl_p(L)$. Assume that $p > 2$. Denote by $J(P)$ the Thompson subgroup of P. Since $C = N_E(P) \le N_E(Z(J(P))) < E$, we have that $N_E(Z(J(P))) = C$.

As $N_L(Z(J(P))) = C \cap L$ is $\{2,q\}$-closed and r-closed for every prime $r \notin \{2,q\}$, $N_L(Z(J(P)))$ is p-nilpotent. Then by Lemma 4.10, L is p-nilpotent, and so $p \nmid |L|$, a contradiction. This implies that $C \cap L$ contains no nontrivial Sylow p-subgroup when $p \notin \{2,q\}$. It follows that $\pi(C \cap L) \subseteq \{2,q\}$ and if $2 \mid |C \cap L|$, then $|C \cap L|_2 = |L|_2$.

If $2 \nmid |C \cap L|$, then $C \cap L$ is a q-group. Since q is odd, L is q-nilpotent as above, which is absurd. Hence $2 \mid |C \cap L|$, and thereby L has a nontrivial Sylow 2-subgroup P contained in $C \cap L$. Let T be a maximal subgroup of G containing $N_G(P)$. By Frattini's argument, $G = N_G(P)L = TL$. As $\Phi_{\mathfrak{X},S}(G) \nleq T$, by hypothesis and Lemma 4.6, there exists a $\hat{\theta}$-completion C' for T such that $(C')^g \nleq T$ for every $g \in G$ and there exists at most one odd prime r such that C' is not r-closed but C' is $\{2,r\}$-closed. With the similar argument as above, $C' \cap L > 1$ and L has a nontrivial Sylow 2-subgroup P' contained in $C' \cap L$ such that $C' \leq N_G(P')$. By Sylow Theorem, $P = (P')^l$ for an element $l \in L$. Hence $(C')^l \leq (N_G(P'))^l = N_G((P')^l) = N_G(P) \leq T$. The final contradiction completes the proof.

Remark   Theorem 4.17 does not hold in general if we replace "maximal" by "c-maximal." For example, let $S = G = A_5$ and $C = A_4$. Then for every c-maximal subgroup M of G, $(C, 1)$ is a $\hat{\theta}$-pair for M such that C is an $\mathfrak{X}$-group. However, G is not solvable.

Corollary 4.18   *Let G be a group. Then G is solvable if and only if for any maximal subgroup M of G, there exists a $\hat{\theta}$-pair (C, D) such that $C^g \nleq M$ for every $g \in G$ and C/D is nilpotent.*

Proof   The sufficiency part directly follows from Theorem 4.17, and the proof of the necessity part is similar as the proof of Corollary 4.16.

Remark   [469, Theorem] directly follows from Corollary 4.18 (The necessity part is evident). Actually, Corollary 4.18 still holds if "maximal" is replaced by "c-maximal." Therefore, similar results e.g., [292, Theorem 3.2], [90, Theorem 4.2], and [291, Theorem 1] can be extended by using the notion of $\hat{\theta}$-pairs.

Theorem 4.19   (Chen, Guo [64]). *Let G be a group and $\mathfrak{F}$ a saturated formation containing $\mathfrak{U}$. Suppose that S is a normal subgroup of G. If for any c-maximal subgroup M of G not containing $S^{\mathfrak{F}}$, the $\mathfrak{F}$-residual of S, there exists a $\hat{\theta}$-pair (C, D) for M such that C/D is cyclic, and for any prime p with $|G : M|_p = p^n$ (where $n \geq 2$ is an integer), $C/C \cap M_G$ satisfies that*
   (i) $|C/C \cap M_G| \nmid p(p^{n-1} - 1)$ when $p \mid |C/C \cap M_G|$;
   (ii) $|C/C \cap M_G| \nmid p^n - 1$ when $p \nmid |C/C \cap M_G|$,
   then either $S \in \mathfrak{F}$ or G has a homomorphic image isomorphic to S_4 and S has a homomorphic image isomorphic to S_4 or A_4.

Proof   Assume that the result is false and let (G, S) be a counterexample for which $|G| + |S|$ is minimal.

First, we claim that for any c-maximal subgroup M of G not containing $S^{\mathfrak{F}}$, there exists a $\hat{\theta}$-completion for M, we still denote it by C, such that C/M_G is cyclic, and for any prime p with $|G:M|_p = p^n$ $(n \geq 2)$, C/M_G satisfies that

(i) $|C/M_G| \nmid p\,(p^{n-1} - 1)$ when $p \mid |C/M_G|$;

(ii) $|C/M_G| \nmid p^n - 1$ when $p \nmid |C/M_G|$.

In fact, since $C/D = C/(C \cap M)_G$ is cyclic, CM_G/M_G is cyclic. With the similar discussion as the proof of Lemma 4.5, either CM_G is a $\hat{\theta}$-completion for M or CM_G/M_G contains a minimal normal subgroup L/M_G of G/M_G such that L is a $\hat{\theta}$-completion for M. In the former case, for any prime p with $|G:M|_p = p^n$ $(n \geq 2)$, by hypothesis, it is obvious that

(i) $|CM_G/M_G| \nmid p\,(p^{n-1} - 1)$ when $p \mid |CM_G/M_G|$;

(ii) $|CM_G/M_G| \nmid p^n - 1$ when $p \nmid |CM_G/M_G|$.

In the latter case, since clearly, $G = ML$ and $(L \cap M)/M_G = 1$, we have that $|G:M| = |L:M_G|$. As L/M_G is cyclic of prime order, for any prime p dividing $|G:M|$, $|G:M| = |L:M_G| = p$. Hence the claim holds.

As $S \notin \mathfrak{F}$, $S^{\mathfrak{F}} > 1$. If G is a simple group, then $S^{\mathfrak{F}} = S = G$. If every maximal subgroup of G has a prime index in G, then $G \in \mathfrak{U}$, and so $S \in \mathfrak{U} \subseteq \mathfrak{F}$, which is impossible. Hence G has at least a c-maximal subgroup M such that there exists a $\hat{\theta}$-completion C for M with C is cyclic. As $C < G$, there exists a subgroup E of G such that C is maximal in E and $E_G \nleq M$. This induces that $E = G$, and so G has a cyclic maximal subgroup C. Then by [248, Chap. IV, Satz 7.4], G is solvable. Therefore, G is cyclic of prime order, and so is S, a contradiction. Thus G is not a simple group.

For any nontrivial normal subgroup N of G, we now show that $(G/N, SN/N)$ satisfies the hypothesis. If M/N is a c-maximal subgroup of G/N not containing $(SN/N)^{\mathfrak{F}}$, then since clearly, $(SN/N)^{\mathfrak{F}} \leq S^{\mathfrak{F}}N/N$, M is a c-maximal subgroup of G not containing $S^{\mathfrak{F}}$. By the above claim, M has a $\hat{\theta}$-completion C such that C/M_G is cyclic, and for any prime p with $|G:M|_p = p^n$ $(n \geq 2)$, we have that

(i) $|C/M_G| \nmid p\,(p^{n-1} - 1)$ when $p \mid |C/M_G|$;

(ii) $|C/M_G| \nmid p^n - 1$ when $p \nmid |C/M_G|$.

Now, it follows from Lemma 4.4 that M/N has a $\hat{\theta}$-completion C/N such that $(C/N)/(M_G/N)$ is cyclic, and for any prime p with $|(G/N):(M/N)|_p = p^n$ $(n \geq 2)$, we also have that

(i) $|(C/N)/(M_G/N)| \nmid p\,(p^{n-1} - 1)$ when $p \mid |(C/N)/(M_G/N)|$;

(ii) $|(C/N)/(M_G/N)| \nmid p^n - 1$ when $p \nmid |(C/N)/(M_G/N)|$.

By the choice of (G, S), either $SN/N \in \mathfrak{F}$ or G/N has a homomorphic image isomorphic to S_4 and SN/N has a homomorphic image isomorphic to S_4 or A_4. But the latter case is contrary to our assumption. Hence $SN/N \in \mathfrak{F}$. We may, therefore, assume that N is the unique minimal normal subgroup of G. Then clearly, $N \leq S$ and $S/N \in \mathfrak{F}$. Since $1 < S^{\mathfrak{F}} \trianglelefteq G$, we have that $S^{\mathfrak{F}} = N$.

If $N \leq \Phi(G)$, then $S \in \mathfrak{F}$ by Lemma 4.11, a contradiction. Hence G has at least one maximal subgroup not containing N. Suppose that for any maximal subgroup T not containing N, T has a prime index in G. Then $N \leq L(G)$, and so N is supersolvable by Lemma 4.12. Evidently, $G = TN$ and $N \cap T = 1$. Since $|G:T|$

is a prime, N is cyclic, and thereby $S \in \mathfrak{F}$ by Chap. 3, Lemma 2.30, a contradiction. Therefore, there exists a c-maximal subgroup M of G not containing N. Obviously $M_G = 1$. By the above claim, M has a $\hat{\theta}$-completion C such that C is cyclic, and for any prime p with $|G : M|_p = p^n$ $(n \geq 2)$, C satisfies that

(i) $|C| \nmid p(p^{n-1} - 1)$ when $p \mid |C|$;

(ii) $|C| \nmid p^n - 1$ when $p \nmid |C|$.

If $C = G$, then it is trivial. So G has a subgroup E such that C is maximal in E and $E_G \nleq M$. This induces that $N \leq E$ by the uniqueness of N. If $N \leq C$, then N is cyclic. By Chap. 3, Lemma 2.30, $S \in \mathfrak{F}$, which contradicts our assumption. Hence $N \nleq C$, and so $E = CN$ for C is maximal in E. It follows from [248, Chap. IV, Satz 7.4] that E is solvable. Consequently, N is an elementary abelian p-subgroup for some prime p. This implies that $|N| = |G : M| = p^n$ and $C_G(N) = N$ by Chap. 1, Proposition 1.4. As clearly, N is noncyclic, $n \geq 2$.

First, we assume that $C \trianglelefteq E$. Then $|E : C| = |N : C \cap N| = p$ by the maximality of C. Since $C \cap N$ is cyclic and $C \cap N > 1$, we have that $|C \cap N| = p$, and so $|N| = p^2$ and $n = 2$. Therefore, $|C| \nmid p(p - 1)$ for $p \mid |C|$. If C is not a p-group, then C has a nontrivial Hall p'-subgroup $C_{p'} \trianglelefteq E$. It follows that $C_{p'} \leq C_E(N) = N$, a contradiction. Hence C is a p-group, and so is E. Note that $E/N = E/C_E(N) \lesssim Aut(Z_p \times Z_p) \cong GL(2, p)$, we have that $|E| \leq p^3$. If $|E| = p^2$, then $E = N$ and $|C| = p$, a contradiction. Thus $|E| = p^3$. Suppose that $p > 2$. Then E is a group of the type (1), (2) or (3) in Lemma 4.13. Since N is noncyclic, E is noncyclic, and thereby E is not a group of the type (1). If E is a group of the type (2), then E is abelian. As $C_E(N) = N$, we have that $E = N$, which is absurd. Now let $E = \langle a, b \rangle$ be a group of the type (3) satisfying that $a^{p^2} = 1, b^p = 1$, and $bab^{-1} = a^{1+p}$. In view of that $[a, b] = a^{-1}b^{-1}ab = (a^p)^{-1} \in Z(E)$, it is easy to calculate that $\Omega_1(E) = \langle g \in E \mid g^p = 1 \rangle = \langle a^p \rangle \times \langle b \rangle = N$. Since $E = N(M \cap E)$ and $N \cap M = 1$, $|M \cap E| = p$. This implies that $M \cap E \leq \Omega_1(E) = N$, a contradiction. Hence $p = 2$, and E is a group of the type (1), (2), (4) or (5) in Lemma 4.13. By arguing similarly as above, we can see that E is not a group of the type (1) or (2). Thus E is either a quaternion group or a dihedral group of order 8. Note that $G \simeq G/M_G \lesssim S_4$ for $|G : M| = |N| = 4$. Since G has a subgroup of order 8, $G = S_4$. It follows from the fact $S \notin \mathfrak{U}$ that $S = S_4$ or A_4, which contradicts the choice of (G, S).

Now consider that $C \ntrianglelefteq E$. Then $N_E(C) = C$. Hence for any element $g \in E \backslash C$, we have that $C \neq C^g$, and so $E = \langle C, C^g \rangle$. Since C and C^g are both cyclic, $E \leq C_E(C \cap C^g)$. Therefore, $C \cap C^g \leq Z(E) \leq C_E(N) = N$. As $E = CN$ and N is abelian, $C \cap N \trianglelefteq E$. Then $(C/C \cap N) \cap (C/C \cap N)^{\bar{g}} = 1$ for any element $\bar{g} \in (E/C \cap N) \backslash (C/C \cap N)$. This shows that $E/C \cap N$ is a Frobenius group with complement $C/C \cap N$. Let $H/C \cap N$ be the Frobenius kernel of $E/C \cap N$ such that

$$E/C \cap N = (C/C \cap N)(H/C \cap N) \text{ and } (C/C \cap N) \cap (H/C \cap N) = 1.$$

By [248, Chap. V, Satz 8.16], either $N \leq H$ or $H \leq N$. But since $|H| = |N|$, we have that $H = N$. It follows that $|C/C \cap N| \mid (|N/C \cap N| - 1)$ and $N/C \cap N \in Syl_p(E/C \cap N)$ by [248, Chap. V, satz 8.3]. Hence $N \in Syl_p(E)$, and thereby $C \cap N \in Syl_p(C)$. Obviously $|C \cap N| \leq p$. If $p \mid |C|$, then $|C \cap N| = p$. Since $|N| = |G : M| = p^n$, we have that $|C/C \cap N| \mid (p^{n-1} - 1)$, and so $|C| \mid p(p^{n-1} - 1)$, a contradiction. If $p \nmid |C|$, then $|C \cap N| = 1$. It follows that $|C| \mid p^n - 1$, also a contradiction. The theorem is thus proved.

Corollary 4.20 *Let G be an S_4-free group and $\mathfrak{F}$ a saturated formation containing $\mathfrak{U}$. Then $G \in \mathfrak{F}$ if and only if for any c-maximal subgroup M of G which is $\mathfrak{F}$-abnormal, there exists a $\hat{\theta}$-pair (C, D) for M such that C/D is cyclic and $|C : C \cap M_G| \geq |G : M|$.*

Proof The necessity part is obvious. Now assume that $|C : C \cap M_G| \geq |G : M|$. If for some prime p with $|G : M|_p = p^n$ $(n \geq 2)$,

  either $|C/C \cap M_G| \mid p(p^{n-1} - 1)$

  or

  $|C/C \cap M_G| \mid p^n - 1$,

then $|C/C \cap M_G| < p^n \leq |G : M|$, which is impossible. Take $S = G$ and Theorem 4.19 applies.

Remark [33, Theorem 1], [292, Theorem 3.5], and [92, Theorem 10] can be directly generalized by Corollarys 4.16 and 4.20 may also be required).

Finally, we focus attention on the case in which assumptions imposed on θ-pairs without any restriction.

Theorem 4.21 (Chen, Guo [64]). *Let G be a group and $\mathfrak{F} = LF(f)$ a saturated formation, where f is an integrated formation function (that is, $f(p) \subseteq \mathfrak{F}$ for all primes). Then $G \in \mathfrak{F}$ if and only if for any $\mathfrak{F}$-abnormal maximal subgroup M of G, there exists a θ-pair (C, D) for M such that C/D is f-central (in sense of Definition 4.14) in G.*

Proof The necessity is clear. We need only to prove the sufficiency. Assume that it is false and let G be a counterexample with minimal order. Note that by Definition 4.14, if C/D is f-central in G and A/B is a section of C/D with $B \trianglelefteq G$, then A/B is f-central in G. Analogously as the proof of Lemma 4.5, we can obtain that for any $\mathfrak{F}$-abnormal maximal subgroup M of G, M has a θ-completion C such that C/M_G is f-central in G. Let N be a minimal normal subgroup of G. It is easy to check that G/N also satisfies the hypothesis of the theorem. Hence by induction, $G/N \in \mathfrak{F}$. Then since $\mathfrak{F}$ is a saturated formation, N is the unique minimal normal subgroup of G and $N \not\leq \Phi(G)$. Thus there exists a maximal subgroup M of G such that $G = MN$ and $M_G = 1$. If M is $\mathfrak{F}$-normal, then $G \in \mathfrak{F}$, contrary to our assumption. We may, therefore, assume that M is $\mathfrak{F}$-abnormal. It follows that M has a θ-completion C such that $C/1$ is f-central in G. Hence $G^{f(p)} \leq C_G(C)$ for each prime divisor p of $|C|$ by definition. If $G^{f(p)} = 1$, then $G \in f(p) \subseteq \mathfrak{F}$ for f is integrated, which is absurd. Now assume that $G^{f(p)} > 1$. Then $N \leq G^{f(p)} \leq C_G(C)$, and so $C \leq C_G(N)$. As $M_G = 1$ and $C > 1$, by [89, Chap. A, Theorem 15.2], N is elementary abelian and $N = C_G(N)$. This induces that p is the unique prime dividing $|N|$. Since $C \leq N \leq G^{f(p)} \leq C_G(C)$, we have that

$$1 < C \leq Z(G^{f(p)}) \trianglelefteq G.$$

Hence $N \leq Z(G^{f(p)})$, and so $G^{f(p)} \leq C_G(N) = N$. This implies that $G^{f(p)} = N$. Therefore, $G/C_G(N) = G/N \in f(p)$. But since $G/N \in \mathfrak{F}$, we obtain that $G \in \mathfrak{F}$. The final contradiction completes the proof.

Remark [25, Theorem 1], and [291, Theorem 8] can be generalized by Theorem 4.21.

Corollary 4.22 *Let G be a group. Suppose that $\mathfrak{F} = LF(f)$ and $\mathfrak{H} = LF(h)$ are saturated formations such that $\mathfrak{F} \subseteq \mathfrak{H}$, where f and h are the canonical local satellites of $\mathfrak{F}$ or the smallest local definitions (see p. 250 below or [89, IV, Definition 3.9]). Then $G \in \mathfrak{H}$ if and only if for any $\mathfrak{H}$-abnormal maximal subgroup M of G, there exists a θ-pair (C, D) for M such that C/D is f-central (in sense of Definition 4.14) in G.*

Proof By [89, Chap. IV, Proposition 3.11] or [135, 3.1.18], $f(p) \subseteq g(p)$ for all $p \in \mathbb{P}$. We may prove the corollary similarly as Theorem 4.21.

Corollary 4.23 *Let G be a group and $\mathfrak{F}$ a saturated formation containing $\mathfrak{N}$. Then $G \in \mathfrak{F}$ if and only if for any $\mathfrak{F}$-abnormal maximal subgroup M of G, there exists a θ-pair (C, D) for M such that $C/D \leq Z(G/D)$.*

Proof In fact, the smallest local definition f of $\mathfrak{N}$ can be defined as follows: $f(p) = 1$ for each prime p. Therefore, the result follows from Corollary 4.22.

Remark [473, Theorem 3.9] can be directly generalized by Corollary 4.23.

Corollary 4.24 *Let G be a group and $\mathfrak{F}$ a saturated formation containing all p-nilpotent groups. Then $G \in \mathfrak{F}$ if and only if for any $\mathfrak{F}$-abnormal maximal subgroup M of G, there exists a θ-pair (C, D) for M such that $O^p(G/D) \leq C_{G/D}(C/D)$.*

Proof Let $\mathfrak{H}$ be the formation of all p-nilpotent groups. Then $\mathfrak{H}$ has the canonical local satellites f such that $f(p) = \mathfrak{S}_p; f(q) = \mathfrak{H}$ for each prime $q \neq p$. As

$$(G/D)/O^p(G/D) \in \mathfrak{S}_p \subseteq \mathfrak{H},$$

we have that

$$(G/D)^{\mathfrak{H}} \leq (G/D)^{\mathfrak{S}_p} \leq O^p(G/D).$$

Since $O^p(G/D) \leq C_{G/D}(C/D)$, C/D is f-central in G, and so Corollary 4.22 applies.

Remark [321, Theorem 4.7] directly follows from Corollary 4.24.

4.5 Additional Information and Some Problems

I. The concept of the Σ-embedded subgroup has more close connections with the theory of CAP-embedded subgroups. Recall that a subgroup H of G is called a CAP-subgroup of G if H either covers or avoids each pair (K, H), where H/K is a chief factor of G. A subgroup H is called a partial CAP-subgroup [36, 37] or a semi cover-avoiding subgroup [226, 100] of G if H either covers or avoids each pair (K, H), where H/K is a factor of some fixed chief series of G. By using the CAP-subgroups and the semi cover-avoiding subgroups, people have obtained many interesting results (see, for example, [36, 37, 226, 227, 100, 101, 224, 326, 297, 433]). It is clear that every CAP-subgroup of G is Σ-embedded in G for each chief series $\Sigma = \{1 = G_0 < G_1 < \ldots < G_{t-1} < G_t = G\}$ of G. On the other hand, every semi cover-avoiding subgroup of G is Σ-embedded in G for some chief series Σ of G.

Since every subgroup of a supersoluble group is a CAP-subgroup of G, one could not say any thing about the classes of groups properly contained in the class $\mathfrak{U}$ of all supersoluble groups by using CAP-subgroups or partial CAP-subgroups. In fact, the concept of Σ-embedded subgroup is a more wide universe because many important classes of groups, for which there are not any description on the base of the concept of CAP-subgroup (for example, the class of the nilpotent groups), may be characterized in the terms of Σ-embedded subgroups. Finally, note that many classification results connected with quasinormal subgroups, CAP-subgroups and partial CAP-subgroups (semi cover-avoiding subgroups) can be nontrivially generalized on the base of the Σ-embedded subgroups.

II. We show that every quasinormal subgroup H of G is a CAP-subgroup of G. Indeed, let K/L be any chief factor of G. Assume that H does not avoid K/L, that is, $H \cap L \neq H \cap K$. We show that H covers K/L, that is, $K = L(H \cap K)$. We may assume that H_G does not cover K/L, so that $H_G \leq L$. If $H_G \neq 1$, then, by induction, H/H_G covers $(K/H_G)/(L/H_G)$, which implies that H covers K/L. Finally, assume that $H_G = 1$. Then $H \leq Z_\infty(G)$ by [32, Corollary 1.5.6]. If $(K/L) \cap Z_\infty(G/L) = 1$, then $(HL/L) \cap (K/L) = 1$ and consequently $H \cap L = H \cap K$, a contradiction. Thus, $K/L \leq Z_\infty(G/L)$ and thereby K/L is of prime order. This implies that H covers K/L. We can also see that quasinormal subgroup H of G has another interesting cover-avoidance property: For any maximal pair (L, K) of G, H either covers (L, K) or avoids (L, K). Thus CAP-subgroups and $CAMP$-subgroups are different generalizations of quasinormality. Nevertheless, we do not know the answer to the following.

Problem 5.1 Is there a group G some $CAMP$-subgroup of which is not a CAP-subgroup of G?

III. In the paper [299], the following generalization of quasinormality was considered.

Definition 5.2 Let p be a prime. A subgroup A of a group G is called a *weak* CAP_p-*subgroup* of G if G has a composition series $1 = G_0 < G_1 < \ldots < G_n = G$ such that A either covers or avoids every maximal pair (K, H) of G such that $G_{i-1} \leq K < H \leq G_i$ for some i, where p divides $|G_i/G_{i-1}|$ and H is not a p-soluble group.

The following theorem was proved.

Theorem 5.3 (Liu, Guo, Kovaleva, Skiba [299]). *Let G be a group and p a prime. The following assertions are equivalent:*

(1) G is p-soluble;

(2) every subgroup of G is a weak CAP_p-subgroup of G;

(3) *every maximal subgroup of G is a weak CAP_p-subgroup of G;*

(4) *every 2-maximal subgroup of G is a weak CAP_p-subgroup of G;*

(5) *every Sylow p-subgroup of G is a weak CAP_p-subgroup of G;*

(6) *either G is a primary group, i.e., a group whose order is a power of some prime, or G contains two p-soluble maximal subgroups M_1 and M_2 such that $(|G : M_1|, |G : M_2|) = r^a q^b$ for some primes r and q and some $a, b \in 0 \cap \mathbb{N}$ and M_1 and M_2 are weak CAP_p-subgroups of G;*

(7) *every nonsupersoluble Schmidt subgroup of G is a weak CAP_p-subgroup of G.*

IV. Despite of the results of this chapter and many other known results about n-maximal subgroups (see, for example, [286, 370, 304, 306, 35, 267]), the fundamental work of Mann [310] still retains its value. In the research of the structure of the groups whose n-maximal subgroups are subnormal, Mann proved that if all n-maximal subgroups of a soluble group G are subnormal and $|\pi(G)| \geq n + 1$, then G is nilpotent; but if $|\pi(G)| \geq n - 1$, then G is ϕ-dispersive for some ordering ϕ of the set of all primes. Finally, in the case $|\pi(G)| = n$, Mann described G completely.

In the paper [271], a supersoluble analog of Mann's theory [310] was developed. In particular, the following theorems were proved.

Theorem 5.4 (V.A. Kovaleva, A.N. Skiba [271]). If every n-maximal subgroup of a soluble group G is $\mathfrak{U}$-subnormal in G and $|\pi(G)| \geq n + 2$, then G is supersoluble.

Theorem 5.5 (V.A. Kovaleva, A.N. Skiba [271]). Let G be a soluble group with $|\pi(G)| \geq n + 1$. Then all n-maximal subgroups of G are $\mathfrak{U}$-subnormal in G if and only if G is a group of one of the following types:

I. G is supersoluble.

II. $G = A \rtimes B$, where $A = G^{\mathfrak{U}}$ and B are Hall subgroups of G, while G is Ore dispersive and satisfies the following:

(1) A is either of the form $N_1 \times \ldots \times N_t$, where each N_i is a minimal normal subgroup of G, which is a Sylow subgroup of G, for $i = 1, \ldots, t$, or a Sylow p-subgroup of G of exponent p for some prime p and the commutator subgroup, the Frattini subgroup, and the center of A coincide, every chief factor of G below $\Phi(G)$ is cyclic, while $A/\Phi(A)$ is a noncyclic chief factor of G;

(2) every n-maximal subgroup H of G is supersoluble and induces on the Sylow p-subgroup of A an automorphism group which is an extension of some p-group by abelian group of exponent dividing $p - 1$ for every prime divisor p of the order of A.

Theorem 5.6 (V.A. Kovaleva, A.N. Skiba [271]). *If every n-maximal subgroup of a soluble group G is $\mathfrak{U}$-subnormal in G and $|\pi(G)| \geq n$, then G is ϕ-dispersive for some ordering ϕ of the set of all primes.*

The proofs of Theorems 5.4, 5.5, and 5.6 rest on the properties of groups whose all 2-maximal subgroups are $\mathfrak{U}$-subnormal. The following theorem gives the description of such groups.

Theorem 5.7 (V.A. Kovaleva, A.N. Skiba [271]). *Every 2-maximal subgroup of a group G is $\mathfrak{U}$-subnormal in G if and only if either G is supersoluble or G is a minimal nonsupersoluble group and $G^{\mathfrak{U}}$ is a minimal normal subgroup of G.*

Note that the restrictions on $|\pi(G)|$ in Theorems 5.4, 5.5, and 5.6 cannot be weakened. For Theorem 5.4 this follows from the description of minimal nonsupersoluble groups (see [359, Chap. VI, Theorems 26.3 and 26.5]). This also applies to Theorem 5.6, as we can see from the example of the symmetric group of degree 4. Now take primes p, q, and r with $p > q > r$ such that r divides $q - 1$, while both q and r divide $p - 1$. Take a non-abelian group $Q \rtimes R$ of order qr, and a simple faithful $F_p(QR)$-module P_1. Finally, put $G = (P_1 \rtimes (Q \rtimes R)) \times P_2$, where P_2 is a group of order p. Then the supersoluble residual $G^{\mathfrak{U}} = P_1$ of G is not a Hall subgroup of G. Furthermore, it is easy to verify that all 3-maximal subgroups of G are $\mathfrak{U}$-subnormal in G.

Theorems 5.4, 5.5, and 5.6 are the motivation for the following

Definition 5.8 We say that a group G is a group of *Spencer $\mathfrak{U}$-height $h_{\mathfrak{U}}(G) = n$* if every maximal chain of G of length n contains a proper K-$\mathfrak{U}$-subnormal entry and there exists at least one maximal chain of G of length $n - 1$ which contains no proper K-$\mathfrak{U}$-subnormal entry.

In view of Theorem 3.5, it seems natural to ask:

Problem 5.9 What is the precise structure of a nonnilpotent group G with Spencer $\mathfrak{U}$-height $h_{\mathfrak{U}}(G) \leq 3$?

Problem 5.9 maybe is too difficult since now we do not know the answer even to the following special case.

Problem 5.10 What is the precise structure of a nonnilpotent group G under the condition that every 3-maximal or every 2-maximal subgroup of G is K-$\mathfrak{U}$-subnormal in G?

V. Theorem 2.5 and 2.13 are a motivation for the following:

Problem 5.11 What is the precise structure of a nonnilpotent group G in which all 3-maximal subgroups are $F(G)$-permutable with all maximal subgroups?

Problem 5.12 What is the structure of a group G in which all 3-maximal subgroups are $F(G)$-permutable with all 2-maximal subgroups?

VI. It is clear that the groups in which every two maximal subgroups are permutable are nilpotent. Lemma 2.21 shows that if every two 2-maximal subgroups of G are permutable then G is a nilpotent or biprimary group, that is, its order is divisible by precisely two primes. Theorem 2.28 shows that if every two 3-maximal subgroups of G are permutable then G is a nilpotent or biprimary group, or its order is divisible by precisely three primes. In connection with these observations, the following open question seems fully natural.

Problem 5.13 Assume that G is a nonnilpotent group and every two of its n-maximal subgroups are permutable $(n > 3)$. Is it true that $|\pi(G)| \leq n$?

VII. Note that the formation analogy of the results of paper [310] were obtained in the paper [272].

VIII. Let A be a proper subgroup of G. We call any chief factor H/A_G of G/H_G a *G-boundary factor* or simply *boundary factor* of A. For any G-boundary factor H/A_G of A, we call the subgroup $(A \cap H)/A_G$ of G/A_G a *G-trace* of A or simply a *trace* of A (see [204]).

In [204], the following theorem is obtained.

Theorem 5.14 (Guo, Skiba [204]). *A nonidentity normal subgroup E is contained $Z_{\mathfrak{U}}(G)$ if and only if for every meet-irreducible subgroup X of E and some G-boundary factor H/X_G of X, where $H \leq E$, the trace $(H \cap X)/X_G$ permutes with some Sylow q-subgroup of G/X_G for every prime q dividing $|G/X_G|$.*

Corollary 5.15 *G is supersoluble if and only if for every meet-irreducible subgroup X of G and some G-boundary factor H/X_G of X, the trace $(H \cap X)/X_G$ permutes with some Sylow q-subgroup of G/X_G for every prime q dividing $|G/X_G|$.*

Problem 5.16 (Guo, Skiba). Is it true that G is p-soluble if and only if every maximal chain of G of length 2 contains a proper subgroup M of G such that either some G-trace of M is subnormal or every G-boundary factor of M is a p'-group?

Problem 5.17 (Guo, Skiba). Suppose that every maximal subgroup of G has a supersoluble G-trace. Is it true that G is soluble?

IX. In [205], Guo and Skiba introduced the following notions.

Let θ be a subgroup functor (see Chap. 3, Section 3.1), and A be a subgroup of a group G. If G has a normal series G

$$\Gamma : 1 = G_0 \leq G_1 \leq \cdots \leq G_t = G$$

such that, for each $i = 1, \ldots, t$, we have

$$(A \cap G_i)G_{i-1}/G_{i-1} \in \theta(G/G_{i-1}),$$

then we say that A Γ*-graduatedly has property* θ (or simply A *graduatedly has property* θ) in G.

For example, we say that a subgroup A of G is: Γ-graduatedly permutable with a Sylow p-subgroup of G, Γ-graduatedly subnormal in G, Γ-graduatedly nilpotent provided, for each $i = 1, \ldots, t$, we respectively have: $(A \cap G_i)G_{i-1}/G_{i-1}$ permutes with a Sylow p-subgroup of G/G_{i-1}; $(A \cap G_i)G_{i-1}/G_{i-1}$ is subnormal in G/G_{i-1}; $(A \cap G_i)G_{i-1}/G_{i-1}$ is nilpotent.

Theorem 5.18 (Guo, Skiba [205]). *G is soluble if and only if some maximal subgroup M of G is graduatedly nilpotent and a Sylow 2-subgroup of M is graduatedly abelian in G.*

Problem 5.19 (Guo, Skiba). Suppose that every maximal subgroup of G is graduatedly supersoluble in G. Is it true then that G is soluble?

Problem 5.20 (Guo, Skiba). Let $\theta(X) = \{A \leq X | A$ be a nilpotent group with Sylow 2-subgroups of the class 2$\}$ for any group X. Suppose that some maximal subgroup of G is a θ-subgroup of G. Is it true then that G is soluble?

Problem 5.21 (Guo, Skiba). Let $\theta(X) = \{A \leq X \mid |X : N_X(A)|$ be a prime or a square of prime$\}$ for any group X. Suppose that every maximal chain of G of length n, for some fixed $1 < n \leq 3$, contains a proper subgroup of G which graduatedly has property θ in G. Is it true then that G is soluble?

Chapter 5
Formations and Fitting Classes

5.1 Generated ω-Composition and ω-Local Formations

We use $R(G)$ to denote the soluble radical of G, that is, $R(G) = G_{\mathfrak{S}}$ is the largest soluble normal subgroup of G. $C^p(G)$ denotes the intersection of the centralizers of all abelian p-chief factors of a group G ($C^p(G) = G$ if G has no such chief factors). For any collection $\mathfrak{X}$ of groups, we denote by $C(\mathfrak{X})$ the class of all simple groups A such that $A \simeq H/K$ for some composition factor H/K of a group $G \in \mathfrak{X}$. Com$(\mathcal{X})$ denotes the class of all groups L such that L is isomorphic to some abelian composition factors of some group in $\mathfrak{X}$. Let $\emptyset \neq \omega \subseteq \mathbb{P}$. $O_\omega(G)$ denotes the maximal normal ω-subgroup of G.

We say that f is a generalized formation function if f is a function of the form

$$f : \omega \cup \{\omega'\} \to \{\text{formation of groups}\}. \tag{5.1}$$

For a generalized formation function f, we put

$$CF_\omega(f) = \{G \text{ is a group} \mid G/R(G) \cap O_\omega(G) \in f(\omega')$$

and

$$G/C^p(G) \in f(p) \text{ for any prime } p \in \omega \cap \pi(\text{Com}(G))\}$$

and we use $LF_\omega(f)$ to denote the class

$$(G \text{ is a group} \mid G/O_\omega(G) \in f(\omega') \text{ and } G/O_{p'p}(G) \in f(p) \text{ for all } p \in \omega \cap \pi(G)).$$

It is not difficult to show that for any generalized formation function f, both classes $CF_\omega(f)$ and $LF_\omega(f)$ are nonempty formations.

Following [389], a formation $\mathfrak{F}$ is said to be a ω-*composition formation* if $\mathfrak{F} = CF_\omega(f)$ for some formation function f of the form (5.1). In this case, f is called a ω-*composition satellite* of $\mathfrak{F}$. If $f(a) \subseteq \mathfrak{F}$ for all $a \in \omega \cup \{\omega'\}$, then the satellite f is called inner or integrated.

If $\mathfrak{F} = LF_\omega(f)$, then $\mathfrak{F}$ is said to be a ω-*local formation* and f is called a ω-*local satellite* of $\mathfrak{F}$. If $f(a) \subseteq \mathfrak{F}$ for all $a \in \omega \cup \{\omega'\}$, then the satellite f is called inner or integrated.

© Springer-Verlag Berlin Heidelberg 2015
W. Guo, *Structure Theory for Canonical Classes of Finite Groups*,
DOI 10.1007/978-3-662-45747-4_5

It is easy to see that a ω-local formation is a ω-composition formation; and if $\omega = \mathbb{P}$, then a ω-local (ω-composition) formation is a local (composition) formation.

If $\omega = p$, then a ω-local formation (ω-composition formation) is called a p-local formation (p-composition formation).

It is easy to see that a ω-local formation (ω-composition formation) is p-local formation (p-composition formation), for any $p \in \omega$.

For two generalized formation functions f and h, we write $f \le h$ if $f(a) \subseteq h(a)$ for all $a \in \omega \cup \{\omega'\}$.

If f is a ω-composition satellite of a ω-composition formation $\mathfrak{F}$ such that $f \le h$ for all ω-composition satellites h of $\mathfrak{F}$, then we call f the smallest ω-composition satellite of $\mathfrak{F}$.

Similarly if f is a ω-local satellite of a ω-local formation $\mathfrak{F}$ such that $f \le h$ for all ω-local satellites h of $\mathfrak{F}$, then we call f the smallest ω-local satellite of $\mathfrak{F}$.

We write $c_\omega \mathrm{form}\mathfrak{X}$ to denote the intersection of all ω-composition formations containing a given set $\mathfrak{X}$ of groups. If G is a group, then we write $c_\omega \mathrm{form}G$ instead of $c_\omega \mathrm{form}\{G\}$.

A formation $\mathfrak{F}$ is said to be a one-generated ω-composition formation if for some group G, we have $\mathfrak{F} = c_\omega \mathrm{form}G$.

Similarly, we write $l_\omega \mathrm{form}\mathfrak{X}$ to denote the intersection of all ω-local formations containing a given set $\mathfrak{X}$ of groups. If G is a group, then we write $l_\omega \mathrm{form}G$ instead of $l_\omega \mathrm{form}\{G\}$. In particular, $l\mathrm{form}\mathfrak{X}$ is the intersection of all local formations containing $\mathfrak{X}$.

A formation $\mathfrak{F}$ is said to be a one-generated ω-local formation if $\mathfrak{F} = l_\omega \mathrm{form}G$ for some group G.

Lemma 1.1 *Let G be a group with $N_1, N_2, \cdots, N_t \trianglelefteq G$ such that $\cap_{i=1}^n N_i = 1$. If $O_p(G/N_i) = 1$, for all $i = 1, 2, \cdots, t$, then $O_p(G) = 1$*

Proof Assume that $O_p(G) \neq 1$. Then $O_p(G) \nsubseteq \cap_{i-1}^n N_i$. Hence there is an index i such that $O_p(G) \nsubseteq N_i$. This implies that $O_p(G/N_i) \neq 1$, a contradiction.

Lemma 1.2 *Let G be a group, p be a prime. Then $O_p(G/C^p(G)) = 1$.*

Proof This follows from Chap. 1, Lemma 2.15 and Lemma 1.1.

Lemma 1.3 *Let $O_p(G) = 1$ and $A = C_p \wr G = K \rtimes G$, where K is the base group of A and C_p a group of order p. Then $K = C^p(A)$.*

Proof Let

$$1 = K_0 < K_1 < \ldots < K_t = K, \qquad (5.2)$$

where K_i/K_{i-1} is a chief factor of A for all $i = 1, \ldots, t$. Let $C_i = C_A(K_i/K_{i-1})$ and $C = C_1 \cap \ldots \cap C_t$. Suppose that $D = C \cap G \neq 1$. Then D stabilizes series (5.2), so D is a p-group by Chap. 1, Lemma 2.13, which implies $D \le O_p(G) = 1$. This contradiction shows that $C = K$. The lemma is proved.

Lemma 1.4 *Let N be a normal subgroup of a group G. Then*

(1) $C^p(N) = C^p(G) \cap N$.

(2) $R(N) \cap O_\omega(N) = (R(G) \cap O_\omega(G)) \cap N.$

Proof

(1) This follows from Chap. 1, Propositions 2.16 and 2.17.
(2) This part is clear.

Lemma 1.5 ([89, IV, 1.14]). *If $F(G)M = G$, then $M \in \text{form}(G)$.*

Lemma 1.6 *Let G be a group and R a minimal normal subgroup of G. If R is an elementary abelian p-group, then $G \in \text{form}(Z_p \wr (G/R))$, where Z_p is a group of order p.*

Proof Let $T = R \wr (G/R) = K \rtimes (G/R)$, where K is the base group of the regular wreath product T. Then there is a subgroup $H \simeq G$ of T such that $T = KH$ by [248, I, (13.9)]. Moreover, $H \in \text{form}(T)$ by Lemma 1.5. On the other hand, T is a subdirect product of groups of the form $Z_p \wr (G/R)$ by [89, A, 18.2]. Hence $G \in \text{form}(Z_p \wr (G/R))$.

Lemma 1.7 *Let $\mathfrak{F} = CF_\omega(f)$ and $p \in \omega$. If $G/O_p(G) \in \mathfrak{F} \cap f(p)$, then $G \in \mathfrak{F}$.*

Proof Without loss of generality, we may assume that $O_p(G)$ is a minimal normal subgroup of G. Let $A = G/O_p(G)$ and $T = Z_p \wr A = K \rtimes A$, where K is the base group of the wreath product T. Since, by Lemma 1.6, $G \in \text{form}T$, we only need to show that $T \in \mathfrak{F}$. By Lemma 1.3, $K = C^p(T)$. Hence $T/C^p(T) \in f(p)$. Let $q \in \pi(\text{Com}(G)) \setminus \{p\}$. Then, clearly, $A/C^q(A) \simeq T/C^q(T)$ and $T/C^q(T) \in f(q)$ since $A \in \mathfrak{F}$. Finally, since $A \in \mathfrak{F}$ and evidently $A/O_\omega(A) \cap R(A) \simeq T/(O_\omega(T) \cap R(T))$, we obtain $T \in \mathfrak{F}$.

Lemma 1.8 *Let $\mathfrak{F} = c_\omega\text{form}(\mathfrak{X})$ and $\pi = \pi(\text{Com}(\mathfrak{F})) \cap \omega$. Then $\mathfrak{F}$ has a smallest ω-composition satellite f and f has the following properties:*

(1) $f(p) = \text{form}(G/C^p(G)|G \in \mathfrak{X})$, *for all* $p \in \pi$.
(2) $f(p) = \varnothing$, *for all* $p \in \omega \setminus \pi$.
(3) $f(\omega') = \text{form}(G/(O_\omega(G) \cap R(G))|G \in \mathfrak{X})$.
(4) *If* $\mathfrak{F} = CF_\omega(h)$, *then* $f(p) = \text{form}(A \mid A \in h(p) \cap \mathfrak{F}, O_p(A) = 1)$ *for all* $p \in \pi$
and

$$f(\omega') = \text{form}(A \mid A \in h(\omega') \cap \mathfrak{F} \text{ and } R(A) \cap O_\omega(A) = 1).$$

Proof (1)–(3). Let t be a satellite such that

$$t(a) = \begin{cases} \text{form}(G/C^p(G)|G \in \mathfrak{X}), & \text{if } a = p \in \pi \\ \varnothing, & \text{if } a = p \in \omega \setminus \pi \\ \text{form}(G/(O_\omega(G) \cap R(G))|G \in \mathfrak{X}), & \text{if } a = \omega'. \end{cases}$$

Let $\mathfrak{M} = CF_\omega(t)$. We will show that $\mathfrak{M} = \mathfrak{F}$. Clearly, $\mathfrak{X} \subseteq \mathfrak{M}$ and so $\mathfrak{F} \subseteq \mathfrak{M}$. On the other hand, if $\mathfrak{F} = CF_\omega(h)$ for some ω-composition satellite h, then by $\mathfrak{X} \subseteq \mathfrak{F}$, we have $t \leq h$. This leads to $\mathfrak{M} \subseteq \mathfrak{F}$, and therefore $\mathfrak{F} = \mathfrak{M}$. It is also clear that t is the smallest ω-composition satellite of the formation $\mathfrak{F}$.

(4) Let h be a ω-composition satellite of $\mathfrak{F}$ and t a satellite such that $t(p) =$ form$(A \mid A \in h(p) \cap \mathfrak{F}$ and $O_p(A) = 1)$, for all $p \in \pi$; $t(p) = \varnothing$, for all $p \in \omega \setminus \pi$ and $t(\omega') = $ form$(A \mid A \in h(\omega') \cap \mathfrak{F}$ and $R(A) \cap O_\omega(A) = 1)$. Then in view of (1)–(3) and by Lemma 1.2, we have $f \le t$. Now we show that $t \le f$. For this purpose, we let $p \in \pi$ and A be a group such that $A \in h(p) \cap \mathfrak{F}$ with $O_p(A) = 1$. Let $T = Z_p \wr A = K \rtimes A$, where K is the base group of the wreath product T. By Lemma 1.7, $T \in \mathfrak{F}$ and so $T/C^p(T) \in f(p)$. However, by Lemma 1.3, we see that $K = C^p(T)$. Hence $A \in f(p)$, and consequently $t(p) \subseteq f(p)$. Finally we note that if $A \in h(\omega') \cap \mathfrak{F}$ and $R(A) \cap O_\omega(A) = 1$, then $A \simeq A/(R(A) \cap O_\omega(A)) \in f(\omega')$, and so $t(\omega') \subseteq f(\omega')$. Thus $f = t$. The proof is completed.

Corollary 1.9 *Let f_i be the smallest ω-composition satellite of the formation $\mathfrak{F}_i$, $i = 1, 2$. Then $\mathfrak{F}_1 \subseteq \mathfrak{F}_2$ if and only if $f_1 \le f_2$.*

Lemma 1.10 *Let $\mathfrak{F} = l_\omega$form$G$ for some group G. Then $\mathfrak{F}$ has a smallest ω-local satellite f and f has the following properties:*

(1) $f(p) = $ form$(G/F_p(G))$, for all $p \in \pi(G) \cap \omega$;
(2) $f(p) = \varnothing$, if $p \in \omega \setminus \pi(G)$;
(3) If $\mathfrak{F} = LF_\omega\langle h \rangle$ for some ω-local satellite h, then

$$f(p) = \text{form}(A \mid A \in h(p) \cap \mathfrak{F} \text{ and } O_p(A) = 1)$$

for all $p \in \pi(G) \cap \omega$ and

$$f(\omega') = \text{form}(A \mid A \in h(\omega') \cap \mathfrak{F} \text{ and } O_\omega(A) = 1).$$

Proof See the proof of Lemma 1.8.

Lemma 1.11 ([362]). *A formation $\mathfrak{F}$ is a p-composition formation if and only if $\mathfrak{F}$ contains each group G with $G/\Phi(O_p(G)) \in \mathfrak{F}$*

5.2 The Criterion of ω-Compositively of the First Factor of One-Generated ω-Composition Product of Two Formations

It is known [96, 382] that if the product $\mathfrak{F} = \mathfrak{M}\mathfrak{H}$ of two formations is a one-generated saturated (composition) formation and $\mathfrak{F} \ne \mathfrak{H}$, then $\mathfrak{M}$ is saturated (composition) formation.

In the paper [389], the following problem was raised:

Problem 2.1 (see Problem 18 in [389]). Let $\mathfrak{F} = \mathfrak{M}\mathfrak{H}$ be one-generated ω-composition formation. Is it true that $\mathfrak{M}$ is a ω-composition formation if $\mathfrak{H} \ne \mathfrak{F}$?

The following theorem not only gives the positive answer to this problem but also a stronger result.

Theorem 2.2   (W. Guo, V.M. Sel'kin and K.P. Shum [162]). *Let* $\mathfrak{M}\mathfrak{H} \subseteq \mathfrak{F}$, *where* $\mathfrak{F}$ *is a one-generated ω-local formation and* $\mathfrak{M}$, $\mathfrak{H}$ *be two nonidentity formations. Suppose that* $\mathfrak{M}\mathfrak{H}$ *is a p-composition formation where* $p \in \omega$. *If* $\mathfrak{H} \neq \mathfrak{M}\mathfrak{H}$, *then* $\mathfrak{M}$ *is a p-local formation.*

In order to prove the theorem, we need the following results.

Lemma 2.3   (see [89, A; 18.2, 18.5] and [381, Lemma 3.1.9]). *Let* $G = A \wr B = K \rtimes B$, *where* $K = \prod_{b\in B} A_1^b$ *is the base group of the wreath product* G *and* A_1 *is the first copy of the group* A *in* K. *Then the following statements hold:*

(i) *If* L *is a minimal normal subgroup of* G, L_1 *is the projection of* L *into* A_1 *and* $L_1 \not\subseteq Z(A_1)$, *then* $L = \prod_{b\in B} (L \cap A_1)^b$.

(ii) *If* R *is a minimal normal subgroup of* A_1 *and* $R \not\subseteq Z(A_1)$, *then* $R_1 = \prod_{b\in B} R^b$ *is a minimal normal subgroup of* G.

(iii) $\mathrm{Soc}(G) \subseteq \prod_{b\in B} M^b$, *where* $M = \mathrm{Soc}(A_1)$.

(iv) *If* $L \trianglelefteq G$, $L \subseteq K$ *and* M *is the projection of* L *into* A_1, *then* $(A_1/M) \wr B$ *is a homomorphic image of the factor group* G/L.

Lemma 2.4   (see [322, Theorem 51.2] or [381, Lemma 3.1.5]). *Let* $A \in \mathrm{form}\,G$ *and* $\exp(A)$ *denotes the exponent of* A. *Then:*

1) $\exp(A) \le \exp(G)$;
2) *every chief factor of* A *is isomorphic to some chief factor of* G;
3) *if* $H \le A$, *then* $c(H/H^{\mathfrak{N}}) \le \max\{c(T/T^{\mathfrak{N}})|T \le G\}$.

Lemma 2.5   (see [381, Lemma 3.1.7].) *Let* A *be a group of prime order* p. *Then the nilpotent class of the group* $A \wr (A^n)$ *is at least* $n + 1$.

Lemma 2.6   (Guo, Shum [170, Lemma 3.16]). *Let* n *be a positive integer,* q *be a prime number and* $T = M \times \cdots \times M$ *be the direct product of* n *copies of a nonabelian group* M *such that* q *does not divide* $|M|$. *Also, let* F_q *be a field with* q *elements. Then there exists a simple* $F_q T$-module V *such that* $\dim_{F_q} V \ge 2^n$.

Proof Let $\overline{F}_q$ be the algebraic closure of F_q. Since M is nonabelian, there exists at least a simple $\overline{F}_q M$-module R with $(T : \overline{F}_q) \ge 2$. Let D be the outer tensor product (cf. [81], Sect. 43) of n copies of the module R. Then by [81, Exercises 2, p. 189], we know that D is a simple $\overline{F}_q(M)$-module and $(D : \overline{F}_q) \ge 2^n$. Now, by Exercises 8 on page 206 of [81] and the formula (12.20) on page 71 of [81], we see that there exists a simple $F_q[M^n]$-module V such that D is a direct summand of the module $V^{\overline{F}_q}$. This leads to $(V : F_q) \ge 2^n$. Thus, the lemma is proved.

Recall that a group G is said to be monolithic if G has a unique minimal normal subgroup N. In this case, N is called the monolith of G.

Lemma 2.7   *Let* $\mathfrak{F} = l_\omega \mathrm{form}\,G$ *be a one-generated ω-local formation and* $\mathfrak{M}\mathfrak{H} \subseteq \mathfrak{F}$, *where* $\mathfrak{M}$ *and* $\mathfrak{H}$ *are nonidentity formations. Then the following statements hold:*

(i) *Every simple group in $\mathfrak{M}$ is abelian;*

(ii) *If some group $B \in \mathfrak{H}$ has an exponent greater than $p^{|G|}$ for some prime p, then $|A| = p$ for all simple groups $A \in \mathfrak{M}$;*

(iii) *If there exists a simple group $A \in \mathfrak{M}$ with $|A| \notin \omega$, then $\mathfrak{H}$ is an abelian formation.*

Proof Let f be the smallest ω-local satellite of the formation $\mathfrak{F}$. Then by Lemma 1.10, we have

$$
f(a) = \begin{cases} \mathrm{form}(G/F_p(G)), & \text{if } a = p \in \omega \cap \pi(G), \\ \varnothing, & \text{if } a = p \in \omega \setminus \pi(G), \\ \mathrm{form}(G/O_\omega(G)), & \text{if } a = \omega'. \end{cases}
$$

Write $B = B_1 \times \ldots \times B_{|G|}$, where $B_1 \simeq \ldots \simeq B_{|G|}$ is nonidentity groups in $\mathfrak{H}$. We now proceed our proof as follows:

(i) Let A be a simple group in $\mathfrak{M}$, and let $D = A \wr B = K \rtimes B$, where K is the base group of the wreath product D. Then, it is clear that $D \in \mathfrak{M}\mathfrak{H}$. Hence $D \in \mathfrak{F}$. Assume that A is a nonabelian group. Then by Lemma 2.3, the group D is monolithic and its monolith is K. Assume that $\pi(K) \cap \omega = \varnothing$. Then $O_\omega(D) = 1$, and so

$$
D/O_\omega(D) \simeq D \in f(\omega') = \mathrm{form}(G/O_\omega(G)).
$$

But $|K| = |A|^{|B|^{|G|}} > |G|$, this contradicts Lemma 2.4. Hence $\pi(K) \cap \omega \neq \varnothing$. Let $q \in \pi(K) \cap \omega$. Then, evidently, $F_q(D) = 1$, and therefore

$$
D/F_q(D) \simeq D \in f(q) = \mathrm{form}(G/F_q(G)),
$$

which is a contradiction. Thus every simple group in $\mathfrak{M}$ must be abelian.

(ii) This part may be proved as (i), we omit the proof.

(iii) In view of (i), we have $|A| = q$ for some prime $q \notin \omega$. Assume that q divides the order $|B|$ of B. Let $D = A \wr B = K \rtimes B$. Clearly $O_\omega(D) = 1$, and so

$$
D \in f(\omega') = \mathrm{form}(G/O_\omega(G)).
$$

Let Q_i be a subgroup of order q in B_i, $i = 1, \cdots, |G|$. Also, let $T = A \wr (Q_1 \times \ldots \times Q_{|G|})$. Then, by Lemma 2.5, we see that the nilpotent class of T is at least $|G| + 1$. By [89, A; (18.2)], we have

$$
T \simeq M \leq D \in \mathrm{form}(G/O_\omega(G)),
$$

which contradicts Lemma 2.4. Hence we may assume that $q \nmid |B_i|$.

Suppose, on the contrary, that the formation $\mathfrak{H}$ is not abelian. Then we may assume that B_1 is a nonabelian group. In this case, by Lemma 2.6, there is a simple $\mathbb{F}_q D$-module V such that $\dim_{\mathbb{F}_q}(V) \geq 2^{|G|}$. Let $H = V \rtimes D$. Then $H \in \mathfrak{M}\mathfrak{H} \subseteq \mathfrak{F}$. Hence $H \simeq H/O_\omega(H) \in \mathrm{form}(G/O_\omega(G))$, which again contradicts Lemma 2.4. Thus $\mathfrak{H}$ is an abelian formation.

Lemma 2.8 *Let* $\mathfrak{F} = \mathfrak{M}\mathfrak{H}$ *be the product of the nonidentity formations* $\mathfrak{M}$ *and* $\mathfrak{H}$. *Assume that every simple group in* $\mathfrak{M}$ *is abelian. If there exists a group* $A \in \mathfrak{M}$ *and a natural number* n *such that, for every group* $B \in \mathfrak{H}$ *with* $|B| \geqslant n$, *the* $\mathfrak{H}$-*residual of the wreath product* $T = A \wr B$ *is not contained subdirectly in the base group of* T, *then there exists a group* Z_p *of prime order* p *and a group* D *having an exponent greater than* p^n *such that* $Z_p \in \mathfrak{M} \cap \mathfrak{H}$ *and* $D \in \mathfrak{H}$.

Proof Let $D_1 \simeq \ldots \simeq D_n$ be nonidentity groups in $\mathfrak{H}$. Let $B_1 = D_1 \times \ldots \times D_n$ and $G_1 = A \wr B_1 = K \rtimes B_1$, where K is the base group of the wreath product G_1. Since, by hypothesis, $G_1^{\mathfrak{H}}$ is not contained subdirectly in K, by Lemma 2.3(iv), we see that there is a normal subgroup $M(B_1)$ of A such that $A/M(B_1)$ is a simple group and $B_2 = (A/M(B_1)) \wr B_1$ is a homomorphic image of the group $G_1/G_1^{\mathfrak{H}} \in \mathfrak{H}$. Analogously, we can also see that there is a normal subgroup $M(B_2)$ of A such that $A/M(B_2)$ is a simple group and the group $B_3 = (A/M(B_2)) \wr B_2$ is a homomorphic image of the group $G_2/G_2^{\mathfrak{H}} \in \mathfrak{H}$, where $G_2 = A \wr B_2$, and so on. Since $A \in \mathfrak{M}$, all groups $A/M(B_1), A/M(B_2), \ldots, A/M(B_n), \ldots$ belong to the formation $\mathfrak{M}$. By the hypothesis, each simple group in $\mathfrak{M}$ is abelian. Since $|A| < \infty$, there exists a prime p and an infinite sequence of indices $i_1, i_1, \cdots, i_n, \ldots$ such that and the order of the group $A/M(B_{i_j})$ is equal to p for all $j = 1, 2, \ldots$.

Let Z_p be a group of order p, and let $T_1 = Z_p, T_2 = Z_p \wr T_1, \cdots, T_n = Z_p \wr T_{n-1}, \ldots$. We show that for any i there exists an index j such that the group T_i is isomorphic to a subgroup of B_{i_j}. If $i = 1$, then the result is evident. Now assume that $i > 1$. Let j be an index such that the group T_{i-1} is isomorphic to a subgroup of B_{i_j}. Then by [89, A, 18.2], $T_i = Z_p \wr T_{i-1}$ is isomorphic to a subgroup of $B_{i_{j+1}} = (A/M(B_{i_j})) \wr B_{i_j}$. Hence for any natural number i, there exists a natural number j such that T_i is isomorphic to a subgroup of $B_j \in \mathfrak{H}$.

Now let P be a p-group and l the length of its composition series. In this case, by [89, A, 18.2] and by induction on l, we see that the group P is isomorphic to a subgroup of some group $T_i \in \mathfrak{H}$. Hence, there is a group $T \in \mathfrak{H}$ such that $\exp(T) \geqslant p^n$. Finally, since $B_2 = (A/M(B_1)) \wr B_1 \in \mathfrak{H}$ and $Z_p(B_2) \neq 1$, we have $Z_p \in \mathfrak{H}$ by Chap. 1, Theorem 2.6(i). Therefore, we have that $Z_p \in \mathfrak{M} \cap \mathfrak{H}$.

Lemma 2.9 *Let* $\mathfrak{F} = \mathfrak{M}\mathfrak{H}$, *where* $\mathfrak{M}$ *and* $\mathfrak{H}$ *are formations and* $\mathfrak{N}_p\mathfrak{H} = \mathfrak{H}$ *for some prime* p. *If for every simple group* $A \in \mathfrak{M}$ *we have* $|A| = p$, *then* $\mathfrak{F} = \mathfrak{H}$.

Proof Assume that $\mathfrak{F} \not\subseteq \mathfrak{H}$ and let G be a group of minimal order in $\mathfrak{F} \setminus \mathfrak{H}$. Let $R = G^{\mathfrak{H}}$ be the monolith of G. Since $G \in \mathfrak{F} = \mathfrak{M}\mathfrak{H}$, we have $R \in \mathfrak{M}$. But $R = A_1 \times \ldots \times A_t$, where $A_1 \simeq \ldots \simeq A_t \simeq A$ is a simple group. Hence $|A| = p$, and so R is a p-group. Therefore $G \in \mathfrak{N}_p\mathfrak{H} = \mathfrak{H}$. The contradiction shows that $\mathfrak{F} \subseteq \mathfrak{H}$. On the other hand, it is trivial that $\mathfrak{H} \subseteq \mathfrak{M}\mathfrak{H} = \mathfrak{F}$, hence $\mathfrak{F} = \mathfrak{H}$.

Lemma 2.10 *Let* p *be a prime number and* $\mathfrak{F} = \mathfrak{M}\mathfrak{H}$, *where every simple group in* $\mathfrak{M}$ *is of order* p, *then* $G = A^{\mathfrak{H}} \wr (A/A^{\mathfrak{H}}) \in \mathfrak{F}$, *for all groups* $A \in \mathfrak{F}$.

Proof Assume that $G \notin \mathfrak{F}$. Then $A^{\mathfrak{H}} \neq 1$. Let $A^{\mathfrak{H}}/N$ be a chief factor of the group A. Since $A \in \mathfrak{F}$, we have $A^{\mathfrak{H}} \in \mathfrak{M}$, and so by our hypothesis, $A^{\mathfrak{H}}/N$ is an elementary abelian p-group. In view of Lemma 1.6, we have

$$A/N \in \text{form}(Z_p \wr (A/A^{\mathfrak{H}})),$$

where Z_p is a group of order p. Since $A/N \notin \mathfrak{H}$, $Z_p \wr (A/A^\mathfrak{H}) \notin \mathfrak{H}$. Now, by using Lemma 2.3, we see that $G^\mathfrak{H}$ is contained subdirectly in the base group K of the wreath product G. Since $A^\mathfrak{H} \in \mathfrak{M}$, we have $K \in \mathfrak{M}$, and so $G \in \mathfrak{F}$. This contradiction completes the proof.

Lemma 2.11 *Let* $\mathfrak{M}\mathfrak{H} \subseteq \mathfrak{F}$, *where* $\mathfrak{F} = l_\omega \mathrm{form} G$ *is a one-generated* ω-*local formation. If* $\mathfrak{M}$ *and* $\mathfrak{H}$ *are formations and* $\mathfrak{H} \neq \mathfrak{M}\mathfrak{H}$. *Then* $\mathfrak{M}$ *is a soluble formation.*

Proof Let f be the smallest ω-local satellite of the formation $\mathfrak{F}$.

Assume that $\mathfrak{M}$ contains some nonsoluble groups and let A be a nonsoluble group with minimal order in $\mathfrak{M}$. Then evidently A has a unique minimal normal subgroup, say P. Clearly P is not abelian and A/P is a soluble group. By Lemma 2.7, we see that $P \neq A$. We now make the following claims:

(1) *For every group* $B \in \mathfrak{H}$ *such that* $|B| > |G|$, *the* $\mathfrak{H}$-*residual of the wreath product* $T = A \wr B$ *is not contained subdirectly in the base group of* T.

Indeed, if we let $T = A \wr B = K \rtimes B$, where K is the base group of the wreath product T. Then, by Lemma 2.3, the group T is monolithic and its monolith L coincides with $P^\natural = \prod_{b\in B} P_1^b$, where P_1 is the monolith of the first copy A_1 of the group A in K. Assume that $T \in \mathfrak{M}\mathfrak{H}$. Then $T \in \mathfrak{F}$. Suppose that $\pi = \pi(P) \cap \omega = \varnothing$. Then $\pi(L) \cap \omega = \varnothing$, and so $O_\omega(T) = 1$. Hence by Lemma 1.10,

$$T \simeq T/O_\omega(T) \in f(\omega') = \mathrm{form}(G/O_\omega(G)).$$

But the group T has a minimal normal subgroup L of order $|P|^{|B|} > |G|$, this contradicts Lemma 2.4. Hence $\pi \neq \varnothing$. Let $p \in \pi$. Then $F_p(T) = 1$, and so

$$T \simeq T/F_p(T) \in f(p) = \mathrm{form}(G/F_p(G))$$

which is absurd. Hence $T \notin \mathfrak{M}\mathfrak{H}$ and thereby the $\mathfrak{H}$-residual of the wreath product $T = A \wr B$ is not contained subdirectly in the base group of T.

(2) *There exist a group* Z_p *of prime order* p *and a group* B *having an exponent greater than* $p^{|G|}$ *such that* $Z_p \in \mathfrak{M} \cap \mathfrak{H}$ *and* $B \in \mathfrak{H}$.

By Lemma 2.7, we know that every simple group in $\mathfrak{M}$ is abelian. Now, let B be a group in $\mathfrak{H}$ such that $|B| > |G|$. Also, let $T = A \wr B = K \rtimes B$, where K is the base group of the wreath product T. Assume that $T^\mathfrak{H}$ is contained subdirectly in K. Then since $A \in \mathfrak{M}$, we have $T^\mathfrak{H} \in \mathfrak{M}$, and so $T \in \mathfrak{M}\mathfrak{H}$ which contradicts to (1). Hence $T^\mathfrak{H}$ is not contained subdirectly in K. Now by Lemma 2.8 and (1) above, we have (2).

(3) *For every group* $T \in \mathfrak{M}\mathfrak{H}$, *we have* $T^\mathfrak{H} \wr (T/T^\mathfrak{H}) \in \mathfrak{F}$.

In fact, by (2) and by Lemma 2.7, we know that $|H| = p$, for every simple group H in $\mathfrak{M}$. Hence by using Lemma 2.10, we see that the claim (3) holds.

(4) $\mathfrak{N}_p\mathfrak{H} = \mathfrak{H}$.

Assume that $\mathfrak{N}_p\mathfrak{H} \not\subseteq \mathfrak{H}$ and let B be a group of minimal order in $\mathfrak{N}_p\mathfrak{H} \setminus \mathfrak{H}$. Let $R = B^\mathfrak{H}$ be the monolith of B. Then, it is clear that R is an abelian p-group. Hence by Lemma 1.6, $B \in \mathrm{form}(Z_p \wr (B/R))$, where Z_p is a group of order p. Therefore $Z_p \wr (B/R) \notin \mathfrak{H}$. Let $T = A \wr (B/R) = K \rtimes (B/R)$, where K

is the base group of the wreath product T. Using Lemma 1.6 and the fact that $A \in \mathfrak{M}$, we see that $T^{\mathfrak{H}}$ is contained subdirectly in $K \in \mathfrak{M}$. This leads to $T \in \mathfrak{F}$. Let $D = T^{|G|} = T_1 \times T_2 \times \ldots \times T_{|G|}$, where $T_1 \simeq T_2 \simeq \ldots \simeq T_{|G|} \simeq T$. Then, it is clear that $D \in \mathfrak{F}$, and so by (3), $E = D^{\mathfrak{H}} \wr (D/D^{\mathfrak{H}}) \in \mathfrak{F}$. Clearly $D^{\mathfrak{H}} \subseteq T_1^{\mathfrak{H}} \times T_2^{\mathfrak{H}} \times \ldots \times T_{|G|}^{\mathfrak{H}}$. Hence $|D/D^{\mathfrak{H}}| \geq |T/T^{\mathfrak{H}}|^{|G|}$. Since $Z_p \in \mathfrak{H}$, we have $R \neq B$. This leads to $|T/T^{\mathfrak{H}}| > 1$, and so $t = |D/D^{\mathfrak{H}}| > |G|$. It is also clear that $T^{\mathfrak{H}} \neq 1$. Now, by Lemma 2.3, the group T is monolithic and its monolith $L = P^{\natural} = \prod_{b \in B/R} P_1^b$, where P_1 is the monolith of the first copy of A in K. Obviously, $\mathrm{Soc}(D) = L_1 \times L_2 \times \ldots \times L_{|G|}$, where L_i is the monolith of the group T_i, and every minimal normal subgroup of $D^{\mathfrak{H}}$ and every minimal normal subgroup X of E are nonabelian groups whose composition factors are isomorphic to the compositions factors of the group P.

Assume that $p \in \pi(P) \cap \omega$. Then $F_p(E) = 1$, and so

$$E \simeq E/F_p(E) \in f(p) = \mathrm{form}(G/F_p(G)).$$

But $|X| \geq |P|' > |G|$, this is clearly impossible by Lemma 2.4. Let $\pi(P) \cap \omega = \varnothing$. Then $O_\omega(E) = 1$, and so

$$E \simeq E/O_\omega(E) \in f(\omega') = \mathrm{form}(G/O_\omega(G)),$$

this contradicts Lemma 2.4. Hence, $\mathfrak{N}_p \mathfrak{H} \subseteq \mathfrak{H}$, and consequently, $\mathfrak{N}_p \mathfrak{H} = \mathfrak{H}$. Now, by using Lemma 2.9, we see that $\mathfrak{M} \mathfrak{H} = \mathfrak{H}$. This contradiction shows that $\mathfrak{M}$ is a soluble formation. The proof is completed.

Proof of Theorem 2.2. In view of Lemma 2.11, $\mathfrak{M}$ is a soluble formation. Hence, we only need to prove that $\mathfrak{M}$ is a p-composition formation. Let $\mathfrak{F} = l_\omega \mathrm{form} G$ and f the smallest ω-local satellite of $\mathfrak{F}$. Suppose on the contrary that the theorem is not true. Then by Lemma 2.11, there exists a group $A \notin \mathfrak{M}$ together with a prime $p \in \omega$ such that $A/\Phi(O_p(A)) \in \mathfrak{M}$. We denote the subgroup $\Phi(O_p(A))$ of A by N. And let $B = A/N$. Let $D \in \mathfrak{H}$ and $T = B \wr D = K \rtimes D$, where K is the base group of the wreath product T. Assume that $T^{\mathfrak{H}}$ is contained subdirectly in K. Then $T^{\mathfrak{H}} \in \mathfrak{M}$, and so $T \in \mathfrak{M} \mathfrak{H}$. Let $T_1 = A \wr D = K_1 \rtimes D$, where K_1 is the base group of the wreath product T_1. Then, there is an epimorphism $\varphi : T_1 \to T$ such that the kernel R of φ coincides with $N^{\natural} = \prod_{d \in D} \Phi(O_p(A_1))^d$, where A_1 is the first copy of A in K_1. Clearly, $R \subseteq \Phi(O_p(K_1)) \subseteq \Phi(O_p(T_1))$. Since $\mathfrak{M}$ is a soluble formation, the group K (and so K_1) is soluble. Since $T_1/R \simeq T \in \mathfrak{M} \mathfrak{H}$ and $\mathfrak{M} \mathfrak{H}$ is a p-composition formation, we have $T_1 \in \mathfrak{M} \mathfrak{H}$ by Lemma 2.11. Hence $T_1^{\mathfrak{H}} \in \mathfrak{M}$. Now we can see that $T^{\mathfrak{H}} = \varphi(T_1^{\mathfrak{H}})$, $K = \varphi(K_1)$ and $T^{\mathfrak{H}}$ is contained subdirectly in K. This shows that $T_1^{\mathfrak{H}}$ is contained subdirectly in K_1. Consequently, the group $T_1^{\mathfrak{H}}$ has a normal subgroup M such that $T_1^{\mathfrak{H}}/M \simeq A$. But $T_1^{\mathfrak{H}} \in \mathfrak{M}$, so $A \in \mathfrak{M}$. This contradiction shows that for all groups $D \in \mathfrak{H}$, the $\mathfrak{H}$-residual of the group $T = B \wr D$ can not be contained subdirectly in the base group of T. Now, by Lemma 2.8, there exists a group Z_p of prime order p and a group D having an exponent greater than $p^{|G|}$ such that $Z_p \in \mathfrak{M} \cap \mathfrak{H}$ and $D \in \mathfrak{H}$. By Lemma 2.7, for all simple groups $X \in \mathfrak{M}$, we have $|X| = p$. In order to show that $\mathfrak{H} = \mathfrak{N}_p \mathfrak{H}$, we first observe that $\mathfrak{H} \subseteq \mathfrak{N}_p \mathfrak{H}$. Assume that $\mathfrak{N}_p \mathfrak{H} \nsubseteq \mathfrak{H}$ and let G be a group of minimal order in $\mathfrak{N}_p \mathfrak{H} \setminus \mathfrak{H}$. Then

$R = G^{\mathfrak{H}}$ is the monolith of G, and so R is a p-group. Using Lemma 1.6, we see that $G \in \mathrm{form}(Z_p \wr (G/R))$, where Z_p is a group of order p.

Form $Y = B \wr (G/R) = K \rtimes (G/R)$, where K is the base group of the wreath product Y. Let B_1 be the first copy of B in K. Since we have already shown that the $\mathfrak{H}$-residual $Y^{\mathfrak{H}}$ is not contained subdirectly in K, by Lemma 2.3, we see that there is a proper normal subgroup M in B_1 such that B_1/M is a simple group and $(B_1/M) \wr (G/R) \in \mathfrak{H}$. Since $B_1 \simeq B \in \mathfrak{M}$, we have $B_1/M \in \mathfrak{M}$ so that $|B_1/M| = p$. But this leads to

$$G \in \mathrm{form}(Z_p \wr (G/R)) = \mathrm{form}((B_1/M) \wr (G/R)) \subseteq \mathfrak{H}.$$

which is clearly a contradiction. Thus, we have $\mathfrak{N}_p \mathfrak{H} \subseteq \mathfrak{H}$, and so $\mathfrak{N}_p \mathfrak{H} = \mathfrak{H}$. Now by Lemma 2.9, we obtain $\mathfrak{M}\mathfrak{H} = \mathfrak{H}$. However, by our hypotheses, $\mathfrak{M}\mathfrak{H} \neq \mathfrak{H}$. This contradiction shows that $\mathfrak{M}$ must be a p-composition formation.

The following corollaries follow directly from Theorem 2.2.

Corollary 2.12 *Let the product $\mathfrak{R}$ of two nonidentity formations $\mathfrak{M}$ and $\mathfrak{H}$ be a composition formation such that $\mathfrak{H} \neq \mathfrak{R}$ and $\mathfrak{R} \subseteq \mathfrak{F}$, for some one-generated local formation $\mathfrak{F}$. Then $\mathfrak{M}$ is a local formation.*

Corollary 2.13 (Skiba [382]). *Let $\mathfrak{F} = \mathfrak{M}\mathfrak{H}$ be a one-generated composition formation. If $\mathfrak{H} \neq \mathfrak{F}$, then $\mathfrak{M}$ is also a composition formation.*

Corollary 2.14 (Guo [134]). *Let $\mathfrak{F} = \mathfrak{M}\mathfrak{H}$ be a composition subformation of some one-generated composition formation. If $(1) \neq \mathfrak{H} \neq \mathfrak{F}$, then $\mathfrak{M}$ is a local formation.*
Note that Corollary 2.14 gave the answer to Problem 12.74 in [316].

Corollary 2.15 (Skiba and Rizhik [390]). *Let $\mathfrak{F} = \mathfrak{M}\mathfrak{H}$ be a one-generated ω-local formation. If $\mathfrak{M}$ is a normal hereditary formation, then $\mathfrak{M}$ is ω-local.*

Corollary 2.16 (Guo, Shum [166]). *Let $\mathfrak{F} = \mathfrak{M}\mathfrak{H}$ be a one-generated ω-saturated formation. If $\mathfrak{H} \neq \mathfrak{F}$, then $\mathfrak{M}$ is also a ω-saturated formation.*

5.3 Noncancellative Factorizations of One-Generated ω-Composition Formations

If

$$\mathfrak{F} = \mathfrak{F}_1 \ldots \mathfrak{F}_t \tag{5.3}$$

is the product of the formations $\mathfrak{F}_1, \cdots, \mathfrak{F}_t$ and $\mathfrak{F} \neq \mathfrak{F}_1 \ldots \mathfrak{F}_{i-1}\mathfrak{F}_{i+1} \ldots \mathfrak{F}_t$ for all $i = 1, \cdots, t$, then we call (5.3) a *noncancellative factorization* of the formation $\mathfrak{F}$.

On the noncancellative factorizations of one-generated ω-composition formations, Shemetkov and Skiba proposed the following Problem in [389].

Problem ([389, Problem 21]). Describe noncancellative factorizations of one-generated ω-composition formations.

The aim of this section is to establish the theory of noncancellative factorizations of one-generated ω-composition formations and give the answer to this problem.

Lemma 3.1 (W. Guo [131, Lemma 4.5.23]). *Let $\mathfrak{X}$ be a nonempty set of groups. If $G \in \text{form}\mathfrak{X}$, then*

$$G/O_p(G) \in \text{form}(A/O_p(A) \mid A \in \mathfrak{X}).$$

Lemma 3.2 *Let f be the smallest ω-composition satellite of a ω-composition formation $\mathfrak{F}$ and let $\pi = \pi(\text{Com}(\mathfrak{F})) \cap \omega$. Then $\mathfrak{F}$ is a one-generated ω-composition formation if and only if $|\pi| \leq \infty$ and $f(a)$ is a one-generated formation, for all $a \in \pi \cup \{\omega'\}$.*

Proof We first assume that for some group G, we have $\mathfrak{F} = c_\omega \text{form}G$. Then, by Lemma 1.8, we have $\text{Com}(\mathfrak{F}) = \text{Com}(G)$, and so $|\pi| < \infty$. On the other hand, we have

$$f(a) = \begin{cases} \text{form}(G/C^p(G)) & \text{if } a = p \in \pi; \\ \text{form}(G/(R(G) \cap O_\omega(G))) & \text{if } a = \omega'. \end{cases}$$

This shows that $f(a)$ is a one-generated formation, for all $a \in \pi \cap \{\omega'\}$.

Now we assume that $|\pi| < \infty$ and that the formations $f(\omega'), f(p_1), \cdots, f(p_n)$, where $\{p_1, \cdots, p_n\} = \pi$, are all one-generated. Let $f(\omega') = \text{form}A$ and $f(p_i) = \text{form}B_i$, $i = 1, \cdots, n$.

Let $T_i = Z_{p_i} \wr (B_i/O_{p_i}(B_i)) = K_i \rtimes (B_i/O_{p_i}(B_i))$, where K_i is the base group of T_i, $i = 1, \ldots, n$. Let $G = A \times T_1 \times \ldots \times T_n$ and $\mathfrak{M} = c_\omega \text{form}G$. In order to complete the proof of the Lemma, it suffices to prove that $\mathfrak{M} = \mathfrak{F}$. Since by Lemma 1.8, f is an inner ω-composition satellite of the formation $\mathfrak{F}$, by Lemma 1.7, we see that $G \in \mathfrak{F}$, and so $\mathfrak{M} \subseteq \mathfrak{F}$. If $\mathfrak{F} \nsubseteq \mathfrak{M}$, then we can let T be a group of minimal order in $\mathfrak{F} \setminus \mathfrak{M}$. Let L be the monolith of T. Then $L = T^{\mathfrak{M}}$. Assume that $R(T) \cap O_\omega(T) = 1$. In this case, we have $T \simeq T/1 \in f(\omega') = \text{form}A$. But $A \simeq G/(T_1 \times \ldots \times T_n) \in \mathfrak{M}$, and so $T \in \mathfrak{M}$, which is a contradiction. Thus $R(T) \cap O_\omega(T) \neq 1$, and hence L must be a p-group, for some $p \in \pi$. Let $p = p_i$. Since $T \in \mathfrak{F}$, we have $T/C^p(T) \in f(p) = \text{form}B_i$. Now, by Lemma 1.2, we have $O_p(T/C^p(T)) = 1$, and so by Lemma 3.1, we see that $T/C^p(T) \in \text{form}(B_i/O_p(B_i))$. Since it is clear that $T_i \in \mathfrak{M}$, and so $T_i/C^p(T_i) \in m(p)$, where m is the smallest ω-composition satellite of $\mathfrak{M}$. But by Lemma 1.3, we have $K_i = C^p(T_i)$, and so $B_i/O_p(B_i) \simeq T_i/K_i \in m(p)$. Hence $T/C^p(T) \in m(p)$. Because $T/L \in \mathfrak{M}$, we have $(T/L)/(R(T/L) \cap O_\omega(T/L)) \in m(\omega')$ and also for all $q \in \pi$, we have $(T/L)/C^q(T/L) \in m(q)$. On the other hand, it is easy to verify that

$$R(T/L) \cap O_\omega(T/L) = (R(T)/L) \cap (O_\omega(T)/L)$$

and $C^q(T/L) = C^q(T)/L$, for all primes $q \neq p$. Hence, we derive that

$$T/(R(T) \cap O_\omega(T)) \simeq (T/L)/((R(T) \cap O_\omega(T))/L) \in m(\omega')$$

and

$$T/C^q(T) \simeq (T/L)/C^q(T/L) \in m(q)$$

This shows that $T \in \mathfrak{M}$, a contradiction. Thus, we have $\mathfrak{F} \subseteq \mathfrak{M}$, and so $\mathfrak{F} = \mathfrak{M}$. The proof is completed.

Lemma 3.3 *Let* $\mathfrak{F} = \mathfrak{M}\mathfrak{H}$, *where* $\mathfrak{M}$ *is a* ω-*composition formation with inner* ω-*composition satellite* m. *If* $\mathfrak{H}$ *is a nonempty formation such that* $\pi(\mathrm{Com}(\mathfrak{H})) \cap \omega \subseteq \pi(\mathrm{Com}(\mathfrak{M}))$, *then* $\mathfrak{F} = CF_\omega(f)$, *where*

$$f(a) = \begin{cases} m(p)\mathfrak{H} & \textit{if } a = p \in \pi(\mathrm{Com}(\mathfrak{M})) \cap \omega; \\ \varnothing & \textit{if } a = p \in \omega \setminus \pi(\mathrm{Com}(\mathfrak{M})); \\ m(\omega')\mathfrak{H} & \textit{if } a = \omega'. \end{cases}$$

Proof Let $\mathfrak{M}_1 = CF_\omega(f)$. Assume that $\mathfrak{M}_1 \not\subseteq \mathfrak{F}$ and let G be a group of minimal order in $\mathfrak{M}_1 \setminus \mathfrak{F}$. Let $L = G^{\mathfrak{F}}$ be the monolith of G. Since $\mathfrak{H} \subseteq \mathfrak{M}\mathfrak{H} = \mathfrak{F}$, we have $G^{\mathfrak{H}} \neq 1$, and so $L \subseteq G^{\mathfrak{H}}$. Since $G/L \in \mathfrak{F}$ and $(G/L)^{\mathfrak{H}} = G^{\mathfrak{H}}/L$, we see that $G^{\mathfrak{H}}/L \in \mathfrak{M}$. Suppose that $R(G) \cap O_\omega(G) = 1$. Then by $G \in \mathfrak{M}_1$, we have

$$G \simeq G/1 \in f(\omega') = m(\omega')\mathfrak{H} \subseteq \mathfrak{M}\mathfrak{H} = \mathfrak{F},$$

which is a contradiction. Thus, L must be a p-group for some $p \in \omega$, and so

$$G/C^p(G) \in f(p) = m(p)\mathfrak{H},$$

that is,

$$(G/C^p(G))^{\mathfrak{H}} \simeq G^{\mathfrak{H}}C^p(G)/C^p(G) \in m(p).$$

But by Lemma 1.4, $G^{\mathfrak{H}} \cap C^p(G) = C^p(G^{\mathfrak{H}})$. This leads to $G^{\mathfrak{H}}/C^p(G^{\mathfrak{H}}) \in m(p)$. Since we also have $G^{\mathfrak{H}}/L \in \mathfrak{M}$, we conclude that $G^{\mathfrak{H}} \in \mathfrak{M}$, i.e., $G \in \mathfrak{F}$. This contradiction shows that $\mathfrak{M}_1 \subseteq \mathfrak{F}$.

Now, suppose that $\mathfrak{F} \not\subseteq \mathfrak{M}_1$ and let G be a group of minimal order in $\mathfrak{F} \setminus \mathfrak{M}_1$ with monolith $L = G^{\mathfrak{M}_1}$. Let $G^{\mathfrak{H}} = 1$. Then $G \in m(\omega')\mathfrak{H}$ and $G/C^p(G) \in m(p)\mathfrak{H}$, for all $p \in \pi(\mathrm{Com}(G)) \cap \omega$, and so $G \in \mathfrak{M}_1$, a contradiction. Hence $G^{\mathfrak{H}} \neq 1$. Since $G \in \mathfrak{F}$, $G^{\mathfrak{H}} \in \mathfrak{M}$ and thereby

$$G^{\mathfrak{H}}/R(G^{\mathfrak{H}}) \cap O_\omega(G^{\mathfrak{H}}) \in m(\omega').$$

By applying Lemma 1.4(2), we obtain

$$(R(G) \cap O_\omega(G)) \cap G^{\mathfrak{H}} = R(G^{\mathfrak{H}}) \cap O_\omega(G^{\mathfrak{H}}).$$

Therefore, we deduce that

$$G^{\mathfrak{H}}/(R(G) \cap O_\omega(G) \cap G^{\mathfrak{H}}) \simeq G^{\mathfrak{H}}(R(G) \cap O_\omega(G))/(R(G) \cap O_\omega(G))$$
$$= (G/(R(G) \cap O_\omega(G)))^{\mathfrak{H}} \in m(\omega'),$$

and consequently

$$G/(R(G) \cap O_\omega(G)) \in m(\omega')\mathfrak{H} = f(\omega').$$

On the other hand, by Lemma 1.4, we have $C^p(G) \cap G^{\mathfrak{H}} = C^p(G^{\mathfrak{H}})$. Thus, for all $p \in \pi(\text{Com}(G)) \cap \omega$, we have

$$(G/C^p(G))^{\mathfrak{H}} = G^{\mathfrak{H}} C^p(G)/C^p(G) \simeq G^{\mathfrak{H}}/G^{\mathfrak{H}} \cap C^p(G) = G^{\mathfrak{H}}/C^p(G^{\mathfrak{H}}) \in m(p),$$

that is,

$$G/C^p(G) \in m(p)\mathfrak{H} = f(p).$$

This implies that $G \in CF_\omega(f) = \mathfrak{M}_1$, which is a contradiction. Thus $\mathfrak{F} \subseteq \mathfrak{M}_1$, and hence $\mathfrak{F} = \mathfrak{M}_1$. This completes the proof.

Let $\mathfrak{H}$ be a class of groups. We now use $\mathfrak{H}(\omega')$ to denote the class

$$\text{form}(A/(R(A) \cap O_\omega(R)) \mid A \in \mathfrak{H}).$$

and use $\mathfrak{H}(p)$ to denote the class

$$\text{form}(A/O_p(A) \mid A \in \mathfrak{H}).$$

Lemma 3.4   Let $\mathfrak{F} = \mathfrak{M}\mathfrak{H}$ be a ω-composition formation, where $\mathfrak{M}$ and $\mathfrak{H}$ are formations with $\mathfrak{M} \subseteq \mathfrak{N}$. Let f be the smallest ω-composition satellite of the formation $\mathfrak{F}$. Assume that $|\pi(\mathfrak{M}) \cap \omega| > 1$. Then

$$\mathfrak{H} = \text{form}(f(p) \cup f(q))$$

for any two different primes $p, q \in \pi(\mathfrak{M}) \cap \omega$.

Proof   Let $\mathfrak{M}_0 = \text{form}(f(p) \cup f(q))$. Then, by our hypothesis, for each group $G \in \mathfrak{F}$, we have $G^{\mathfrak{H}} \subseteq F(G)$, so $G/C^p(G), G/C^q(G) \in \mathfrak{H}$. Using Lemma 1.8, we see that $f(p) \cup f(q) \subseteq \mathfrak{H}$, and so $\mathfrak{M}_0 \subseteq \mathfrak{H}$. Assume that $\mathfrak{H} \not\subseteq \mathfrak{M}_0$ and let A be a group of minimal order in $\mathfrak{H} \setminus \mathfrak{M}_0$. Let R be the monolith of A. Then, it is clear that either $R \not\subseteq O_p(A)$ or $R \not\subseteq O_q(A)$. Let $R \not\subseteq O_p(A)$ and let $T = Z_p \wr A = K \rtimes A$, where K is the base group of the wreath product T. It is clear that $T \in \mathfrak{F}$. Hence

$$T/C^p(T) \in f(p) = \text{form}(G/C^p(G) \mid G \in \mathfrak{F}) \subseteq \mathfrak{M}_0.$$

But by Lemma 1.3, $C^p(T) = K$, and so $T/C^p(T) \simeq A \in \mathfrak{M}_0$, which is a contradiction. Hence $\mathfrak{H} \subseteq \mathfrak{M}_0$, and therefore $\mathfrak{H} = \mathfrak{M}_0$.

For any nonempty class $\mathfrak{H}$ of groups, we write $\mathfrak{H}/O_p(\mathfrak{H})$ to denote the formation generated by the set $\{G/O_p(G) \mid G \in \mathfrak{H}\}$.

Lemma 3.5   Let $\mathfrak{F} = \mathfrak{M}\mathfrak{H}$ be the product of nonidentity formations $\mathfrak{M}$ and $\mathfrak{H}$ such that $\mathfrak{H} \neq \mathfrak{F}$. Also let $\mathfrak{M} \subseteq \mathfrak{N}_\omega$. Then $\mathfrak{F}$ is a one-generated ω-composition formation if and only if the following conditions are true:

(1) $\mathfrak{M}$ is a local formation such that $|\pi(\mathfrak{M})| < \infty$;
(2) $\pi(\text{Com}(\mathfrak{H})) \cap \omega \subseteq \pi(\mathfrak{M})$;
(3) $\mathfrak{H}(\omega')$ is a one-generated formation;
(4) If $|\pi(\mathfrak{M})| > 1$, then $\mathfrak{H}$ is a one-generated formation;

(5) If $\mathfrak{M} = \mathfrak{N}_p$ for some prime p, then $\mathfrak{H}(p)$ is a one-generated formation.

Proof First, we assume that the above conditions (1)–(5) hold. We need to show that $\mathfrak{F}$ is a one-generated ω-composition formation. First of all, by Lemma 3.3, $\mathfrak{F}$ is a ω-composition formation. In view of Lemma 3.2, we only need to prove that $|\pi(\mathrm{Com}(\mathfrak{F})) \cap \omega| < \infty$ and that the formations $f(\omega')$ and $f(p)$ (for all $p \in \pi(\mathrm{Com}(\mathfrak{F})) \cap \omega$) are one-generated formations, where f is the smallest ω-composition satellite of $\mathfrak{F}$.

Let $G \in \mathfrak{F} = \mathfrak{M}\mathfrak{H}$ and $p \in \pi(\mathrm{Com}(G)) \cap \omega$. Since $\pi(\mathrm{Com}(\mathfrak{H})) \cap \omega \subseteq \pi(\mathfrak{M})$, we have $p \in \pi(\mathrm{Com}(\mathfrak{M}))$. But $|\pi(\mathfrak{M})| < \infty$, and so $|\pi(\mathrm{Com}(\mathfrak{F})) \cap \omega| < \infty$. Now we prove that $f(\omega') = \mathfrak{H}(\omega')$. Since $\mathfrak{H} \subseteq \mathfrak{F}$, we have $A/(R(A) \cap O_\omega(A)) \in f(\omega')$, for every group $A \in \mathfrak{H}$, and therefore $\mathfrak{H}(\omega') \subseteq f(\omega')$. Let $A \in \mathfrak{F}$. Then $A^{\mathfrak{H}} \subseteq F(A)$. Hence $A/F(A) \in \mathfrak{H}$, and thus $A/(R(A) \cap O_\omega(A)) \in \mathfrak{H}$. But

$$(R(A) \cap O_\omega(A))/F(A) = R(A/F(A)) \cap O_\omega(A/F(A)),$$

so $A/(R(A) \cap O_\omega(A)) \in \mathfrak{H}(\omega')$. By invoking Lemma 1.8, we see that $f(\omega') \subseteq \mathfrak{H}(\omega')$. Therefore $f(\omega') = \mathfrak{H}(\omega')$ is a one-generated formation.

Now let $p \in \pi(\mathrm{Com}(\mathfrak{F})) \cap \omega$. Then $p \in \pi(\mathfrak{M})$. In view of Lemma 3.3, the formation $\mathfrak{F}$ has an inner ω-composition satellite t such that $t(p) = \mathfrak{H}$. Now, by using Lemma 1.8, we see that

$$f(p) = \mathrm{form}(A \mid A \in \mathfrak{H} \text{ and } O_p(A) = 1) = \mathrm{form}(A/O_p(A) \mid A \in \mathfrak{H}) = \mathfrak{H}(p).$$

Let $|\pi(\mathfrak{M})| > 1$ and B be a group such that $\mathfrak{H} = \mathrm{form}B$. Also, let $A \in \mathfrak{H}$. Then by Lemma 3.1

$$A/O_p(A) \in \mathrm{form}(B/O_p(B)),$$

and so $f(p) \subseteq \mathrm{form}(B/O_p(B))$. Next let $T = Z_p \wr (B/O_p(B))$. Clearly $T \in \mathfrak{F}$, and hence $T/C^p(T) \in f(p)$. But by Lemma 1.3, we have

$$T/C^p(T) \simeq B/O_p(B).$$

This shows that $B/O_p(B) \in f(p)$, and hence $\mathrm{form}(B/O_p(B)) \subseteq f(p)$. Thus, $f(p) = \mathrm{form}(B/O_p(B))$ is a one-generated formation. It follows that $\mathfrak{F}$ is a one-generated ω-composition formation.

To prove the converse of the lemma, we let $\mathfrak{F} = c_\omega \mathrm{form}G$ for some group G. By applying Theorem 2.2, we see that $\mathfrak{M}$ is a ω-local formation. But since $\mathfrak{M} \subseteq \mathfrak{N}_\omega$, we have that $\mathfrak{M}$ is a local formation. By Lemma 1.5, we see that $\mathfrak{M} \subseteq \mathfrak{F}$, and hence $\pi(\mathfrak{M}) \subseteq \pi(\mathfrak{F}) = \pi(G)$. This proves that $|\pi(\mathfrak{M})| < \infty$.

Assume that there exists a prime p such that $p \in \pi(\mathrm{Com}(\mathfrak{H})) \cap \omega$ and $p \notin \pi(\mathfrak{M})$. Since $\mathfrak{H} \subseteq \mathfrak{F}$, we have $\mathfrak{N}_p \subseteq \mathfrak{F}$ and $Z_p \notin \mathfrak{M}$. Let Z_q be a simple group in $\mathfrak{M}$. Then $q \neq p$. Let B be a cyclic group of order p^m, where $m = |G|$. Let $D = Z_q \wr B = K \rtimes B$, where K be the base group of D. Then, it is clear that $D \in \mathfrak{F}$, and hence $D/C^q(D) = D/K \simeq B \in f(q)$. But by Lemma 1.8, we have

$$f(q) = \text{form}(G/C^q(G))$$

which contradicts Lemma 2.4. Hence $\pi(\text{Com}(\mathfrak{H})) \cap \omega \subseteq \pi(\mathfrak{M})$. Same as above, we can easily prove that $\mathfrak{H}(\omega') = f(\omega')$. Thus, by Lemma 1.8, we have

$$\mathfrak{H}(\omega') = \text{form}(G/(R(G) \cap O_\omega(G)))$$

which is a one-generated formation.

Let $|\pi(\mathfrak{M})| > 1$. Then, by Lemma 3.4, $\mathfrak{H}$ is also a one-generated formation.

Finally, we assume that $\mathfrak{M} = \mathfrak{N}_p$, for some prime p. We show that $\mathfrak{H}(p) = f(p)$. Indeed, if $A \in \mathfrak{H}$ and $T = Z_p \wr (A/O_p(A)) = K \rtimes (A/O_p(A))$, where K is the base group of the wreath product T, then

$$A/O_p(A) \simeq T/K = T/C^p(T) \in f(p)$$

Consequently, we have $\mathfrak{H}(p) \subseteq f(p)$. Next if $A \in \mathfrak{F}$, then $A/C^p(A) \in \mathfrak{H}$ and by Lemma 1.2, we have $O_p(A/C^p(A)) = 1$. This leads to $A/C^p(A) \in \mathfrak{H}(p)$, and so

$$f(p) = \text{form}(A/C^p(A) \mid A \in \mathfrak{F}) \subseteq \mathfrak{H}(p).$$

Thus we have proved that $f(p) \subseteq \mathfrak{H}(p)$, and therefore

$$\mathfrak{H}(p) = f(p) = \text{form}(G/C^p(G))$$

is a one-generated formation.

Lemma 3.6 ([346, Corollary 4.11]). *Let $\mathfrak{F}$ be a ω-local formation. Assume that $\mathfrak{F} \not\subseteq \mathfrak{N}_\omega\mathfrak{N}$ but $\mathfrak{F}_1 \subseteq \mathfrak{N}_\omega\mathfrak{N}$ for all its proper ω-local subformations $\mathfrak{F}_1$ of $\mathfrak{F}$. Then $\mathfrak{F} = l_\omega\text{form}G$, where G is a monolithic group with monolith $R = G^{\mathfrak{N}_\omega\mathfrak{N}}$ such that either $\pi = \pi(R) \cap \omega = \varnothing$ or $\pi \neq \varnothing$ and one of the following statements hold:*

(1) R is a nonabelian group and $R = G^{\mathfrak{N}_p\mathfrak{N}}$, for all $p \in \pi(R) \cap \omega$;
(2) $G = R \rtimes H$, where $R = C_G(R)$ is a p-group and $H = Q \rtimes N$ is a monolithic group with monolith $Q = C_H(Q) = O_q(H)$ such that $p \neq q \in \omega$ and N is a nonidentity nilpotent group.

Lemma 3.7 [89, VII, Lemma 1.7]. *Let $\mathfrak{F} = \text{form}G$, where G is a soluble group. Then the set of all subformations of the formation $\mathfrak{F}$ is a finite set.*

Lemma 3.8 *Let $\mathfrak{M}\mathfrak{H} \subseteq \mathfrak{F}$, where $\mathfrak{F}$ is a one-generated ω-local formation and $\mathfrak{M}\mathfrak{H}$ be a ω-composition formation such that both formations $\mathfrak{M}, \mathfrak{H}$ are nonidentity. If $\mathfrak{H} \neq \mathfrak{M}\mathfrak{H}$, then $\mathfrak{M} \subseteq \mathfrak{N}_\omega\mathfrak{N}$.*

Proof Let $\mathfrak{F} = l_\omega\text{form}G$. Suppose that $n = |G|$ and f is the smallest ω-local satellite of $\mathfrak{F}$.

Assume that $\mathfrak{M} \not\subseteq \mathfrak{N}_\omega\mathfrak{N}$, and let D be a group in $\mathfrak{M} \setminus \mathfrak{N}_\omega\mathfrak{N}$. Then, by Lemma 2.11, we see that D is a soluble group. Let $\mathfrak{M}_0 = c_\omega\text{form}D$. Then as an application of Theorem 2.2, we see that $\mathfrak{M}$ is a ω-local formation so that $\mathfrak{M}_0 \subseteq \mathfrak{M}$. By using Lemma 3.7, Corollary 1.9, and Lemma 1.10, we see that the set of all ω-composition subformations of the formation $\mathfrak{M}_0$ must be finite. Hence $\mathfrak{M}_0$ has a ω-composition subformation $\mathfrak{H}_0$ such that $\mathfrak{H}_0 \not\subseteq \mathfrak{N}_\omega\mathfrak{N}$ but $\mathfrak{H}_1 \subseteq \mathfrak{N}_\omega\mathfrak{N}$ for any proper ω-composition subformation $\mathfrak{H}_1$ of $\mathfrak{H}_0$. Since D is a soluble group, $\mathfrak{M}_0$ is a soluble formation. It follows that $\mathfrak{M}_0$ and every ω-composition subformation of $\mathfrak{M}_0$ are ω-local formations. In view of Lemma 3.6, we conclude that $\mathfrak{H}_0 = l_\omega \text{form} A$, where A is a monolithic group with monolith $P = A^{\mathfrak{N}_\omega\mathfrak{N}}$ and either $\pi = \pi(P) \cap \omega = \varnothing$ or P is a p-group, where $p \in \omega$, and $A = P \rtimes H$, where $P = C_G(P) = F(A) = F_p(A)$ and $H = Q \rtimes N$ is a monolithic group, where $Q = C_H(Q) = O_q(H)$ is the monolith of H, with $p \neq q \in \omega$ and N is a nonidentity nilpotent group.

Assume that for every group $B \in \mathfrak{H}$ such that $|B| > n$, the $\mathfrak{H}$-residual $D^{\mathfrak{H}}$ of $D = A \wr B$ is not contained subdirectly in the base group of the wreath product D. Then, by Lemmas 2.7, 2.8 and 2.10, we see that for every group $T \in \mathfrak{M}\mathfrak{H}$, we always have $T^{\mathfrak{H}} \wr (T/T^{\mathfrak{H}}) \in \mathfrak{M}\mathfrak{H}$ and there is a prime p such that for all simple group A in $\mathfrak{M}$, we have $|A| = p$. Now, using the same arguments as in the proof of Lemma 2.11, we can show that $\mathfrak{H} = \mathfrak{M}\mathfrak{H}$, which contradicts our hypothesis. This shows that there is a group B in $\mathfrak{H}$ such that $|B| > n$ and the $\mathfrak{H}$-residual $D^{\mathfrak{H}}$ of the wreath product $D = A \wr B$ is contained subdirectly in K. Therefore $D \in \mathfrak{M}\mathfrak{H}$. By Lemma 2.3, the group D is monolithic whose monolith R coincides with $P^\natural = \prod_{b \in B} P_1^b$, where P_1 is the monolith of the first copy of A in K. If $\pi = \varnothing$, then $O_\omega(D) = 1$, and so by Lemma 1.10, we have

$$D \simeq D/O_\omega(D) \in f(\omega') = \text{form}(G/O_\omega(G)).$$

On the other hand, we can easily see that the group D has a minimal normal subgroup R of order $|P|^{|B|} \geq |P|^n > n$, this contradicts Lemma 2.4. Hence $\pi \neq \varnothing$. It is not difficult to see that $R \not\subseteq \Phi(D)$. Hence there exists a maximal subgroup M of D such that $RM = D$ and $C = C_D(R)$. Thus, $C = C \cap RM = R(C \cap M)$. Clearly, $C \cap M \trianglelefteq D$. This leads to $C = R$ so that $F_p(D) = R$. Since $D \in \mathfrak{M}\mathfrak{H} \subseteq \mathfrak{F}$, by Lemma 1.10, we have

$$D/F_p(D) = D/R \in f(p) = \text{form}(G/F_p(G)).$$

Now by using [89, A; (18.2)], we see that $D/R \simeq (A/P) \wr B \simeq H \wr B$. Thus, $H \wr B \in \text{form}(G/F_p(G))$. However, the group $H \wr B$ has a minimal normal subgroup of order $|Q|^{|B|} \geq |Q|^n > n$, which is a contradiction. Thus, we conclude that $\mathfrak{M} \subseteq \mathfrak{N}_\omega\mathfrak{N}$.

Lemma 3.9 *Let $\mathfrak{F} = \mathfrak{M}\mathfrak{H}$, where $\mathfrak{M}$ and $\mathfrak{H}$ are nonidentity formations. Then $\mathfrak{F}$ is neither abelian formation nor a nilpotent formation with $|\pi(\mathfrak{F})| > 1$.*

Proof Let A be a simple group in $\mathfrak{M}$ and B a simple group in $\mathfrak{H}$. Let $G = A \wr B = K \rtimes B$, where K is the base group of G. Clearly $G \in \mathfrak{F}$. Assume that $\mathfrak{F}$ is abelian. Then $B \subseteq C_G(K)$. But $C_G(K) \subseteq K$, we obtain a contradiction. This shows that $\mathfrak{F}$ is not abelian. If we assume that $\mathfrak{F}$ is nilpotent and $|\pi(\mathfrak{F})| \geq 2$, then, clearly, $A \in \mathfrak{F}$, and so $|A| = p$ is a prime. Let $q \in \pi(\mathfrak{F}) \setminus \{p\}$, Z_q be a group of order q. Assume that $|B| = p$. Let $Z_q \in \mathfrak{M}$ and let $G = Z_q \wr B = K \rtimes B$ where K is the base group of G. Then $G \in \mathfrak{F}$ and hence G is nilpotent. Clearly, B is a Sylow p-subgroup of G, and hence B is normal in G. But then, we have $B \subseteq C_G(K)$, a contradiction. Analogously, we can obtain contradictions in the cases $|B| \neq p$ or $Z_q \in \mathfrak{H}$.

The following lemma can be similarly proved as Lemma 2.7(iii). We omit the details.

Lemma 3.10 Let $\mathfrak{M}\mathfrak{H} \subseteq \mathfrak{F}$, where $\mathfrak{F}$ is a one-generated ω-local formation and $\mathfrak{M}$ is a ω-local formation. Let m be the smallest ω-local satellite of $\mathfrak{M}$. Suppose that there are a prime $p \in \omega$ and a simple group A such that $A \in m(p)$ and $|A| \neq p$. Then $A \notin \mathfrak{H}$ and the formation $\mathfrak{H}$ is abelian.

Lemma 3.11 Let $\mathfrak{F}$ be a ω-composition formation. If $\mathfrak{F} \subseteq \mathfrak{N}_\omega \mathfrak{N}$, then $\mathfrak{F}$ is a hereditary formation.

Proof Let $\mathfrak{F} = CF_\omega(f)$, where f is the smallest ω-composition satellite of $\mathfrak{F}$. Since $\mathfrak{F} \subseteq \mathfrak{N}_\omega \mathfrak{N}$, then by Lemma 1.8, for any $a \in \omega \cup \{\omega'\}$ the formation $f(p)$ is nilpotent, so it is hereditary by Lemma 1.5. Now let $H \leq G \in \mathfrak{F}$. Since G is soluble, $C^p(G) = O_{p'p}(G)$ and so $C^p(G) \cap H \leq O_{p'p}(H) = C^p(H)$. Hence $H/C^p(H) \in f(p)$. Similarly we have $H/O_\omega(H) \in f(p)$. Hence $H \in \mathfrak{F}$.

We now prove the following technical lemma.

Lemma 3.12 Let

$$\mathfrak{F} = \mathfrak{F}_1 \mathfrak{F}_2 \ldots \mathfrak{F}_t$$

be a noncancellative factorization of a formation $\mathfrak{F}$ and assume that $\mathfrak{F} \subseteq \mathfrak{R}$ for some one-generated ω-local formation $\mathfrak{R}$. If $\mathfrak{F}$ is a ω-composition formation, then $t \leq 3$.

Proof Assume on the contrary that $t \geq 4$. Let $\mathfrak{M} = \mathfrak{F}_1 \mathfrak{F}_2$ and $\mathfrak{H} = \mathfrak{F}_3 \ldots \mathfrak{F}_t$. Then, by Theorem 2.2, we see that $\mathfrak{F}_1$ and $\mathfrak{M}$ are ω-local formations and by Lemma 3.8, $\mathfrak{F}_1, \mathfrak{M} \subseteq \mathfrak{N}_\omega \mathfrak{N}$. Suppose that $\mathfrak{M} \subseteq \mathfrak{N}$. Then by Lemma 3.9, we have $\mathfrak{M} = \mathfrak{N}_p$, for some prime p. Also, by Lemma 3.11, we know that $\mathfrak{F}_1$ contains every subgroup of every group in $\mathfrak{F}_1$. Hence $\mathfrak{F}_1 \subseteq \mathfrak{M}$, and so $\mathfrak{F}_1 \subseteq \mathfrak{N}_p$. This leads to $\mathfrak{M} = \mathfrak{F}_1$ and $\mathfrak{F} = \mathfrak{F}_1 \mathfrak{F}_3 \ldots \mathfrak{F}_t$. However, this is a contradiction and hence $\mathfrak{M} \not\subseteq \mathfrak{N}$. Let A be a group with minimal order in $\mathfrak{M} \setminus \mathfrak{N}_\omega$. And let $R = A^{\mathfrak{N}_\omega}$ be the monolith of A. Assume that $R = A$. Then A is a simple group and $|A| \notin \omega$. Now, by Lemma 2.7, we see that $\mathfrak{H}$ is an abelian formation which contradicts Lemma 3.9. Hence $R \subset A$, and consequently, $R = F_p(A) = O_p(A)$, for some prime p. Since $A \in \mathfrak{M} \subseteq \mathfrak{N}_\omega \mathfrak{N}$, we have $p \in \omega$. Evidently, $C_A(R) = R$, and so $O_p(A/R) = 1$ by Chap. 1, Lemma 2.15. Thus there exists a simple group B such that $|B| \neq p$ and $B \simeq A/N$ for some normal subgroup N of A such that $R \subseteq N$. Let m be the smallest ω-local satellite of $\mathfrak{M}$. Then $A/F_p(A) \in m(p)$, and so $B \in m(p)$. Hence by Lemma 3.10, $\mathfrak{H}$ is an abelian formation, which is a contradiction. Thus $t \leq 3$.

We are now able to answer the open Problem 21 in [367].

Theorem 3.13 (Guo, Selkin and Shum [162]). *The product*

$$\mathfrak{F}_1 \mathfrak{F}_2 \ldots \mathfrak{F}_t \tag{5.4}$$

is a noncancellative factorization of some one-generated ω-composition formation $\mathfrak{F}$ if and only if the following conditions hold.

(1) $t \leq 3$ *and every factor in (5.4) is a nonidentity formation;*

(2) $\mathfrak{F}_1$ *is a one-generated ω-local subformation in $\mathfrak{N}_\omega\mathfrak{N}$ and $\pi(\mathrm{Com}(\mathfrak{F})) \cap \omega \subseteq \pi(\mathfrak{F}_1)$;*

(3) *If $\mathfrak{F}_1 \not\subseteq \mathfrak{N}_\omega$, then $t = 2$, $\mathfrak{F}_2$ is an abelian one-generated formation and for all groups $A \in \mathfrak{F}_1$ and $B \in \mathfrak{F}_2$, we have $(|A/F_\omega(A)|, |B|) = 1$ and $(|A/O_\omega(A)|, |B|) = 1$;*

(4) *If $\mathfrak{F}_1 \subseteq \mathfrak{N}_\omega$ and $t = 3$, then $\mathfrak{F}_3$ is a one-generated abelian formation, for every $p \in \pi(\mathfrak{F}_1)$, the formation $\mathfrak{F}_2(p)$ is a one-generated nilpotent formation, and $\pi(A/O_p(A)) \cap \pi(B) = \varnothing$, for all groups $A \in \mathfrak{F}_2$ and $B \in \mathfrak{F}_3$;*

(5) *If $\mathfrak{F}_1 \subseteq \mathfrak{N}_\omega$, $t = 2$ and $|\pi(\mathfrak{F}_1)| > 1$, then $\mathfrak{F}_2(\omega')$ and $\mathfrak{F}_2$ are one-generated formations;*

(6) *If $\mathfrak{F}_1 = \mathfrak{N}_p$, for some prime $p \in \omega$, then there is a group $B \in \mathfrak{F}_2$ such that for all groups $A \in \mathfrak{F}_1$ the $\mathfrak{F}_2$-residual $T^{\mathfrak{F}_2}$ of the wreath product $T = A \wr B$ is contained subdirectly in the base group of T. Moreover, if $t = 2$, then $\mathfrak{F}_2(\omega'), \mathfrak{F}_2(p)$ are one-generated formations and $\mathfrak{F}_2 \not\subseteq \mathfrak{F}_1$.*

Proof We first assume that $\mathfrak{F} = \mathfrak{F}_1\mathfrak{F}_2\ldots\mathfrak{F}_t = c_\omega\mathrm{form}G$ for some group G. Let f be the smallest ω-composition satellite of the formation $\mathfrak{F}$. We will show that the factors of the noncancellative factorization in (5.4) all satisfy conditions (1)–(6).

It is clear that every factor in (5.4) is a nonidentity formation. In additions, by Lemma 3.12, we have $t \leq 3$. Thus, (1) is true.

Now, by Theorem 2.2, we see that $\mathfrak{F}_1$ is a ω-local formation and by Lemma 3.8 we have $\mathfrak{F}_1 \subseteq \mathfrak{N}_\omega\mathfrak{N}$. By using Corollary 1.9 and Lemma 3.7, we can easily show that the set of all ω-local subformations of the formation $\mathfrak{F}_1$ is finite. Hence, there is a chain

$$(1) = \mathfrak{M}_0 \subseteq \mathfrak{M}_1 \subseteq \ldots \subseteq \mathfrak{M}_{n-1} \subseteq \mathfrak{M}_n = \mathfrak{F}_1,$$

where $\mathfrak{M}_i$ is a maximal ω-local subformation of $\mathfrak{M}_{i+1}$, $i = 0, 1, \cdots, n - 1$. Let $A_i \in \mathfrak{M}_i \setminus \mathfrak{M}_{i-1}$, $i = 1, 2, \cdots, n$. Then $l_\omega\mathrm{form}(\mathfrak{M}_{i-1} \cup \{A_i\}) = \mathfrak{M}_i$, and so

$$\mathfrak{F}_1 = l_\omega\mathrm{form}(A_1, \cdots, A_n) = l_\omega\mathrm{form}(A_1 \times \ldots \times A_n)$$

is a one-generated ω-local formation. By using the same arguments as in the proof of Lemma 3.5, one can easily see that $\pi(\mathrm{Com}(\mathfrak{F})) \cap \omega \subseteq \pi(\mathfrak{F}_1)$. This shows that (2) holds.

Assume that $\mathfrak{F}_1 \not\subseteq \mathfrak{N}_\omega$. Let $\mathfrak{H} = \mathfrak{F}_2\ldots\mathfrak{F}_t$ and A be a group of minimal order in $\mathfrak{F}_1 \setminus \mathfrak{N}_\omega$. Let R be the monolith of A. Then $R = A^{\mathfrak{N}_\omega} = C_A(R)$ and $R \not\subseteq \Phi(A)$. If R is a p-group, then $R = O_p(A) = F_p(A)$. By (2), A/R is a nilpotent group. But by [135, Lemma 1.7.11], we have $O_p(A/C_A(R)) = 1$, and so $p \notin \pi(A/R)$. Assume that $R = A$. Then $|R| = p$ and since $R = A^{\mathfrak{N}_\omega}$, we can deduce that $p \notin \omega$. Hence $\mathfrak{H}$ is an abelian formation by Lemma 2.7. Let $R \neq A$ and let $R \leq M \leq A$, where M is a maximal subgroup of A. Then A/M is a simple group with $|A/M| \neq p$. Let m be the smallest ω-local satellite of $\mathfrak{F}_1$. Since $A \in \mathfrak{F}_1$, we have $A/F_p(A) = A/R \in m(p)$, and so $A/M \in m(p)$. Now, by using Lemma 3.10, we see that $t = 2$ and that $\mathfrak{F}_2$ is an abelian formation. It follows that $\mathfrak{F} = \mathfrak{F}_1\mathfrak{F}_2$ is a soluble ω-composition formation and hence, in this case, $\mathfrak{F}$ is a ω-local formation.

Now we prove that $\mathfrak{F}_2$ is a one-generated formation. Assume that $\pi(\mathfrak{F}_1) \cap \omega = \{p\}$. Since $\mathfrak{F}_1 \nsubseteq \mathfrak{N}_\omega$, we may choose in $\mathfrak{F}_1$ a group A of prime order $q \notin \omega$. Let $B \in \mathfrak{F}_2$ and $T = A \wr B$. Then, it is clear that $T \in \mathfrak{F}$ and $O_\omega(T) = O_p(T) = 1$. Hence by Lemma 1.8, we have

$$T \simeq T/O_\omega(T) \in f(\omega') = \text{form}(G/O_\omega(G)).$$

In follows that $\mathfrak{F}_2 \subseteq \text{from}(G/O_\omega(G))$. Since the formation $\mathfrak{F}_2$ is soluble, $\mathfrak{F}_2$ is one-generated formation. Now let $|\pi(\mathfrak{F}_1) \cap \omega| > 1$ and let p, q be two different primes in $\pi(\mathfrak{F}_1) \cap \omega$. Let $B \in \mathfrak{F}_2$. Then by Lemma 3.3, we have $\mathfrak{F} = CF_\omega(t)$, where

$$t(a) = \begin{cases} m(p)\mathfrak{F}_2 & \text{if } a = p \in \pi(\text{Com}(\mathfrak{F}_1)) \cap \omega; \\ \varnothing & \text{if } a = p \in \omega \setminus \pi(\text{Com}(\mathfrak{F}_1)); \\ m(\omega')\mathfrak{F}_2 & \text{if } a = \omega'. \end{cases}$$

We note that the formation function m is an inner ω-composition satellite of $\mathfrak{F}_1$. Now, using Lemma 1.8, we deduce that

$$B/O_p(B) \in f(\omega') = \text{form}(G/C^p(G))$$

and

$$B/O_q(B) \in \text{form}(q) = \text{form}(G/C^q(G)).$$

Hence

$$B \in \text{form}((G/C^p(G)) \times (G/C^q(G))).$$

This shows that $\mathfrak{F}_2 \subseteq \text{form}(G/C^p(G) \times (G/C^q(G))$ and so $\mathfrak{F}_2$ is a one-generated formation.

Let $A \in \mathfrak{F}_1$ and $B \in \mathfrak{F}_2$. Assume that there is a prime p such that $p \in \pi(A/F_\omega(A))$ and $p \in \pi(B)$. Then there is a prime $q \in \omega$ such that $p \in \pi(A/F_q(A))$. Since $A \in \mathfrak{F}_1 \subseteq \mathfrak{N}_\omega \mathfrak{N}$, we have $p \neq q$. Let Z_p be a group of order p, $n = |G|$. Let $D = Z_p \wr (A_1 \times \ldots \times A_n)$, where $A_1 \simeq \ldots \simeq A_n \simeq Z_p$ and let $T = Z_q \wr D$, where Z_q is a group of order q. Clearly, $D \in m(q)\mathfrak{F}_2$. Thus, by Lemma 1.7, we have

$$T \in \mathfrak{N}_q(m(q)\mathfrak{F}_2) = (\mathfrak{N}_q m(a))\mathfrak{F}_2 \subseteq \mathfrak{F}_1 \mathfrak{F}_2.$$

It is not difficult to see that $C^q(T) = F_q(T) = O_q(T)$, and so in the factor group $T/F_q(T)$ there is a subgroup $L \simeq D$. But

$$T/C^q(T) \in f(q) = \text{form}(G/F_q(G))$$

and by Lemma 2.5, we see that the nilpotent class of the group D is at least $n + 1$, which contradicts Lemma 2.4. This shows that the exponents of the groups $A/F_\omega(A)$ and B must be coprime.

Now assume that there is a prime p such that $p \in \pi(A/O_\omega(A)) \cap \pi(B)$. Let $T = A \wr (B_1 \times \ldots \times B_n)$ where $B_1 \simeq \ldots \simeq B_m \simeq Z_p$. Since $\mathfrak{F}_2$ is an abelian formation, we have $V = B_1 \times \ldots \times B_n \in \mathfrak{F}_2$, and so $T \in \mathfrak{F}_1\mathfrak{F}_2$. In view of Lemma 1.10, we have

$$T/O_\omega(T) \in f(\omega') = \text{form}(G/O_\omega(G)).$$

Let $K = A_1 \times A_2 \times \ldots \times A_{|V|}$ be the base group of the wreath product T. Then, it is clear that

$$O_\omega(K) = O_\omega(A_1) \times O_\omega(A_2) \times \ldots \times O_\omega(A_{|V|}) = (O_\omega(A))^\natural = \prod_{b \in V_|} (O_\omega(A_1))^b$$

Hence, by [89, A, (18.2)], we have

$$T/O_\omega(T) \simeq (A/O_\omega(A)) \wr V \in \text{form}(G/O_\omega(G)).$$

But since $A \in \mathfrak{N}_\omega\mathfrak{N}$, $p \notin \omega$, and so there is a subgroup $L \simeq Z_p \wr V$ in $T/O_\omega(T)$, this contradicts Lemma 2.4. Hence the exponents of the groups $A/O_\omega(A)$ and B must be coprime. This shows that the condition (3) holds.

Next we assume that $\mathfrak{F}_1 \subseteq \mathfrak{N}_\omega$ and $t = 3$. First we consider the case that $|\pi(\mathfrak{F}_1)| > 1$. We claim that $\mathfrak{F}_2(p)$ is nilpotent formation, for all $p \in \omega$. In fact, if the claim is not true, then we can let A be a nonnilpotent group of smallest order in $\mathfrak{F}_2(p)$. Let R be the monolith of A and q a prime in $\pi(\mathfrak{F}_1)$ such that $R \nsubseteq O_q(A)$. Let $T = Z_q \wr A$, where Z_q is a group of order q. Then, it is clear that

$$T \in \mathfrak{F}_1(\mathfrak{F}_2(p)) \subseteq \mathfrak{F}_1\mathfrak{F}_2 \subseteq \mathfrak{N}_\omega\mathfrak{N}.$$

Also, it is not difficult to show that $F(T) = K$, where K is the base group of T. Clearly, $T/K \simeq A$ is not nilpotent, and so $T \notin \mathfrak{N}_\omega\mathfrak{N}$. This contradiction shows that $\mathfrak{F}_2(p)$ is a nilpotent formation, and our claim is hence established. If $\mathfrak{F}_1\mathfrak{F}_2 \subseteq \mathfrak{N}_\omega$, then by Lemma 3.9, we have $\mathfrak{F}_1\mathfrak{F}_2 = \mathfrak{F}_1 = \mathfrak{N}_p$, which is impossible. Hence $\mathfrak{F}_1\mathfrak{F}_2 \nsubseteq \mathfrak{N}_\omega$ and so by (3), $\mathfrak{F}_3$ is an abelian one-generated formation. Let $p \in \pi(\mathfrak{F}_1)$, $A \in \mathfrak{F}_2$, $T_{\omega'} = A/O_p(A)$ and $B \in \mathfrak{F}_3$. Assume that there is a prime q such that $q \in \pi(T_{\omega'}) \cap \pi(B)$. Let Z_q be a group of order q. Let $T = Z_q \wr (Z_1 \times \ldots \times Z_n)$, where $Z_1 \simeq Z_2 \simeq \ldots \simeq Z_n \simeq Z_q$ and $n = |G|$. Form $Y = T_{\omega'} \wr (B_1 \times \ldots \times B_n)$, where $B_i \simeq B$. Then, by Lemma 2.5, the nilpotent class $c(T)$ of the group T is at least $n + 1$. If $q \notin \omega$, then by [89, A, (18.2)] and by Lemma 1.8, we have

$$T/O_\omega(T) \simeq T \simeq E \leq Y/O_\omega(Y) \in \text{form}(G/O_\omega(G)).$$

This clearly contradicts Lemma 2.4. Let $q \in \omega$. Then, in view of (2), we have $q \in \pi(\mathfrak{F}_1) \setminus \{p\}$. Let Z_p be a group of order p and $D = Z_p \wr T = K \rtimes T$, where K is the base group of D. Then, clearly, $D \in \mathfrak{F}_1\mathfrak{F}_2\mathfrak{F}_3 = \mathfrak{F}$, and so by Lemma 1.8, we have

$$D/C^p(D) = D/K \simeq T \in \text{form}(G/C^p(G)),$$

this again contradicts Lemma 2.4. Thus, for all groups $A \in \mathfrak{F}_2$ and $B \in \mathfrak{F}_3$, we have $\pi(A/O_p(A)) \cap \pi(B) = \varnothing$. Now we claim that $\mathfrak{F}_2(p)$ is a one-generated formation. Indeed, by (2), $\mathfrak{F}_1\mathfrak{F}_2$ is a one-generated ω-local formation. By using Lemma 3.5, we see that $\mathfrak{F}_2$ or $\mathfrak{F}_2(p)$ is a one-generated formation. But $\mathfrak{F}_2$ is a soluble formation and $\mathfrak{F}_2(p) \subseteq \mathfrak{F}_2$. Hence $\mathfrak{F}_2(p)$ is a one-generated formation in view of Lemma 3.7.

Now, we consider $\mathfrak{F}_1 = \mathfrak{N}_p$, for some $p \in \omega$. By Lemma 3.5, $\mathfrak{F}_2(p)$ is a one-generated nilpotent formation since $\mathfrak{F}_1\mathfrak{F}_2$ is a one-generated ω-composition formation by (2). Thus, condition (4) holds.

Condition (5) and the first two conditions of (6) follow directly from Lemma 3.5. It is clear that $\mathfrak{F}_2 \not\subseteq \mathfrak{F}_1$. Now we assume that for every group $B \in \mathfrak{F}_2$ there is a group $A \in \mathfrak{F}_1$ such that the $\mathfrak{F}_2$-residual $T^{\mathfrak{F}_2}$ of the wreath product $T = A \wr B$ is not contained subdirectly in the base group of T. And let B be a group of minimal order in $\mathfrak{N}_p\mathfrak{F}_2 \setminus \mathfrak{F}_2$. Then the group B is monolithic and its monolith is $R = B^{\mathfrak{F}_2}$. Now let $T = A \wr (B/R)$, where A is a group in $\mathfrak{F}_1$ such that the $\mathfrak{F}_2$-residual $T^{\mathfrak{F}_2}$ of T is not contained subdirectly in the base group K of T. Then the formation $\mathfrak{F}_2$ contains the group $Z_p \wr (B/R)$, where $|Z_p| = p$. Evidently, R is an elementary abelian p-group, so by Lemma 1.6, $B \in \mathfrak{F}_2$. This contradiction shows that $\mathfrak{N}_p\mathfrak{F}_2 \subseteq \mathfrak{F}_2$. Hence $\mathfrak{N}_p\mathfrak{F}_2 = \mathfrak{F}_2 = \mathfrak{F}$. This contradiction shows that condition (6) holds.

Now, to shows the converse statement, we suppose that

$$\mathfrak{F} = \mathfrak{F}_1\mathfrak{F}_2\ldots\mathfrak{F}_t$$

and the condition (1)–(6) hold. We first show that $\mathfrak{F}$ is one-generated ω-composition formation. In view of Lemma 3.2, we only need to show that $f(a)$ is a one-generated formation for all $a \in (\pi(Com(\mathfrak{F}))\cap\omega)\cup\{\omega'\}$, where f is the smallest ω-composition satellite of the formation $\mathfrak{F}$ and $|\pi(Com(\mathfrak{F})) \cap \omega| < \infty$.

By the condition (2), we see that $|\pi(Com(\mathfrak{F})) \cap \omega| < \infty$. Let $\mathfrak{H} = \mathfrak{F}_2\ldots\mathfrak{F}_t$, and m be the smallest ω-composition satellite of the formation $\mathfrak{F}_1$. Then by Lemma 3.3, we have $\mathfrak{F} = CF_\omega(t)$, where

$$t(a) = \begin{cases} m(p)\mathfrak{H}, & \text{if } a = p \in \pi(\mathrm{Com}(\mathfrak{F}_1)) \cap \omega \\ \varnothing, & \text{if } a = p \in \omega \setminus \pi(\mathrm{Com}(\mathfrak{F}_1)) \\ m(\omega')\mathfrak{H}, & \text{if } a = \omega'. \end{cases}$$

Assume that $\mathfrak{F}_1 \not\subseteq \mathfrak{N}_\omega$. Then by hypothesis, $\mathfrak{H} = \mathfrak{F}_2$ is an one-generated abelian formation. Let $p \in \pi(\mathrm{Com}(\mathfrak{F}_1)) \cap \omega$. Since $\mathfrak{F}_1$ is a one-generated ω-composition formation and $\mathfrak{F}_1 \subseteq \mathfrak{N}_\omega\mathfrak{N}$, $m(p)$ is a nilpotent one-generated formation by Lemma 3.2. Since for any groups $A \in \mathfrak{F}_1$ and $B \in \mathfrak{F}_2$, we have $(|A/F_p(A)|, |B|) = 1$, then by Lemma 1.8 $m(p) \cap \mathfrak{H} = (1)$. Now by using the similar arguments as in the proof of [135, Theorem 4.5.8], we see that $m(p)\mathfrak{H}$ is a one-generated formation. But $f(p) \subseteq t(p) = m(p)\mathfrak{H}$, so by Lemma 3.7, $f(p)$ is also a one-generated formation. Similarly, we can show that $f(\omega')$ is a one-generated formation. Hence, $\mathfrak{F}$ is indeed a one-generated ω-composition formation.

Let $\mathfrak{F}_1 \subseteq \mathfrak{N}_\omega$. In this case, by Lemma 3.5, we only need to show that if $t = 3$, then $\mathfrak{H}(\omega')$ and $\mathfrak{H}(p)$ (for all $p \in \pi(\mathfrak{F}_1)$) are one-generated formations and $\mathfrak{H}$ is a one-generated formation if $|\pi(\mathfrak{F})| > 1$.

Let $p \in \pi(\mathfrak{F}_1) \cap \omega$. Consider the formation $\mathfrak{F}_0 = \mathfrak{F}_2(p)\mathfrak{F}_3$. In view of condition (4), we have $\pi(\mathfrak{F}_2(p))\cap\pi(\mathfrak{F}_3) = \varnothing$. Besides, $\mathfrak{F}_2(p)$ is a nilpotent one-generated formation. In these cases, the product $\mathfrak{F}_2(p)\mathfrak{F}_3$ is a one-generated formation. But evidently $\mathfrak{F}_2(p)\mathfrak{F}_3$ is a soluble formation, so every subformation of $\mathfrak{F}_2(p)\mathfrak{F}_3$ is also one-generated. Thus, in order to prove that $(\mathfrak{F}_2\mathfrak{F}_3)(p)$ is a one-generated formation, we only need to show that it is indeed a subformation of $\mathfrak{F}_2(p)\mathfrak{F}_3$. Let $A \in \mathfrak{F}_2\mathfrak{F}_3$. Then $A^{\mathfrak{F}_3} \in \mathfrak{F}_2$, and so $O_p(A^{\mathfrak{F}_3})$ is a characteristic subgroup of $A^{\mathfrak{F}_3}$ such that

$$A^{\mathfrak{F}_3}/O_p(A^{\mathfrak{F}_3}) \in \mathfrak{F}_2(p).$$

But since $A^{\mathfrak{F}_3}/O_p(A^{\mathfrak{F}_3}) = (A/O_p(A^{\mathfrak{F}_3}))^{\mathfrak{F}_3}$, we have $A/O_p(A^{\mathfrak{F}_3}) \in \mathfrak{F}_2(p)\mathfrak{F}_3$, and so $A/O_p(A) \in \mathfrak{F}_2(p)\mathfrak{F}_3$. Thus $(\mathfrak{F}_2\mathfrak{F}_3)(p) \subseteq \mathfrak{F}_2(p)\mathfrak{F}_3$. This shows that the formation $(\mathfrak{F}_2\mathfrak{F}_3)(p)$ is one-generated.

Assume that $|\pi(\mathfrak{F})| > 1$ and let $p, q \in \pi(\mathfrak{F}_1)$. Let $A \in \mathfrak{F}_2$ and $B \in \mathfrak{F}_3$. Since $|A/O_p(A)|, |B|) = 1$ and $(|A/O_q(A)|, |B|) = 1$, we have $(|A|, |B|) = 1$. This shows that the exponents of the formations $\mathfrak{F}_2$ and $\mathfrak{F}_3$ are coprime. Same as above, one can show that $\mathfrak{F}_2$ is a nilpotent formation. Hence $\mathfrak{F}_2\mathfrak{F}_3$ is a one-generated formation.

Consider the formation $\mathfrak{F}_0 = \mathfrak{F}_2(\omega')\mathfrak{F}_3$. Clearly, $\mathfrak{F}_2(\omega')$ is a one-generated nilpotent formation and $\pi(\mathfrak{F}_2(\omega')) \cap \pi(\mathfrak{F}_3) = \varnothing$. Hence $\mathfrak{F}_0$ is a soluble one-generated formation. Now, in order to prove that $(\mathfrak{F}_2\mathfrak{F}_3)(\omega')$ is a one-generated formation, we only need to show that $(\mathfrak{F}_2\mathfrak{F}_3)(\omega') \subseteq \mathfrak{F}_2(\omega')\mathfrak{F}_3$. Let $A \in \mathfrak{F}_2\mathfrak{F}_3$. Since $O_\omega(A^{\mathfrak{F}_3})$ is a characteristic subgroup of $A^{\mathfrak{F}_3}$ such that $A^{\mathfrak{F}_3}/O_\omega(A^{\mathfrak{F}_3}) \in \mathfrak{F}_2(\omega')$, we have $A/O_\omega(A) \in \mathfrak{F}_2(\omega')\mathfrak{F}_3$. Hence $(\mathfrak{F}_2\mathfrak{F}_3)(\omega') \subseteq \mathfrak{F}_2(\omega')\mathfrak{F}_3$.

Finally, we still need to show that the factorization (5.4) is noncancellative. For this purpose, we first take $t = 2$. Assume that $\mathfrak{F}_1 \not\subseteq \mathfrak{N}_\omega$. Then by conditions (3) $\mathfrak{F}_2$ is an abelian formation, and so by Lemma 3.9, $\mathfrak{F} \neq \mathfrak{F}_2$. Suppose that $\mathfrak{F} = \mathfrak{F}_1$, and let A be a group of minimal order in $\mathfrak{F}_1 \setminus \mathfrak{N}_\omega$. Let R be the monolith of A. Then $R = A^{\mathfrak{N}_\omega}$. It is clear that $R \not\subseteq \Phi(A)$. Let B be a simple group in $\mathfrak{F}_2$ and $T = A \wr B = K \rtimes B$ where K is the base group of T. Since $\mathfrak{F}_2$ is abelian, $T^{\mathfrak{F}_2}$ is contained subdirectly in K, and therefore

$$T \in \mathfrak{F} = \mathfrak{F}_1 \subseteq \mathfrak{N}_\omega\mathfrak{N}.$$

It is also clear that $R = F(A)$ and by Lemma 2.3, we deduce that

$$L = R^{\natural} = \prod_{b\in B} R_1^b = F(T)$$

is the monolith of T, where R_1 is the monolith of the first copy A in K. Since $T \in \mathfrak{N}_\omega\mathfrak{N}$, we have $T^{\mathfrak{N}} \in \mathfrak{N}_\omega$, i.e., $T^{\mathfrak{N}} \subseteq L$. Let R be an ω'-group. Then L is an ω'-group, and so $O_\omega(T) = 1$. Since $T \in \mathfrak{N}_\omega\mathfrak{N}$, T must be a nilpotent group. But $F(T) = L \neq T$, this is a contradiction. Hence R is a p-group, for some $p \in \omega$. However, since $A \notin \mathfrak{N}_\omega$, we have $R = F_\omega(A)$ and consequently, $(|A/R|, |B|) = 1$. Let B be a q-group. Then B is a Sylow q-subgroup of $T_1 = (A/R) \wr B$. By [89, A, (18.2)], we have $T_1 \simeq T/L$. This proves that T_1 is a nilpotent group. Thus, $B \trianglelefteq T$, and so $B \cap K_1 \neq 1$, where K_1 is the base group of T_1, a contradiction. This shows that $\mathfrak{F}_1 \neq \mathfrak{F} \neq \mathfrak{F}_2$.

Now we assume that $\mathfrak{F}_1 \subseteq \mathfrak{N}_\omega$. Let $|\pi(\mathfrak{F}_1)| > 1$ and p, q be different primes in $\pi(\mathfrak{F}_1)$. Also we let B be a group such that $\mathfrak{F}_2 = \text{form} B$. Since $\mathfrak{F}_1$ is an ω-local formation, we have $\mathfrak{N}_{\{p,q\}} \subseteq \mathfrak{F}_1$. Hence $\mathfrak{F} \neq \mathfrak{F}_2$ by Lemma 2.4. In view of Lemma 3.9, we conclude that $\mathfrak{F} \neq \mathfrak{F}_1$.

Let $\pi(\mathfrak{F}_1) = \{p\}$, for some $p \in \omega$. Then $\mathfrak{F}_1 = \mathfrak{N}_p$. Let B be a group in $\mathfrak{F}_2$ such that for every group $A \in \mathfrak{F}_1$ the $\mathfrak{F}_2$-residual $T^{\mathfrak{F}_2}$ of the wreath product $T = A \wr B$ is contained subdirectly in the base group of T. Assume that $\mathfrak{F} = \mathfrak{F}_2 = \mathfrak{N}_p \mathfrak{F}_2$ and let A be a nonidentity group in $\mathfrak{N}_p$. If $T = A \wr B$, then $T \in \mathfrak{F} = \mathfrak{F}_2$, and so $T^{\mathfrak{F}_2} = 1$ is not contained subdirectly in the base group of T. This contradiction shows that $\mathfrak{F} \neq \mathfrak{F}_2$. Also by condition (6), $\mathfrak{F}_2 \not\subseteq \mathfrak{F}_1$, and thereby, we have $\mathfrak{F} \neq \mathfrak{F}_1$. This shows that the factorization (5.4) is noncancellative.

Now let $t = 3$. Consider the case $\pi(\mathfrak{F}_1) = \{p\}$. Then, it is clear that $\mathfrak{F}_2 \not\subseteq \mathfrak{N}_p$. Let A be a group of minimal order in $\mathfrak{F}_2 \setminus \mathfrak{N}_p$. Then $O_p(A) = 1$. Thus, if $B \in \mathfrak{F}_3$, we have $(|A|, |B|) = 1$. Let $T = Z_p \wr (A \wr B)$. Evidently, $T \in \mathfrak{F}$ and T is not a metanilpotent group. Hence $\mathfrak{F} \not\subseteq \mathfrak{N}^2$. Since the formations $\mathfrak{F}_1 \mathfrak{F}_2, \mathfrak{F}_1 \mathfrak{F}_3$ are all metanilpotent, $\mathfrak{F} \neq \mathfrak{F}_1 \mathfrak{F}_2, \mathfrak{F}_1 \mathfrak{F}_3$. By the condition (6), we can let B be a group in $\mathfrak{F}_2$ such that for all nonidentity groups $A \in \mathfrak{F}_1$ the $\mathfrak{F}_2$-residual $D^{\mathfrak{F}_2}$ of the group $D = A \wr B$ is contained subdirectly in the base group of the wreath product D. Let C be a nonidentity group in $\mathfrak{F}_3$. Assume that $\mathfrak{F}_2 \mathfrak{F}_3 = \mathfrak{F}_1 \mathfrak{F}_2 \mathfrak{F}_3$ and $T = D \wr C = K \rtimes C$, where K is the base group of D. Then since $\mathfrak{F}_3$ is abelian formation, we see that $T^{\mathfrak{F}_3}$ is contained subdirectly in K, and so $D \in \mathfrak{F}_2$, that is, $D^{\mathfrak{F}_2} = 1$, a contradiction. Hence $\mathfrak{F} \neq \mathfrak{F}_2 \mathfrak{F}_3$. Now assume that $|\pi(\mathfrak{F}_1)| > 1$. Let $\{p, q\} \subseteq \pi(\mathfrak{F}_1)$. Then, by (4), $\mathfrak{F}_2(p)$ and $\mathfrak{F}_2(q)$ are one-generated nilpotent formations. Hence, $\mathfrak{F}_2$ is also one-generated nilpotent formation by Lemma 3.4. Therefore, $\mathfrak{F} \neq \mathfrak{F}_2 \mathfrak{F}_3$ since $\mathfrak{F}$ is not a metanilpotent formation. For the case $|\pi(\mathfrak{F}_1)| > 1$, the proof is similar. Thus, the theorem is proved.

Let $\mathfrak{F}$ be a formation. We call $\mathfrak{F}$ is a limited formation if $\mathfrak{F}$ is a subformation of some one-generated formation. As an analogy, we call a ω-composition formation $\mathfrak{F}$ limited if it is a subformation of some one-generated ω-composition formation.

The following theorem can be proved analogously as Theorem 3.13, we omit the details.

Theorem 3.14 *The product*

$$\mathfrak{F}_1 \mathfrak{F}_2 \dots \mathfrak{F}_t \qquad (5.5)$$

is a noncancellative factorization of some limited ω-composition formation $\mathfrak{F}$ if and only if the following conditions hold:

(1) $t \leq 3$ and every factor in (5.5) is a nonidentity formation;

(2) $\mathfrak{F}_1$ is a one-generated ω-local subformation in $\mathfrak{N}_\omega \mathfrak{N}$ and $\pi(\mathrm{Com}(\mathfrak{F})) \cap \omega \subseteq \pi(\mathfrak{F}_1)$;

(3) If $\mathfrak{F}_1 \not\subseteq \mathfrak{N}_\omega$, then $t = 2$, $\mathfrak{F}_2$ is an abelian one-generated formation and for all groups $A \in \mathfrak{F}_1$ and $B \in \mathfrak{F}_2$, $(|A/F_\omega(A)|, |B|) = 1$ and $(|A/O_\omega(A)|, |B|) = 1$;

(4) If $\mathfrak{F}_1 \subseteq \mathfrak{N}_\omega$ and $t = 3$, then $\mathfrak{F}_3$ is a one-generated abelian formation and for every $p \in \pi(\mathfrak{F}_1)$, the formation $\mathfrak{F}_2(p)$ is one-generated nilpotent and for all groups $A \in \mathfrak{F}_2$ and $B \in \mathfrak{F}_3$, $\pi(A/O_p(A)) \cap \pi(B) = \varnothing$;

(5) If $\mathfrak{F}_1 \subseteq \mathfrak{N}_\omega$, $t = 2$ and $|\pi(\mathfrak{F}_1)| > 1$, then $\mathfrak{F}_2(\omega')$, $\mathfrak{F}_2$ are limited formations;

(6) If $\mathfrak{F}_1 = \mathfrak{N}_p$ for some prime p, then $\mathfrak{F}_2(\omega')$ and $\mathfrak{F}_2(p)$ (if $p \in \omega$) are limited formations, $\mathfrak{F}_2 \not\subseteq \mathfrak{F}_1$, and there is a group $B \in \mathfrak{F}_2$ such that for all groups $A \in \mathfrak{F}_1$, the $\mathfrak{F}_2$-residual $T^{\mathfrak{F}_2}$ of the wreath product $T = A \wr B$ is contained subdirectly in the base group of T.

Some applications of Theorem 3.13. Theorem 3.13 not only gave the solution of Problem 21 in [367] but also generalizes many known results. Some of these results we give below.

First of all, from Lemma 3.5(5) and Theorem 3.13 we get

Corollary 3.15 ([186]). *Let $\mathfrak{F} = \mathfrak{M}\mathfrak{H}$ be a uncancellative factorization of a Baer-local formation $\mathfrak{F}$ and $\mathfrak{H} = \mathfrak{M}_2 \cdots \mathfrak{M}_t$. Then $\mathfrak{F}$ is one-generated Baer-local formation if and only if the following conditions are satisfied:*

1) $\mathfrak{M}$ *is a metanilpotent one-generated saturated formation and* $C(\mathfrak{H}) \cap \mathfrak{A} \subseteq \mathfrak{M}$;
2) *If the formation $\mathfrak{M}$ is not primary (i.e., $|\pi(\mathfrak{M})| > 1$), then $\mathfrak{H}$ is a one-generated formation; besides, if $\mathfrak{M}$ is not nilpotent, then $\mathfrak{H}$ is abelian;*
3) *For all groups $A \in \mathfrak{M}$ and $B \in \mathfrak{H}$, the groups $A/F(A)$ and B have the coprime exponents;*
4) *If $\mathfrak{M} = \mathfrak{N}_p$, for some prime p, then $\mathfrak{H}/O_p(\mathfrak{H})$ is a one-generated formation.*

Corollary 3.16 (Guo, Shum [170]). *Let*

$$\mathfrak{F} = \mathfrak{M}_1 \mathfrak{M}_2 \cdots \mathfrak{M}_t \tag{5.6}$$

be a uncancellative factorization of a Baer-local formation $\mathfrak{F}$ and $\mathfrak{H} = \mathfrak{M}_2 \cdots \mathfrak{M}_t$. Then $\mathfrak{F}$ is one-generated Baer-local formation if and only if the following conditions are satisfied:

1) $\mathfrak{M}_1$ *is a metanilpotent one-generated local formation and* $C(\mathfrak{H}) \cap \mathfrak{A} \subseteq \mathfrak{M}_1$;
2) *If the formation $\mathfrak{M}_1$ is not primary (i.e., $|\pi(\mathfrak{M})| > 1$), then $\mathfrak{H}$ is a one-generated formation; besides, if $\mathfrak{M}_1$ is not nilpotent, then $\mathfrak{H}$ is abelian;*
3) *For all groups $A \in \mathfrak{M}_1$ and $B \in \mathfrak{H}$, the groups $A/F(A)$ and B have the coprime exponents;*
4) *If $\mathfrak{M}_1 = \mathfrak{N}_p$ for some prime p, then $\mathfrak{H}/O_p(\mathfrak{H})$ is a one-generated formation;*
5) $t \leq 3$, *and if $t = 3$, then $\mathfrak{M}_1$ and $\mathfrak{M}_2$ are nilpotent, $\mathfrak{M}_3$ is abelian and the exponents of $\mathfrak{M}_2$ and $\mathfrak{M}_3$ are coprime.*

Corollary 3.17 *(Skiba and Shemetkov [388]). Let p be a prime. Then the formation $\mathfrak{N}_p$ can not be expressed in the form $\mathfrak{N}_p = \mathfrak{M}\mathfrak{H}$, where $\mathfrak{M}$ and $\mathfrak{H}$ are proper subformations in $\mathfrak{N}_p$.*

Proof Assume that the corollary is not true. Then, by $\mathfrak{M} \neq \mathfrak{N}_p \neq \mathfrak{H}$, we see that $\mathfrak{M}\mathfrak{H}$ is a noncancellative factorization of $\mathfrak{N}_p$. Since $\mathfrak{N}_p = l_p \mathrm{form} A$, where A is a nonidentity group in $\mathfrak{N}_p$, by Theorem 3.13, $\mathfrak{M}$ is a p-local formation. However, since $\mathfrak{M} \neq \mathfrak{N}_p$, we have $\mathfrak{M} = (1)$, which is the formation of all identity groups. This leads to $\mathfrak{F} = (1)\mathfrak{H} = \mathfrak{H}$, a contradiction.

Corollary 3.18 *(Skiba [96]). The product $\mathfrak{F}_1 \ldots \mathfrak{F}_t$ of formations $\mathfrak{F}_1, \cdots, \mathfrak{F}_t$ is a non-cancellative factorization of a one-generated formation $\mathfrak{F}$ if and only if the following conditions hold:*

(i) $t = 2$ *and both formations $\mathfrak{F}_1, \mathfrak{F}_2$ are one-generated;*

(ii) $\mathfrak{F}_1$ *is nilpotent,* $\mathfrak{F}_2$ *is abelian and* $(|A|, |B|) = 1$ *for all groups* $A \in \mathfrak{F}_1$ *and*
$\quad B \in \mathfrak{F}_2$.

Proof We note that if p is a prime and $\mathfrak{F}$ is a formation such that $p \notin \pi(\text{Com}(\mathfrak{F}))$,
then $\mathfrak{F} = CF_p(f)$, where $f(p) = \varnothing$ and $f(\omega') = \mathfrak{F}$.

Now, we assume that $\mathfrak{F}_1 \ldots \mathfrak{F}_t$ is a noncancellative factorization of a one-generated
formation $\mathfrak{F} = \text{form}\,G$. Since evidently $\text{Com}(\mathfrak{F}) = \text{Com}(G)$, we may choose a prime
p such that $p \notin \pi(\text{Com}(\mathfrak{F}))$. In this case, $\mathfrak{F} = c_p \text{form}\,G$ is a one-generated p-
composition formation. Consequently, conditions (1)–(6) of Theorem 3.13 all hold
for the formations $\mathfrak{F}_1, \cdots, \mathfrak{F}_t$. Thus, $\mathfrak{F}_1$ is a one-generated p-local formation such that
$\mathfrak{F}_1 \subseteq \mathfrak{N}_p \mathfrak{N}$. By Lemma 3.11, we have $\mathfrak{F}_1 \subseteq \mathfrak{F}$. This leads to $p \notin \pi(\text{Com}(\mathfrak{F}_1)) =
\pi(\mathfrak{F}_1)$ and so $\mathfrak{F}_1 \subseteq \mathfrak{N}$. Hence $\mathfrak{F}_1 \nsubseteq \mathfrak{N}_p$, and therefore by Theorem 3.13(3), the
product $\mathfrak{F}_2 \ldots \mathfrak{F}_t$ is an abelian formation. By using Lemma 3.9, we can easily see
that $t = 2$. Since $p \notin \pi(\mathfrak{F}_1)$ for every groups $A \in \mathfrak{F}_1$ and $B \in \mathfrak{F}_2$, we have
$(|A/O_p(A)|, |B|) = (|A|, |B|) = 1$. Moreover, we have $\mathfrak{F}_1 = l_p \text{form}\,A = \text{form}\,A$,
for some group A.

Now assume that $\mathfrak{F} = \mathfrak{F}_1\mathfrak{F}_2$, where the above conditions (i), (ii) of the corollary
hold for the formations $\mathfrak{F}_1$ and $\mathfrak{F}_2$. Let A, B be the groups such that $\mathfrak{F}_1 = \text{form}\,A$ and
$\mathfrak{F}_2 = \text{form}\,B$. Let $\pi_1 = \pi(A)$, $\pi_2 = \pi(B)$. Then for every group $T \in \mathfrak{F}$, we have
$\pi(T) \subseteq \pi_1 \cup \pi_2$. This implies that there exists a prime p such that $p \notin \pi(\text{Com}(\mathfrak{F}))$.
It is not difficult to check that the conditions (1)–(3) of Theorem 3.13 are satisfied
for $\mathfrak{F}_1, \mathfrak{F}_2$ and so $\mathfrak{F} = c_p \text{form}\,G$ for some group G. But since $p \notin \pi(\text{Com}(G))$, we
can immediately see that

$$c_p \text{form}\,G = \text{form}\,G$$

is a one-generated formation.

5.4   On Two Problems of the Lattice Theory of Formations

Complete lattices of formations. Let θ be any nonempty set of formation. If $\mathfrak{F} \in \theta$,
then we say that $\mathfrak{F}$ is a θ-formation.

We say that θ is a complete lattice of formations if the intersection of any set
of θ-formations is again a θ-formation, and there exists a θ-formation $\mathfrak{F}$ such that
$\mathfrak{M} \subseteq \mathfrak{F}$ for any θ-formation $\mathfrak{M}$.

If $\mathfrak{X} \subseteq \mathfrak{H} \in \theta$, then we use the symbol $\theta \text{form}\,\mathfrak{X}$ to denote the intersection of all
θ–formations containing $\mathfrak{X}$, and we say that $\theta \text{form}\,\mathfrak{X}$ is the θ-formation generated
by $\mathfrak{X}$. If $\mathfrak{X} = \{G\}$, then $\theta \text{form}\,\mathfrak{X}$ is said to be the θ-formation generated by the group
G, and denoted by $\theta \text{form}\,G$.

A formation $\mathfrak{F}$ is called a one-generated θ-formation if for some group G we have
$\mathfrak{F} = \theta \text{form}\,G$.

For any set of θ-formations $\{\mathfrak{F}_i \mid i \in I\}$, we write $\vee_\theta(\mathfrak{F}_i | i \in I)$ to denote the
intersection of all θ-formations containing $\cup_{i \in I} \mathfrak{F}_i$. In particular, we use $\mathfrak{F}_1 \vee_\theta \mathfrak{F}_2$ to
denote the intersection of all θ-formations containing $\mathfrak{F}_1 \cup \mathfrak{F}_2$.

The complete lattice τ^l. A formation function f is called θ-valued if every nonempty value $f(p)$ of f is a θ-formation. For any complete lattice θ of formations, we write θ^ℓ to denote the set of all saturated formations $\mathfrak{F}$ such that $\mathfrak{F} = LF(f)$ for at leas one θ-valued local satellite f. We write $\theta^l \mathrm{form} \mathfrak{X}$ to denote the intersection of all θ^l-formations containing $\mathfrak{X}$.

For any set $\{f_i \mid i \in I\}$ of formation functions f_i, we write $\cap(f_i \mid \in I)$ to denote the formation function f such that $f(p) = \cap_{i \in I} f_i(p)$ for all primes p.

The following lemma is evident.

Lemma 4.1 *Let f_i be a local satellite of the formation $\mathfrak{F}_i, i \in I$. Then $\cap(f_i \mid \in I)$ is a local satellite of the intersection $\cap_{i \in I} \mathfrak{F}_i$.*

Lemma 4.2 *For any complete lattice θ of formations, the set θ^l is also a complete lattice of formations.*

Proof In view of Lemma 4.1, the intersection of any set of formations in θ^l belongs is still in θ^l. Now let $\mathfrak{F}$ be a θ-formation such that $\mathfrak{M} \subseteq \mathfrak{F}$ for any θ-formation $\mathfrak{M}$, and let $f(p) = \mathfrak{F}$ for all primes p. Then $\mathfrak{F} = LF(f) \in \theta^l$ and for any θ^l-formation $\mathfrak{H}$ we have $\mathfrak{H} \subseteq \mathfrak{F} = LF(f)$.

If $\{f_i \mid i \in I\}$ is the set of all θ-valued local satellites of $\mathfrak{F}$, then by Lemma 4.1, the formation function $\cap(f_i \mid \in I)$ is still a θ-valued local satellite of $\mathfrak{F}$. Such a satellite is called the smallest θ-valued local satellite of $\mathfrak{F}$.

Lemma 4.3 *Let θ be a complete lattice of formations, $\mathfrak{X} \subseteq \mathfrak{H} \in \theta$ and $\mathfrak{F} = \theta^\ell \mathrm{form} \mathfrak{X}$. Let f be the smallest θ-valued local satellite of $\mathfrak{F}$. Then the following assertions hold.*

(1) $\pi(\mathfrak{X}) = \pi(\mathfrak{F})$;
(2) $f(p) = \phi$ if $p \in \mathbf{P} \backslash \pi(\mathfrak{X})$; $f(p) = \theta \mathrm{form}(A/F_p(A) | A \in \mathfrak{X})$ if $p \in \pi(\mathfrak{X})$.
(3) If h is any θ-valued local satellite of $\mathfrak{F}$, then $f(p) = \theta \mathrm{form}(A | A \in \mathfrak{F} \cap h(p)$ and $O_p(A) = 1)$ for all $p \in \pi(\mathfrak{X})$.

Proof (2) Let t be a formation function such that $t(p) = \phi$ if $p \in \mathbf{P} \backslash \pi(\mathfrak{X})$ and $t(p) = \theta \mathrm{form}(A/F_p(A) | A \in \mathfrak{X})$ if $p \in \pi(\mathfrak{X})$. We show that $f = t$. Let $\mathfrak{M} = LF(t)$. Then, evidently, $\mathfrak{X} \subseteq \mathfrak{M}$. Let f_1 be any θ-valued local satellite of $\mathfrak{F}$. Since $\mathfrak{X} \subseteq \mathfrak{F}$, for any prime $p \in \pi(\mathfrak{X})$ we have $\theta \mathrm{form}(A/F_p(A) | A \in \mathfrak{X}) \subseteq \mathfrak{F}$, and so $t(p) \subseteq f_1(p)$ for all primes p. Hence $\mathfrak{M} \subseteq \mathfrak{F}$, which implies that $\mathfrak{M} = \mathfrak{F}$ and $t = f$.

(1) follows from (2).

(3) Let f_1 be a formation function such that

$$f_1(p) = \theta \mathrm{form}(A | A \in \mathfrak{F} \cap h(p), O_p(A) = 1)$$

for all $p \in \pi(\mathfrak{X})$ and $f_1(p) = \emptyset$ for all primes $p \notin \pi(\mathfrak{X})$. Then $f \le f_1$. Indeed, if $B \in \mathfrak{F}$ and $p \in \pi(B)$, then $B/F_p(B) \in \mathfrak{F} \cap h(p)$ and $O_p(B/F_p(B)) = 1$. Hence by (2), $f(p) \subseteq f_1(p)$. Now let $p \in \pi(\mathcal{X})$, $A \in \mathfrak{F} \cap h(p)$ and $O_p(A) = 1$. Let $G = P \wr A = K \rtimes A$, where K is the base group of the regular wreath product G. Then $K = F_p(G)$ and so $A \simeq G/F_p(G)$. But by Chap. 1, Proposition 1.15, $G \in \mathfrak{F}$. Hence $A \in f(p)$, and so $f_1(p) \subseteq f(p)$. Thus $f = f_1$. The lemma is proved.

Corollary 4.4 *Let f_i be the smallest θ–valued local satellite of $\mathfrak{F}_i, i = 1, 2$. Then $\mathfrak{F}_1 \subseteq \mathfrak{F}_2$ if and only if $f_1 \leqslant f_2$, that is, $f_1(p) \subseteq f_2(p)$ for all primes p.*

For any set $\{f_i \mid i \in I\}$ of θ-valued formation functions f_i, we write $\vee_\theta(f_i | i \in I)$ to denote the formation function f such that $f(p)$ is the θ-formations generated by $\cup_{i \in I} f_i(p)$ for all primes p.

From Lemma 4.3 we get

Lemma 4.5 *Let f_i be the smallest θ-valued local satellite of the formation $\mathfrak{F}_i, i \in I$. Then $\vee_\theta(f_i | i \in I)$ is the smallest θ-valued local satellite of the formation $\mathfrak{F} = \vee_{\theta^t}(\mathfrak{F}_i | i \in I)$.*

The lattice l_n, l_n^τ and l_∞. A subgroup functor τ is said to be a *Skiba subgroup functor* [259] if the following conditions are satisfied:

(1) $G \in \tau(G)$ for every group G;
(2) for any epimorphism $\varphi : A \to B$, where A, B are two groups, and for any group $H \in \tau(A)$ and $T \in \tau(B)$, we have that $H^\varphi \in \tau(B)$ and $T^{\varphi^{-1}} \in \tau(A)$.

In the following, τ denotes an arbitrary Skiba subgroup functor.

A formation $\mathfrak{F}$ is called τ-closed if $\tau(G) \subseteq \mathfrak{F}$ for every group G in $\mathfrak{F}$.

Every formation can be considered as 0-multiply local. Let $n > 0$. Then a formation $\mathfrak{F}$ is called n-multiply local [381] if it has a local satellite f such that every nonempty value $f(p)$ of f is $n - 1$-multiply local. A formation $\mathfrak{F}$ is said to be totally local if it is n-multiply local for all natural number n. n-multiply composition formation and totally composition formation are defined analogously.

We use l_n, l_n^τ and l_∞ to denote the sets of all n-multiply local, τ-closed n-multiply local and total local formations, respectively. The formation $\mathfrak{G}$ of all groups is, evidently, total local and τ-closed for any subgroup functor τ; in particular, it is τ-closed n-multiply local for any n. Hence, in view of Lemma 4.1, l_n, l_n^τ and l_∞ are all complete lattices of formations.

Let $\mathfrak{H}$ and $\mathfrak{F}$ be two θ-formations. If $\mathfrak{H}$ is a known θ-formation, then we can obtain some information about the θ-formation $\mathfrak{F}$ because $\mathfrak{F}$ contains a part of the formation $\mathfrak{H}$, that is, $\mathfrak{H} \cap \mathfrak{F}$. For example, when we study a local formation $\mathfrak{F}$, the nilpotent subformation $\mathfrak{F} \cap \mathfrak{N}$ is often used. However, in general, it would be rather difficult to describe the formation $\mathfrak{F}$ without additional restrictions on $\mathfrak{F} \cap \mathfrak{H}$. One possible restriction on $\mathfrak{F} \cap \mathfrak{H}$ is to consider the θ–lattice of formations $\mathfrak{F}/_\theta(\mathfrak{H} \cap \mathfrak{F}) = \{\mathfrak{M}|\mathfrak{M}$ is a θ–formation such that $\mathfrak{H} \cap \mathfrak{F} \subseteq \mathfrak{M} \subseteq \mathfrak{F}\}$.

In this respect, A.N. Skiba raised the following interesting questions (see p. 192 in [381]):

Problem 4.6 [381, Problem 5.2.2]. Describe the soluble totally local formations $\mathfrak{F}$ satisfying the lattice $\mathfrak{F}/_\infty(\mathfrak{F} \cap \mathfrak{N})$ of all totally local formations between $\mathfrak{F} \cap \mathfrak{N}$ and $\mathfrak{F}$ is Boolean (or a complemented lattice).

Problem 4.7 [381, Problem 5.2.1]. Describe the τ-closed n-multiply local formations $\mathfrak{F}$ satisfying the lattice $\mathfrak{F}/_n^\tau(\mathfrak{F} \cap \mathfrak{N})$ of all τ-closed n-multiply local formations between $\mathfrak{F} \cap \mathfrak{N}$ and $\mathfrak{F}$ is Boolean (or a complemented lattice).

Our next goal is to give the answers to these Problems.

Lemma 4.8 *Let* $\mathfrak{F}$ *be a local formation and suppose that* $\phi \neq \pi \subseteq \mathbf{P}$. *Then* $\mathfrak{F} = \mathfrak{S}_\pi \mathfrak{F}$ *if and only if* $\mathfrak{F}$ *has a screen* f *such that* $f(p) = \mathfrak{F}$ *for all* $p \in \pi$.

Proof Let $\mathfrak{F} = LF(f)$ with $f(p) = \mathfrak{F}$ for all $p \in \pi$. Then $\mathfrak{N}_p \mathfrak{F} \subseteq \mathfrak{F}$ for all $p \in \pi$ by Lemma 1.7. Assume that $\mathfrak{S}_\pi \mathfrak{F} \not\subseteq \mathfrak{F}$. Then we can find a group G with minimal order in $\mathfrak{S}_\pi \mathfrak{F} \backslash \mathfrak{F}$. Clearly, this group G has a unique minimal normal subgroup $R = G^{\mathfrak{F}}$ which is a p-group for some $p \in \pi$. This implies that $G \in \mathfrak{F}$ since $\mathfrak{N}_p \mathfrak{F} \subseteq \mathfrak{F}$, a contradiction. Hence $\mathfrak{S}_\pi \mathfrak{F} = \mathfrak{F}$.

Suppose that $\mathfrak{F} = \mathfrak{S}_\pi \mathfrak{F}$ and f is the canonical local satellite of the formation $\mathfrak{F}$. If $\mathfrak{F} \not\subseteq f(p)$ for some $p \in \pi$. Then there is a group G with minimal order in $\mathfrak{F} \backslash f(p)$. We denote by R the unique minimal normal subgroup of G. Since $\mathfrak{N}_p f(p) = f(p)$, $O_p(G) = 1$. It follows from Chap. 1, Lemma 5.9 that there exists a simple $\mathbb{F}_p G$-module which is faithful for G. Let $F = P \rtimes G$ be the semidirect product of P and G, where $|P| = p$. Then, it is easy to see that $F \in \mathfrak{F}$. So we have

$$F/F_p(F) \simeq G \in f(p),$$

which is a contradiction. Therefore, $\mathfrak{F} \subseteq f(p)$ and hence $\mathfrak{F} = f(p)$.

Lemma 4.9 (see [135, Theorem 3.1.20]). *If* $\mathfrak{F} = \mathfrak{M}\mathfrak{H}$, *where* $\mathfrak{M} = LF(m), \mathfrak{H} = LF(h)$, *and* h *is an inner satellite of* $\mathfrak{H}$, *then* $\mathfrak{F} = LF(f)$, *where* f *is defined by*

(1) $f(p) = m(p)\mathfrak{H}$, *if* $p \in \pi(\mathfrak{M})$,
(2) $f(p) = h(p)$, *if* $p \in \pi(\mathfrak{H}) \backslash \pi(\mathfrak{M})$,
(3) $f(p) = \phi$, *if* $p \notin \pi(\mathfrak{M}) \cup \pi(\mathfrak{H})$.

The following lemma is evident.

Lemma 4.10 *Let* $\mathfrak{S}_\pi$ *be the class of all soluble* π-*groups. Then* $\mathfrak{S}_\pi$ *is a local formation with a local satellite* f *such that* $f(p) = \mathfrak{S}_\pi$ *for all* $p \in \pi$ *and* $f(p) = \phi$ *for all* $q \in \mathbf{P} \backslash \pi$.

Lemma 4.11 *Let* G *be a group with a nonabelian unique minimal normal subgroup* $R = G^{\mathfrak{N}}$. *If* $\pi = \pi(R)$ *and* $\mathfrak{F} = \ell_\infty \mathrm{form} G$, *then*

$$\mathfrak{F} = \mathfrak{S}_\pi \mathfrak{F} = \mathfrak{S}_\pi \ell \mathrm{form} G.$$

Proof Let f be the smallest ℓ_∞-valued local satellite of the formation $\mathfrak{F}$. Then by Lemma 4.3, we have

$$f(p) = \ell_\infty \mathrm{form}(G/F_p(G)) = \ell_\infty \mathrm{form} G$$

for all $p \in \pi$. Hence, by Lemma 4.10, $\mathfrak{F} = \mathfrak{S}_\pi \mathfrak{F}$. As it is clear that $\ell \mathrm{form} G \subseteq \ell_\infty \mathrm{form} G = \mathfrak{F}$, thereby $\mathfrak{S}_\pi \ell \mathrm{form} G \subseteq \mathfrak{F}$.

To prove the opposite inclusion, it suffices for us to show that the formation $\mathfrak{S}_\pi \ell \mathrm{form} G$ is totally local. By Lemma 4.9, we know that $\mathfrak{S}_\pi$ has a local satellite s such that $s(p) = \mathfrak{S}_\pi$ for all $p \in \pi$ and $s(q) = \phi$ for all $q \in \mathbf{P} \backslash \pi$. On the other hand, by Lemma 4.3, we know that $\ell\mathrm{form}G$ has a local satellite h such that

$$h(q) = \begin{cases} \mathrm{form}(G/F_q(G)) = \mathrm{form}G, & \text{if } q \in \pi, \\ \mathrm{form}(G/F_q(G)) = (1), & \text{if } q \in \pi(G)\backslash\pi, \\ \phi, & \text{if } q \in \mathbf{P}\backslash\pi(G). \end{cases}$$

Then, it follows from Lemma 4.9 that the formation $\mathfrak{S}_\pi\ell\mathrm{form}G$ has the following satellite m such that

$$m(p) = \begin{cases} \mathfrak{S}_\pi\ell\mathrm{form}G, & \text{if } p \in \pi, \\ (1), & \text{if } p \in \pi(G)\backslash\pi, \\ \phi, & \text{if } p \in \mathbf{P}\backslash\pi(G). \end{cases}$$

Assume that the formation $\mathfrak{S}_\pi\ell\mathrm{form}G$ is n-multiple local. Then it follows from the description of the screen m that the formation $\mathfrak{S}_\pi\ell\mathrm{form}G$ is $(n+1)$-multiple local. So, by induction, we can see that $\mathfrak{S}_\pi\ell\mathrm{form}G$ is totally local. Therefore $\mathfrak{F} \subseteq \mathfrak{S}_\pi\ell\mathrm{form}G$ and hence $\mathfrak{F} = \mathfrak{S}_\pi\ell\mathrm{form}G$. The lemma is proved.

For a totally local formation $\mathfrak{F}$, we write $L_\infty(\mathfrak{F})$ to denote the lattice of all totally local formations which are in $\mathfrak{F}$.

Lemma 4.12 ([366, Lemma 8.10]). *Let* $G = NM$, *where* $N \lhd G, M \leqslant G$ *and* $N \in \mathfrak{N}^{n+1}$. *Then* $M \in \ell_n\mathrm{form}G$.

Lemma 4.13 *Let* $\mathfrak{F} = \ell_\infty\mathrm{form}G$ *be the totally local formation generated by a group* G. *Then the following statements hold:*

1) If G *is soluble, then* $L_\infty(\mathfrak{F})$ *is finite;*
2) If G *is nonsoluble, then* $L_\infty(\mathfrak{F})$ *is infinite.*

Proof

1) Suppose G is a soluble group with its nilpotent length $\ell(G) = n$. We prove that $L_\infty(\mathfrak{F})$ is finite by using induction on n.
   For $n = 1$, we have $\mathfrak{F} = \mathfrak{N}_\pi$, where $\pi = \pi(G)$. Since $|\pi| < |G| < \infty$ and every nilpotent local formation $\mathfrak{H}$ has the form $\mathfrak{N}_\varphi$, where $\varphi = \pi(\mathfrak{H})$, this implies that the lattice $L_\infty(\mathfrak{F})$ is finite.
   For $n > 1$, we suppose $L_\infty(\ell_\infty\mathrm{form}A)$ is finite for every soluble group A with $\ell(A) < n$. Let $\mathfrak{M}$ be an arbitrary totally local subformation of $\mathfrak{F}$ and let m, f be the smallest ℓ_∞-valued local satellite of $\mathfrak{M}$ and $\mathfrak{F}$ respectively. Then, by Corollary 4.4, we have $m \leqslant f$. By applying Lemma 4.3, we have

$$f(p) = \begin{cases} \ell_\infty\mathrm{form}(G/F_p(G)), & \text{if } p \in \pi(G), \\ \phi, & \text{if } P \in \mathbf{P}\backslash\pi(G). \end{cases}$$

Obviously, $\ell(G/F_p(G)) < \ell(G)$. This shows that the lattice $L_\infty(\ell_\infty\mathrm{form}$ $(G/F_p(G)))$ is finite for every $p \in \pi(G)$. It follows from $|\pi(G)| = |\pi(\mathfrak{F})| < \infty$ that the number of totally local subformations in $\mathfrak{F}$ is finite, i.e., $L_\infty(\mathfrak{F})$ is finite.

2) Suppose that G is nonsoluble. Let A be a group of minimal order in $\mathfrak{F}\backslash\mathfrak{S}$. Clearly, A has a unique minimal normal subgroup $R = A^{\mathfrak{S}}$ which is nonabelian. Let $\mathfrak{M} = \ell_\infty \text{form} A$ and $\pi = \pi(R)$. Then $\mathfrak{N} = \mathfrak{S}_\pi \mathfrak{M}$ (cf. the proof of Lemma 4.11). By Lemma 4.9, we know that the formations $\mathfrak{N}_\pi, \mathfrak{N}_\pi^2 = \mathfrak{N}_\pi \mathfrak{N}_\pi, \cdots, \mathfrak{N}_\pi^n = \mathfrak{N}_\pi \mathfrak{N}_\pi \cdots \mathfrak{N}_\pi, \cdots$ are all totally local and they are contained in $\mathfrak{M} \subseteq \mathfrak{F}$. Hence the number of totally local subformations in $\mathfrak{F}$ must be infinite. The lemma is proved.

A totally local formation $\mathfrak{F}$ is called a minimal totally local non-$\mathfrak{H}$-formation (or $\mathfrak{H}_\infty$-critical formation) if $\mathfrak{F} \not\subseteq \mathfrak{H}$ but $\mathfrak{F}_1 \subseteq \mathfrak{H}$ for every proper totally local subformation $\mathfrak{F}_1$ of $\mathfrak{F}$. In general, a formation $\mathfrak{F}$ is said to be $\mathfrak{H}$-critical if $\mathfrak{F} \not\subseteq \mathfrak{H}$ but $\mathfrak{F}_1 \subseteq \mathfrak{H}$ for every proper subformation $\mathfrak{F}_1$ of $\mathfrak{F}$. The following lemma characterizes the minimal totally local nonnilpotent formations.

Lemma Lemma 4.14 $\mathfrak{F}$ is a minimal totally local nonnilpotent formation if and only if $\mathfrak{F} = \mathfrak{N}_p \mathfrak{N}_q$ for different primes p and q.

Proof Let $\mathfrak{F}$ be a minimal total local nonnilpotent formation and G a group of minimal order in $\mathfrak{F}\backslash\mathfrak{N}$. Then G has a unique minimal normal subgroup $R = G^{\mathfrak{N}}$. If R itself is nonabelian, then $|\pi| > 1$, where $\pi = \pi(R)$. It follows from Lemma 4.11 that $\mathfrak{F} = \ell_\infty \text{form} G = \mathfrak{S}_\pi \ell_\infty \text{form} G$ and hence $\mathfrak{F}$ has a nonnilpotent totally local soluble subformation $\mathfrak{S}_\pi$. This shows that $\mathfrak{F} = \mathfrak{S}_\pi$ and G is soluble, a contradiction. Thus, R must be a p-group for a prime p. It is clear that $R \not\subseteq \Phi(G)$ and so $C_G(R) = R$. Since G/R is nilpotent, $p \notin \pi(G/R)$ by Chap. 1, Lemma 2.15. Let $\varphi = \{p_1, p_2, \cdots, p_t\} = \pi(G/R)$. Suppose that $|\varphi| > 1$. Let P_i be a group of order $p_i, i = 1, 2, \cdots, t$. Also, we let P be a group of order p. Then, by Lemma 4.12, we know that $M \in \ell \text{form} G \subseteq \ell_\infty \text{form} G = \mathfrak{F}$ for any subgroup M of G, i.e., $\mathfrak{F}$ is subgroup-closed. This leads to $D = P \wr P_1 \in \mathfrak{F}$ and hence $\mathfrak{M} = \ell_\infty \text{form} D \subseteq \mathfrak{F}$. But it is clear that the formation $\mathfrak{M} = \ell_\infty \text{form} D$ is not nilpotent. Therefore $\mathfrak{M} = \mathfrak{F}$ and thereby, by Lemma 4.3, we have

$$\ell_\infty \text{form}(G/F_p(G)) = \ell_\infty \text{form}(G/R)$$
$$= \ell_\infty \text{form}(D/F_p(D)) = \ell_\infty \text{form} P_1.$$

Since $\pi(G/R) \neq \pi(P_1)$, we have $\ell_\infty \text{form}(G/R) \neq \ell_\infty \text{form} P_1$. However, this is clearly a contradiction and so the only choice for t is $t = 1$. Let $q = p_1$. Obviously, $F_{p_1}(G) = R$. It follows from Lemma 4.3 that $\mathfrak{F}$ must have the smallest ℓ_∞-valued local satellite f such that

$$f(r) = \begin{cases} \ell_\infty \text{form}(G/R) = \mathfrak{N}_q, & \text{if } r = p, \\ (1), & \text{if } r = q, \\ \phi, & \text{if } r \in \mathbf{P}\backslash\{p, q\}. \end{cases}$$

It is now easy to check, by using Lemma 4.3, that the formation $\mathfrak{N}_p\mathfrak{N}_q$ has the smallest ℓ_∞-valued local satellite m such that

$$m(r) = \begin{cases} \mathfrak{N}_q, & \text{if } r = p, \\ (1), & \text{if } r = q, \\ \phi, & \text{if } r \in \mathbf{P}\backslash\{p,q\}. \end{cases}$$

Therefore $\mathfrak{F} = \mathfrak{N}_p\mathfrak{N}_q$.

Now let $\mathfrak{F} = \mathfrak{N}_p\mathfrak{N}_q$. By using the same arguments as above, we can also show that $\mathfrak{F}$ has the smallest ℓ_∞-valued local satellite f such that

$$f(r) = \begin{cases} \mathfrak{N}_q, & \text{if } r = p, \\ (1), & \text{if } r = q, \\ \phi, & \text{if } r \in \mathbf{P}\backslash\{p,q\}. \end{cases}$$

Let $\mathfrak{M}$ be any proper totally local subformation of $\mathfrak{F}$ and m the smallest ℓ_∞-valued local satellite of $\mathfrak{M}$. Then, by Corollary 4.4, we know that $m \leqslant f$. Let r be a prime number such that $m(r) \subset f(r)$, consequently, it can be seen that $r \in \{p,q\}$. If $r = q$, then $m(q) = \phi$ and hence $q \notin \pi(\mathfrak{M})$. This leads to $m(p) = 1$ since m is inner, and so $\mathfrak{M} = \mathfrak{N}_p$ is a nilpotent formation. If $m(p) \subset f(p)$, then $m(p) \subseteq (1)$ and hence $\mathfrak{M} \subseteq \mathfrak{N}_{\{p,q\}}$. This implies that every proper totally local subformation of $\mathfrak{F}$ is nilpotent. Obviously, $\mathfrak{F} \not\subseteq \mathfrak{N}$. Therefore $\mathfrak{F}$ is a minimal totally local nonnilpotent formation. The lemma is proved.

Lemma 4.15 *Let $\mathfrak{F}$ be a nonnilpotent totally local formation. Then $\mathfrak{F}$ contains an $\mathfrak{N}_\infty$-critical subformation.*

Proof Let G be a group of minimal order in $\mathfrak{F}\backslash\mathfrak{N}$. Then G has a unique minimal normal subgroup $R = G^{\mathfrak{N}}$. If R itself is abelian, then the formation $\mathfrak{F}$ is, of course, soluble, i.e., $\mathfrak{F} \subseteq \mathfrak{S}$. It follows from Lemma 4.13 that the set of totally local subformations of $\mathfrak{M} = \ell_\infty \text{form} G$ is finite. Consequently, there exists a totally local subformation $\mathfrak{H}$ of $\mathfrak{F}$ such that $\mathfrak{H} \not\subseteq \mathfrak{N}$ and every proper totally local subformation of $\mathfrak{H}$ is contained in $\mathfrak{N}$.

On the other hand, if R is a nonabelian group and $\pi = \pi(R)$, then we let $\mathfrak{M} = \ell_\infty \text{form} G$. Hence, by Lemma 4.11, we have $\mathfrak{M} = \mathfrak{S}_\pi \ell \text{form} G$. Obviously $|\pi| > 1$. Let p, q be different primes in π. Then by Lemma 4.14, $\mathfrak{N}_p\mathfrak{N}_q$ is a minimal totally local nonnilpotent formation. Furthermore, it is clear that $\mathfrak{N}_p\mathfrak{N}_q \subseteq \mathfrak{M} \subseteq \mathfrak{F}$. The lemma is hence proved.

In the following, for any formation $\mathfrak{F}$, we denote by $\mathfrak{F}_0$ the intersection $\mathfrak{F} \cap \mathfrak{N}$.

Lemma 4.16 *If $\mathfrak{F}/_\infty\mathfrak{F}_0$ be a complemented lattice, then $\mathfrak{F}$ is a soluble formation.*

Proof Suppose, on the contrary, that the formation $\mathfrak{F}$ is not soluble. Then, we have $\mathfrak{F} \not\subseteq \mathfrak{N}$, and hence by Lemma 4.15, $\mathfrak{F}$ has a minimal totally local nonnilpotent subformation $\mathfrak{H}_1$. Let $\{\mathfrak{H}_i | i \in I\}$ be the collection of all minimal totally local nonnilpotent subformations of $\mathfrak{F}$. Let $\mathfrak{M} = \mathfrak{F}_0 \vee_\infty (\vee_\infty(\mathfrak{H}_i | i \in I))$. Then, by Lemma 4.14, we know that every $\mathfrak{H}_i (i \in I)$ is soluble and thereby this leads to $\mathfrak{M} \neq \mathfrak{F}$.

Now, let $\mathfrak{H}$ be the complement of $\mathfrak{M}$ in $\mathfrak{F}/_\infty\mathfrak{F}_0$. Then, clearly, $\mathfrak{H} \neq \mathfrak{F}_0$. Hence, by Lemma 4.15, we know that $\mathfrak{H}$ has at least a minimal totally local nonnilpotent subformation $\mathfrak{H}_i$. Clearly, $\mathfrak{H}_i \subseteq \mathfrak{M} \cap \mathfrak{H} = \mathfrak{F}_0 \subseteq \mathfrak{N}$. This contradiction shows that $\mathfrak{F}$ is a soluble formation.

Lemma 4.17 *Every soluble totally local formation generated by one element must be a compact element in the lattice of all totally local formations.*

Proof Let $\mathfrak{M} = \ell_\infty \text{form} G$, where G is a soluble group with nilpotent length $\ell(G) = n$. We need is to prove that $\mathfrak{M}$ is compact element of the lattice ℓ_∞ of all totally local formations. We prove this result by induction. Of course, the result holds trivially if $\ell(G) = 1$.

Let $\mathfrak{M} \subseteq \mathfrak{F} = \vee_\infty(\mathfrak{F}_i | i \in I)$, where $\mathfrak{F}_i$ is a totally local formation. Let m, f, and f_i be the smallest ℓ_∞-valued local satellites of $\mathfrak{M}$, $\mathfrak{F}$, and $\mathfrak{F}_i$, respectively. Then, by Lemma 4.5, we have

$$f(p) = \begin{cases} \vee_\infty(f_i(p) | i \in I), & \text{if } p \in \pi(\mathfrak{F}), \\ \phi, & \text{if } f \in \mathbf{P} \backslash \pi(\mathfrak{F}). \end{cases}$$

Also, by Corollary 4.4, we know that $m \leqslant f$. Now, by Lemma 4.3, we can see that

$$m(p) = \begin{cases} \ell_\infty \text{form}(G/F_p(G)), & \text{if } p \in \pi(G), \\ \phi, & \text{if } p \in \mathbf{P} \backslash \pi(G). \end{cases}$$

Trivially, $m(p) \leqslant f(p)$ for all $p \in \pi(G)$ since $m \leqslant f$. This leads to

$$m(p) = \ell_\infty \text{form}(G/F_p(G)) \subseteq \vee_\infty(f_i(p) | i \in I).$$

As $\ell(G/F_p(G)) < \ell(G)$, so by induction hypothesis, there are $i_1, i_2, \cdots, i_t \in I$ such that

$$G/F_p(G) \in f_{i_1}(p) \vee_\infty \cdots \vee_\infty f_{i_t}(p).$$

Since $|\pi(G)| < \infty$, there are some $j_1, j_2, \cdots, j_k \in I$ such that

$$G \in \mathfrak{F}_{j_1} \vee_\infty \cdots \vee_\infty \mathfrak{F}_{j_k}.$$

This implies that

$$\mathfrak{M} = \ell_\infty \text{form} G \subseteq \mathfrak{F}_{j_1} \vee_\infty \cdots \vee_\infty \mathfrak{F}_{j_k}.$$

and hence $\mathfrak{M}$ is a compact element of ℓ_∞.

Lemma 4.18 ([381, Theorem 4.2.12]). *The lattice of soluble totally local formations is distributive.*

Lemma 4.19 ([135, Theorem 5.1.2]). *For any two elements a and b of a modular lattice, the sublattices $(a \vee b)/a$ and $b/(a \wedge b)$ are isomorphic.*

Lemma 4.20 *Let $\mathfrak{F}$ be a soluble totally local formation. Then the lattice $\mathfrak{F}/_\infty\mathfrak{F}_0$ is algebraic and its elements with finite height are all compact elements.*

Proof Let $\mathfrak{M}$ be an arbitrary element with finite height in the lattice $\mathfrak{F}/_{\infty}\mathfrak{F}_0$. Consider the following ascending chain of totally local formations

$$\mathfrak{F}_0 = \mathfrak{M}_0 \subseteq \mathfrak{M}_1 \subseteq \cdots \subseteq \mathfrak{M}_{t-1} \subseteq \mathfrak{M}_t = \mathfrak{M},$$

where each $\mathfrak{M}_{i-1}$ is a maximal totally local subformation of $\mathfrak{M}_i (i = 1, 2, \cdots, t)$. Let A_i be a group with minimal order in $\mathfrak{M}_i \backslash \mathfrak{M}_{i-1}(i = 1, 2, \cdots, t)$. Then, we have

$$\mathfrak{M} = \ell_{\infty}\text{form}(\mathfrak{F}_0 \cup \{A_1, A_2, \cdots, A_t\})$$

$$= \mathfrak{F}_0 \vee_{\infty} (\ell_{\infty}\text{form}(A_1 \times A_2 \times \cdots \times A_t)).$$

This means that every element with finite height in the lattice $\mathfrak{F}/_{\infty}\mathfrak{F}_0$ has the form $\mathfrak{F}_0 \vee_{\infty} (\ell_{\infty}\text{form}G)$ for some groups $G \notin \mathfrak{F}_0$. Conversely, if $\mathfrak{M} = \mathfrak{F}_0 \vee_{\infty} (\ell_{\infty}\text{form}G)$ for some groups $G \in \mathfrak{F}$, then $\mathfrak{M}$ must be an element with finite height in $\mathfrak{F}/_{\infty}\mathfrak{F}_0$. In fact, since $G \in \mathfrak{F}$, we know that G is a soluble group by the given condition of the lemma. Hence, it follows from Lemma 4.13 that in the formation $\ell_{\infty}\text{form}G$ the number of totally local subformations is finite. By Lemma 4.18, the lattice of all soluble totally local formations must be distributive, and consequently, it is a modular lattice. Thus, by Lemma 4.19, we can obtain the following isomorphism between the lattices:

$$(\mathfrak{F}_0 \vee_{\infty} (\ell_{\infty}\text{form}G))/_{\infty}\mathfrak{F}_0 \simeq (\ell_{\infty}\text{form}G)/_{\infty}((\ell_{\infty}\text{form}G) \cap \mathfrak{F}_0).$$

Thus, $\mathfrak{M} = \mathfrak{F}_0 \vee_{\infty} \ell_{\infty}\text{form}G$ is an element with finite height in $\mathfrak{F}/_{\infty}\mathfrak{F}_0$.

Now, let

$$\mathfrak{M} = \mathfrak{F}_0 \vee_{\infty} (\ell_{\infty}\text{form}G) \subseteq \vee_{\infty}(\mathfrak{F}_i | i \in I),$$

where $\mathfrak{F}_i$ is in the lattice $\mathfrak{F}/_{\infty}\mathfrak{F}_0$ for all $i \in I$. Let f_i be the smallest ℓ_{∞}-valued local satellite of $\mathfrak{F}_i$, m the smallest ℓ_{∞}-valued local satellite of $\mathfrak{M}$ and f a minimal ℓ_{∞}-valued local satellite of $\vee_{\infty}(\mathfrak{F}_i | i \in I)$. Then, by Lemma 4.5, we know that $f(p) = \vee_{\infty}(f_i(p) | i \in I)$ if $p \in \pi(\bigcup_{i \in I} \mathfrak{F}_i)$, and $f(p) = \phi$ if $p \notin \pi(\bigcup_{i \in I} \mathfrak{F}_i)$.

It is also clear that

$$\mathfrak{M} = \ell_{\infty}\text{form}(\mathfrak{F}_0 \cup \{G\}).$$

Hence, by Lemma 4.3, $m(p) = \ell_{\infty}\text{form}(G/F_p(G))$ if $p \in \pi(G)$; $m(p) = (1)$ if $p \in \pi(\mathfrak{F}) \backslash \pi(G)$ and $m(p) = \phi$ if $p \in \mathbf{P} \backslash \pi(\mathfrak{F})$. By Corollary 4.4, we know that $m \leqslant f$. Hence $G/F_p(G) \in \vee_{\infty}(f_i(p) | i \in I)$ for all $p \in \pi(G)$. Therefore, by Lemma 4.17, there exist indices $i_1, i_2, \cdots, i_t \in I$ such that

$$G/F_p(G) \in f_{i_1}(p) \vee_{\infty} \cdots \vee_{\infty} f_{i_t}(p).$$

Since $|\pi(G)| < \infty$, there are indices $j_1, j_2, \cdots, j_k \in I$ such that $G \in \mathfrak{F}_{j_1} \vee_{\infty} \cdots \vee_{\infty} \mathfrak{F}_{j_k}$. Because $\mathfrak{F}_{j_1}, \mathfrak{F}_{j_2}, \cdots, \mathfrak{F}_{j_k} \in \mathfrak{F}/_{\infty}\mathfrak{F}_0$, and $\mathfrak{F}_0 \subseteq \bigcap_{i=1}^{k} \mathfrak{F}_{j_i}$, we have

$$\mathfrak{M} = \mathfrak{F}_0 \vee_{\infty} (\ell_{\infty}\text{form}G) \subseteq \mathfrak{F}_{j_1} \vee_{\infty} \cdots \vee_{\infty} \mathfrak{F}_{j_k}.$$

This shows that every element with finite height in $\mathfrak{F}/_{\infty}\mathfrak{F}_0$ is a compact element in the lattice $\mathfrak{F}/_{\infty}\mathfrak{F}_0$.

We now show that the lattice $\mathfrak{F}/_\infty\mathfrak{F}_0$ is algebraic. In fact, if $\mathfrak{H} \in \mathfrak{F}/_\infty\mathfrak{F}_0$, then we have

$$\mathfrak{H} = \ell_\infty\mathrm{form}(G | G \in \mathfrak{H}) = \ell_\infty\mathrm{form}(\mathfrak{F}_0 \bigcup (\bigcup_{G \in \mathfrak{H}} \ell_\infty\mathrm{form} G)).$$

This implies that every element $\mathfrak{H}$ of the lattice $\mathfrak{F}/_\infty\mathfrak{F}_0$ is the join of all elements with finite heights in $\mathfrak{F}/_\infty\mathfrak{F}$. Hence, we conclude that the lattice $\mathfrak{F}/_\infty\mathfrak{F}_0$ is algebraic and the lemma is proved.

Lemma 4.21 *Let $\mathfrak{F}_0 \subseteq \mathfrak{M}_1 \subseteq \mathfrak{M} \subseteq \mathfrak{F}$, where $\mathfrak{M}_1, \mathfrak{M}, \mathfrak{F}$ are soluble totally local formations. If $\mathfrak{H}$ is a complement of $\mathfrak{M}_1$ in $\mathfrak{F}/_\infty\mathfrak{F}_0$, then $\mathfrak{H} \cap \mathfrak{M}$ is a complement of $\mathfrak{M}_1$ in $\mathfrak{M}/_\infty\mathfrak{M}_0$.*

Proof By the given conditions, we know that $\mathfrak{M}_1 \cap \mathfrak{H} = \mathfrak{F}_0$ and $\mathfrak{M} \vee_\infty \mathfrak{H} = \mathfrak{F}$. Hence, by Lemma 4.18, we have

$$\mathfrak{M} = \mathfrak{M} \cap (\mathfrak{M}_1 \vee_\infty \mathfrak{H}) = \mathfrak{M}_1 \vee_\infty (\mathfrak{M} \cap \mathfrak{H}).$$

Moreover, it is clear that $(\mathfrak{M} \cap \mathfrak{H}) \cap \mathfrak{M}_1 = \mathfrak{F}_0$. This implies that $\mathfrak{M}_1 \cap \mathfrak{H}$ is the complement of $\mathfrak{M}_1$ in $\mathfrak{M}/_\infty\mathfrak{M}_0$.

Lemma 4.22 *Let $\mathfrak{M}, \mathfrak{F}$ be totally local formations with $\mathfrak{M} \subseteq \mathfrak{F}$. If $\mathfrak{F}/_\infty\mathfrak{F}_0$ is a complemented lattice, then $\mathfrak{M}/_\infty\mathfrak{M}_0$ is a complemented lattice.*

Proof By Lemmas 4.16, 4.18 and 4.19, we can see that the following lattices are isomorphic:

$$(\mathfrak{F}_0 \vee_\infty \mathfrak{M})/_\infty\mathfrak{F}_0 \simeq \mathfrak{M}/_\infty(\mathfrak{M} \cap \mathfrak{F}_0) = \mathfrak{M}/_\infty(\mathfrak{M} \cap \mathfrak{N}) = \mathfrak{M}/_\infty\mathfrak{M}_0.$$

However, since $(\mathfrak{F}_0 \vee_\infty \mathfrak{M})/_\infty\mathfrak{F}_0$ is a sublattice of the lattice $\mathfrak{F}/_\infty\mathfrak{F}_0$, by Lemma 4.21 we know that $(\mathfrak{F}_0 \vee_\infty \mathfrak{M})/_\infty\mathfrak{F}_0$ is a complemented lattice. Consequently, $\mathfrak{M}/_\infty\mathfrak{M}_0$ is the required complemented lattice.

Lemma 4.23 *Let $\mathfrak{M}$ be a nilpotent local formation and $\{\mathfrak{H}_i | i \in I\}$ a set of $\mathfrak{N}_\infty$-critical formations. If $\mathfrak{H}$ is an $\mathfrak{N}_\infty$-critical formation with $\mathfrak{H} \subseteq \mathfrak{F} = \mathfrak{M} \vee_\infty (\vee_\infty(\mathfrak{H}_i | i \in I))$, then $\mathfrak{H} = \mathfrak{H}_i$ for some $i \in I$.*

Proof Let f, m be the smallest ℓ_∞-valued local satellites of the formations $\mathfrak{F}$ and $\mathfrak{M}$ respectively. If h_i is the smallest ℓ_∞-valued local satellites of the formation $\mathfrak{H}_i$, then by Lemma 4.5, we have

$$f(p) = \begin{cases} \ell_\infty\mathrm{form}(m(p) \bigcup (\bigcup_{i \in I} h_i(p))), & \text{if } p \in \pi(\mathfrak{F}), \\ \phi, & \text{if } p \notin \pi(\mathfrak{F}). \end{cases}$$

Now, let h be the smallest ℓ_∞-valued local satellite of $\mathfrak{H}$. Then, by Lemma 4.14, we know that $\mathfrak{H} = \mathfrak{N}_p\mathfrak{N}_q$, where p, q are different primes. By Lemma 4.9, the formation $\mathfrak{H}$ has the following inner screen f_1:

$$f_1(r) = \begin{cases} \mathfrak{N}_p\mathfrak{N}_q, & \text{if } r = p, \\ \mathfrak{N}_q, & \text{if } r = q, \\ \phi, & \text{if } r \in \mathbf{P}\backslash\{p,q\}. \end{cases}$$

It follows from Lemma 4.3 that

$$h(r) = \begin{cases} \mathfrak{N}_q, & \text{if } r = p, \\ (1), & \text{if } r = q, \\ \phi, & \text{if } r \in \mathbf{P}\backslash\{p,q\}. \end{cases}$$

Since the group Z_q of order q belongs to $h(p)$, by Corollary 4.4, we know that $Z_q \in f(p)$. Hence, by the above description of the screen f, we have $(1) \subset h_i(p)$ for some $i \in I$. Consequently, among the $\{\mathfrak{H}_i | i \in I\}$, there is at least a formation $\mathfrak{H}_i$ of the form $\mathfrak{N}_p\mathfrak{N}_s$. Let $\{\mathfrak{N}_p\mathfrak{N}_{r_1}, \mathfrak{N}_p\mathfrak{N}_{r_2}, \cdots, \mathfrak{N}_p\mathfrak{N}_{r_n}, \cdots\}$ be the set of all formations in $\{\mathfrak{H}_i | i \in I\}$. Then, it is clear that

$$f(p) = \ell_\infty \text{form}(\mathfrak{N}_{r_1} \cup \mathfrak{N}_{r_2} \cup \cdots \cup \mathfrak{N}_{r_n} \cup \cdots) = \mathfrak{N}_\pi,$$

where $\pi = \{r_1, r_2, \cdots, r_n, \cdots\}$. Since $Z_q \in f(p)$, we have $q \in \pi$ and thereby

$$\mathfrak{N}_p\mathfrak{N}_q \in \{\mathfrak{N}_p\mathfrak{N}_{r_1}, \mathfrak{N}_p\mathfrak{N}_{r_2}, \cdots, \mathfrak{N}_p\mathfrak{N}_{r_n}, \cdots\}.$$

The lemma is proved.

Lemma 4.24 *Let $\mathfrak{F} = \ell_\infty \text{form} G$ be a totally local formation generated by a group G. Then there are only finitely many $\mathfrak{N}_\infty$-critical subformations of $\mathfrak{F}$.*

Proof Let $\mathfrak{H}$ be an $\mathfrak{N}_\infty$-critical subformation of $\mathfrak{F}$. Then by Lemma 4.14, we have $\mathfrak{H} = \mathfrak{N}_p\mathfrak{N}_q$ for some primes $p, q \in \pi(\mathfrak{F})$. However, by Lemma 4.3, we know that $\pi(G) = \pi(\mathfrak{F})$, and hence $\pi(\mathfrak{F})$ is a finite set. It follows that the set of all prime pairs of the form (p,q), where $p, q \in \pi(\mathfrak{F})$, is a finite set. In other words, we have shown that $\mathfrak{F}$ has only a finite number of subformations of the form $\mathfrak{N}_p\mathfrak{N}_q$, where $p, q \in \pi(\mathfrak{F})$. The lemma is proved.

By the above lemmas, we are now able to conclude the following theorem which answers Problem 4.6. It is worthwhile to note that in our statement, we do not assume that $\mathfrak{F}$ is a soluble formation.

Theorem 4.25 (Guo [164, Theorem 3.13]). *Let $\mathfrak{F}$ be a nonnilpotent totally local formation. Then the following statements are equivalent:*

1) *$\mathfrak{F}/_\infty\mathfrak{F}_0$ is a complemented lattice;*
2) *$\mathfrak{F}$ is soluble formation and $\mathfrak{F}/_\infty\mathfrak{F}_0$ is a algebraic lattice; moreover, $\mathfrak{F} = \mathfrak{F}_0 \vee_\infty$ $(\vee_\infty(\mathfrak{H}_i | i \in I))$, where $\{\mathfrak{H}_i | i \in I\}$ is the set of all $\mathfrak{N}_\infty$-critical subformations of $\mathfrak{F}$;*
3) *$\mathfrak{F}/_\infty\mathfrak{F}_0$ is a Boolean lattice.*

Proof 1)$\Rightarrow$2). By Lemmas 4.16 and 4.20, we know that $\mathfrak{F}$ is soluble and $\mathfrak{F}/_\infty\mathfrak{F}_0$ is algebraic. Now we prove that $\mathfrak{F} = \mathfrak{F}_0 \vee_\infty (\vee_\infty(\mathfrak{H}_i | i \in I))$, where $\{\mathfrak{H}_i | i \in I\}$ is the set of all $\mathfrak{N}_\infty$-critical subformations of $\mathfrak{F}$. We first suppose that $\mathfrak{F}$ is a totally local formation generated by one element. Then by Lemma 4.24, the number t of all $\mathfrak{N}_\infty$-critical subformations of $\mathfrak{F}$ is finite. We now prove condition 2) by using induction on the number t.

Let $\mathfrak{H}_i$ be any $\mathfrak{N}_\infty$-critical subformation of $\mathfrak{F}$ and $\mathfrak{M} = \mathfrak{F}_0 \vee_\infty \mathfrak{H}_i$. Then, by condition 1), there exists a totally local subformation $\mathfrak{H}$ such that $\mathfrak{M} \cap \mathfrak{H} = \mathfrak{F}_0$ and $\mathfrak{M} \vee_\infty \mathfrak{H} = \mathfrak{F}$. It is clear that $\mathfrak{H}_i \not\subseteq \mathfrak{H}$ (for if otherwise, we have $\mathfrak{H}_i \subseteq \mathfrak{H} \cap \mathfrak{M} = \mathfrak{F}_0 = \mathfrak{F} \cap \mathfrak{N} \subseteq \mathfrak{N}$, which contradicts the fact that $\mathfrak{H}_i$ is $\mathfrak{N}_\infty$-critical). Suppose that $\mathfrak{M} \subset \mathfrak{F}$. Then by Lemma 4.22 and by induction, we have $\mathfrak{H} = \mathfrak{H}_0 \vee_\infty (\vee_\infty \mathfrak{F}_j | j \in I)$, where $\{\mathfrak{F}_j | j \in J\}$ is the set of all $\mathfrak{N}_\infty$-critical subformations of $\mathfrak{H}$. But since $\mathfrak{H}_0 = \mathfrak{F}_0$, by Lemma 4.23, we have

$$\mathfrak{F} = \mathfrak{M} \vee_\infty \mathfrak{H} = (\mathfrak{F}_0 \vee_\infty \mathfrak{H}_i) \vee_\infty (\mathfrak{F}_0 \vee_\infty (\vee_\infty(\mathfrak{F}_j | j \in J)))$$
$$= (\mathfrak{F}_0 \vee_\infty \mathfrak{H}_i) \vee_\infty (\mathfrak{F}_j | j \in J)$$
$$= \mathfrak{F}_0 \vee_\infty (\vee_\infty \mathfrak{H}_i | i \in I)).$$

where $\{\mathfrak{H}_i | i \in I\}$ is the set of all $\mathfrak{N}_\infty$-critical subformations of $\mathfrak{F}$.

Now we consider the general case. Let G be an arbitrary nonnilpotent group in $\mathfrak{F}$ and $\mathfrak{M} = \ell_\infty \text{form} G$. Then, by Lemma 4.22, we know that the condition 1) also holds on the lattice $\mathfrak{M}/_\infty\mathfrak{M}_0$. This leads to $\mathfrak{M}/_\infty\mathfrak{M}_0$ is a complemented lattice. Hence, by the result proved in the previous paragraph, we have $\mathfrak{M} = \mathfrak{M}_0 \vee_\infty (\vee_\infty(\mathfrak{M}_j | j \in J))$, where $\{\mathfrak{M}_j | j \in J\}$ is the set of all $\mathfrak{N}_\infty$-critical subformations of $\mathfrak{M}$. Thus, we obtain

$$\mathfrak{F} = \ell_\infty \text{form}(G | G \in \mathfrak{F})$$
$$= \ell_\infty \text{form} G_1 \vee_\infty \ell_\infty \text{form} G_2 \vee_\infty \cdots \vee_\infty \ell_\infty \text{form} G_n \vee_\infty \cdots$$
$$= \mathfrak{F}_0 \vee_\infty (\vee_\infty(\mathfrak{H}_i | i \in I)),$$

where $\{\mathfrak{H}_i | i \in I\}$ is the set of all $\mathfrak{N}_\infty$-critical subformations of $\mathfrak{F}$. Therefore, condition 2) holds.

2)$\Rightarrow$ 3). We first prove that $\mathfrak{F}/_\infty\mathfrak{F}_0$ is a complemented lattice. In doing so, we let $\mathfrak{M}$ be a nonnilpotent formation belonging to the lattice $\mathfrak{F}/_\infty\mathfrak{F}_0$. Let $\mathfrak{X}_1$ be the set of all $\mathfrak{N}_\infty$-critical subformations of $\mathfrak{M}$ and let $\mathfrak{X}_2 = \{\mathfrak{H}_i | i \in I\}\backslash\mathfrak{X}_1$. Denote $(\ell_\infty \text{form}(\bigcup_{\mathfrak{H}_i \in \mathfrak{X}_2} \mathfrak{H}_i)) \vee_\infty \mathfrak{F}_0$ by $\mathfrak{H}$. Then, it is trivial to see that $\mathfrak{M} \vee_\infty \mathfrak{H} = \mathfrak{F}$. Assume that $\mathfrak{M} \cap \mathfrak{H} \not\subseteq \mathfrak{F}_0$. Then, by Lemma 4.15, we know that $\mathfrak{M} \cap \mathfrak{H}$ has at least an $\mathfrak{N}_\infty$-critical subformation $\mathfrak{H}^*$. If $\mathfrak{H}^* \in \mathfrak{X}_1$, then $\mathfrak{H}^* \notin \mathfrak{X}_2$. However, by Lemma 4.23 we know that in the formation $\mathfrak{F}_0 \vee_\infty \ell_\infty \text{form}(\bigcup_{\mathfrak{H}_i \in \mathfrak{X}_2} \mathfrak{H}_i)$, there is no $\mathfrak{N}_\infty$-critical subformation which is not included in $\mathfrak{X}_2$, which is a contradiction. Thus, we must have $\mathfrak{M} \cap \mathfrak{H} = \mathfrak{F}_0$ and therefore, the lattice $\mathfrak{F}/_\infty\mathfrak{F}_0$ must be complemented. By Lemma 4.16, the formation $\mathfrak{F}$ is also soluble. Thus, it follows from Lemma 4.18 that $\mathfrak{F}/_\infty\mathfrak{F}_0$ is a distributive lattice. This shows that the lattice $\mathfrak{F}/_\infty\mathfrak{F}_0$ is indeed a Boolean lattice.

3)$\Rightarrow$ 1) This part is clear. Thus, the theorem is proved.

The solution of Problem 4.7. We first need some information on τ-closed n-multiply local formations.

A group G is said to be a formation-critical group, if there are two formations $\mathfrak{M}$ and $\mathfrak{H}$ such that G is a group with minimal order in $\mathfrak{M} \backslash \mathfrak{H}$.

A subgroup functor τ is said to be closed if $\tau(H) \subseteq \tau(G)$ for any group G and $H \in \tau(G)$.

We denote by $\overline{\tau}$ the intersection of all such closed subgroup functor τ_i that $\tau \leq \tau_i$. The functor $\overline{\tau}$ is called the closure of functor τ.

Let $\mathfrak{X}$ be a set of groups. We denote by $S_\tau \mathfrak{X}$ the set of all such groups H that $H \in \tau(G)$ for some group $G \in \mathfrak{X}$. Recall that for any class $\mathfrak{X}$ of groups, $H\mathfrak{X} = \{G | G$ is an epimorphic image of an $\mathfrak{X}$-group$\}$, $R_0 \mathfrak{X} = \{G |$ there are normal subgroups $N_1, N_2, \cdots, N_t$ of G such that $\bigcap_{i=1}^{t} N_i = 1$ and $G/N_i \in \mathfrak{X}, i = 1, 2, \cdots, t\}$.

We use $\tau \text{form} \mathfrak{X}$ to denote the intersection of all τ-closed formations containing $\mathfrak{X}$.

Lemma 4.26 (Guo, Shum [165, Theorem 3.3]). *Let $\mathfrak{X}$ be a set of groups. Then the following equality holds:*

$$\tau \text{form} \mathfrak{X} = H R_0 S_{\overline{\tau}}(\mathfrak{X}).$$

Lemma 4.27 ([366, Theorem 3.47].) *Let $\mathfrak{S}(n)$ denote the class of all such finite soluble groups G that exponents of G, orders of chief factors of G and nilpotent classes of nilpotent factors of G are no more than n. Then for any natural number n, the number of non-isomorphic formation-critical group in $\mathfrak{S}(n)$ is finite.*

Lemma 4.28 *Let $\mathfrak{F}$ be a formation. If $\mathfrak{X}$ is a set of all formation-critical $\mathfrak{F}$-groups, then $\mathfrak{F} = \text{form} \mathfrak{X}$.*

Proof Suppose that $\mathfrak{F} \neq \text{form } \mathfrak{X}$. Let G be a group with minimal order in $\mathfrak{F} \backslash \text{form} \mathfrak{X}$. Then G is a formation-critical $\mathfrak{F}$-group. It follows that $G \in \mathfrak{X}$ and hence $G \in \text{form} \mathfrak{X}$, a contradiction. Therefore $\mathfrak{F} = \text{form} \mathfrak{X}$.

Lemma 4.29 *If the number of non-isomorphic formation-critical group in $\mathfrak{F}$ is finite, then the number of subformations of $\mathfrak{F}$ is finite.*

Proof Let $\mathfrak{M}$ be a subformation of $\mathfrak{F}$ and let $\{A_1, A_2, \cdots, A_t\}$ is the set of all non-isomorphic formation-critical group in $\mathfrak{F}$. Suppose that $\{A_{i_1}, A_{i_2}, \cdots, A_{i_m}\}$ is the set of all non-isomorphic formation-critical groups in $\mathfrak{M}$. Then by Lemma 4.28 we know that $\mathfrak{M} = \text{form}\{A_{i_1}, \cdots, A_{i_m}\}$, and hence, obviously, the numbers of subformations of $\mathfrak{F}$ is finite.

A class $\mathfrak{F}$ is called a semiformation if every homomorphic image of every group in $\mathfrak{F}$ is still in $\mathfrak{F}$.

Lemma 4.30 ([381, Lemma 2.1.6].) *Let A have a nonabelian unique minimal normal subgroup, $\mathfrak{M}$ be a τ-closed semiformation and $A \in l_n^\tau \text{form} \mathfrak{M}$. Then $A \in \mathfrak{M}$.*

Lemma 4.31 ([381, Lemma 1.2.21].) *Let $\mathfrak{F}$ be a τ-closed semiformation generated by $\mathfrak{X}$. Then*

$$\mathfrak{F} = H S_{\overline{\tau}} \mathfrak{X}.$$

Let $\mathfrak{H}$ be a nonempty class of groups. A τ-closed n-multiple local formation $\mathfrak{F}$ is called a minimal τ-closed n-multiple local non-$\mathfrak{H}$-formation (or $\mathfrak{H}_n^\tau$-critical formation) if $\mathfrak{F} \nsubseteq \mathfrak{H}$ but $\mathfrak{F}_1 \subseteq \mathfrak{H}$ for every proper τ-closed n-multiple local subformation $\mathfrak{F}_1$ of $\mathfrak{F}$.

A group $G \notin \mathfrak{F}$ is said to be τ-minimal non-$\mathfrak{F}$-group if all proper τ-subgroup of G belongs to $\mathfrak{F}$.

Lemma 4.32   ([381, Corollary 2.1.13].) $\mathfrak{F}$ is a minimal τ-closed n-multiple local nonnilpotent formation (where $n \geq 1$) if and only if $\mathfrak{F} = l_n^\tau \text{form} G$, where G is a τ-minimal nonnilpotent group and one of the following conditions holds:

1) G is a Schmidt group;
2) $n = 1$, G is a nonabelian simple group and $\tau(G) \subseteq \{G, 1\}$.

A formation $\mathfrak{F}$ is called τ-irreducible if $\mathfrak{F} \neq \tau\text{form}(\cup_{i \in I} \mathfrak{X}_i)$, where $\{\mathfrak{X}_i | i \in I\}$ is the set of all proper τ-closed subformation of $\mathfrak{F}$.

Lemma 4.33   ([381, Lemma 2.1.5].) *Suppose that a group A have a unique minimal normal subgroup $R \nsubseteq \Phi(A)$. Then the formation $\mathfrak{F} = \tau\text{form} A$ is τ-irreducible and $\mathfrak{F}$ has a unique maximal τ-closed subformation $\mathfrak{M} = \tau\text{form}(\{A/R\} \cup \mathfrak{X})$, where $\mathfrak{X}$ is the set of all proper τ-subgroup of A.*

Lemma 4.34   *Let Z_q be a group of order q. Then*

$$\tau\text{form} Z_q = \text{form} Z_q.$$

Proof   Since any subformation of the formation $\mathfrak{N}$ of all nilpotent groups is hereditary by Lemma 1.5, $\text{form} Z_q$ is hereditary. It follows that $\text{form} Z_q$ is τ-closed, and hence $\tau\text{form} Z_q = \cap\{\mathfrak{H}_i | Z_q \in \mathfrak{H}_i$ and $\mathfrak{H}_i$ is τ-closed$\} \subseteq \text{form} Z_q$. On the other hand, it is clear that $\text{form} Z_q \subseteq \tau\text{form} Z_q$. Therefore $\tau\text{form} Z_q = \text{form} Z_q$.

Lemma 4.35   ([381, Lemma 1.2.24].) *For any set $\{\mathfrak{M}_i | i \in I\}$ of τ-closed formations, we have*

$$\tau\text{form}(\bigcup_{i \in I} \mathfrak{M}_i) = \text{form}(\bigcup_{i \in I} \mathfrak{M}_i).$$

Lemma 4.36   ([381, Lemma 5.2.15].) *Let A be a nonabelian simple group and $\mathfrak{X}$ be the set of all proper $\bar{\tau}$-subgroups of A. If $\mathfrak{F} = \tau\text{form} A$, $\mathfrak{M} = \text{form} A$ and $\mathfrak{H} = \text{form} \mathfrak{X}$, then $\mathfrak{F} = \mathfrak{M} \oplus \mathfrak{H}$ (Here $\mathfrak{M} \oplus \mathfrak{H}$ denotes that $\mathfrak{M} \cap \mathfrak{H} = (1)$ and for every $G \in \mathfrak{F}$ we have $G = A_1 \times A_2$, where $A_1 \in \mathfrak{M}$ and $A_2 \in \mathfrak{H}$).*

Lemma 4.37   ([381, Theorem 1.2.29].) *Let $\mathfrak{X}$ be a set of groups, $\mathfrak{F}$ be a nonempty formation and for every group $G \in \mathfrak{X}$ the $\mathfrak{F}$-residual $G^{\mathfrak{F}}$ have no Frattini G–chief factor. If $A \in \text{form} \mathfrak{X} \backslash \mathfrak{F}$ and A have a unique minimal normal subgroup, then $A \in H(\mathfrak{X})$.*

Lemma 4.38   ([381, Corollary 4.2.8].) *The lattice l_n^τ is modular.*

Lemma 4.39   ([381, Corollary 4.2.9].) *Let $\mathfrak{M}, \mathfrak{H}$ be two τ-closed n-multiple local formations. Then*

$$(\mathfrak{M} \vee_n^\tau \mathfrak{H})/_n^\tau \mathfrak{M} \simeq \mathfrak{H}/_n^\tau(\mathfrak{H} \cap \mathfrak{M}).$$

Lemma 4.40 ([381, Lemma 2.1.11].) *Let one of two formations $\mathfrak{H}$ and $\mathfrak{F}$ is soluble and $\mathfrak{F} \in l_n^\tau$. If $\mathfrak{F} \nsubseteq \mathfrak{H}$, then $\mathfrak{F}$ has a $\mathfrak{H}_n^\tau$-critical subformation.*

Lemma 4.41 *Let $\mathfrak{F}$ be a τ-closed n-multiply local formation generated by one element. Then the number of all soluble τ-closed n-multiply local subformations of $\mathfrak{F}$ is finite.*

Proof Let $\mathfrak{F} = l_n^\tau \mathrm{form}\, G$, for some group $G \in \mathfrak{F}$. We prove the lemma by induction on n.

Suppose that $n = 0$, i.e., $\mathfrak{F} = \tau \mathrm{form}\, G$. By Lemma 4.26 we know that $\mathfrak{F} = H R_0 S_{\overline{\tau}}(G)$. If $m = |G|$, then by Corollary 2.4 we know that the exponents, orders of chief factors and nilpotent classes of nilpotent factors of all groups in $\mathfrak{F}$ are no more than m. It follows from Lemma 4.27 that the number of non-isomorphic formation-critical groups in $\mathfrak{F} \cap \mathfrak{G}$ is finite. Hence by Lemma 4.29, we have that in $\mathfrak{F}$ the number of soluble subformations is finite.

Suppose that $n > 0$ and the lemma is true for $n - 1$. Let $\mathfrak{M}$ be any soluble τ-closed n-multiply local subformation of $\mathfrak{F}$. Let m and f are the smallest l_{n-1}^τ-value local satellite of $\mathfrak{M}$ and $\mathfrak{F}$ respectively. Then by Corollary 4.4 we know that $m \leq f$. On the other hand, by Lemma 4.3 we have that

$$f(p) = \begin{cases} l_{n-1}^\tau \mathrm{form}(G/F_p(G)), & \text{if } p \in \pi(G), \\ \phi, & \text{if } p \in \mathbf{P} \backslash \pi(G). \end{cases}$$

By our assumption, $f(p) = l_{n-1}^\tau \mathrm{form}(G/F_p(G))$ has only finite number of soluble τ-closed $(n-1)$-multiply local subformations ($m(p)$ is one of them). Since $\pi(G)$ is a finite set, the number of soluble τ-closed n-multiply local subformation of $\mathfrak{F}$ is finite. The lemma is proved.

Let $\mathfrak{F}$ be a τ-closed n-multiply local formation. Then we denote by $L_n^\tau(\mathfrak{F})$ the lattice of all τ-closed n-multiply local formations which is contained in $\mathfrak{F}$.

Lemma 4.42 *Let $\mathfrak{F}$ be a τ-closed n-multiply local formation generated by one element. Then the set of all formations each of which is either a atom of the lattice $L_n^\tau(\mathfrak{F})$ or an $\mathfrak{N}_n^\tau$-critical subformation of $\mathfrak{F}$ is finite.*

Proof Let $\mathfrak{F} = l_n^\tau \mathrm{form}\, G$ and $\mathfrak{M}$ a atom of the lattice $l_n^\tau(\mathfrak{F})$. Then $\mathfrak{M} = l_n^\tau \mathrm{form}\, A$, for some simple group A. If A is nonabelian, then by Lemma 4.30, $A \in \mathfrak{H}$, where $\mathfrak{H}$ is a τ-closed semiformation generated by G. But by Lemma 4.31 we have $\mathfrak{H} = H S_{\overline{\tau}}(G)$. This means that the number of nonsoluble atoms in $L_n^\tau(\mathfrak{F})$ is finite.

By Lemma 4.41, the number of soluble atoms in $L_n^\tau(\mathfrak{F})$ is also finite.

Now suppose that $\mathfrak{M}$ is an $\mathfrak{N}_n^\tau$-critical subformation of $\mathfrak{F}$. Then by Lemma 4.32, $\mathfrak{M} = l_n^\tau \mathrm{form}\, A$, where A is a τ-minimal nonnilpotent group and one of the following conditions holds:

1) A is a Schmidt group;
2) $n = 1$, A is a nonabelian simple group and $\tau(A) \subseteq \{A, 1\}$.

It follows that $\mathfrak{F}$ has only finite number of non-soluble $\mathfrak{N}_n^\tau$-critical subformations.

If A is a Schmidt group, then A is a soluble group by Chap. 1, Proposition 1.9. But by Lemma 4.41, we know that every τ-closed n-multiply local formation generated by one element has only finite number of soluble τ-closed n-multiply local subformations. The lemma is proved.

Lemma 4.43 $\mathfrak{F}$ is a minimal τ-closed nonnilpotent formation if and only if $\mathfrak{F} = \tau \mathrm{form} G$, where G is a τ-minimal nonnilpotent group with a unique minimal normal subgroup $R = G^{\mathfrak{N}} \not\subseteq \Phi(G)$.

Proof Necessity. Let G be a group with minimal order in $\mathfrak{F} \backslash \mathfrak{N}$. Then G is a τ-minimal nonnilpotent group with a unique minimal normal subgroup $R = G^{\mathfrak{N}} \not\subseteq \Phi(G)$. Obviously, $\mathfrak{F} = \tau \mathrm{form} G$.

Sufficiency. Let $\mathfrak{F} = \tau \mathrm{form} G$, where G is a τ-minimal nonnilpotent group with a unique minimal normal subgroup $R = G^{\mathfrak{N}} \not\subseteq \Phi(G)$. Then by Lemma 4.33, the formation $\mathfrak{F}$ is τ-irreducible and that if $\mathfrak{M}$ is a maximal τ-closed subformation of $\mathfrak{F}$, then $\mathfrak{M} = \tau \mathrm{form}(\{G/R\} \cup \mathfrak{X})$, where $\mathfrak{X}$ is the set of all proper τ-subgroup of G. But G is a τ-minimal nonnilpotent group, so $\mathfrak{X} \subseteq \mathfrak{N}$. Hence $\mathfrak{M} \subseteq \mathfrak{N}$. On the other hand, $\mathfrak{F} \not\subseteq \mathfrak{N}$. Therefore, $\mathfrak{F}$ is a minimal τ-closed nonnilpotent formation. The lemma is proved.

Lemma 4.44 Let $\mathfrak{M}$ be a nilpotent τ-closed n-multiply local formation and $\{\mathfrak{H}_i | i \in I\}$ be a set of $\mathfrak{N}_n^\tau$-critical formations. If $\mathfrak{H}$ is an $\mathfrak{N}_n^\tau$-critical formation and $\mathfrak{H} \subseteq \mathfrak{F} = \mathfrak{M} \vee_n^\tau (\vee_n^\tau(\mathfrak{H}_i | i \in I))$, then $\mathfrak{H} = \mathfrak{H}_i$ for some $i \in I$.

Proof Suppose that $n \geq 1$. Then by Lemma 4.32, $\mathfrak{H} = l_n^\tau \mathrm{form} G$, where G is a τ-minimal nonnilpotent group and one of the following conditions holds: (1) G is a Schmidt group; (2) $n = 1$ and G is a nonabelian simple group with $\tau(G) \subseteq \{G, 1\}$.

Let f, m be the smallest l_{n-1}^τ-valued local satellites of formations $\mathfrak{F}$ and $\mathfrak{M}$ respectively, h_i be the smallest l_{n-1}^τ-value local satellite of $\mathfrak{H}_i$. Then by Lemma 4.3, we have that

$$f(p) = \begin{cases} l_{n-1}^\tau \mathrm{form}((m(p) \cup (\cup_{i \in I} h_i(p))), & \text{if } p \in \pi(\mathfrak{F}), \\ \phi, & \text{if } p \notin \pi(\mathfrak{F}). \end{cases} \tag{5.7}$$

Let h be the smallest l_{n-1}^τ-valued local satellite of the formation $\mathfrak{H}$. By Corollary 4.4, $h \leq f$. Assume G is a Schmidt group. Then by Chap. 1, Prop. 1.9, G is a soluble group with $|\pi(G)| = 2$, G has a normal p-Sylow subgroup and $G/F_p(G) \simeq Z_q$ where Z_q is a group of order q, for $q \in \pi(G) \setminus \{p\}$. It follows from Lemma 4.3 that

$$h(r) = \begin{cases} l_{n-1}^\tau \mathrm{form} Z_q, & \text{if } r = p, \\ (1), & \text{if } r = q, \\ \phi, & \text{if } r \in \mathbf{P} \backslash \{p, q\}. \end{cases}$$

Since $Z_q \in h(p) \subseteq f(p)$, there exists $i \in I$ such that $(1) \subset h_i(p)$. Suppose that $n = 1$. By Lemmas 4.36 and 4.3, we see that $f(p) = \mathrm{form} \mathfrak{X}$, where $\mathfrak{X}$ is some set of simple groups. On the other hand, if $A \in \mathfrak{X}$ and A is a nonabelian group, then there exists $i \in I$ such that $\mathfrak{H} = l_n^\tau \mathrm{form} A$ and $p \mid |A|$. If A is abelian group of order r, then there exists $j \in I$ such that $\mathfrak{H}_j = l_n^\tau \mathrm{form} B$, where B is a Schmidt group with $\pi(B) = \{p, q\}$ and a Sylow p-subgroup of B is normal. Since $Z_p \in \mathrm{form}\mathfrak{X}$, Z_q coincide with an abelian group in $\mathfrak{X}$. Hence, there exists $i \in I$ such that $\mathfrak{H}_i = l_n^\tau \mathrm{form} B$, where B is a Schmidt group with $\pi(B) = \{p, q\}$ and B has a normal Sylow p-subgroup. Therefore, by Lemma 4.3, $\mathfrak{H} = \mathfrak{H}_i$.

Now suppose that $n > 1$. Then by Lemma 4.32 and Lemma 4.3, we have that $f(p) = \mathfrak{N}_\pi$ and $r \in \pi$ if and only if there exists $\mathfrak{H} \in \{\mathfrak{H}_i | i \in I\}$ such that $\mathfrak{H} = l_n^\tau \mathrm{form} A$ where A is a Schmidt group with $\pi(A) = \{p, r\}$ and A has a normal Sylow p-subgroup. But $Z_q \in f(p)$, hence $\mathfrak{H} = \mathfrak{H}_i$ for some $i \in I$.

Now assume $n = 1$ and G is a nonabelian simple group such that $\tau(G) \subseteq \{1, G\}$. Then, for any $p \in \pi(G)$, we have that $h(p) = \tau \mathrm{form} G \subseteq f(p)$. Hence in $\{\mathfrak{H}_i | i \in I\}$ there is an $\mathfrak{H}_i$ such that $\mathfrak{H}_i = l_1^\tau \mathrm{form} G = \mathfrak{H}$.

Suppose that $n = 0$. Then by Lemma 4.43, $\mathfrak{H} = l_0^\tau \mathrm{form} A = \tau \mathrm{form} A$, $\mathfrak{H}_i = \tau \mathrm{form} A_i$, where A and A_i are τ-minimal nonnilpotent groups and A has a unique minimal normal subgroup $R = A^\mathfrak{N} \not\subseteq \Phi(A)$ and A_i has a unique minimal normal subgroup $R_i = A_i^\mathfrak{N} \not\subseteq \Phi(A_i)$.

Let $\mathfrak{X}_i$ is the set of all proper τ-subgroups of A_i. Then by Lemma 4.26, we have

$$\mathfrak{H}_i = \tau \mathrm{form} A_i = \mathrm{form}(\mathfrak{X}_i \cup \{A_i\}).$$

Since every nilpotent formation is hereditary by Lemma 4.35, we have that

$$\mathfrak{F} = \mathfrak{M} \vee^\tau (\vee^\tau(\mathfrak{H}_i | i \in I)) = \tau \mathrm{form}(\mathfrak{M} \cup (\cup_{i \in I}(\mathfrak{X}_i \cup \{A_i\})))$$
$$= \mathrm{form}(\mathfrak{X} \cup (\cup_{i \in I}(\{A_i\} \cup \mathfrak{X}_i))),$$

where $\mathfrak{X}$ is the set of all subgroups of all groups from $\mathfrak{M}$. It is clear that for every group T in $\mathfrak{X} \cup (\cup_{i \in I}(\{A_i\} \cup \mathfrak{X}_i))$, the $\mathfrak{N}$–residual $T^\mathfrak{N}$ has no Frattini T-chief factor. So by Lemma 4.37, the nonnilpotent group A is a homomorphism image of a group in $\mathfrak{X} \cup (\cup_{i \in I}(\{A_i\} \cup \mathfrak{X}_i))$. Since every group in $\mathfrak{X} \cup (\cup_{i \in I}\mathfrak{X}_i)$ is nilpotent, there exists $i \in I$ such that A is a homomorphism image of A_i. But A_i has a unique minimal normal subgroup $R_i = A_i^\mathfrak{N}$, so $A \simeq A_i$, and hence $\mathfrak{H} = \mathfrak{H}_i$. The lemma is proved.

Lemma 4.45 *Let* $\mathfrak{F}_0 \subseteq \mathfrak{M}_1 \subseteq \mathfrak{M} \subseteq \mathfrak{F}$, *where* $\mathfrak{M}_1, \mathfrak{M}, \mathfrak{F}$ *are* τ-closed n-multiply local formations. If $\mathfrak{H}$ is a complement of $\mathfrak{M}_1$ in the lattice $\mathfrak{F}/_n^\tau \mathfrak{F}_0$, then $\mathfrak{H} \cap \mathfrak{M}$ is a complement of $\mathfrak{M}_1$ in the lattice $\mathfrak{M}/_n^\tau \mathfrak{M}_0$.

Proof By the hypothesis, $\mathfrak{M}_1 \cap \mathfrak{H} = \mathfrak{F}_0$ and $\mathfrak{M}_1 \vee_n^\tau \mathfrak{H} = \mathfrak{F}$. It follows from Lemma 4.38 that

$$\mathfrak{M} = \mathfrak{M} \cap (\mathfrak{M}_1 \vee_n^\tau \mathfrak{H}) = \mathfrak{M}_1 \vee_n^\tau (\mathfrak{M} \cap \mathfrak{H}).$$

Moreover, it is clear that $(\mathfrak{M} \cap \mathfrak{H}) \cap \mathfrak{M}_1 = \mathfrak{F}_0$. Hence $\mathfrak{M}_1 \cap H$ is a complement of $\mathfrak{M}_1$ in $\mathfrak{M}/_n^\tau \mathfrak{M}_0$.

Lemma 4.46 *Let* $\mathfrak{F}, \mathfrak{M}$ *be two* τ-closed n-multiply local formations and $\mathfrak{M} \subseteq \mathfrak{F}$. If $\mathfrak{F}/_n^\tau \mathfrak{F}_0$ is a complemented lattice, then $\mathfrak{M}/_n^\tau \mathfrak{M}_0$ is also a complemented lattice.

Proof By Lemma 4.39 we have the following lattice isomorphism:

$$(\mathfrak{F}_0 \vee_n^\tau \mathfrak{M})/_n^\tau \mathfrak{F}_0 \simeq \mathfrak{M}/_n^\tau(\mathfrak{M} \cap \mathfrak{F}_0) = \mathfrak{M}/_n^\tau(\mathfrak{M} \cap \mathfrak{N})$$
$$= \mathfrak{M}/_n^\tau \mathfrak{M}_0.$$

Since $(\mathfrak{F}_0 \vee_n^\tau \mathfrak{M})/_n^\tau \mathfrak{F}_0$ is a sublattice of $\mathfrak{F}/_n^\tau \mathfrak{F}_0$, by Lemma 4.45, $(\mathfrak{F}_0 \vee_n^\tau \mathfrak{M})/_n^\tau \mathfrak{F}_0$ is a complemented lattice. Consequently, $\mathfrak{M}/_n^\tau \mathfrak{M}_0$ is complemented lattice. The lemma is proved.

Theorem 4.47 *(Guo [137]). Let $\mathfrak{F}$ be a nonnilpotent τ-closed n-multiply local formation. Then the following statements are equivalent:*

1) $\mathfrak{F}/_n^\tau \mathfrak{F}_0$ *is a complemented lattice;*
2) $\mathfrak{F} = \mathfrak{F}_0 \vee_n^\tau (\vee_n^\tau \mathfrak{H}_i | i \in I)$, *where $\{\mathfrak{H}_i | i \in I\}$ is the set of all $\mathfrak{N}_n^\tau$-critical subformations of $\mathfrak{F}$;*
3) $\mathfrak{F}/_n^\tau \mathfrak{F}_0$ *is a Boolean lattice.*

Proof Suppose that the statement 1) holds and $\{\mathfrak{H}_i | i \in I\}$ be the set of all $\mathfrak{N}_n^\tau$-critical subformations of $\mathfrak{F}$. Suppose first that $\mathfrak{F}$ is nonnilpotent τ-closed n-multiply local formation generated by one element. Then by Lemma 4.44 we know that the number t of all $\mathfrak{N}_n^\tau$-critical subformations of $\mathfrak{F}$ is finite. We prove the statement 2) by induction on t.

Let $\mathfrak{H}_i$ be some $\mathfrak{N}_n^\tau$-critical subformation of $\mathfrak{F}$ and $\mathfrak{M} = \mathfrak{F}_0 \vee_n^\tau \mathfrak{H}_i$. By the statement 1), there exists a τ-closed n-multiply local subformation $\mathfrak{H}$ of $\mathfrak{F}$ such that $\mathfrak{M} \cap \mathfrak{H} = \mathfrak{F}_0$ and $\mathfrak{M} \vee_n^\tau \mathfrak{H} = \mathfrak{F}$. If $\mathfrak{M} \subset \mathfrak{F}$, then by Lemma 4.46 and by induction, $\mathfrak{H} = \mathfrak{H}_0 \vee_n^\tau (\vee_n^\tau(\mathfrak{F}_j | j \in J))$, where $\{\mathfrak{F}_j | j \in J\}$ is the set of all $\mathfrak{N}_n^\tau$-critical subformations of $\mathfrak{H}$. Since $\mathfrak{H}_0 = \mathfrak{F}_0$, by Lemma 4.44 we have

$$\mathfrak{F} = (\mathfrak{F}_0 \vee_n^\tau \mathfrak{H}) \vee_n^\tau (\mathfrak{F}_0 \vee_n^\tau (\vee_n^\tau(\mathfrak{F}_j | j \in J)))$$
$$= (\mathfrak{F}_0 \vee_n^\tau \mathfrak{H}_i) \vee_n^\tau (\vee_n^\tau(\mathfrak{F}_j | j \in J))$$
$$= \mathfrak{F}_0 \vee_n^\tau (\vee_n^\tau(\mathfrak{H}_i | i \in I)),$$

where $\{\mathfrak{H}_i \mid i \in I\}$ is the set of all $\mathfrak{N}_n^\tau$-critical subformations of $\mathfrak{F}$.

Now we consider the general case. Let G be some nonnilpotent group in $\mathfrak{F}$ and $\mathfrak{M} = l_n^\tau \text{form} G$. By Lemma 4.46, the statement 1) hold on $\mathfrak{M}/_n^\tau \mathfrak{M}_0$, i.e., $\mathfrak{M}/_n^\tau \mathfrak{M}_0$ is also a complemented lattice. So by the result proved in previous paragraph we have that $\mathfrak{M} = \mathfrak{M}_0 \vee_n^\tau (\vee_n^\tau(\mathfrak{M}_j | j \in J))$ where $\{\mathfrak{M}_j | j \in J\}$ is the set of all $\mathfrak{N}_n^\tau$-critical subformations of $\mathfrak{M}$. Thus, $\mathfrak{F} = l_n^\tau \text{form}(G | G \in \mathfrak{F}) = \mathfrak{F}_0 \vee_n^\tau (\vee_n^\tau(\mathfrak{H}_i | i \in I))$, where $\{\mathfrak{H}_i | i \in I\}$ is the set of all $\mathfrak{N}_n^\tau$-critical subformations of $\mathfrak{F}$.

Suppose that the statement 2) holds. We first prove that $\mathfrak{F}/_n^\tau \mathfrak{F}_0$ is a complemented lattice. Let a nonnilpotent formation $\mathfrak{M}$ belongs to the lattice $\mathfrak{F}/_n^\tau \mathfrak{F}_0$. Let $\mathfrak{X}_1$ be a set of all $\mathfrak{N}_n^\tau$-critical subformations of the formation $\mathfrak{M}$ and $\mathfrak{X}_2 = \{\mathfrak{H}_i | i \in I\} \backslash \mathfrak{X}_1$. Let $\mathfrak{H} = (l_n^\tau \text{form}(\cup_{\mathfrak{H}_i \in \mathfrak{X}_2} \mathfrak{H}_i)) \vee_n^\tau \mathfrak{F}_0$. Then $\mathfrak{M} \vee_n^\tau \mathfrak{H} = \mathfrak{F}$. If $\mathfrak{M} \cap \mathfrak{H} \not\subseteq \mathfrak{F}_0$, then by Lemma 4.40, $\mathfrak{M} \cap \mathfrak{H}$ has at least an $\mathfrak{N}_n^\tau$-critical subformation, say $\mathfrak{H}^*$. Suppose that $\mathfrak{H}^* \in \mathfrak{X}_1$. Then $\mathfrak{H}^* \notin \mathfrak{X}_2$. It follows from Lemma 4.44 that in $\mathfrak{F}_0 \vee_n^\tau (l_n^\tau \text{form}(\cup_{\mathfrak{H}_i \in \mathfrak{X}_2} \mathfrak{H}_i))$ there is no $\mathfrak{N}_n^\tau$-critical subformation which does not belong to $\mathfrak{X}_2$. This contradiction shows that $\mathfrak{H}$ is a complement of $\mathfrak{M}$ in the lattice $\mathfrak{F}/_n^\tau \mathfrak{F}_0$.

Now we ascertain that the lattice $\mathfrak{F}/_n^\tau \mathfrak{F}_0$ is Boolean. Let $\mathfrak{M}_1, \mathfrak{M}_2$, and $\mathfrak{M}_3$ are any elements of the lattice $\mathfrak{F}/_n^\tau \mathfrak{F}_0$. We prove that

$$\mathfrak{M}_1 \cap (\mathfrak{M}_2 \vee_n^\tau \mathfrak{M}_3) = (\mathfrak{M}_1 \cap \mathfrak{M}_2) \vee_n^\tau (\mathfrak{M}_1 \cap \mathfrak{M}_3). \tag{5.8}$$

Since $\mathfrak{F}/_n^\tau \mathfrak{F}_0$ is a complemented lattice, by Lemma 4.46 we know that for every formation $\mathfrak{M}$ belonging to the lattice $\mathfrak{F}/_n^\tau \mathfrak{F}_0$, the lattice $\mathfrak{M}/_n^\tau \mathfrak{M}_0$ is also a complemented lattice. It follows that $\mathfrak{M} = \mathfrak{M}_0 \vee_n^\tau (\vee_n^\tau(\mathfrak{H}_i | i \in I))$ where $\{\mathfrak{H}_i | i \in I\}$ is the set of all $\mathfrak{N}_n^\tau$-critical subformation of $\mathfrak{M}$. Therefore, to prove the inclusion

$$\mathfrak{M}_1 \cap (\mathfrak{M}_2 \vee_n^\tau \mathfrak{M}_3) \subseteq (\mathfrak{M}_1 \cap \mathfrak{M}_2) \vee_n^\tau (\mathfrak{M}_1 \cap \mathfrak{M}_3), \tag{5.9}$$

suffices it to prove that every $\mathfrak{N}_n^\tau$-critical subformation $\mathfrak{H}$ of $\mathfrak{M} = \mathfrak{M}_1 \cap (\mathfrak{M}_2 \vee_n^\tau \mathfrak{M}_3)$ belongs to the formation $(\mathfrak{M}_1 \cap \mathfrak{M}_2) \vee_n^\tau (\mathfrak{M}_1 \cap \mathfrak{M}_3)$. Let $\mathfrak{X}_i$ be the set of all $\mathfrak{N}_n^\tau$-critical subformation of the formation $\mathfrak{M}_i, i = 1, 2, 3$. Then $\mathfrak{H} \subseteq \mathfrak{M}_1$ and

$$\mathfrak{H} \subseteq ((\mathfrak{M}_2)_0 \vee_n^\tau (l_n^\tau \text{form}(\cup_{\mathfrak{H}_i \in \mathfrak{X}_2} \mathfrak{H}_i))) \vee_n^\tau ((\mathfrak{M}_3)_0 \vee_n^\tau (l_n^\tau \text{form}(\cup_{\mathfrak{H}_i \in \mathfrak{X}_3} \mathfrak{H}_i)))$$

$$= (\mathfrak{M}_2 \vee_n^\tau \mathfrak{M}_3)_0 \vee_n^\tau (l_n^\tau \text{form}(\cup_{\mathfrak{H}_i \in \mathfrak{X}_2 \cup \mathfrak{X}_3} \mathfrak{H}_i)).$$

It follows from Lemma 4.44 that $\mathfrak{H} \in \mathfrak{X}_2 \cup \mathfrak{X}_3$. Therefore either $\mathfrak{H} \subseteq \mathfrak{M}_1 \cap \mathfrak{M}_2$ or $\mathfrak{H} \subseteq \mathfrak{M}_1 \cap \mathfrak{M}_3$. This shows that the inclusion (5.9) holds. The inverse inclusion of (5.9) is clear. So the equality (5.8) holds. Hence $\mathfrak{F}/_n^\tau \mathfrak{F}_0$ is a Boolean lattice. 3)→1) is clear. The theorem is thus proved.

5.5 On a Question of the Theory of Graduated Formation

Let

$$f : \mathfrak{G} \rightarrow \{\text{formations of groups}\}$$

be a map of the class of all groups $\mathfrak{G}$ into the set of all formations with the following properties:

(i) $f(1) = \mathfrak{G}$;
(ii) $f(G_1) = f(G_2)$ whenever $G_1 \simeq G_2$;
(iii) $f(G) \subseteq f(N) \cap f(G/N)$ for any group G and any its normal subgroup N.

Following L. A. Shemetkov [357], the above function f is called a screen. For any screen f, the symbol $\langle f \rangle$ denotes the class of all groups G such that either $G = 1$ or $G \neq 1$ and $G/C_G(H/K) \in f(H/K)$ for every chief factor H/K of G. A formation $\mathfrak{F}$ is said to be a *graduated formation* if $\mathfrak{F} = \langle f \rangle$ for some screen f.

In 1995, O. V. Melnikov posed the following problem at Gomel algebraic seminar.

Melnikov's Problem 5.1. Whether it is true that every graduated formation is a Baer-local formation?

In this section we give a negative answer to this problem.

Let f be a function such that $f(G)$ is a formation for any elementary group G. Then we say that f is an e-function if $f(H) = f(T)$ for any two elementary groups H and T with $H \simeq T$. Let $EF(f)$ be the class of all groups G such that $G/C_G(H/K) \in f(H/K)$ for any chief factor of G. By the definition, all identity groups are in $EF(f)$.

Proposition 5.2 (Guo [136]). *For any e-function f, the class $EF(f)$ is an nonempty formation.*

Proof Let $G \in EF(f)$ and N an arbitrary normal subgroup of G. Let $(H/N)/(K/N)$ be an chief factor of G/N. Then H/K is a chief factor of G and $G/C_G(H/K) \in f(H/K)$ by assumption on G. It is clear that $C_{G/N}((H/N)/(K/N)) = C_G(H/K)/N$. Hence,

$$(G/N)/C_{G/N}((H/N)/(K/N)) = (G/N)/(C_G(H/K)/N)$$

$$\simeq G/C_G(H/K) \in f(H/K) = f((H/N)/(K/N)).$$

Consequently, $G/N \in EF(f)$, which means that the class $EF(f)$ is closed under taking homomorphic images.

Now let $G/N_1 \in EF(f)$ and $G/N_2 \in EF(f)$, we show $G/N_1 \cap N_2 \in EF(f)$. Let $L = N_1 \cap N_2$ and $(H/L)/(K/L)$ be a chief factor of G/N. Then, H/K is a chief factor of G. Assume that $HN_1 = KN_1$, then $H = H \cap KN_1 = K(H \cap N_1)$. Hence, $H \cap N_1 \nleq K$ and $H \cap N_1 \supset K \cap N_1$. But $H \cap N_1/K \cap N_1$ is G-isomorphic to $K(H \cap N_1)/K$. This implies that $K(H \cap N_1)/K = H/K$. So $H \cap N_1/K \cap N_1$ is a chief factor of G, which is G-isomorphic to H/K. On the other hand, the factor $((H \cap N_1)N_2)/((K \cap N_1)N_2)$ is G-isomorphic to

$$(H \cap N_1)/(K \cap N_1)(H \cap N_1 \cap N_2) = (H \cap N_1)/((K \cap N_1)(N_1 \cap N_2)) = (H \cap N_1)/(K \cap N_1).$$

This shows that $((H \cap N_1)N_2)/((K \cap N_1)N_2)$ is a chief factor of G and it is G-isomorphic to H/K. It follows that $((H \cap N_1)N_2)/((K \cap N_1)N_2)$ is G-isomorphic to the factor $(((H \cap N_1)N_2)/N_2)/((((K \cap N_1)N_2))/N_2)$ of $G/N_2 \in EF(f)$. Therefore

$$G/C_G(H/K) \simeq (G/N_2)/C_G(H/K)/N_2)$$

$$= (G/N_2)/(C_{G/N_2}(((H \cap N_1)N_2)/N_2)/(((K \cap N_1)N_2)/N_2))$$

$$\in f(((H \cap N_1)N_2)/((K \cap N_1)N_2)) = f(H/K).$$

Thus

$$(G/L)/C_{G/L}((H/L)/(K/L)) = (G/L)/(C_G(H/K)/L) \simeq G/C_G(H/K)$$

$$\in f((H/L)/(K/L)).$$

Assume that $HN_1 \neq KN_1$. Since $(HN_1)/(KN_1) \simeq H/(K(H \cap N_1)) = H/K$, as above, we have

$$(G/L)/C_{G/L}((H/L)/(K/L)) \in f((H/L)/(K/L)).$$

This implies that $G/N_1 \cap N_2 \in EF(f)$. Thus, $EF(f)$ is a formation.

We will say that a formation $\mathfrak{F}$ is a *weak graduated formation* if $\mathfrak{F} = EF(f)$ for some e-function f. Let f be an e-function. We say that f is a (primarily) similar e-function if $f(H) = f(T)$ for any two elementary (abelian, respectively) groups H and K with isomorphic composition factors.

Theorem 5.3 (Guo [136]). *A formation $\mathfrak{F}$ is a Baer-local formation if and only if $\mathfrak{F} = EF(f)$ for some primarily similar e-function f.*

Proof Let $\mathfrak{F} = EF(f)$ for some primarily similar e-function f. We prove that $\mathfrak{F}$ is a Baer-local formation. Let $\overline{f}$ be a formation function such that $\overline{f}(p) = f(H)$ if H is an abelian elementary group p-group; and $\overline{f}(o) = \mathfrak{F}$. Let $\mathfrak{M} = CLF(\overline{f})$. We prove that $\mathfrak{M} = \mathfrak{F} = EF(f)$.

Assume that $\mathfrak{M} \not\subseteq EF(f)$. Let G be a group with minimal length of composition series and $G \in \mathfrak{M} \setminus EF(f)$. Since the class $EF(f)$ is an formation by Prop. 5.2, G has an unique minimal normal subgroup R and $R = G^{EF(f)}$. Assume that R is a nonabelian. Then $G_{\mathfrak{S}} = 1$, and so $G \simeq G/1 \in \overline{f}(o) = \mathfrak{F}$, a contradiction. Hence R is an abelian p-group for some prime p. Consequently, $G/C^p(G) \in \overline{f}(p) = f(R)$, which implies that $G/C_G(R) \in f(R)$. But $G/R \in EF(f)$ and so $G \in EF(f) = \mathfrak{F}$, a contradiction. Hence $\mathfrak{M} \subseteq EF(f)$.

Assume that $\mathfrak{F} \not\subseteq \mathfrak{M}$. Let G be a group in $\mathfrak{F} \setminus \mathfrak{M}$ with minimal length of composition series, $R = G^{\mathfrak{M}}$ is the unique minimal normal subgroup of G. If R is nonabelian, then

$$G \simeq C/C_G(R) \in \mathfrak{F} = \overline{f}(o).$$

Since $G/R \in LCF(\overline{f})$, we have that $G \in \mathfrak{M} = LCF(\overline{f})$, a contradiction. Hence R is an abelian p-group for some prime p. Hence $G/C_G(R) \in f(R) = \overline{f}(p)$, which implies that $G \in CLF(\overline{f})$, a contradiction again. Thus $\mathfrak{F} = \mathfrak{M}$.

Conversely, assume that $\mathfrak{F}$ is a Baer-local formation and $\mathfrak{F} = CLF(f)$, where f is a composition satellite of $\mathfrak{F}$. Let $\overline{f}$ be a primarily similar e-function such that $\overline{f}(H) = \mathfrak{F}$ for any elementary nonabelian group H and $\overline{f}(H) = f(p)$ if H is elementary abelian p-group. Let $\mathfrak{M} = EF(\overline{f})$. Then as above it is may be shown that $\mathfrak{M} = \mathfrak{F}$. The theorem is proved.

If H/K is a chief factor of a group G and $H/K = A_1 \times \cdots A_t$, where $A_1 \cong A_2 \cong \cdots \cong A_t \cong A$, A is a simple group, then we say that H/K is an A-chief factor of G.

If $H = A_1 \times \cdots A_t$, where $A_1 \cong A_2 \cong \cdots \cong A_t \cong A$, A is a simple group, then number t is said to be the rank of the group H.

Suppose that a function h associates to every simple group A with some systems (possibly empty) of natural numbers $h(A)$. Then we say that h is a *rang function*.

For every rank function h, we let $\mathfrak{F}(h) = \{G \mid$ the rank of A-chief factor of G in $h(A)$, for any simple group $A\}$.

By the definition, all identity groups are in $\mathfrak{F}(h)$, for every rank function h. Hence the class $\mathfrak{F}(h)$ is nonempty. In addition, the following holds.

Proposition 5.4 (Guo [136]). *For any rank function h, then class $\mathfrak{F}(h)$ is a weak graduated formation.*

Proof Let $G \in \mathfrak{F}(h)$ and N be a normal subgroup of G. Let $(H/N)/(K/N)$ be a chief factor of G/N. Then

$$(H/N)/(K/N) \simeq H/K \simeq A_1 \times \cdots \times A_t,$$

where $A_1 \cong A_2 \cong \cdots \cong A_t \cong A$, A is a simple group. Since $G \in \mathfrak{F}(h)$, $t \in h(A)$. Hence for any A-chief factor $(H/N)/(K/N)$ of G/N, where $(H/N)/(K/N) = |A|^t$, we have $t \in h(A)$. This shows that $G/N \in \mathfrak{F}(h)$, that is, $\mathfrak{F}(h)$ is closed under taking homomorphic images.

Let G/N_1, $G/N_2 \in \mathfrak{F}(h)$, $L = N_1 \cap N_2$ and $(H/L)/(K/L)$ be any chief A-factor of G/L. Then H/K is chief A-factor of G. Assume that $HN_1 = KN_1$. Then $(H/L)/(K/L)$ is isomorphic to the chief factor $(((H \cap N_1)N_2)/N_2)/((((K \cap N_1)N_2)/N_2)$ of $G/N_2 \in \mathfrak{F}(h)$ (see proof of Proposition 5.2). As above, if $|(H/L)/(K/L)| = |A|^t$, then by $|(((H \cap N_1)N_2)/N_2)/(((K \cap N_1)N_2)/N_2)| = |A|^t$, we have $t \in h(A)$. Now, assume that $HN_1 \neq KN_1$. Since HN_1/KN_1 is G-isomorphic to $H/K(H \cap N_1) = H/K$, we also have that $t \in h(A)$. Hence, $\mathfrak{F}(h)$ is a formation.

Now we show that $\mathfrak{F}(h)$ is a weak graduated formation. Suppose that for any elementary group T, which is isomorphic to a chief factor H/K of some $G \in \mathfrak{F}(h)$, we have $f(T) = \mathfrak{G}$; and for any elementary group T, which is not isomorphic to a chief factor H/K of any $G \in \mathfrak{F}(h)$, we have $f(T) = \emptyset$. We show that $\mathfrak{F}(h) = EF(f)$. Assume that $\mathfrak{F}(h) \not\subseteq EF(f)$, and let G be a group with minimal length of composition series in $\mathfrak{F}(h) \setminus EF(f)$. Since $EF(f)$ is nonempty formation by Prop. 5.2, G is a monolithic group with monolith $R = G^{EF(f)}$. Since $G \in \mathfrak{F}(h)$, $f(R) = \mathfrak{G}$, and hence $G/C_G(R) \in f(R)$. But as $R = G^{EF(f)}$, $G/R \in EF(f)$. This implies that $G \in EF(f)$. This contradiction shows that $\mathfrak{F}(h) \subseteq EF(f)$. Now assume that $EF(f) \not\subseteq \mathfrak{F}(h)$ and let G be a group with minimal length of composition series in $EF(f) \setminus \mathfrak{F}(h)$. Since $\mathfrak{F}(h)$ is a nonempty formation, G is a monolithic group with monolith $R = G^{\mathfrak{F}(h)}$. Since $G \in EF(f)$, $G/C_G(R) \in f(R)$. So, $f(R) \neq \emptyset$. If $R = A_1 \times \cdots A_t$, where $A_1 \cong A_2 \cong \cdots \cong A_t \cong A$, A is a simple group, then by definition of function of f, we have $t \in h(A)$ and thereby $G \in \mathfrak{F}(h)$, a contradiction. Thus $EF(f) \subseteq \mathfrak{F}(h)$, which induces that $\mathfrak{F}(h) = EF(f)$ is weak graduated formation.

We say that an e-function f is graduated if $f(H) \subseteq f(N)$ for any normal subgroup N of any elementary group H.

Proposition 5.5 (Guo [136]). *A nonempty formation $\mathfrak{F}$ is graduated if and only if $\mathfrak{F} = EF(f)$ for some graduated e-function f.*

Proof Let $\mathfrak{F} = EF(f)$, where f is a graduated e-function. Let $\bar{f}$ be the function such that every nonelementary group G associates with the empty formation $\emptyset$, $\bar{f}(H) = f(H)$ for any nonidentity elementary group H, and $\bar{f}(1) = \mathfrak{G}$. We show that $\bar{f}$ is a screen. Let G be an arbitrary group and N be a nontrivial normal subgroup of G. If G is an elementary group, then by definition of the function $\bar{f}$ we have $\bar{f}(G) = f(G)$. In this case, N and G/N are also elementary groups. Hence $\bar{f}(N) = f(N)$ and $\bar{f}(G/N) = f(G/N)$. It is easy to see G has a normal subgroup such that

$L \simeq G/N$. Hence $\overline{f}(L) = f(L) = f(G/N)$. It follows that $\overline{f}(G) \subseteq \overline{f}(N) \cap \overline{f}(L) = \overline{f}(N) \cap \overline{f}(G/N)$. Therefore $\overline{f}$ is a screen.

We now show that $\mathfrak{F} = \langle \overline{f} \rangle$. Assume that $\mathfrak{F} \nsubseteq \langle \overline{f} \rangle$ and let G be a group in $\mathfrak{F} \backslash \langle \overline{f} \rangle$ with minimal length of composition series. Since $\langle \overline{f} \rangle$ is a formation, G has a unique minimal normal subgroup $R = G^{\langle \overline{f} \rangle}$. Since $G \in \mathfrak{F} = EF(f)$, $G/C_G(R) \in f(R) = \overline{f}(R)$. But $G/R \in \langle \overline{f} \rangle$, so $G \in \langle \overline{f} \rangle$. This contradiction shows that $\mathfrak{F} \subseteq \langle \overline{f} \rangle$. Now assume that $\langle \overline{f} \rangle \nsubseteq \mathfrak{F}$ and let G be a group in $\langle \overline{f} \rangle \backslash \mathfrak{F}$ with minimal length of composition series. Then $R = G^{\mathfrak{F}}$ is the unique minimal normal subgroup of G. Since R is an elementary group, $\overline{f}(R) = f(R)$. But since $G \in \langle \overline{f} \rangle$, $G/C_G(R) \in \overline{f}(R)$. Hence $G/C_G(R) \in f(R)$, and consequently $G \in EF(f)$. This contradiction shows that $\mathfrak{F} = \langle \overline{f} \rangle$ is a graduated formation.

Conversely, if $\mathfrak{F} = \langle f \rangle$, where f is a screen, then, clearly, $\mathfrak{F} = EF(\overline{f})$, where $\overline{f}$ is a graduated e-function such that $\overline{f}(H) = f(H)$, for any elementary group H.

Let A be a simple group, $\{H_i/K_i \mid i \in I\}$ the set of all chief factors of a nonelementary nonidentity group G such that composition factors of H_i/K_i are isomorphic to A and $|H_i/K_i| = |A|^{n_i} (i \in I)$. Then the symbol $r_A(G)$ denotes $\max\{n_i \mid i \in I\}$ and is called A-rank of G. If G has no composition factors isomorphic to A, then we put $r_A(G) = 0$. If A is a group of prime order p, then the A-rank of G is equal to r_p-rank of G, in sense [248, p. 685]. Finally, we use $r(G)$ to denote $\max\{r_A(G) \mid A$ is a composition factor of $G\}$. If G is an identity group, then we put $r(G) = r_A(G) = 0$.

Now we construct an example to show that there is a graduated formation, which is not a Baer-local formation.

Example 5.6 (Guo, [136]). Let p, q and r be primes such that $p \mid q - 1$ and $pq \mid r - 1$ (for example, $p = 2, q = 3, r = 7$). Let H be a transitive permutation group of degree q with $|H| = pq$. Let $k = r - 1$ and $S = H_1 \times \cdots \times H_k$, where $H_1 \cong H_2 \cong \cdots \cong H_k \cong H$, F be a free group of rang qk with free generating a_{ij}, $i = 1, \ldots, k$, $j = 1, \ldots, q$. For any $s \in S$, where $s = h_1 \ldots h_k$, $h_i \in H_i$, put

$$(a_{ij})^s = a_{i(j^{h^i})}.$$

Then S may be considered as an automorphisms group of F.

Let $K_r(F)$ be the r-th term of the lower central series of F and $F^r = \langle a^r \mid a \in F \rangle$. Let $R = F/K_r(F)F^r$. Then R is a r-group of exponent r and its class $\leq r - 1$. The above defined acting of S on F allows us to consider S as a group operators of R. Let $G = R \rtimes S$.

Let f be an e-function such that $f(Z_p) = (1)$, $f(Z_q) = \mathrm{form}\, Z_p$,

$$f(Z_r) = f(Z_r \times Z_r) = \cdots = f(\underbrace{Z_r \times Z_r \times \cdots \times Z_r}_{p}) = \mathrm{form}\, S,$$

$$f(\underbrace{Z_p \times Z_p \times \cdots \times Z_p}_{n}) = f(\underbrace{Z_q \times Z_q \times \cdots \times Z_q}_{m})$$

$$= f(\underbrace{Z_r \times Z_r \times \cdots \times Z_r}_{t}) = \varnothing$$

for all $n \geq 2$, $m \geq 2$, $t \geq p + 1$, and $f(H) = \emptyset$ for all elementary groups H such that every group Z_p, Z_q, Z_r is not isomorphic to a composition factor of H. Then, clearly, f is a graduated e-function.

Consider the formation $\mathfrak{F} = EF(f)$. It is easy to see from Proposition 5.5 that $\mathfrak{F}$ is a graduated formation. We now show that this formation is not a Baer-local formation. Since G is soluble, it is enough to show that $G/\Phi(G) \in \mathfrak{F}$ and $G \notin \mathfrak{F}$. Let H/K be any chief factor of G such that $\Phi(G) \subseteq K$. If H/K is a p-group, then by the construction of G, we see that $H/K = G/K$ is of order p and hence $G/C_G(H/K) = G/G \in f(H/K) = f(Z_p)$. If H/K is a q-group, then $R \subseteq K$ and $G/C_G(H/K) \simeq Z_p \in f(H/K) = f(Z_q)$. Let H/K be a r-factor. Then according to the proof in [248, p. 714], we have $|H/K| = Z_r^t$, where $t = 1$ or $t = p$. So

$$G/C_G(H/K) \in \mathrm{form}(G/R) = formS = f(Z_r) = f(\underbrace{Z_r \times Z_r \times \cdots \times Z_r}_{p}).$$

This means that $G/\Phi(G) \in EF(f)$. But, as the proof in [248, p. 715], if t is the r_r-rank of G, then $t \geq p^k$. Thus, G has a Frattini chief r-factor H/K such that $H/K \cong \underbrace{Z_r \times Z_r \times \cdots \times Z_r}_{p}$, where $t \geq p + 1$. Hence $f(H/K) = \emptyset$ and consequently $G \notin EF(f)$.

5.6 $\mathfrak{F}$-Injectors and $\mathfrak{F}$-Covering Subgroups

$\mathfrak{F}$-injector and $\mathfrak{F}$-covering subgroups. Let $\mathfrak{F}$ be a class of groups. A subgroup V of a group G is said to be:

(1) An $\mathfrak{F}$-*injector* of G if $V \cap K$ is an $\mathfrak{F}$-maximal subgroup of K for every subnormal subgroup K of G.
(2) An $\mathfrak{F}$-*covering subgroup* of G if $V \in \mathfrak{F}$ and $T = KH$ whenever $H \leq T \leq G$ and $T/K \in \mathfrak{F}$.

The theory of the F-injectors and $\mathfrak{F}$-covering subgroups well represented in many books (see, for example, [135, 89]). In this section, we give some new results about the subgroups. In particular, we give the solutions of some open questions.

Description of $\mathfrak{F}$-injectors of finite soluble groups. In this part, all groups are finite and soluble.

Concerning Fitting classes and injectors, there was the following problem.

Problem 6.1 (Hartley [236]). Let $\mathfrak{F}$ be a local Fitting class (see Sect. 5.7), could we describe the $\mathfrak{F}$-injectors of a group?

We will give the solution of this problem in the important case when $\mathfrak{F}$ is Hartey class.

For any two Fitting classes $\mathfrak{F}$ and $\mathfrak{H}$, if $\mathfrak{H} = Q\mathfrak{H}$, that is, $\mathfrak{H}$ is factor group closed, then their product $\mathfrak{F} \diamond \mathfrak{H} = (G : G$ has a normal subgroup $N \in \mathfrak{F}$ with $G/N \in \mathfrak{H})$ (see [89, p. 566]). In this case, in order to write concise, in the following, we use $\mathfrak{F}\mathfrak{H}$ to denote the product $\mathfrak{F} \diamond \mathfrak{H}$ of two Fitting classes $\mathfrak{F}$ and $\mathfrak{H}$.

Let $\Sigma = \{\pi_i | i \in I\}$ be the set of pairwise disjoint subsets of $\mathbb{P}$, the set of all prime numbers, and $\mathbb{P} = \cup_{i \in I} \pi_i$. Then, a function $h : \Sigma \longrightarrow \{$nonempty Fitting classes$\}$ is called a Hartley function (or in brevity, H-function). Let

$$LH(h) = \cap_{i \in I} h(\pi_i) \mathfrak{S}_{\pi_i'} \mathfrak{S}_{\pi_i},$$

where $\mathfrak{S}_\pi$ is the class of all soluble π-groups and $h(\pi_i) \mathfrak{S}_{\pi_i'} \mathfrak{S}_{\pi_i}$ is the usual Fitting product.

A Fitting class $\mathfrak{H}$ is called a Hartley class if $\mathfrak{H} = LH(h)$, for some Hartley function h. In this case, we also say that $\mathfrak{H}$ is defined by the local H-function h or h is a local Hartley function of the Hartley class $\mathfrak{H}$.

For any two H-functions f and h, we write $f \leq h$ if $f(\pi_i) \subseteq h(\pi_i)$ for all $i \in I$.

Let f be a local H-function of a Hartley class $\mathfrak{F}$. f is called integrated if $h(p) \subseteq \mathfrak{F}$ for all $p \in \mathbb{P}$.

If $\mathfrak{X}$ is a set of groups, then we use Fit$(\mathfrak{X})$ to denote the Fitting class generated by $\mathfrak{X}$, that is, Fit$(\mathfrak{X})$ is the smallest Fitting class containing $\mathfrak{X}$.

The following lemma is clear.

Lemma 6.2 *Every Hartley class can be defined by a local integrated H-function.*

Lemma 6.3 (Guo, Vorob'ev [211]). *Every Hartley class $\mathfrak{H}$ can be defined by a local integrated H-function h such that $h(\pi_i) \subseteq h(\pi_j) \mathfrak{S}_{\pi_j'}$, for all $i \neq j \in I$.*

Proof Let $\mathfrak{H}$ be a Hartley class. By Lemma 6.2, $\mathfrak{H} = LH(h_1)$, for some integrated h-function h_1. We define

$$\psi(\pi_i) = \{G \mid G \simeq H^{\mathfrak{S}_{\pi_i'}}, \text{ for some } H \in h_1(\pi_i)\},$$

for all $i \in I$.

Let X be a group in $\psi(\pi_i)(i \in I)$. Then $X \simeq Y^{\mathfrak{S}_{\pi_i'}}$, for some group $Y \in h_1(\pi_i)$. Since every Fitting class is closed with respect to normal subgroup, $Y^{\mathfrak{S}_{\pi_i'}} \in h_1(\pi_i)$ and so $X \in h_1(\pi_i)$. This shows that $\psi \leq h_1$, and consequently

$$\psi(\pi_i) \mathfrak{S}_{\pi_i'} \subseteq h_1(\pi_i) \mathfrak{S}_{\pi_i'}.$$

If $Y_1 \in h_1(\pi_i) \mathfrak{S}_{\pi_i'}$, then $Y_1/(Y_1)_{h_1(\pi_i)} \in \mathfrak{S}_{\pi_i'}$. Since $(Y_1^{\mathfrak{S}_{\pi_i'}})^{\mathfrak{S}_{\pi_i'}} = Y_1^{\mathfrak{S}_{\pi_i'}}$, we have $Y_1^{\mathfrak{S}_{\pi_i'}} \in \psi(\pi_i)$, that is, $Y_1 \in \psi(\pi_i) \mathfrak{S}_{\pi_i'}$. Therefore, we obtain the following equation:

$$\psi(\pi_i) \mathfrak{S}_{\pi_i'} = h_1(\pi_i) \mathfrak{S}_{\pi_i'}. \tag{5.10}$$

Now, let h be a function such that $h(\pi_i) = \text{Fit}(\psi(\pi_i))$, for all $i \in I$. Let

$$\mathfrak{M} = \cap_{i \in I} h(\pi_i) \mathfrak{S}_{\pi_i'} \mathfrak{S}_{\pi_i}.$$

We now prove that $\mathfrak{M} = \mathfrak{H}$. In fact, since $\psi \leq h_1$, we have $h \leq h_1$ and hence $h(\pi_i) \mathfrak{S}_{\pi_i'} \subseteq h_1(\pi_i) \mathfrak{S}_{\pi_i'}$, for all $i \in I$. Then, by the Eq. (5.10), we see that

$$\text{Fit}(h_1(\pi_i) \mathfrak{S}_{\pi_i'}) = h_1(\pi_i) \mathfrak{S}_{\pi_i'} = \text{Fit}(\psi(\pi_i) \mathfrak{S}_{\pi_i'}).$$

Therefore

$$h_1(\pi_i)\mathfrak{S}_{\pi_i'} = \mathrm{Fit}(\psi(\pi_i)\mathfrak{S}_{\pi_i'}) \subseteq (\mathrm{Fit}(\psi(\pi_i))\mathfrak{S}_{\pi_i'} = h(\pi_i)\mathfrak{S}_{\pi_i'}.$$

This shows that

$$h(\pi_i)\mathfrak{S}_{\pi_i'} = h_1(\pi_i)\mathfrak{S}_{\pi_i'},$$

for all $i \in I$. Thus, this induces that $\mathfrak{M} = \mathfrak{H}$ and h is a local H-function of $\mathfrak{H}$. Since $h \leq h_1$ and h_1 is a integrated local H-function of $\mathfrak{H}$, obviously h is also a integrated H-function.

Now, in order to complete the proof of the lemma, we only need to prove that $h(\pi_i) \subseteq h(\pi_j)\mathfrak{S}_{\pi_j'}$, for all $i \neq j \in I$. In fact, let $L \in h_1(\pi_i)$ and $i \neq j$. Since $\pi_i \cap \pi_j = \emptyset$, $\mathfrak{S}_{\pi_j} \subseteq \mathfrak{S}_{\pi_i'}$ and so $L^{\mathfrak{S}_{\pi_i'}} \subseteq L^{\mathfrak{S}_{\pi_j}}$. However, since $L \in \mathfrak{H}$, we have $L/L_{h_1(\pi_j)\mathfrak{S}_{\pi_j'}} \in \mathfrak{S}_{\pi_j}$, and consequently, $L^{\mathfrak{S}_{\pi_j}} \subseteq L_{h_1(\pi_j)\mathfrak{S}_{\pi_j'}}$. Thus, $L^{\mathfrak{S}_{\pi_j}} \in h_1(\pi_j)\mathfrak{S}_{\pi_j'}$. This shows that, for all groups $G \in h_1(\pi_i)$, the $\mathfrak{S}_{\pi_i'}$-residual of G is contained in $h_1(\pi_j)\mathfrak{S}_{\pi_j'}$. Therefore, if $R \in \psi(\pi_i)$, then $R \simeq V^{\mathfrak{S}_{\pi_i'}}$ for some group $V \in h_1(\pi_i)$ and $R \in h_1(\pi_j)\mathfrak{S}_{\pi_j'}$. This induces that

$$\psi(\pi_i) \subseteq h_1(\pi_j)\mathfrak{S}_{\pi_j'}.$$

Thus

$$h(\pi_i) = \mathrm{Fit}\,\psi(\pi_i) \subseteq \mathrm{Fit}(h_1(\pi_j)\mathfrak{S}_{\pi_j'}) = h_1(\pi_j)\mathfrak{S}_{\pi_j'} = h(\pi_j)\mathfrak{S}_{\pi_j'}.$$

The proof is completed.

By Lemma 6.3, for any Hartley class $\mathfrak{H}$, we may always assume that $\mathfrak{H}$ is defined by a local integrated H-function h such that $h(\pi_i) \subseteq h(\pi_j)\mathfrak{S}_{\pi_j'}$ for all $i, j \in I$ and $i \neq j$. We call the subgroup $G_h = \Pi_{i \in I} G_{h(\pi_i)}$ the h-radical of G.

Lemma 6.4 *If V is a subgroup of a group G such that $V/G_h \in \cap_{j \in I} \mathfrak{S}_{\pi_j'}\mathfrak{S}_{\pi_j}$, then $V \in \mathfrak{H}$.*

Proof Since $G_h \trianglelefteq V$, we have

$$G_{h(\pi_j)} = (G_h)_{h(\pi_j)} = G_h \cap V_{h(\pi_j)} \subseteq V_{h(\pi_j)}, \text{ for all } j \in I.$$

We first prove that $G_h/(G_h)_{h(\pi_j)}$ is a π_j'-group. Since $G_{h(\pi_j)}G_{h(\pi_i)}/G_{h(\pi_j)} \simeq G_{h(\pi_j)}/G_{h(\pi_i)} \cap G_{h(\pi_j)}$, where $i \neq j$, we have that $G_{h(\pi_j)}G_{h(\pi_i)}/G_{h(\pi_j)} \simeq G_{h(\pi_i)}/(G_{h(\pi_i)})_{h(\pi_j)}$. However, by Lemma 6.3, $h(\pi_i) \subseteq h(\pi_j)\mathfrak{S}_{\pi_j'}$. Therefore, $G_{h(\pi_i)} \in h(\pi_j)\mathfrak{S}_{\pi_j'}$ and thereby $G_{h(\pi_i)}/(G_{h(\pi_i)})_{h(\pi_j)} \in \mathfrak{S}_{\pi_j'}$. This means that

$$G_{h(\pi_i)}G_{h(\pi_j)}/G_{h(\pi_j)} \in \mathfrak{S}_{\pi_j'},$$

for any $i \neq j$. Hence $G_h/G_{h(\pi_j)} \in \mathfrak{S}_{\pi_j'}$. Then, by the isomorphism

$$V_{h(\pi_j)}G_h/V_{h(\pi_j)} \simeq G_h/G_h \cap V_{h(\pi_j)} \simeq (G_h/G_{h(\pi_j)})/(G_h \cap V_{h(\pi_j)}/G_{h(\pi_j)}),$$

we see that $V_{h(\pi_j)}G_h/V_{h(\pi_j)}$ is a π'_j-group. Since $V/G_h \in \cap_{j \in I}\mathfrak{S}_{\pi'_j}\mathfrak{S}_{\pi_j}$, by using the isomorphism

$$V/V_{h(\pi_j)}G_h \simeq (V/G_h)/(V_{h(\pi_j)}G_h/G_h),$$

we have $V/V_{h(\pi_j)}G_h \in \mathfrak{S}_{\pi'_j}\mathfrak{S}_{\pi_j}$. It follows that

$$(V/V_{h(\pi_j)})/(V_{h(\pi_j)}G_h/V_{h(\pi_j)}) \in \mathfrak{S}_{\pi'_j}\mathfrak{S}_{\pi_j},$$

and consequently, $V/V_{h(\pi_j)} \in \mathfrak{S}_{\pi'_j}(\mathfrak{S}_{\pi'_j}\mathfrak{S}_{\pi_j}) = \mathfrak{S}_{\pi'_j}\mathfrak{S}_{\pi_j}$, for all $j \in I$. Therefore, $V \in \cap_{j \in I}h(\pi_j)\mathfrak{S}_{\pi'_j}\mathfrak{S}_{\pi_j} = \mathfrak{H}$. This completes the proof.

Lemma 6.5 *Let* $\mathfrak{H} = LH(h) = \cap_{i \in I}h(\pi_i)\mathfrak{S}_{\pi'_i}\mathfrak{S}_{\pi_i}$ *and* G *a group. If* $\mathfrak{D} = \cap_{i \in I}\mathfrak{S}_{\pi'_i}\mathfrak{S}_{\pi_i}$, *then* $G_{\mathfrak{H}}/G_h = (G/G_h)_{\mathfrak{D}}$.

Proof Let $(G/G_h)_{\mathfrak{D}} = L/G_h$. Then, we only need to prove that $L = G_{\mathfrak{H}}$. Since $G_{\mathfrak{H}} \in \mathfrak{H}$ and $(G_{\mathfrak{H}})_{h(\pi_i)} = G_{h(\pi_i)}$, we have $G_{\mathfrak{H}}/G_{h(\pi_i)} \in \mathfrak{S}_{\pi'_i}\mathfrak{S}_{\pi_i}$ for all $i \in I$. Consequently, $G_{\mathfrak{H}}/G_h \in \mathfrak{D}$. It follows that $G_{\mathfrak{H}}/G_h \subseteq L/G_h$ and so $G_{\mathfrak{H}} \subseteq L$. On the other hand, since $L/G_h \in \mathfrak{D}$, by Lemma 6.4, we have $L \in \mathfrak{H}$ and so $L \subseteq G_{\mathfrak{H}}$. Thus, the lemma holds.

Lemma 6.6 *Let* $\mathfrak{H} = LH(h) = \cap_{i \in I}h(\pi_i)\mathfrak{S}_{\pi'_i}\mathfrak{S}_{\pi_i}$ *and* G *a group. Then, a subgroup* V *of* G *containing* $G_{\mathfrak{H}}$ *is an* $\mathfrak{H}$-*subgroup if and only if* $V/G_h \in \cap_{i \in I}\mathfrak{S}_{\pi'_i}\mathfrak{S}_{\pi_i}$.

Proof Let $\mathfrak{D} = \cap_{i \in I}\mathfrak{S}_{\pi'_i}\mathfrak{S}_{\pi_i}$. If $V/G_h \in \mathfrak{D}$, then $V \in \mathfrak{H}$ by Lemma 6.4.

Now, assume that $V \in \mathfrak{H}$ and $V \supseteq G_{\mathfrak{H}}$. Since the H-function h is integrated,

$$V_{h(\pi_i)} \cap G_{\mathfrak{H}} = (G_{\mathfrak{H}})_{h(\pi_i)} = G_{h(\pi_i)},$$

for every $i \in I$. It follows that $[V_{h(\pi_i)}, G_{\mathfrak{H}}] \subseteq G_{h(\pi_i)}$ and so $V_{h(\pi_i)} \subseteq C_G(G_{\mathfrak{H}}/G_{h(\pi_i)})$. We now prove that

$$C_G(G_{\mathfrak{H}}/G_{h(\pi_i)}) \subseteq G_{\mathfrak{H}}.$$

Assume that $C = C_G(G_{\mathfrak{H}}/G_{h(\pi_i)})$ is not a subgroup of $G_{\mathfrak{H}}$. Then, there exists a normal subgroup K of G such that $K \subseteq C$ and $K/K \cap G_{\mathfrak{H}}$ is a chief factor of G. Obviously, $C \cap G_{\mathfrak{H}} \neq K$ and $C \cap G_{\mathfrak{H}} = K \cap G_{\mathfrak{H}}$. Hence

$$K/C \cap G_{\mathfrak{H}} = K/K \cap G_{\mathfrak{H}} \simeq KG_{\mathfrak{H}}/G_{\mathfrak{H}},$$

and so $K/K \cap G_{\mathfrak{H}}$ is an elementary abelian p-group. This means that $(K/K \cap G_{\mathfrak{H}})^{\mathfrak{A}} = 1$, where $\mathfrak{A}$ is the formation of all abelian groups, and so $K^{\mathfrak{A}}(K \cap G_{\mathfrak{H}})/K \cap G_{\mathfrak{H}} = 1$ by Chap. 1, Lemma 1.1. It follows that $K^{\mathfrak{A}} \subseteq K \cap G_{\mathfrak{H}}$. Since $K \subseteq C_G(G_{\mathfrak{H}}/G_{h(\pi_i)})$,

$$K \subseteq C_G(K \cap G_{\mathfrak{H}}/G_{h(\pi_i)}).$$

So $[K^{\mathfrak{A}}, K] \subseteq [K \cap G_{\mathfrak{H}}, K] \subseteq G_{h(\pi_i)}$. Therefore, $K/G_{h(\pi_i)}$ is a nilpotent group with nilpotent class at most 2. Let $P/G_{h(\pi_i)}$ be a nonidentity normal Sylow p-subgroup of $K/G_{h(\pi_i)}$. By [135, Theorem 2.6.7], we know that every Sylow p-subgroup covers the p-chief factor $K/K \cap G_{\mathfrak{H}}$. Thus, $P(K \cap G_{\mathfrak{H}}) \supseteq K$, and consequently $PG_{\mathfrak{H}} = KG_{\mathfrak{H}}$. Now, we prove that $P \in \mathfrak{H}$.

If $p \in \pi_i$, where $i \in I$, then $P \in h(\pi_i)\mathfrak{S}_{\pi_i}$. By Lemma 6.3, h is the H-function of the class $\mathfrak{H}$ such that $h(\pi_i) \subseteq h(\pi_j)\mathfrak{S}_{\pi_j'}$, for $j \neq i$ and $j \in I$. This shows that

$$h(\pi_i)\mathfrak{S}_{\pi_i} \subseteq h(\pi_j)\mathfrak{S}_{\pi_j'}\mathfrak{S}_{\pi_i}.$$

Hence $P \in h(\pi_j)\mathfrak{S}_{\pi_j'}\mathfrak{S}_{\pi_j}$, for all $j \neq i$. If $p \in \pi_i'$, then $P \in h(\pi_i)\mathfrak{S}_{\pi_i'}$ and so $P \in h(\pi_i)\mathfrak{S}_{\pi_i'}\mathfrak{S}_{\pi_i}$. This shows that, in any case, $P \in \cap_{i \in I}h(\pi_i)\mathfrak{S}_{\pi_i'}\mathfrak{S}_{\pi_i} = \mathfrak{H}$. It is clear that $P \trianglelefteq G$. Hence $KG_{\mathfrak{H}} = PG_{\mathfrak{H}} \in \mathfrak{H}$. It follows that $KG_{\mathfrak{G}} = G_{\mathfrak{H}}$ and so $K \subseteq G_{\mathfrak{H}}$. Consequently, $K \cap G_{\mathfrak{G}} = C \cap G_{\mathfrak{H}} = K$. This contradiction proves that $C \subseteq G_{\mathfrak{H}}$.

The above argument shows that $V_{h(\pi_i)} \subseteq C \subseteq G_{\mathfrak{H}}$ and so $V_{h(\pi_i)} = V_{h(\pi_i)} \cap G_{\mathfrak{H}} = G_{h(\pi_i)}$ for all $i \in I$. It follows from $V \in \mathfrak{H}$ that $V/G_{h(\pi_i)} = V/V_{h(\pi_i)} \in \mathfrak{S}_{\pi_i'}\mathfrak{S}_{\pi_i}$, for $i \in I$, and hence $V/G_h = V/\Pi G_{h(\pi_i)} \in \cap\mathfrak{S}_{\pi_i'}\mathfrak{S}_{\pi_i} = \mathfrak{D}$. This completes the proof.

Lemma 6.7 (D'Arcy [82]). *Let* $\mathfrak{D} = \cap_{i \in I}\mathfrak{S}_{\pi_i'}\mathfrak{S}_{\pi_i}$. *If* V *is a* $\mathfrak{D}$-*maximal subgroup of* G *containing the* $\mathfrak{D}$-*radical of* G, *then* V *is a* $\mathfrak{D}$-*injector of* G.

Theorem 6.8 (Guo, Vorob'ev [211]). *Let* G *be a soluble group,* V *a subgroup of* G *and* $\mathfrak{H} = LH(h)$ *a Hartley class. Then:*

1) V *is an* $\mathfrak{H}$-*injector of* G *if and only if* V/G_h *is a* $\mathfrak{D}$-*injector of* G/G_h, *where* $\mathfrak{D} = \cap_{i \in I}\mathfrak{S}_{\pi_i'}\mathfrak{S}_{\pi_i}$;
2) *The set of all* $\mathfrak{H}$-*injectors of* G *is exactly the set of the* $\mathfrak{H}$-*maximal subgroups of* G *containing the* $\mathfrak{H}$-*radical of* G.

Proof

1) Let V be an $\mathfrak{H}$-injector of G. Then, obviously, $G_{\mathfrak{H}} \subseteq V$ and V is an $\mathfrak{H}$-maximal subgroup of G. By Lemma 6.6, we have $V/G_h \in \mathfrak{D}$. Then, by the $\mathfrak{H}$-maximality of V in G, we see that V/G_h is a $\mathfrak{D}$-maximal subgroup of G/G_h by Lemma 6.6. Since $G_{\mathfrak{H}}/G_h = (G/G_h)_{\mathfrak{D}}$ by Lemma 6.5 and $G_{\mathfrak{H}} \subseteq V$, V/G_h is a $\mathfrak{D}$-maximal subgroup of G/G_h containing the $\mathfrak{D}$-radical of G/G_h. Hence, by Lemma 6.7, V/G_h is a $\mathfrak{D}$-injector of G/G_h.

Now we prove the converse by induction on the order of G. By Fischer, Gaschutz and Hatrly theorem (see [106] or [135, 2.5.1]), we know that G/G_h has a $\mathfrak{D}$-injector, V/G_h say. Let M be a maximal normal subgroup of G.

We claim that the subgroup $M_h = \Pi_{i \in I}M_{h(\pi_i)}$ of M is equal to the intersection $G_h \cap M$. In order to prove the claim, we first establish that $G_h/G_{h(\pi_i)}$ is a π_j'-group, for all $j \in I$. In fact, by Lemma 6.3, $h_{(\pi_i)} \subseteq h_{(\pi_j)}\mathfrak{S}_{\pi_j'}$, for all different $i, j \in I$. Then, by the isomorphism

$$G_{h(\pi_j)}G_{h(\pi_i)}/G_{h(\pi_j)} \simeq G_{h(\pi_i)}/G_{h(\pi_i)} \cap G_{h(\pi_j)} = G_{h(\pi_i)}/(G_{h(\pi_i)})_{h(\pi_j)},$$

we see that $G_{h(\pi_j)}G_{h(\pi_i)}/G_{h(\pi_j)}$ is a π_j'-group. It follows that $G_h/G_{h(\pi_j)}$ is also a π_j'-group.

Now, since

$$(G_h \cap M)G_{h(\pi_j)}/G_{h(\pi_j)} \simeq (G_h \cap M)/(G_h \cap M) \cap G_{h(\pi_j)}$$
$$= (G_h \cap M)/(M \cap G_{h(\pi_j)}) = (G_h \cap M)/M_{h(\pi_j)},$$

$(G_h \cap M)/M_{h(\pi_j)}$ is a π'_j-group, for all $j \in I$. This induces that $G_h \cap M/M_h \in \cap_{j \in I} \mathfrak{S}_{\pi'_j}$. However, obviously,

$$\cap_{i \in I} \mathfrak{S}_{\pi'_j} = \mathfrak{S}_{\cap_{j \in I} \pi'_j} = \mathfrak{S}_{(\cup_{j \in I} \pi_j)'} = (1).$$

Hence $G_h \cap M = M_h$.

Assume that G_h is a subgroup of M. Then, $M_h = G_h$. Since V/G_h is a $\mathfrak{D}$-injector of G/G_h, $V \cap M/G_h$ is a $\mathfrak{D}$-injector of M/G_h and consequently $V \cap M/M_h$ is a $\mathfrak{D}$-injector of M/M_h. Hence, by induction, $V \cap M$ is an $\mathfrak{H}$-injector of M. Since $V/G_h \in \mathfrak{D}$, by Lemma 6.4, we see that $V \in \mathfrak{H}$. We now prove that V is an $\mathfrak{H}$-maximal subgroup of G. Assume that $V \subset V_1$, where V_1 is an $\mathfrak{H}$-maximal subgroup of G. Then, obviously, $V \cap M = V_1 \cap M$. This shows that V_1 is an $\mathfrak{H}$-maximal subgroup of G and $V \cap M$ is an $\mathfrak{H}$-injector of M, for any maximal normal subgroup M of G. Hence V_1 is an $\mathfrak{H}$-injector of G, and consequently $V_1 \supseteq G_{\mathfrak{H}}$. Then, by Lemma 6.6, we have $V_1/G_h \in \mathfrak{D}$, contrary to the fact that V/G_h is $\mathfrak{D}$-maximal in G/G_h. Therefore, $V = V_1$ and hence V is an $\mathfrak{H}$-injector of G.

Now assume that G_h is not contained in M. In this case, by the maximality of M, we have that $G = G_h M$. Since

$$G/G_h \simeq M/G_h \cap M = M/M_h,$$

$V \cap M/M_h$ is a $\mathfrak{D}$-injector of M/M_h. Then, by induction, $V \cap M$ is an $\mathfrak{H}$-injector of M. By Lemma 6.4, we know that $V \in \mathfrak{H}$. If $V \subset F_1$, where F_1 is an $\mathfrak{H}$-maximal subgroup of G, then $V \cap M = F_1 \cap M$. Since $V \supseteq G_h$, $VM = G$. Consequently,

$$F_1 = F_1 \cap VM = V(F_1 \cap M) = V(V \cap M) = V$$

and so V is an $\mathfrak{H}$-maximal subgroup of G. Therefore V is an $\mathfrak{H}$-injector of G. Thus, the statement 1) holds.

2) If V is an $\mathfrak{H}$-injector of G, then by the definition of $\mathfrak{H}$-injector, we see that $V \supseteq G_{\mathfrak{H}}$ and V is an $\mathfrak{H}$-maximal subgroup of G.

Now, assume that V is any $\mathfrak{H}$-maximal subgroup of G containing the $\mathfrak{H}$-radical $G_{\mathfrak{H}}$. We prove that V is an $\mathfrak{H}$-injector of G. Since h is an integrated H-function of $\mathfrak{H}$, we have that $G_h \subseteq G_{\mathfrak{H}}$ and hence $V \supseteq G_h$. Then, by Lemma 6.6, $V/G_h \in \mathfrak{D} = \cap_{i \in I} \mathfrak{S}_{\pi'_i} \mathfrak{S}_{\pi_i}$. By the $\mathfrak{H}$-maximality of V in G and Lemma 6.6, we see that V/G_h is a $\mathfrak{D}$-maximal subgroup of G/G_h. On the other hand, By the condition and Lemma 6.5, we know that $V/G_h \supseteq (G/G_h)_{\mathfrak{D}}$. It follows from Lemma 6.7 that V/G_h is a $\mathfrak{D}$-injector of $G/G_{\mathfrak{H}}$. Thus, by using the statement 1), we obtain that V is an $\mathfrak{H}$-injector of G. This completes the proof.

Now we consider some applications of Theorem 6.8.

Firstly, since the $\mathfrak{N}$-radical of a group G is the Fitting subgroup $F(G)$ of G, we have:

Corollary 6.9 (Fischer [105]). *A subgroup V of a soluble group G is an $\mathfrak{N}$-injectors of G if and only if V is a nilpotent maximal subgroups of G containing $F(G)$.*

The following result follows immediately from Theorem 6.8.

Corollary 6.10 *Let π be some set of prime numbers. Then, a subgroup V of a soluble group G is a π-nilpotent injector of G if and only if V is a maximal π-nilpotent subgroup of G containing the π-nilpotent radical of G.*

Let $\mathfrak{H} = \mathfrak{X}\mathfrak{N}$, where $\mathfrak{X}$ is any nonempty Fitting class and $\mathfrak{N}$ is the Fitting class of all nilpotent groups. Let h be an H-function such that $h(p) = \mathfrak{X}$ for any $p \in \mathbb{P}$, where $\mathfrak{X}$ is any nonempty Fitting class. Then, $LH(h) = \cap_{p\in\mathbb{P}}h(p)\mathfrak{S}_{p'}\mathfrak{S}_p = \mathfrak{X}(\cap_{p\in\mathbb{P}} \mathfrak{S}_{p'}\mathfrak{N}_p) = \mathfrak{X}\mathfrak{N} = \mathfrak{H}$. Hence, by Theorem 6.8, we have:

Corollary 6.11 (Hartley [236]). *Let $\mathfrak{H} = \mathfrak{X}\mathfrak{N}$, where $\mathfrak{X}$ is any nonempty Fitting class and $\mathfrak{N}$ is the Fitting class of all nilpotent groups, then the set of all $\mathfrak{H}$-injectors of a soluble group G is exactly the set of the subgroups V of G such that $V/G_\mathfrak{X}$ is a nilpotent injector of $G/G_\mathfrak{X}$.*

Corollary 6.12 *Let $\mathfrak{H} = \cap_{p\in\mathbb{P}}h(p)\mathfrak{S}_{p'}\mathfrak{S}_p$ be a Hartley class. Then, a subgroup V of a soluble group G is an $\mathfrak{H}$-injector of a soluble group G if and only if V/G_h is a nilpotent injector of G/G_h.*

Corollary 6.13 *Let $\mathfrak{N}^k$ $(k > 0)$ be the class of all groups with nilpotent length at most k and G a soluble group. Then the set of all $\mathfrak{N}^k$-injectors of G is exactly the set of all subgroups V of G such that $V/G_{\mathfrak{N}^{k-1}}$ is a nilpotent injector of $G/G_{\mathfrak{N}^{k-1}}$. In particular, a subgroup V of a soluble group G is a metanilpotent injector of G if and only if $V/F(G)$ is a nilpotent injector of $G/F(G)$.*

Proof Let h be an H-function such that $h(p) = \mathfrak{N}^{k-1}$, for all primes p. Then $\cap_{p\in\mathbb{P}}h(p)\mathfrak{S}_{p'}\mathfrak{N}_p = \cap_{p\in\mathbb{P}}\mathfrak{N}^{k-1}\mathfrak{S}_{p'}\mathfrak{N}_p = \mathfrak{N}^{k-1}(\cap_{p\in\mathbb{P}} \mathfrak{S}_{p'}\mathfrak{N}_p) = \mathfrak{N}^{k-1}\mathfrak{N} = \mathfrak{N}^k$. This shows that $\mathfrak{N}^k = LH(h)$ is an Hartley class and obviously, $G_h = \mathfrak{N}^{k-1}$. Thus, by Theorem 6.8, the corollary holds.

Two problems on existence and conjugacy of $\mathfrak{N}$-injectors in nonsoluble groups.

The main result of the theory of Fitting classes is the following theorem proved by Fischer, Gaschütz and Hartley [106].

Theorem 6.14 (see [89, IX, Theorem 1.4] and [135, Theorem 2.5.2]) *Let $\mathfrak{F}$ be a Fitting class. Then a soluble group G has at least one $\mathfrak{F}$-injector, and any two $\mathfrak{F}$-injectors of G are conjugate in G.*

To develop this important result of Fischer, Gaschütz and Hartley is one of efforts of mathematicians. In connection with this, the following problem was proposed in [361] and [316]

Problem 6.15 ([361] and [316, Problem 12.96]). To weaken the condition of the Theorem of Fischer, Gaschütz and Hartley or to find the nonempty Fitting class $\mathfrak{F}$ such that every nonsoluble finite group has an $\mathfrak{F}$-injector.

In 1948, Čunikhin [76] introduced the concept of π-selected group: A group G is called π-selected if $|\pi(H/K) \cap \pi| \leq 1$ for every chief factor H/K of G.

Obviously, a π-soluble group is a π-selected group, but the converse does not hold in general. In connection with the above results, the following open problem was proposed by W. Guo [135] and L. A. Shemetkov [361]:

Problem 6.16 ([361], [135, Problems 5,6]). Let $\mathcal{F}$ be a nonempty Fitting class, $\pi = \pi(\mathfrak{F})$, G a group and $G/G_{\mathfrak{F}}$ be a π-selected group.

1) Has G an $\mathfrak{F}$-injector?
2) If G has $\mathfrak{F}$-injectors, are any two $\mathfrak{F}$-injectors of G conjugate in G?

In this direction, in 1984, V. G. Sementovskii proved the following theorem:

Theorem 6.17 (Sementovskii [347]). *Let $\mathfrak{F}$ be a Fitting class and G a group. If $G/G_{\mathfrak{F}}$ is a soluble group, then G has an $\mathfrak{F}$-injector and any two $\mathfrak{F}$-injectors of G are conjugate in G.*

In 1997, Guo extended Sementovskii's result and proved the following theorem.

Theorem 6.18 (Guo [130]). *Let $\mathfrak{F}$ be a Fitting class, $\pi = \pi(\mathfrak{F})$ and G a group. If $G/G_{\mathfrak{F}}$ is a π-soluble group, then G has an $\mathfrak{F}$-injector and any two $\mathfrak{F}$-injectors of G are conjugate in G.*

Our next goal is to resolve Problem 6.16. We prove that the answer to this Problem is affirmative, and consequently, a new progress on Problem 6.15 is also obtained.

Lemma 6.19 *Let $\mathfrak{F}$ be a nonempty Fitting class and H an $\mathfrak{F}$-injector of G. Then the following statements hold:*

1) *H is an $\mathfrak{F}$-maximal subgroup of G;*
2) *$G_{\mathfrak{F}} \leq H$;*
3) *For every $\alpha \in G$, H^α is also an $\mathfrak{F}$-injector of G;*
4) *If K is a subnormal subgroup of G, then $H \cap K$ is an $\mathfrak{F}$-injector of K.*

Proof This is immediate from the definition of $\mathfrak{F}$-injector.

Lemma 6.20 *Let $\mathfrak{F}$ be a nonempty Fitting class. If N is a subnormal subgroup of G, then $N_{\mathfrak{F}} = N \cap G_{\mathfrak{F}}$.*

Proof See [89, Chapter IX, Lemma 1.1].

Lemma 6.21 ([135, Lemma 2.5.8]). *Let $\mathfrak{F}$ be a nonempty Fitting class and G a group satisfying $G/G_{\mathfrak{F}}$ is π-selected, where $\pi = \pi(\mathfrak{F})$. Moreover, suppose that $N \trianglelefteq G$, G/N is abelian, and V_1, V_2 are $\mathfrak{F}$-maximal subgroups of G such that $V_1 \cap N = V_2 \cap N$ is an $\mathfrak{F}$-maximal subgroups of N which contains $N_{\mathfrak{F}}$. Then V_1 and V_2 are conjugate in G.*

Lemma 6.22 (Hall [233] or see [135, Theorem 1.10.6]). *If G has a subnormal series such that every factor of which is an E_{π}^N-group, then G is a D_π-group and every Hall π-subgroup of G is soluble.*

Lemma 6.23 *Let $\mathfrak{F}$ be a Fitting class and G a group. Suppose that $G/G_\mathfrak{F}$ is soluble and G/N is nilpotent. If V is an $\mathfrak{F}$-maximal subgroup of G and $V \cap N$ is an $\mathfrak{F}$-injector of N, then V is an $\mathfrak{F}$-injector of G.*

Proof Assume that the lemma is not true and G is a counterexample of minimal order. By Theorem 6.18, we know that G has an $\mathfrak{F}$-injector, say V_0. Then, by Lemma 6.19(1), V_0 is a $\mathfrak{F}$-maximal subgroup of G.

Assume that $NV \neq G$. Since G/N is nilpotent, NV is a subnormal subgroup of G. Clearly, V is also an $\mathfrak{F}$-maximal subgroup of NV. By Lemma 6.20, $(NV)_\mathfrak{F} = NV \cap G_\mathfrak{F}$. Hence

$$NV/(NV)_\mathfrak{F} = NV/(NV \cap G_\mathfrak{F}) \cong NVG_\mathfrak{F}/G_\mathfrak{F} \le G/G_\mathfrak{F}$$

is soluble. Since $NV/N \le G/N$ and G/N is nilpotent, NV/N is nilpotent. Therefore, by the choice of G, we have that V is an $\mathfrak{F}$-injector of NV. By using Lemma 6.19(4), we see that $V_0 \cap NV$ is an $\mathfrak{F}$-injector of NV. Since any two $\mathfrak{F}$-injectors of NV are conjugate in NV by Theorem 6.17, there exists an element $x \in NV$ such that $(V_0 \cap NV)^x = V$, and consequently, $V \subseteq V_0^x$. Since V is an $\mathfrak{F}$-maximal subgroup of G, $V = V_0^x$. Then by Lemma 6.19, we obtain that V is an $\mathfrak{F}$-injector of G, a contradiction.

Now, assume that $NV = G$ and M is a maximal normal subgroup of G containing N. Because G/N is nilpotent, M/N is nilpotent and G/M is abelian. Obviously, $V \cap M \lhd V$ and so $V \cap M \in \mathfrak{F}$. Let V_1 be an $\mathfrak{F}$-maximal subgroup of M with $V \cap M \subseteq V_1$. Then $V_1 \cap N = V \cap N$ is an $\mathfrak{F}$-injector of N and, by Lemma 6.20, $M/M_\mathfrak{F} = M/M \cap G_\mathfrak{F} \cong MG_\mathfrak{F}/G_\mathfrak{F} \le G/G_\mathfrak{F}$ is soluble. By the choice of G, we see that V_1 is an $\mathfrak{F}$-injector of M. Since $M = NV \cap M = N(V \cap M) = NV_1$, we have $|N(V \cap M)| = |NV_1| = |M|$. It follows that $\frac{|N||V \cap M|}{|N \cap V|} = \frac{|N||V_1|}{|N \cap V_1|}$ and hence $V \cap M = V_1$. This shows that $V \cap M$ is an $\mathfrak{F}$-injector of M. On the other hand, by using Lemma 6.19(4), we know that $V_0 \cap M$ is also an $\mathfrak{F}$-injector of M. Hence, by Lemma 6.18, $V_0 \cap M$ and $V \cap M$ are conjugate in M, and consequently, there exists an element $y \in M$ such that $V^y \cap M = V_0 \cap M$. Moreover, by Lemma 6.19, V_0, V^y are $\mathfrak{F}$-maximal subgroup of M and $M_\mathfrak{F} \subseteq V_0 \cap V$. Therefore, by using Lemma 6.21, we obtain that V, V_0 are conjugate in G. This induces that V is an $\mathfrak{F}$-injector of G. This contradiction completes the proof.

The following lemma is obvious.

Lemma 6.24 *Let $\mathfrak{F}$ be a nonempty Fitting class and G a group. If every chief factor of $G/G_\mathfrak{F}$ is an E_π^N-group and $N \unlhd G$, then every chief factor of $N/N_\mathfrak{F}$ is also an E_π^N-group.*

In order to resolve Problem 6.16, we prove the following more general result.

Theorem 6.25 (Guo, Li [152]). *Let $\mathfrak{F}$ be a nonempty Fitting class, $\pi = \pi(\mathfrak{F})$ and G a group such that every chief factor of $G/G_\mathfrak{F}$ is an E_π^N-group. Then G has at least one $\mathfrak{F}$-injector and any two $\mathfrak{F}$-injectors are conjugate in G.*

Proof We proceed the proof via the following steps:

1) G is a D_π-group and if H is a Hall π-subgroup of G, then $H/H_\mathfrak{F}$ is soluble.

This is immediate from Lemma 6.22 and $G_{\mathfrak{F}} \leq H_{\mathfrak{F}}$.

2) *If G has an 𝔉-injector, then an 𝔉-injector of G is also an 𝔉-injector of a Hall π-subgroup of G.*

Assume that G has an $\mathfrak{F}$-injector V. By 1) we know that G has a Hall π-subgroup, say H, and $H/H_{\mathfrak{F}}$ is soluble. Since V is a π-group and G is a D_{π}-group, we may assume, without loss of generality, that $V \subseteq H$. Now we prove that V is also an $\mathfrak{F}$-injector of H. Assume that V is not an $\mathfrak{F}$-injector of H. By the hypothesis of the theorem, G has the following series

$$1 \leq G_{\mathfrak{F}} = G_0 \leq G_1 \leq G_2 \leq \cdots \leq G_n = G$$

such that G_i/G_{i-1} is a G-chief factor and an E_{π}^N-group, for $i = 1, \cdots, n$. It is easy to see that $G_{\mathfrak{F}} \subseteq H$. Hence H possesses the following series:

$$1 \subseteq G_{\mathfrak{F}} = G_0 \cap H \leq G_1 \cap H \leq \cdots \leq G_{n-1} \cap H \leq G_n \cap H = H.$$

By Lemma 6.19(2), $(G_0 \cap H) \cap V = G_{\mathfrak{F}}$ is an $\mathfrak{F}$-injector of $G_{\mathfrak{F}}$. Since V is not an $\mathfrak{F}$-injector of H, we may assume that there exists $l \in \{1, \cdots, n\}$ such that $V \cap G_l$ is not an $\mathfrak{F}$-injector of $H \cap G_l$, but $V \cap G_{l-1}$ is an $\mathfrak{F}$-injector of $H \cap G_{l-1}$. Because V is an $\mathfrak{F}$-injector of G, we have $V \cap T_l$ is an $\mathfrak{F}$-maximal subgroup in $H \cap T_l$. Since H is soluble and $(H \cap G_l)/(H \cap G_{l-1}) \cong (H \cap G_l)G_{l-1}/G_{l-1}$ is nilpotent, by Lemma 6.23 the subgroup $V \cap G_l$ is an $\mathfrak{F}$-injector of $H \cap G_l$. This contradiction shows that V is an $\mathfrak{F}$-injector of H.

3) *If G has 𝔉-injectors, then any two 𝔉-injectors are conjugate in G.*

Suppose that V_1, V_2 are two $\mathfrak{F}$-injectors of G. Since $\pi = \pi(\mathfrak{F})$, V_i is a π-group, for $i \in \{1, 2\}$. Since G is a D_{π}-group, we may assume that $V_1 \subseteq H_1$ and $V_2 \subseteq H_2$, where H_i is a Hall π-subgroup of G. Then, V_i is an $\mathfrak{F}$-injector of H_i. By the conjugacy of Hall π-subgroups and 1), we obtain that V_1 and V_2 are conjugate in G.

4) *G has an 𝔉-injector.*

Let H be a Hall π-subgroup of G. Then, $H/H_{\mathfrak{F}}$ is soluble by 1). Hence, H has an $\mathfrak{F}$-injector by Theorem 6.17. In order to prove that G has an $\mathfrak{F}$-injector, we only need to prove that an arbitrary $\mathfrak{F}$-injector of H is an $\mathfrak{F}$-injector of G.

Let V be an $\mathfrak{F}$-injector of H. Suppose that K is an arbitrary subnormal subgroup of G. Then, obviously, $H \cap K$ is a subnormal subgroup of H. By Lemma 6.19(4), $V \cap K = V \cap H \cap K$ is an $\mathfrak{F}$-injector of $H \cap K$. We claim that $H \cap K$ is a Hall π-subgroup of K. In fact, let

$$1 \leq K = K_0 \leq K_1 \leq \cdots \leq K_{t-1} \leq K_t = G$$

be a subnormal series of G. Then $|K_{t-1} : K_{t-1} \cap H| = |HK_{t-1}|/|H| \mid |G/H|$ is a π'-number. Hence $H \cap K_{t-1}$ is a Hall π-subgroup of K_{t-1}. By continuing this process, we obtain that $H \cap K$ is a Hall π-subgroup of K. If $K < G$, then by induction on the order $|G|$ of G, we may assume that $V \cap K$ is an $\mathfrak{F}$-injector of K, and hence it is an $\mathfrak{F}$-maximal subgroup of K.

Now, suppose that $K = G$. Let V_1 be an $\mathfrak{F}$-maximal subgroup of G containing V. By the above, we have seen that, for every proper subnormal subgroup K_1 of

G, $V_1 \cap K_1 = V \cap K_1$ is an $\mathfrak{F}$-maximal subgroup of K_1. Thus, by the definition of $\mathfrak{F}$-injector, we know that V_1 is an $\mathfrak{F}$-injector of G. Since $G \in D_\pi$, there exists an element $x \in G$ such that $V \leq V_1 \leq H^x$. On the other hand, since V is an $\mathfrak{F}$-injector of H, V^x is an $\mathfrak{F}$-injector of H^x. However, by 2), we know that V_1 is also an $\mathfrak{F}$-injector of H^x. Since $H/H_{\mathfrak{F}}$ is a soluble group, by Theorem 6.17, V_1 and V^x are conjugate in G. Therefore, $V = V_1$ is an $\mathfrak{F}$-injector of G. This completes the Proof.

It is easy to see that every chief factor of a π-selected group is an E_π^N-group. Hence, form Theorem 6.25 we obtain the following corollary which gives a positive answer to Problem 6.16, and a new progress on Problem 6.15 is also obtained.

Corollary 6.26 (Guo, Li [152]). *Let $\mathfrak{F}$ be a nonempty Fitting class and $\pi = \pi(\mathfrak{F})$. If $G/G_{\mathfrak{F}}$ is π-selected, then G has at least one $\mathfrak{F}$-injector and any two $\mathfrak{F}$-injectors are conjugate in G.*

From Theorem 6.25, a generalized result of Lemma 6.23 is also obtained.

Corollary 6.27 *Suppose $\mathfrak{F}$ and G satisfy the hypothesis in Theorem 6.25. Let N be a normal subgroup of G and G/N an E_π^N-group. If V is an $\mathfrak{F}$-maximal subgroup of G and $V \cap N$ is an $\mathfrak{F}$-injector of N, then V is an $\mathfrak{F}$-injector of G.*

Proof Let H be a Hall π-subgroup of G containing V. By Theorem 6.25 and its proof, we see that $H/H_{\mathfrak{F}}$ is a soluble group and G has an $\mathfrak{F}$-injector V_1 which is also an $\mathfrak{F}$-injector of H. Since any two $\mathfrak{F}$-injectors are conjugate, we only need to prove that V is an $\mathfrak{F}$-injector of H. Since $N \lhd G$, by Lemma 6.24, every chief factor of $N/N_{\mathfrak{F}}$ is also a E_π^N-group. Since $H \cap N$ is a Hall π-subgroup of N and $V \cap N$ is an $\mathfrak{F}$-injector of N, $V \cap N$ is an $\mathfrak{F}$-injector of $H \cap N$ by using the proof of Theorem 6.25. Since $H/H \cap N \cong HN/N$ is a Hall π-subgroup of G/N, $H/H \cap N$ is nilpotent. Therefore, V is an $\mathfrak{F}$-injector of H by Lemma 6.23. This completes the proof.

Corollary 6.28 *Let $\mathfrak{F}$ be a nonempty Fitting class, $\pi = \pi(\mathfrak{F})$, G a group and $V \leq G$. Suppose that G has a normal series:*

$$1 = G_0 \leq G_1 \leq \cdots \leq G_{t-1} \leq G_t = G$$

such that $G_1 \leq G_{\mathfrak{F}}$ and $G_i/G_{i-1} \in E_\pi^N$. If $G_i \cap V$ is an $\mathfrak{F}$-maximal subgroup of G_i, $i = 1, \cdots, t$, then V is an $\mathfrak{F}$-injector of G.

Proof Since $G_1 \unlhd G_{\mathfrak{F}} \in \mathfrak{F}$, $V \cap G_1 = G_1$ is an $\mathfrak{F}$-injector of G_1. Hence we can assume that $t > 1$ and by using induction on t, we may suppose that $V \cap G_{t-1}$, for $t \geq 2$, is an $\mathfrak{F}$-injector of G_{t-1}. Therefore the corollary follows from Corollary 6.27.

Theorem 6.29 (Guo, Li [152, Theorem 3.6]). *Let $\mathfrak{F}$ be a nonempty Fitting class, $\pi = \pi(\mathfrak{F})$ and suppose that every chief factor of $G/G_{\mathfrak{F}}$ is a E_π^N-group. Let A be a subgroup of G and V an $\mathfrak{F}$-injector of G. If $V \leq A \leq G$ and A possesses a Hall π-subgroup, then V is an $\mathfrak{F}$-injector of A.*

Proof Since every chief factor of $G/G_{\mathfrak{F}}$ is a E_π^N-group, G has a normal series

$$1 = G_0 \leq G_1 \leq \cdots \leq G_{t-1} \leq G_t = G,$$

where $G_1 = G_{\mathfrak{F}}$, and for each $i \geq 1$, G_i/G_{i-1} is an E_{π}^N-group. Since V is an $\mathfrak{F}$-injector of G, $V \cap G_i$ is an $\mathfrak{F}$-injector of G_i, for every i. Put $A_i = G_i \cap A, i = 1, \cdots, t$. Since $G_1 = G_{\mathfrak{F}} \leq V \leq A$, we have that $A_1 = G_1 \cap A = G_1 \subseteq G_{\mathfrak{F}} \subseteq A_{\mathfrak{F}}$. Because $A \cap G_i/A \cap G_{i-1} \cong (A \cap G_i)G_{i-1}/G_{i-1}$ and A possesses a Hall π-subgroup, A_i/A_{i-1} is an E_{π}^N-group. Therefore, the series

$$1 = A_0 \leq A_1 \leq \cdots \leq A_{t-1} \leq A_t = A$$

satisfies the hypothesis of Corollary 6.28. Moreover, it is easy to see that $V \cap A_i = V \cap A \cap G_i = V \cap G_i$ is an $\mathfrak{F}$-maximal subgroup of A_i. Hence V is an $\mathfrak{F}$-injector in A. The proof is completed.

One problem on existence and conjugacy of $\mathfrak{F}$-covering subgroups in non-soluble groups. Gaschütz proved in [113] a fundamental theorem on $\mathfrak{F}$-covering subgroups which yields numerous families of conjugate subgroups in a soluble group.

Theorem 6.30 (Gaschütz). *Let $\mathfrak{F}$ be a nonempty saturated formation of finite groups. Then each soluble group possesses exactly one conjugacy class of $\mathfrak{F}$-covering subgroups.*

In [341] and [374] P. Schmid and E. Shmighirev extended Theorem 6.30 by considering the groups with soluble $\mathfrak{F}$-residuals.

Theorem 6.31 (P. Schmid, E. Shmighirev). *Let $\mathfrak{F}$ be a nonempty saturated formation and G be a group whose $\mathfrak{F}$-residual is soluble. Then G possesses exactly one conjugacy class of $\mathfrak{F}$-covering subgroups.*

In [359], Theorem 6.31 was strengthened in the following way.

Theorem 6.32 *Let $\mathfrak{F}$ be a nonempty saturated formation and G a group. Assume that $G^{\mathfrak{F}}$ is $\pi(\mathfrak{F})$-soluble. Then G possesses exactly one conjugacy class of $\mathfrak{F}$-covering subgroups.*

In [135] and [361], the following problem was posed:

Problem 6.33. Let $\mathfrak{F}$ be a nonempty saturated formation, G is a group and the $\mathfrak{F}$-residual $G^{\mathfrak{F}}$ is a $\pi(\mathfrak{F})$-selected group.

1) Has G an $\mathfrak{F}$-covering subgroup?
2) If G has $\mathfrak{F}$-covering subgroups, are any two $\mathfrak{F}$-covering subgroups of G conjugate?

A positive answer to the problem was obtained in Li and Guo [278]. Here we shall prove a stronger result.

Theorem 6.34 (Guo, Al-Sharo, Shemetkov [144]). *Let $\mathfrak{F}$ be a nonempty saturated formation, $\pi = \pi(\mathfrak{F})$. Let K be a normal subgroup of a group G such that every G-chief factor of K is E_{π}^N-selected. Assume that G/K possesses exactly one conjugacy class of $\mathfrak{F}$-covering subgroups. Then G possesses exactly one conjugacy class of $\mathfrak{F}$-covering subgroups.*

Corollary 6.35 *Let $\mathfrak{F}$ be a nonempty saturated formation. Let K be a $\pi(\mathfrak{F})$-selected normal subgroup of a group G. Assume that G/K possesses exactly one conjugacy class of $\mathfrak{F}$-covering subgroups. Then G possesses exactly one conjugacy class of $\mathfrak{F}$-covering subgroups.*

Corollary 6.36 [278]. *Let $\mathfrak{F}$ be a nonempty saturated formation. If $G^{\mathfrak{F}}$ is $\pi(\mathfrak{F})$-selected, then G possesses exactly one conjugacy class of $\mathfrak{F}$-covering subgroups.*

Lemma 6.37 (see [135, Theorem 2.2.4]). *Let $\mathfrak{F}$ be a formation and K a normal subgroup of a group G. Then the following statements hold:*

1) *If H is an $\mathfrak{F}$-covering subgroup of G, then HK/K is an $\mathfrak{F}$-covering subgroup of G/K;*
2) *If B/K is an $\mathfrak{F}$-covering subgroup of G/K and H is an $\mathfrak{F}$-covering subgroup of B, then H is an $\mathfrak{F}$-covering subgroup of G.*

Lemma 6.38 ([233], [237]). *If K is a normal subgroup of G such that $K \in E_\pi^N$ and $G/K \in D_\pi$, then $G \in D_\pi$.*

A new proof of Lemma 6.38 and its generalizations can be found in [365].

Lemma 6.39 ([444]). *If $G \in E_\pi^N$, then $G \in D_\pi$.*

Lemma 6.40 ([60], see also [359, Theorem 15.9]). *Let $\mathfrak{F}$ be a saturated formation. If $HF(G) = G$ and $H \in \mathfrak{F}$, then H is contained in some $\mathfrak{F}$-covering subgroup of G.*

Proof of Theorem 6.34. Let G be a counterexample of minimal order. By the condition, G/K has an $\mathfrak{F}$-covering subgroup R/K. Assume that $R \neq G$. It is clear that every R-chief factor of K is E_π^N-selected. By the inductive hypothesis, there exists an $\mathfrak{F}$-covering subgroup S_1 of R. By Lemma 6.37, S_1 is an $\mathfrak{F}$-covering subgroup of G. Let S_2 be another $\mathfrak{F}$-covering subgroup of G. Then by Lemma 6.37 again, $S_1 K/K$ and $S_2 K/K$ are $\mathfrak{F}$-covering subgroups of G/K. By the condition,

$$R/K = S_1 K/K = S_2^x K/K$$

for some $x \in G$. Clearly, S_1 and S_2^x are $\mathfrak{F}$-covering subgroups of R. Therefore, S_1 and S_2^x are conjugate in R. Hence the theorem is true for the case $R \neq G$.

Now we assume that $G/K \in \mathfrak{F}$. Let L be a minimal normal subgroup of G contained in K. Assume that $K \neq L$. Since the theorem is true for G/L, we have that G/L possesses exactly one conjugacy class of $\mathfrak{F}$-covering subgroups. Let T/L be one of them. Evidently, $T \neq G$. Hence, by the inductive hypothesis, T has an $\mathfrak{F}$-covering subgroup F_1 and $F_1 L = T$. By Lemma 6.37, F_1 is an $\mathfrak{F}$-covering subgroup of G. If F_2 is another $\mathfrak{F}$-covering subgroup of G, then by Lemma 6.37, $F_1 L/L$ and $F_2 L/L$ are $\mathfrak{F}$-covering subgroups of G/L. Therefore, $F_1 L/L$ and $F_2 L/L$ are conjugate in G/L. We now have that $T = F_1 L = F_2^x L$ for some $x \in G$. Since the theorem is true for T, we see that the theorem is also true for G.

In order to complete the proof, we now assume that K is a minimal normal subgroup of G. Actually, K is the $\mathfrak{F}$-residual of G. By using the condition, we see that $K \in E_\pi^N$. Since G/K is a π-group, it follows from Lemma 6.38 that $G \in D_\pi$. Let H be a Hall π-subgroup of G. Evidently, $HK = G$ and $H \cap K$ is a Hall π-subgroup of K. As $K \in E_\pi^N$, $H \cap K$ is nilpotent. Since

$$H/H \cap K \simeq G/K \in \mathfrak{F},$$

we have that the $\mathfrak{F}$-residual of H is nilpotent. Hence by Theorem 6.31, H possesses exactly one conjugacy class of $\mathfrak{F}$-covering subgroups. Let S be an $\mathfrak{F}$-covering subgroup of H. Then $H = S(H \cap K)$. Therefore,

$$SK = S(H \cap K)K = HK = G.$$

We note that S is an $\mathfrak{F}$-maximal subgroup of G. In fact, we can suppose that S is a proper subgroup of an $\mathfrak{F}$-maximal subgroup S_1 of G. Since $G \in D_\pi$, we have that $S_1^x \subseteq H$ for some $x \in G$. Clearly, S_1^x is $\mathfrak{F}$-maximal in G. Since $SK = G = S_1K$, we have $S_1^x K = G$. Hence,

$$S_1^x(H \cap K) = H \cap S_1^x K = H \cap G = H.$$

Moreover, since S_1^x is $\mathfrak{F}$-maximal in H, by Lemma 6.40, S_1^x is an $\mathfrak{F}$-covering subgroup of H. By Theorem 6.31, we can see that the $\mathfrak{F}$-covering subgroups of H are conjugate, hence we obtain that $|S_1^x| = |S|$. This contradicts $S \subset S_1$. Now we only need to show that S is an $\mathfrak{F}$-covering subgroup of G. Let U be a subgroup of G such that $S \subseteq U \subset G$. Since $SK = UK = G$, $U = S(U \cap K)$ and

$$U/U \cap K \simeq S/S \cap U \cap K = S/S \cap K.$$

This leads to that $U/U \cap K$ is a π-group. Since $U \cap K$ is contained in a E_π^N-selected group K, we have $U \cap K \in E_\pi^N$. Hence by Lemma 6.38, $U \in D_\pi$. This shows that S is contained in a Hall π-subgroup U_π of U. From $U = S(U \cap K)$ and $S \subseteq U_\pi$ we have $U_\pi = S(U_\pi \cap U \cap K) = S(U_\pi \cap K)$. Since $K \in E_\pi^N$, $U_\pi \cap K$ is nilpotent by Lemma 6.39. As S is $\mathfrak{F}$-maximal in G, S is $\mathfrak{F}$-maximal in $U_\pi = S(U_\pi \cap K)$. By Lemma 6.40, S is an $\mathfrak{F}$-covering subgroup of U_π. By the inductive hypothesis, U possesses an $\mathfrak{F}$-covering subgroup S_1. Since $U \in D_\pi$, we can assume that $S_1 \subseteq U_\pi$. Hence, we obtain that S and S_1 are $\mathfrak{F}$-covering subgroups of U_π. By Theorem 6.31, S and S_1 are conjugate in U_π. From the above results, we deduce that S is an $\mathfrak{F}$-covering subgroup of U. Hence, $SU^{\mathfrak{F}} = U$.

Thus, we have obtained that $G/K \in \mathfrak{F}$ and G possesses at least one $\mathfrak{F}$-covering subgroup. Let S_1 and S_2 be two $\mathfrak{F}$-covering subgroups of G. Since $G \in D_\pi$, we can assume that S_1 and S_2 are contained in a Hall π-subgroup H of G. By Lemma 6.39, $H \cap K$ is nilpotent and $H/H \cap K \simeq G/K \in \mathfrak{F}$. Therefore, by Theorem 6.31, S_1 and S_2 are conjugate in H. The theorem is proved.

5.7 Classes Determined by Hall Subgroups

Local Fitting classes. Let π and ω be two nonempty sets of primes. We use $\mathfrak{S}^\pi$ to denote the class of all π-soluble groups. $\mathfrak{X}_\omega$ denotes the class of all finite ω-groups lying in $\mathfrak{X}$. $\mathfrak{S}_\omega^\pi$ denotes the class of all π-soluble ω-groups.

A function f of the form

$$f : \mathbb{P} \to \{\text{Fitting classes}\}$$

is an H-function. The set $\sigma = \text{Supp}(f) = \{p \in \mathbb{P} : f(p) \neq \emptyset\}$ is called the support of the function f.

For above an H-function f and a class $\mathfrak{X}$ of groups, let $LR(f) = \mathfrak{X}_\sigma^\pi \cap (\cap_{p \in \sigma} f(p)\mathfrak{N}_p\mathfrak{X}_{p'})$ where $\sigma = \text{Supp}(f)$.

A Fitting class $\mathfrak{F}$ is called local in the universe $\mathfrak{X}$ if there exists an H-function f such that $\mathfrak{F} = LR(f)$. In this case, we say that $\mathfrak{F}$ is locally defined by f or f is an H-function of $\mathfrak{F}$.

In this section, all groups considered are finite π-soluble groups, where π is some given subset of the set $\mathbb{P}$.

Shemetkov problem for Fitting classes.

For any class $\mathfrak{F}$ of groups and any set of primes π, we put

$$K_\pi(\mathfrak{F}) = \{G : G \text{ is a group and a Hall } \pi - \text{subgroup of } G \text{ is in } \mathfrak{F}\},$$

$$C_\pi(\mathfrak{F}) = \{G \mid G \in K_\pi(\mathfrak{F}) \text{ and any two Hall } \pi - \text{subgroups of } G \text{ are conjugate in } G\}.$$

In the group theory, the following problem is known.

Problem 7.1 (L. A. Shemetkov [359, Problem 19]). Let $\mathfrak{F}$ be a local formation. Is $C_\pi(\mathfrak{F})$ a local formation?

It is natural to consider the dual problem of Problem (7.1):

Problem 7.2. Let $\mathfrak{F}$ be a local Fitting class. Is $C_\pi(\mathfrak{F})$ a local Fitting class?

L. A. Shemetkov and A. F. Vasil'ev in [369] proved that the formation $C_{\{3,5\}}(\mathfrak{N})$ is not local. Thus, Problem 7.1 has a negative solution. However, it was proved that $C_\pi(\mathfrak{F})$ is a formation if $\mathfrak{F}$ is a formation [425, Theorem 1] and $C_\pi(\mathfrak{F})$ is solubly saturated if $\mathfrak{F}$ is solubly saturated [425, Corollary 2.2].

Moreover, Problem 7.2 has also a negative solution in general case since $C_{\{2,3\}}(\mathfrak{G}) = C_{\{2,3\}}$ is not closed under taking normal subgroups (see, for instance, [119, p. 7], [421, example], [422, example 5.3]) and hence it is not a Fitting class.

However, it makes sense to consider Problems 7.1 and 7.2 for some special cases (for example, only for solvable, π-solvable and so on).

It was proved that

(I) *For any local formation $\mathfrak{F}$, the class $C_\pi(\mathfrak{F}) \cap \mathfrak{G}$ is a local formation* (Blessenohl [54])

(II) *For any local formation $\mathfrak{F}$ and for the class $\mathfrak{X}$ of all π-separated groups, the class $C_\pi(\mathfrak{F}) \cap \mathfrak{X}$ is a local formation* (Slepova [392])

In the paper [463], Problem 7.2 was solved in the class $\mathfrak{G}$ of all soluble groups, that is, it was proved that for any local Fitting class $\mathfrak{F}$, the class $C_\pi(\mathfrak{F}) \cap \mathfrak{G}$ is a local Fitting class.

Our next goal is to prove the analogous result in the class $\mathfrak{G}^\pi$.

Let f be an H-function of a local Fitting class $\mathfrak{F}$. Then f is called

(i) Integrated if $f(p) \subseteq \mathfrak{F}$ for all $p \in \mathbb{P}$; and

(ii) Full if $f(p) = f(p)\mathfrak{N}_p$ for all $p \in \mathbb{P}$.

(iii) A canonical local H-function if f is integrated and full as well.

A Fitting class $\mathfrak{F}$ is said be a Lockett class if $\mathfrak{F} = \mathfrak{F}^*$, where $\mathfrak{F}^*$ is the smallest Fitting class containing $\mathfrak{F}$ such that the $\mathfrak{F}^*$-radical of the direct product $G \times H$ of any two groups G and H is equal to the direct product of the $\mathfrak{F}^*$-radical of G and the $\mathfrak{F}^*$-radical of H, that is, $(G \times H)_{\mathfrak{F}^*} = G_{\mathfrak{F}^*} \times H_{\mathfrak{F}^*}$. We use $\text{Hall}_\pi(G)$ to denote the set of all Hall π-subgroups of G.

Lemma 7.3 [431]. *Every local Fitting class $\mathfrak{F}$ can be defined by an integrated H-function F such that $F(p)\mathfrak{N}_p = F(p)$ for all $p \in \mathbb{P}$ and each nonempty value $F(p)$ are Lockett class.*

Lemma 7.4 *Let $\mathfrak{F}$ be a nonempty Fitting class.*

(a) $K_\pi(\mathfrak{F})$ *is a Fitting class for any* $\pi \subseteq \mathbb{P}$.

(b) *If H is a Hall subgroup of a group G, then $G_{K_\pi(\mathfrak{F})} \cap H = H_{\mathfrak{F}}$.*

(c) *If $\mathfrak{H}$ be Fitting class, then $K_\pi(\mathfrak{F}\mathfrak{H}) = K_\pi(\mathfrak{F})K_\pi(\mathfrak{H})$.*

Proof

(a) It is clear by the definition of $K_\pi(\mathfrak{F})$.

(b) Put $\mathfrak{R} = K_\pi(\mathfrak{F})$ and $K = G_{\mathfrak{R}}$. Since $K \trianglelefteq G$, we have $H \cap K \in \text{Hall}_\pi(K)$ and $H \cap K \trianglelefteq H$. Hence $H \cap K \subseteq H_{\mathfrak{F}}$. Let $F/K = F_\pi(G/K)$, then $F/K \in \mathfrak{N}_\pi$ since $F \in K_\pi(\mathfrak{F})\mathfrak{E}_{\pi'}\mathfrak{N}_\pi = K_\pi(\mathfrak{F})\mathfrak{N}_\pi$. Therefore, $F/K \leq HK/K \in \text{Hall}_\pi(G/K)$. Obviously, $H_{\mathfrak{F}} \in \text{Hall}_\pi(H_{\mathfrak{F}}K)$. So $H_{\mathfrak{F}}K \in \mathfrak{R}$. On the other hand, $F \cap H_{\mathfrak{F}}K \, sn \, G$ by $F/K \in \mathfrak{N}$, so $F \cap H_{\mathfrak{F}}K \leq G_{\mathfrak{R}} = K$. Therefore, $[F, H_{\mathfrak{F}}K] \leq F \cap H_{\mathfrak{F}}K \leq K$, and consequently, $H_{\mathfrak{F}}K \leq C_G(F/K) \leq F$ (cf. [135, Theorem 1.8.19]. It follows that $H_{\mathfrak{F}} \leq H \cap F \cap H_{\mathfrak{F}}K \leq H \cap K$. Thus, (b) holds.

(c) Let H be a Hall π-subgroup of G. If $G \in K_\pi(\mathfrak{F}\mathfrak{H})$, then $H \in \mathfrak{F}\mathfrak{H}$, that is, $H/H_{\mathfrak{F}} \in \mathfrak{H}$. By (b), we know that $H_{\mathfrak{F}} = G_{K_\pi(\mathfrak{F})} \cap H$, so $H/H_{\mathfrak{F}} \simeq HG_{K_\pi(\mathfrak{F})}/G_{K_\pi(\mathfrak{F})} \in \mathfrak{H}$ and hence $G/G_{K_\pi(\mathfrak{F})} \in K_\pi(\mathfrak{H})$. This shows that $K_\pi(\mathfrak{F}\mathfrak{H}) \leq K_\pi(\mathfrak{F})K_\pi(\mathfrak{H})$. On the other hand, if $G \in K_\pi(\mathfrak{F})K_\pi(\mathfrak{H})$, then $G/G_{K_\pi(\mathfrak{F})} \in K_\pi(\mathfrak{H})$. It follows from (b) that $H/H_{\mathfrak{F}} \simeq HG_{K_\pi(\mathfrak{F})}/G_{K_\pi(\mathfrak{F})} \in \mathfrak{H}$. Hence $H \in \mathfrak{F}\mathfrak{H}$ and consequently $G \in K_\pi(\mathfrak{F}\mathfrak{H})$. Thus, (c) holds.

Remark The statements (b) and (c) in Lemma 7.4 maybe not true in the class $\mathfrak{G}$ of all finite groups. For example, put $\mathfrak{F} = \mathfrak{N} = \mathfrak{H}$, $G = A_5$ and $\pi = \{2, 3\}$. Then $H \simeq A_4$ is a Hall π-subgroup of G. Clearly $G_{K_\pi(\mathfrak{F})} \cap H = G_{K_\pi(\mathfrak{F})} = 1$, but $H_{\mathfrak{F}} \neq 1$. Hence (b) is not true. Since $H \in \mathfrak{N}^2$, we know $G \in K_\pi(\mathfrak{F}\mathfrak{H})$. But $G \notin K_\pi(\mathfrak{F})K_\pi(\mathfrak{H})$ since $G_{K_\pi(\mathfrak{F})} = 1$. Hence (c) is not true.

The following lemma is evident.

Lemma 7.5 *Let $\mathfrak{F}$ and $\mathfrak{H}$ be two Fitting classes. Then the following statements hold:*

(a) *If $\mathfrak{F} \subseteq \mathfrak{H}$, then $K_\pi(\mathfrak{F}) \subseteq K_\pi(\mathfrak{H})$.*

(b) $K_\pi(\mathfrak{F} \cap \mathfrak{H}) = K_\pi(\mathfrak{F}) \cap K_\pi(\mathfrak{H})$.

Lemma 7.6 *Let $\mathfrak{F}$ be a nonempty Fitting class. Then the following statements hold:*

(a) *If $p \in \pi$ and $\mathfrak{F} = \mathfrak{S}_{p'}^{\pi}$, then $K_{\pi}(\mathfrak{F}) = \mathfrak{F}$;*
(b) *If $\mathfrak{F}\mathfrak{N}_p = \mathfrak{F}$ for some prime p, then $K_{\pi}(\mathfrak{F})\mathfrak{N}_p = K_{\pi}(\mathfrak{F})$.*

Proof

(a) Since a subgroup of a soluble p'-group is a soluble p'-group, it is easy to see that $\mathfrak{F} \subseteq K_{\pi}(\mathfrak{F})$. Let $G \in K_{\pi}(\mathfrak{F})$ and H is a Hall π-subgroup of G. Then $H \in \mathfrak{F}$, and so $|H|$ is a p'-number. On the other hand, since $p \in \pi$, we have $\pi' \subseteq p'$. Hence $|G : H|$ is also a p'-number. It follows that $|G|$ is a p'-number and $G \in \mathfrak{F}$. Therefore $\mathfrak{F} = K_{\pi}(\mathfrak{F})$.

(b) Obviously, $K_{\pi}(\mathfrak{F}) \subseteq K_{\pi}(\mathfrak{F})\mathfrak{N}_p$. Now assume that $G \in K_{\pi}(\mathfrak{F})\mathfrak{N}_p$. Then $G/G_{K_{\pi}(\mathfrak{F})}$ is a p-group. Let H is a Hall π-subgroup of G. Then, by Lemma 7.4(b), $H/H_{\mathfrak{F}} = H/H \cap G_{K_{\pi}(\mathfrak{F})} \simeq HG_{K_{\pi}(\mathfrak{F})}/G_{K_{\pi}(\mathfrak{F})} \leq G/G_{K_{\pi}(\mathfrak{F})}$ is a p-group, that is, $H/H_{\mathfrak{F}} \in \mathfrak{N}_p$. This means that $H \in \mathfrak{F}\mathfrak{N}_p = \mathfrak{F}$, that is, $G \in K_{\pi}(\mathfrak{F})$. Therefore, $K_{\pi}(\mathfrak{F})\mathfrak{N}_p = K_{\pi}(\mathfrak{F})$.

Lemma 7.7 *Let $\mathfrak{F}$ be a nonempty Fitting class, and π, σ be two sets of prime numbers such that $\pi \cap \sigma = \emptyset$. Then the following statements hold:*

(a) *$K_{\pi}(\mathfrak{F})\mathfrak{S}_{\sigma}^{\pi} = K_{\pi}(\mathfrak{F})$. In particular, $K_{\pi}(\mathfrak{F})\mathfrak{S}_{\pi'}^{\pi} = K_{\pi}(\mathfrak{F})$;*
(b) *$K_{\pi}(\mathfrak{F})\mathfrak{N}_{\sigma} = K_{\pi}(\mathfrak{F})$.*

Proof

(a) First, it is clear that $K_{\pi}(\mathfrak{F}) \subseteq K_{\pi}(\mathfrak{F})\mathfrak{S}_{\sigma}^{\pi}$. Assume that $G \in K_{\pi}(\mathfrak{F})\mathfrak{S}_{\sigma}^{\pi}$ and H is a Hall π-subgroup of G. Then $G/G_{K_{\pi}(\mathfrak{F})} \in \mathfrak{S}_{\sigma}^{\pi}$ and the Hall π-subgroup $HG_{K_{\pi}(\mathfrak{F})}/G_{K_{\pi}(\mathfrak{F})}$ of $G/G_{K_{\pi}(\mathfrak{F})}$ is a σ-group. Since $HG_{K_{\pi}(\mathfrak{F})}/G_{K_{\pi}(\mathfrak{F})} \simeq H/H \cap G_{K_{\pi}(\mathfrak{F})}$, $H/H_{\mathfrak{F}}$ is a σ-group by Lemma 7.4(b). Hence $H/H_{\mathfrak{F}} \in \mathfrak{S}_{\pi}^{\pi} \cap \mathfrak{S}_{\sigma}^{\pi} = (1)$, where (1) is the class consists of identity groups. Consequently, $H = H_{\mathfrak{F}}$ and hence $G \in K_{\pi}(\mathfrak{F})$.

(b) By the statement (a) of the lemma, we see that $K_{\pi}(\mathfrak{F})\mathfrak{N}_{\sigma} \subseteq K_{\pi}(\mathfrak{F})\mathfrak{S}_{\sigma}^{\pi} = K_{\pi}(\mathfrak{F})$. Thus, the statement (b) hold.

Theorem 7.8 (Guo, Li [153]). *For any local Fitting class $\mathfrak{F}$, the class $K_{\pi}(\mathfrak{F})$ is a local Fitting class in the universe $\mathfrak{S}^{\pi}$.*

Proof Since $\mathfrak{F}$ is a local Fitting class, by Lemma 7.3, there exists an H-function F such that $\mathfrak{F} = LR(F)$ and $F(p)\mathfrak{N}_p = F(p) \subseteq \mathfrak{F}$ for all $p \in \mathbb{P}$ and each value $F(p)$ are a Lockett class, for every $p \in \sigma = Supp(F)$. Then, we have that

$$\mathfrak{F} = \mathfrak{S}_{\sigma}^{\pi} \cap (\cap_{p \in \sigma} F(p)\mathfrak{N}_p\mathfrak{S}_{p'}^{\pi}) = \mathfrak{S}_{\sigma}^{\pi} \cap (\cap_{p \in \sigma} F(p)\mathfrak{S}_{p'}^{\pi}). \tag{5.11}$$

If $\pi = \mathbb{P}$, then $K_{\pi}(\mathfrak{F}) = \mathfrak{F}$ and so the theorem holds. Assume that $\pi = \emptyset$, then $K_{\pi}(\mathfrak{F}) = \mathfrak{S}^{\pi}$. However, it is easy to see that the class of all π-soluble groups $\mathfrak{S}^{\pi} = LR(h)$, where h is the H-function such that $h(p) = \mathfrak{S}^{\pi}$, for all $p \in \mathbb{P}$. This shows that, in this case, $K_{\pi}(\mathfrak{F})$ is a local Fitting class.

We now assume that $\emptyset \subsetneqq \pi \subsetneqq \mathbb{P}$ and define an H-function as follows:

$$f(p) = \begin{cases} K_{\pi \cap \sigma}(F(p)), & \text{if } p \in \pi \cap \sigma, \\ K_\pi(\mathfrak{F}), & \text{if } p \in \pi', \\ \emptyset, & \text{if } p \in \pi \cap \sigma'. \end{cases}$$

Then $\omega = Supp(f) = \sigma \cup \pi'$, and so

$$LR(f) = \mathfrak{S}^\pi_{\sigma \cup \pi'} \cap ((\cap_{p \in \pi \cap \sigma} K_{\pi \cap \sigma}(F(p)) \mathfrak{N}_p \mathfrak{S}^\pi_{p'}) \cap (\cap_{p \in \pi'} K_\pi(\mathfrak{F}) \mathfrak{N}_p \mathfrak{S}^\pi_{p'}). \tag{5.12}$$

In order to prove the theorem, we only need to ascertain that $K_\pi(\mathfrak{F}) = LR(f)$.

For this purpose, we let $\mathfrak{M} = \cap_{p \in \pi \cap \sigma} K_{\pi \cap \sigma}(F(p)) \mathfrak{N}_p \mathfrak{S}^\pi_{p'}$. Since the H-function F is Full, by Lemma 7.6(b), we see that $\mathfrak{M} = \cap_{p \in \pi \cap \sigma} K_{\pi \cap \sigma}(F(p)) \mathfrak{S}^\pi_{p'}$. We now prove

$$\mathfrak{M} = K_{\pi \cap \sigma}(\mathfrak{F}). \tag{5.13}$$

Indeed, by the equality (5.11), we have that $\mathfrak{F} \subseteq \cap_{\pi \cap \sigma} F(p) \mathfrak{S}^\pi_{p'}$. Then, by Lemmas 7.5, 7.4(c) and 7.6(a), we see that $K_{\pi \cap \sigma}(\mathfrak{F}) \subseteq K_{\pi \cap \sigma}(\cap_{\pi \cap \sigma} F(p) \mathfrak{S}^\pi_{p'}) \subseteq \cap_{p \in \pi \cap \sigma} K_{\pi \cap \sigma}(F(p) \mathfrak{S}^\pi_{p'}) = \cap_{p \in \pi \cap \sigma} K_{\pi \cap \sigma}(F(p)) K_{\pi \cap \sigma}(\mathfrak{S}^\pi_{p'}) = \cap_{p \in \pi \cap \sigma} K_{\pi \cap \sigma}(F(p)) \mathfrak{S}^\pi_{p'}$. Therefore, $K_{\pi \cap \sigma}(\mathfrak{F}) \subseteq \mathfrak{M}$. On the other hand, since the H-function F is integrated, by Lemma 7.5(a), we have $K_{\pi \cap \sigma}(F(p)) \subseteq K_{\pi \cap \sigma}(\mathfrak{F})$ for every $p \in \pi \cap \sigma$. It follows that $K_{\pi \cap \sigma}(F(p)) \mathfrak{S}^\pi_{p'} \subseteq K_{\pi \cap \sigma}(\mathfrak{F}) \mathfrak{S}^\pi_{p'}$ for all $p \in \pi \cap \sigma$, and consequently, $\mathfrak{M} \subseteq \cap_{p \in \pi \cap \sigma} K_{\pi \cap \sigma}(\mathfrak{F}) \mathfrak{S}^\pi_{p'} = K_{\pi \cap \sigma}(\mathfrak{F}) \mathfrak{S}^\pi_{(\pi \cap \sigma)'}$. However, by Lemma 7.7(a), we see that $K_{\pi \cap \sigma}(\mathfrak{F}) \mathfrak{S}^\pi_{(\pi \cap \sigma)'} = K_{\pi \cap \sigma}(\mathfrak{F})$, thus, the equality (5.13) hold.

Let $\mathfrak{M}_1 = \cap_{p \in \pi'} K_\pi(\mathfrak{F}) \mathfrak{N}_p \mathfrak{S}^\pi_{p'}$. We prove

$$\mathfrak{M}_1 = K_\pi(\mathfrak{F}) \mathfrak{S}^\pi_\pi. \tag{5.14}$$

In fact, by Lemma 7.7(b), we have $\mathfrak{M}_1 = \cap_{p \in \pi'} K_\pi(\mathfrak{F}) \mathfrak{S}^\pi_{p'} = K_\pi(\mathfrak{F})(\cap_{p \in \pi'} \mathfrak{S}^\pi_{p'}) = K_\pi(\mathfrak{F}) \mathfrak{S}^\pi_\pi$. Hence the equality (5.14) hold.

Now, by the equalities (5.12), (5.13), and (5.14), we obtain that

$$LR(f) = \mathfrak{S}^\pi_{\sigma \cup \pi'} \cap \mathfrak{M} \cap \mathfrak{M}_1 = \mathfrak{S}^\pi_{\sigma \cup \pi'} \cap K_{\pi \cap \sigma}(\mathfrak{F}) \cap K_\pi(\mathfrak{F}) \mathfrak{S}^\pi_\pi. \tag{5.15}$$

Let $\mathfrak{D} = \mathfrak{S}^\pi_{\sigma \cup \pi'} \cap K_{\pi \cap \sigma}(\mathfrak{F})$. We prove that $\mathcal{D} = K_\pi(\mathfrak{F})$. Assume that $G \in K_\pi(\mathfrak{F})$ and H is a Hall π-subgroup of G. Then, $H \in \mathfrak{F}$. Since $\mathfrak{F} \subseteq \mathfrak{S}^\pi_\sigma$, $|H|$ is a $(\pi \cap \sigma)$-number. It follows that $|G|$ is a $(\sigma \cup \pi')$-number, that is, $G \in \mathfrak{S}^\pi_{\sigma \cup \pi'}$. In addition, since $\pi' \subseteq (\sigma \cap \pi)'$, we see that H is a $(\sigma \cap \pi)$-Hall subgroup of G. This shows that $G \in K_{\pi \cap \sigma}(\mathfrak{F})$, and hence $G \in \mathfrak{D}$. On the other hand, assume that $G \in \mathfrak{D}$ and H is a $(\pi \cap \sigma)$-Hall subgroup of G. Then, $|G|$ is a $(\sigma \cup \pi')$-number and $H \in \mathfrak{F}$. It is clear that the index $|G : H|$ is a $(\pi' \cup \sigma')$-number. Hence $|G : H|$ is a μ-number, where $\mu = (\pi' \cup \sigma') \cap (\sigma \cup \pi')$. Obviously, $\mu \subseteq \pi'$. Thus, H is a Hall π-subgroup of G. This means that $G \in K_\pi(\mathfrak{F})$. Therefore $\mathfrak{D} = K_\pi(\mathfrak{F})$.

Finally, by using the above results and Lemma 7.7, we have that $LR(f) = \mathfrak{D} \cap \mathfrak{M}_1 = K_\pi(\mathfrak{F}) \cap K_\pi(\mathfrak{F})\mathfrak{S}_\pi^\pi = K_\pi(\mathfrak{F})\mathfrak{S}_{\pi'}^\pi \cap K_\pi(\mathfrak{F})\mathfrak{S}_\pi^\pi = K_\pi(\mathfrak{F})(\mathfrak{S}_{\pi'}^\pi \cap \mathfrak{S}_\pi^\pi) = K_\pi(\mathfrak{F})$. This completes the proof of the theorem.

Formations determined by Hall subgroups.

Definition 7.9 Let $\mathfrak{X}$ be a local formation. We define the classes $\check{C}_\pi(\mathfrak{X})$ and $\check{C}'_\pi(\mathfrak{X})$ of groups as follows:

$$\check{C}_\pi(\mathfrak{X}) = \{G : G \text{ is a group, and every } \mathfrak{X} - \text{covering subgroup of } G$$

$$\text{contains some Hall } \pi - \text{subgroup of } G\}.$$

$$\check{C}'_\pi(\mathfrak{X}) = \{G : G \text{ is a group, and every } \mathfrak{X} - \text{covering subgroup of } G$$

$$\text{contains some Hall } \pi - \text{subgroup of } G \text{ as its normal subgroup}\}.$$

Lemma 7.10 (cf. [135, Theorem 2.2.4]). *Let $\mathfrak{F}$ be a formation and G a group. Then the following statements hold.*

1) *If H is an $\mathfrak{F}$-covering subgroup of G and $H \leq K \leq G$, then H is also an $\mathfrak{F}$-covering subgroup of K.*
2) *If H is an $\mathfrak{F}$-covering subgroup of G and $N \trianglelefteq G$, then HN/N is an $\mathfrak{F}$-covering subgroup of $G/N/$.*
3) *If R/N is an $\mathfrak{F}$-covering subgroup of G/N and H is an $\mathfrak{F}$-covering subgroup of R, then H is an $\mathfrak{F}$-covering subgroup of G.*

From Lemma 7.11 to Theorem 7.16, we assume that all groups are soluble.

Lemma 7.11 *Let $\mathfrak{X}$ be a formation. Then, $\check{C}'_\pi(\mathfrak{X})$ is also a formation, and the classes $\mathfrak{X} \cap \mathfrak{S}_\pi \mathfrak{S}_{\pi'}$ and $\mathfrak{S}_{\pi'}$ are subformations of $\check{C}'_\pi(\mathfrak{X})$.*

Proof By using Lemma 7.10(2), it is easy to see that $\check{C}'_\pi(\mathfrak{X})$ is a formation. The rest is obvious.

The following three lemmas are obvious.

Lemma 7.12 *Let $\mathfrak{F}$ and $\mathfrak{H}$ be nonempty formations. Then, $\mathfrak{F}$ is a subformation of $\mathfrak{H}$ if and only if $G^\mathfrak{H} \subseteq G^\mathfrak{F}$ for any group G.*

Lemma 7.13 *Let $\mathfrak{F}$ be a nonempty subformation of a formation $\mathfrak{H}$, G be a group and H an $\mathfrak{H}$-covering subgroup of G. If $H \in \mathfrak{F}$, then H is an $\mathfrak{F}$-covering subgroup of G.*

Lemma 7.14 *For any formation $\mathfrak{X}$, the following equality hold:*

$$\check{C}'_\pi(\mathfrak{X} \cap \mathfrak{S}_\pi \mathfrak{S}_{\pi'}) = \check{C}_\pi(\mathfrak{X} \cap \mathfrak{S}_\pi \mathfrak{S}_{\pi'}).$$

Lemma 7.15 *Let $\mathfrak{X}$ be a local formation. Then $\mathfrak{S}_{\pi'}\check{C}_\pi(\mathfrak{X}) = \check{C}_\pi(\mathfrak{X})$.*

Proof Let $G \in \mathfrak{S}_{\pi'}\check{C}_\pi(\mathfrak{X})$ and $N = G^{\check{C}_\pi(\mathfrak{X})}$. Then, by the definition of the product of two formations, we see that $N \in \mathfrak{S}_{\pi'}$ and $G/N \in \check{C}_\pi(\mathfrak{X})$. This means that N is a π'-group and every $\mathfrak{X}$-covering subgroup of G/N contains some Hall π-subgroup R/N. Since N is a π'-group, we can assume that $R/N = HN/N$, where H is a Hall π-subgroup of R. Let P is a covering subgroup of G. Then, by Lemma 7.10, we know that PN/N is an $\mathfrak{X}$-covering subgroup of G/N. Since any two $\mathfrak{X}$-covering subgroup of a soluble group are conjugate by Theorem 6.30, we may assume, with no loss of generality, that PN/N contains HN/N. This shows that every $\mathfrak{X}$-covering subgroup of G contains a Hall π-subgroup of G, that is, $G \in \check{C}_\pi(\mathfrak{X})$. The inverse inclusion is obvious. This completes the proof.

The following theorem shows that the class $\check{C}_\pi(\mathfrak{F})$ is a local formation, for every local formation $\mathfrak{F}$. The theorem also give the structure of the formation function of $\check{C}'_\pi(\mathfrak{F})$.

We use $Cov_{\mathfrak{H}}G$ to denote the set of all $\mathfrak{H}$-covering subgroups of the group G.

Theorem 7.16 (Guo [139]). *Let $\mathfrak{F}$ be a local formation. Then the following statements hold:*

1) $\check{C}'_\pi(\mathfrak{F}) = \check{C}_\pi(\mathfrak{F} \cap \mathfrak{S}_\pi \mathfrak{S}_{\pi'})$;
2) $\check{C}'_\pi(\mathfrak{F})$ *is a local formation defined by the formation function f such that*

$$f(p) = \begin{cases} \{G \ : Cov_{\mathfrak{H}}G \subseteq h(p)\}, & \text{if } p \in \pi; \\ \mathfrak{S}, & \text{if } p \in \pi', \end{cases}$$

where $\mathfrak{H} = \mathfrak{F} \cap \mathfrak{S}_\pi \mathfrak{S}_{\pi'}$ is a formation with the canonical local satellite h.

Proof

1) Let $\mathfrak{H} = \mathfrak{F} \cap \mathfrak{S}_\pi \mathfrak{S}_{\pi'}$. Since $\check{C}'_\pi(\mathfrak{H}) = \check{C}_\pi(\mathfrak{H})$ by Lemma 7.14, we only need to prove that $\check{C}'_\pi(\mathfrak{F}) \subseteq \check{C}'_\pi(\mathfrak{H})$ for proving $\check{C}'_\pi(\mathfrak{F}) \subseteq \check{C}_\pi(\mathfrak{H})$.

In order to prove $\check{C}'_\pi(\mathfrak{F}) \subseteq \check{C}'_\pi(\mathfrak{H})$, we only need to prove that if $G \in \check{C}'_\pi(\mathfrak{F})$, then an $\mathfrak{F}$-covering subgroup of the group G is also an $\mathfrak{H}$-covering subgroup of G. In fact, let $G \in \check{C}'_\pi(\mathfrak{F})$ and F be some $\mathfrak{F}$-covering subgroup of G, then there exists a Hall π-subgroup G_π of G such that $G_\pi \trianglelefteq F$. Obviously, $F_\pi \subseteq G_\pi$. On the other hand, since $G_\pi \subseteq F$, $G_\pi \subseteq F_\pi$. Hence $F_\pi = G_\pi$. However, since $G_\pi \trianglelefteq F$, we have $F_\pi \trianglelefteq F$. This means that $F \in \mathfrak{S}_\pi \mathfrak{S}_{\pi'}$. Thus, $F \in \mathfrak{F} \cap \mathfrak{S}_\pi \mathfrak{S}_{\pi'} = \mathfrak{H}$. According to the definition of $\mathfrak{F}$-covering subgroup of a group G, for any subgroup U of G containing the covering subgroup F, we have that $FU^{\mathfrak{F}} = U$. Since $\mathfrak{H} \subseteq \mathfrak{F}$, $U^{\mathfrak{F}} \subseteq U^{\mathfrak{H}}$ by Lemma 7.11. It follows that $FU^{\mathfrak{H}} = U$. This shows that F is an $\mathfrak{H}$-covering subgroup of G.

Now, we prove the inverse inclusion $\check{C}_\pi(\mathfrak{H}) \subseteq \check{C}'_\pi(\mathfrak{F})$. Analogous to the above, by Lemma 7.14, we only need to prove that $\check{C}'_\pi(\mathfrak{H}) \subseteq \check{C}'_\pi(\mathfrak{F})$. Assume that it is false and $G \in \check{C}'_\pi(\mathfrak{H}) \setminus \check{C}'_\pi(\mathfrak{F})$ of minimal order. Let K be a minimal normal subgroup of G. If K_1 is another minimal normal subgroup of G, then $K \cap K_1 = 1$. Since $\check{C}'_\pi(\mathfrak{H})$ is a formation by Lemma 7.11, we have $G/K \in \check{C}'_\pi(\mathfrak{H})$ and $G/K_1 \in \check{C}'_\pi(\mathfrak{H})$. Then, by the choice of G, we have that $G/K \in \check{C}'_\pi(\mathfrak{F})$ and $G/K_1 \in \check{C}'_\pi(\mathfrak{F})$. Now, since

$\check{C}'_\pi(\mathfrak{F})$ is a formation, we have $G/K \cap K_1 = G \in \check{C}'_\pi(\mathfrak{F})$. This contradiction shows that G has a unique minimal normal subgroup K.

Let V be an $\mathfrak{F}$-covering subgroup of G and H an $\mathfrak{H}$-covering subgroup of G. Then, by Lemma 7.10, we have VK/K is an $\mathfrak{F}$-covering subgroup of G/K and HK/K is an $\mathfrak{H}$-covering subgroup of G/K. Since $G/K \in \check{C}'_\pi(\mathfrak{H})$, we have $G_\pi K/K \trianglelefteq HK/K$, and consequently, $G_\pi = (HK)_\pi$. Now, by the choice of G, we also have $G/K \in \check{C}'_\pi(\mathfrak{F})$. This means that $G_\pi K/K \trianglelefteq VK/K$, where G_π is a Hall π-subgroup of G. On the other hand, obviously, $G_\pi K/K = V_\pi K/K$, where V_π is a Hall π-subgroup of V, so $V_\pi K \trianglelefteq VK/K$ and consequently, $VK/K \in \mathfrak{S}_\pi\mathfrak{S}_{\pi'}$. It follows that $VK/K \in \mathfrak{F} \cap \mathfrak{S}_\pi\mathfrak{S}_{\pi'} = \mathfrak{H}$. Since $\mathfrak{H}$ is a subformation of $\mathfrak{F}$, by Lemma 7.13 we see that VK/K is also an $\mathfrak{H}$-covering subgroup of G/K. However, because any two $\mathfrak{H}$-covering subgroups of a soluble group are conjugate, there exists an element $x \in G$ such that $V^x K/K = HK/K$. Thus, $V^x K = HK$, and so HK/K is an $\mathfrak{F}$-covering subgroup of G/K.

Suppose that $HK \neq G$. Then, since H is an $\mathfrak{H}$-covering subgroup of G and by the hypotheses $G \in \check{C}'_\pi(\mathfrak{H})$, we have $(HK)_\pi = G_\pi \trianglelefteq H$. However, by Lemma 7.10, we see that H is also an $\mathfrak{H}$-covering subgroup of HK. Hence $HK \in \check{C}'_\pi(\mathfrak{H})$. Since $|HK| < |G|$, by the choice of G, we have $HK \in \check{C}'_\pi(\mathfrak{F})$. Then, as above we have seen that if V_1 is an $\mathfrak{F}$-covering subgroup of HK, then V_1 is an $\mathfrak{H}$-covering subgroup of HK. Since two $\mathfrak{H}$-covering subgroup of a soluble group are conjugate, we have that V_1 and H are conjugate in HK, that is, there exists an element $y \in HK$ such that $V_1^y = H$. Hence H is also an $\mathfrak{F}$-covering subgroup of HK. Since HK/K is an $\mathfrak{F}$-covering subgroup of G/K, we have H is an $\mathfrak{F}$-covering subgroup of G by Lemma 7.10. This shows that $G \in \check{C}'_\pi(\mathfrak{F})$, a contradiction.

Now, we assume that $HK = G$. Then $G/K \in \mathfrak{F}$ and G/K is itself an $\mathfrak{F}$-covering subgroup of G/K and $G^\mathfrak{F} \subseteq K$. By the minimality of K, we have $G^\mathfrak{F} = K$.

If $K \subseteq \Phi(G)$, Then, since $G/\Phi(G) \simeq (G/K)/(\Phi(G)/K)$ and $G/K \in \mathfrak{F}$, we have that $G/\Phi(G) \in \mathfrak{F}$. Since $\mathfrak{F}$ is a local formation and a local formation is a saturated formation (cf. [135, Theorem 3.1.11]), we obtain that $G \in \mathfrak{F}$. Analogously, since $G/K = HK/K \in \mathfrak{S}_\pi\mathfrak{S}_{\pi'}$ and the formation of all π-closed groups is local formation, we have $G \in \mathfrak{S}_\pi\mathfrak{S}_{\pi'}$. This induces that $G \in \mathfrak{F} \cap \mathfrak{S}_\pi\mathfrak{S}_{\pi'} = \mathfrak{H}$. Moreover, by Lemma 7.10, $\mathfrak{H}$ is a subformation of $\check{C}'_\pi(\mathfrak{F})$. Therefore, $G \in \check{C}'_\pi(\mathfrak{F})$, a contradiction.

Assume that $K \not\subseteq \Phi(G)$. Since K is a minimal normal subgroup of the soluble group G and $K = G^\mathfrak{F}$, we have that $G^\mathfrak{F}$ is abelian subgroup of G. It follows that $G^\mathfrak{F} \cap H \trianglelefteq G$ and so $G^\mathfrak{F} \cap H = 1$ or $G^\mathfrak{F}$. If $G^\mathfrak{F} \cap H = G^\mathfrak{F}$, then $G = H \in \mathfrak{H} \subseteq \mathfrak{F}$, a contradiction. Hence H is a complement of $G^\mathfrak{F}$ in G. It follows from [135, Theorem 3.3.4] that H is a $\mathfrak{F}$-covering subgroup of G. On the other hand, since H is a $\mathfrak{H}$-covering subgroup of G and $G \in \check{C}'_\pi(\mathfrak{H})$, we have $G_\pi \trianglelefteq H$. This shows that $G \in \check{C}'_\pi(\mathfrak{F})$. The final contradiction complete the proof of 1).

2) In order to prove that $\check{C}'_\pi(\mathfrak{F})$ is a local formation, by 1), we only need to prove that $\check{C}_\pi(\mathfrak{H})$ is a local formation. Since $\mathfrak{H} = \mathfrak{F} \cap \mathfrak{S}_\pi\mathfrak{S}_{\pi'}$ is a intersection of two local formations, we know that $\mathfrak{H}$ is also a local formation. Hence, by [135, Corollary 3.1.17], $\mathfrak{H}$ has a canonical local satellite h such that $h(p) = \mathfrak{N}_p h(p)$, for every prime p.

Now let f be a formation function such that

$$f(p) = \begin{cases} \{G \ : Cov_{\mathfrak{H}}G \subseteq h(p)\}, & \text{if } p \in \pi; \\ \mathfrak{S}, & \text{if } p \in \pi', \end{cases}$$

In order to prove that $\check{C}_\pi(\mathfrak{H})$ is a local formation, we only need to prove that $\check{C}_\pi(\mathfrak{H}) = LF(f)$.

Let G be a group in the class $LF(f) \setminus \check{C}_\pi(\mathfrak{H})$ of minimal order and K a minimal normal subgroup of G. Then, obviously, K is the unique minimal normal subgroup of G. Since $|G/K| < |G|$, by the choice of G, we see that $G/K \in \check{C}_\pi(\mathfrak{H})$. If F is a $\mathfrak{H}$-covering subgroup of G and H is a Hall π-subgroup of G, then FK/K is a $\mathfrak{H}$-covering subgroup of G/K and HK/K is a Hall π-subgroup of G. Hence $HK/K \subseteq FK/K$ and consequently, $HK \subseteq HK$. Since G is a soluble group, K is an abelian p-group for some prime p.

Assume $p \in \pi$. Then, since $G \in LF(f)$, we have

$$G/C_G(K) \in f(p) = \{G \ : Cov_{\mathfrak{H}}G \subseteq h(p)\}.$$

Let $h^*(p) = \{G \ : Cov_{\mathfrak{H}}G \subseteq h(p)\}$. Then, by [89, Theorem IV.5.19 (c)], we know that the $\mathfrak{H}$-covering subgroup F covers every h^*-central chief factor of G. Hence $K \subseteq F$, and so $F = FK \supseteq HK \supseteq H$. This shows that $G \in \check{C}_\pi(\mathfrak{H})$, a contradiction.

Assume that $p \in \pi'$. Then $K \in \mathfrak{S}_{\pi'}$, and consequently $G \in \mathfrak{S}_{\pi'}\check{C}_\pi(\mathfrak{H})$. It follows from Lemma 7.15 that $G \in \check{C}_\pi(\mathfrak{H})$. This contradiction induces that $LF(f) \subseteq \check{C}_\pi(\mathfrak{H})$.

Now, we prove the inverse inclusion $\check{C}_\pi(\mathfrak{H}) \subseteq LF(f)$. If it is false, we may assume that L is a group of minimal order in $\check{C}_\pi(\mathfrak{H}) \setminus LF(f)$. Then, it is easy to know that L has a unique minimal normal subgroup R. By the choice, we have that $L/R \in LF(f)$. Since L is soluble, R is an abelian p-group for some prime p.

Suppose that $p \in \pi'$. Then, obviously, $L/C_G(R) \in \mathfrak{S} = f(p)$. On the other hand, since $L/R \in LF(f)$, every chief factor between L and R is f-central. This shows that that $L \in LF(f)$, a contradiction.

Assume $p \in \pi$. Let H^* be a Hall π-subgroup of L and F^* an $\mathfrak{H}$-covering subgroup of L. Since R is a normal p-group, we have $R \subseteq H^*$. Since $L \in \check{C}_\pi(\mathfrak{H})$, we can assume $H^* \subseteq F^*$. Then $R \subseteq F^*$, that is, F^* covers R. Hence, by [89, Theorem IV.5.19 c)], we see that R is an h^*-central chief factor. Since $h^*(p) = f(p)$, for every $p \in \pi$, R is also an f-central chief factor. Moreover, since $L/R \in LF(f)$, we obtain that $L \in LF(f)$. The final contradiction completes the proof.

5.8   The Theory of $\mathfrak{F}$-Cocentrality of Chief Factors and $\mathfrak{F}$-Cohypercenter for Fitting Classes

It is well known that the $\mathfrak{F}$-hypercenter in the theory of local formations plays an important role in the research of groups (see also the previous chapters). In this connection, naturally, the following problem arose (this problem was proposed at Gomel seminar many years ago).

Problem 8.1. Would we establish the theory of $\mathfrak{F}$-centrality of chief factors and $\mathfrak{F}$-hypercenter of a group for local Fitting classes $\mathfrak{F}$ in the universe $\mathfrak{S}$?

In this section, we will establish the theory.

$\mathfrak{F}$-cocentral chief factors. In this section, all groups are soluble.

Definition 8.2 [212]. Let $\mathfrak{F} = LR(f) = \mathfrak{S}_\pi \cap (\cap_{p \in \pi} f(p)\mathfrak{N}_p\mathfrak{S}_{p'})$ be a local Fitting class and $\pi = \mathrm{Supp}(f)$. Then we say that a p-chief factor H/K of a group G is f-cocentral in G if $p \in \pi$ and $G_{f(p)\mathfrak{N}_p}$ covers H/K; otherwise we say that H/K is f-coeccentric in G.

Lemma 8.3 [89, VIII.2.4(d)]. *If $\mathfrak{F}$ is a nonempty Fitting class and N is a subnormal subgroup of G, then $N_\mathfrak{F} = G_\mathfrak{F} \cap N$.*

Lemma 8.4 *Let $\mathfrak{F} = LR(f)$ and $p \in \pi = \mathrm{Supp}(f)$. Then a p-chief factor H/K of a group G is f-cocentral in G if and only if $H = KH_{f(p)\mathfrak{N}_p}$.*

Proof Assume that the p-chief factor H/K of G is f-cocentral. Then $H \subseteq KG_{f(p)\mathfrak{N}_p}$. By Dedekind modular law, we have

$$H = H \cap KG_{f(p)\mathfrak{N}_p} = K(H \cap G_{f(p)\mathfrak{N}_p}).$$

It follow from Lemma 8.3 that $H = KH_{f(p)\mathfrak{N}_p}$. The converse is clear.

Lemma 8.5 (see [432] or [389]). *The product of any two local Fitting classes is a local Fitting class.*

Remark 8.6 The definition of f-cocentral chief factors depends on the choice of H-functions of $\mathfrak{F}$. In order to show it, we give the following .

Example 8.7 Let $\mathfrak{F} = \mathfrak{N}_3\mathfrak{S}_{3'}$. Since $\mathfrak{N}_3$ and $\mathfrak{S}_{3'}$ are local Fitting classes, by Lemma 8.5, $\mathfrak{F}$ is also a local Fitting class. By a direct check, we can see that $\mathfrak{F}$ can be defined by each of the following two H-functions φ and ψ, where

$$\varphi(p) = \begin{cases} \mathfrak{N}_3 & \text{if } p = 3, \\ \mathfrak{F} & \text{if } p \neq 3. \end{cases}$$

$$\psi(p) = \begin{cases} \mathfrak{N}_3 & \text{if } p = 3, \\ \mathfrak{S} & \text{if } p \neq 3. \end{cases}$$

for every prime p.

Let $G = S_4$ be the symmetric group of degree 4 and $A = A_4$ the alternating group of degree 4. Then G/A is a chief factor of G, $G_{\varphi(2)\mathfrak{N}_2} = K$ is Klein 4-group and $G_{\psi(2)\mathfrak{N}_2} = G$. Clearly, G has a unique chief series $1 \lhd K \lhd A \lhd G$. Since $AG_{\varphi(2)\mathfrak{N}_2} = AK = A < G$, the 2-factor G/A is not φ-cocentral by Lemma 8.4. But, since $AG_{\psi(2)\mathfrak{N}_2} = AG = G$, G/A is a ψ-cocentral 2-chief factor of G.

However, the following theorem shows that the concept of f-cocentral chief factor does not depend on the choice of H-functions f of $\mathfrak{F}$ if the H-functions f are integrated.

Theorem 8.8 (Guo, Vorobev [212]). *Suppose that $\mathfrak{F} = LR(\varphi) = LR(\psi)$ for some integrated H-functions φ and ψ, and H/K be a p-chief factor of G. Then the following two statements are equivalent.*

(i) H/K is φ-cocentral in G;
(ii) H/K is ψ-cocentral in G.

Proof Without loss of generality, we may assume that $\varphi \leq \psi$. In fact, since $\mathfrak{F} = LR(\varphi \cap \psi)$ and $\varphi \cap \psi$ is an integrated H-function of $\mathfrak{F}$, we can always choose two integrated H-functions φ and ψ such that $\varphi \leq \psi$. Let $\pi = \mathrm{Supp}(\varphi) = \mathrm{Supp}(\psi)$. Then we have $G_{\varphi(p)} \subseteq G_{\psi(p)}$ for all $p \in \pi$ and so every φ-cocentral chief factor is a ψ-cocentral p-chief factor.

Now we only need to show that $(ii) \Longrightarrow (i)$. Let H/K be a ψ-cocentral p-chief factor of G. Then by Lemma 8.4, we have

$$H/K = (H \cap G_{\psi(p)\mathfrak{N}_p})K/K. \tag{5.16}$$

We fist prove that $\psi(p)\mathfrak{N}_p \subseteq \mathfrak{F}$. For the purpose, we construct an H-function f of $\mathfrak{F}$ such that $f(r) = \psi(r)\mathfrak{N}_r$ if $r \in \pi$ and $f(r) = \varnothing$ if $r \notin \pi$. Obviously, $Supp(\psi) = Supp(f) = \pi$. Since $\mathfrak{F} = LR(\psi) = \mathfrak{S}_\pi \cap (\bigcap_{p \in \pi} \psi(p)\mathfrak{N}_p\mathfrak{S}_{p'})$, by the multiplicative associative law of Fitting classes, we have $LR(f) = \mathfrak{S}_\pi \cap (\bigcap_{p \in \pi} f(p)\mathfrak{N}_p\mathfrak{S}_{p'}) = \mathfrak{S}_\pi \cap (\bigcap_{p \in \pi} \psi(p)\mathfrak{N}_p\mathfrak{N}_p\mathfrak{S}_{p'}) = \mathfrak{S}_\pi \cap (\bigcap_{p \in \pi} \psi(p)\mathfrak{N}_p\mathfrak{S}_{p'}) = \mathfrak{F}$. This means that f is also an H-function of $\mathfrak{F}$. Let $L \in f(p)$. Since $f(p) \subseteq f(p)\mathfrak{N}_p\mathfrak{S}_{p'}$, we have $L \in f(p)\mathfrak{N}_p\mathfrak{S}_{p'}$. Assume that q is an arbitrary prime in π different from p. Then $\mathfrak{N}_p \subseteq \mathfrak{S}_{q'}$ and so $L^{\mathfrak{S}_{q'}} \subseteq L^{\mathfrak{N}_p}$. But since $L \in f(p) = \psi(p)\mathfrak{N}_p$, $L/L_{\psi(p)} \in \mathfrak{N}_p$. Hence $L^{\mathfrak{N}_p} \subseteq L_{\psi(p)}$. Since ψ is an integrated H-function of $\mathfrak{F}$, $L^{\mathfrak{N}_p} \in \mathfrak{F}$. Hence $L^{\mathfrak{S}_{q'}} \in \mathfrak{F}$. Then by $\mathfrak{F} = LR(f)$, we obtain that $L^{\mathfrak{S}_{q'}}/(L^{\mathfrak{S}_{q'}})_{f(q)} \in \mathfrak{N}_q\mathfrak{S}_{q'}$. It follows that $(L^{\mathfrak{S}_{q'}})^{\mathfrak{N}_q\mathfrak{S}_{q'}} \in f(q)$. Since $(L^{\mathfrak{S}_{q'}})^{\mathfrak{N}_q\mathfrak{S}_{q'}} = L^{\mathfrak{N}_q\mathfrak{S}_{q'}\mathfrak{S}_{q'}}$ (see [89, IV.1.8(b)], we have $L^{\mathfrak{N}_q\mathfrak{S}_{q'}} \in f(q)$. Hence

$$L/L_{f(q)} \simeq (L/L^{\mathfrak{N}_q\mathfrak{S}_{q'}})/(L_{f(q)}/L^{\mathfrak{N}_q\mathfrak{S}_{q'}}) \in \mathfrak{N}_q\mathfrak{S}_{q'}$$

and so $L \in f(q)\mathfrak{N}_q\mathfrak{S}_{q'}$ for $q \neq p$ and $q \in \pi$. This implies that $L \in \mathfrak{S}_\pi \cap (\bigcap_{p \in \pi} f(p)\mathfrak{N}_p\mathfrak{S}_{p'})$. Thus $\psi(p)\mathfrak{N}_p \subseteq \mathfrak{F}$. It follows that

$$G_{\varphi(p)\mathfrak{N}_p} \subseteq G_{\psi(p)\mathfrak{N}_p} \subseteq G_{\mathfrak{F}}. \tag{5.17}$$

Consequently, $G_{\psi(p)\mathfrak{N}_p} \in \mathfrak{F}$ and thereby $G_{\psi(p)\mathfrak{N}_p}/(G_{\psi(p)\mathfrak{N}_p})_{\varphi(p)\mathfrak{N}_p} \in \mathfrak{S}_{p'}$. But since $(G_{\psi(p)\mathfrak{N}_p})_{\varphi(p)\mathfrak{N}_p} = G_{\psi(p)\mathfrak{N}_p \cap \varphi(p)\mathfrak{N}_p}$ and $\varphi \leq \psi$, we have $G_{\psi(p)\mathfrak{N}_p}/G_{\varphi(p)\mathfrak{N}_p} \in \mathfrak{S}_{p'}$. This means that every Sylow p-subgroup P of $G_{\varphi(p)\mathfrak{N}_p}$ is a Sylow p-subgroup of $G_{\psi(p)\mathfrak{N}_p}$. Now because H/K is a p-factor of G, by comparing the orders of $K(H \cap G_{\varphi(p)\mathfrak{N}_p})$ and $K(H \cap P)$, we see that

$$K(H \cap G_{\varphi(p)\mathfrak{N}_p}) = K(H \cap P). \tag{5.18}$$

Thus by the equalities (5.16) and (5.17), we obtain that

$$H/K = (G_{\psi(p)\mathfrak{N}_p} \cap H)K/K = (P \cap H)K/K = (G_{\varphi(p)\mathfrak{N}_p} \cap H)K/K.$$

This induces that $KH_{\varphi(p)\mathfrak{N}_p} \supseteq H$. Therefore, H/K is φ-cocentral in G. This completes the proof.

This theorem shows that the f-cocentrality of chief factors in the theory of Fitting classes does not depend on the choice of integrated H-functions of the Fitting class $\mathfrak{F}$. In connection with this, we call an f-cocentral chief factor an $\mathfrak{F}$-cocentral chief factor when f is an integrated H-function of $\mathfrak{F}$.

In the theory of formations, there exists an analogue of Theorem 8.8. In fact, Carter and Hawkes [60] proved that if f_1 and f_2 are two integrated formation functions of a formation $\mathfrak{F}$, then a chief factor is f_1-central in G if and only if it is f_2-central in G. Moreover, Carter and Hawkes [60] proved that for a local formation $\mathfrak{F} = LF(f)$, a group $G \in \mathfrak{F}$ if and only if every chief factor of G is f-central in G (see also [89, Theorem IV.3.2]. Now we give the following theorem which is an analogue of the result of Carter and Hawkes.

Theorem 8.9 (Guo, Vorob'ev [212]). *Let $\mathfrak{F} = LR(f)$ be a local Fitting class. Then $G \in \mathfrak{F}$ if and only if every chief factor of G is f-cocentral.*

Proof Assume that $G \in \mathfrak{F}$ and let $\pi = Supp(f)$. Then $G \in f(p)\mathfrak{N}_p\mathfrak{S}_{p'}$ for all primes $p \in \pi$. Let H/K be a p-chief factor of G and $H_{f(p)\mathfrak{N}_p}$ be an $f(p)\mathfrak{N}_p$-radical of H. Since $H \in \mathfrak{F}$, $H/H_{f(p)\mathfrak{N}_{p'}} \in \mathfrak{S}_{p'}$. Let P be a Sylow p-subgroup of H. Then $PH_{f(p)\mathfrak{N}_p}/H_{f(p)\mathfrak{N}_p}$ is a Sylow subgroup of $H/H_{f(p)\mathfrak{N}_p}$. Hence $PH_{f(p)\mathfrak{N}_p}/H_{f(p)\mathfrak{N}_p} \in \mathfrak{S}_{p'} \cap \mathfrak{N}_p = (1)$. It follows that $PH_{f(p)\mathfrak{N}_p} = H_{f(p)\mathfrak{N}_p}$ and so $P \subseteq H_{f(p)\mathfrak{N}_p}$. This implies that P is also a Sylow p-subgroup of $H_{f(p)\mathfrak{N}_p}$. But since $H/K \in \mathfrak{N}_p$, $H/K = PK/K \leq H_{f(p)\mathfrak{N}_p}K/K \leq H/K$. It follows that $H = H_{f(p)\mathfrak{N}_p}K$. Therefore, by Lemma 8.4, the p-chief factor H/K is f-cocentral in G.

Now assume that $G \notin \mathfrak{F}$. Then there exists some prime $p \in \pi$ such that $G \notin f(p)\mathfrak{N}_p$. Hence there exists a p-chief factor H/K above $G_{f(p)\mathfrak{N}_p}$ such that $G_{f(p)\mathfrak{N}_p}$ does not cover H/K. This shows H/K is not f-cocentral. The theorem is proved.

$\mathfrak{F}$-**cohypercenter.** We now establish the theory of $\mathfrak{F}$-cohypercenter for local Fitting classes $\mathfrak{F}$.

Definition 8.10 Let $\mathfrak{F}$ be a local Fitting class.

i) Suppose that L is a normal subgroup of G. Then the factor group G/L is said to be $\mathfrak{F}$-cohypercentral in G if there exists a series

$$L = L_0 \lhd L_1 \lhd \cdots \lhd L_k = G$$

such that every factor L_{i+1}/L_i is a G-chief factor and it is $\mathfrak{F}$-cocentral in G, for all $i \in \{0, 1, \cdots, k - 1\}$.

ii) The intersection of all normal subgroups K of G such that G/K is $\mathfrak{F}$-cohypercentral is called the $\mathfrak{F}$-cohypercenter of G and denoted by $Z^{\mathfrak{F}}(G)$.

In general, $G/Z^{\mathfrak{F}}(G)$ is not $\mathfrak{F}$-cohypercentral (see Example 8.12 below).

In the theory of formations, it is well known that if $\mathfrak{F}$ is a local formation with canonical local satellite F and $\mathfrak{F}$ is closed with respect to normal subgroups, then $Z_{\mathfrak{F}}(G) \in \mathfrak{F}$ (see Chapter I). We now establish an analogous result in the theory of Fitting classes.

Theorem 8.11 (Guo, Vorobev [212]). *Let $\mathfrak{F} = LR(f)$ be a local Fitting class, where f is an integrated H-function. Then for any group G and any normal subgroup L of G, the following statements hold:*

(a) *If G/L is $\mathfrak{F}$-cohypercentral in G, then $G = LG_{\mathfrak{F}}$;*

(b) *If $\mathfrak{F}$ is closed under homomorphic images and G/L is $\mathfrak{F}$-cohypercentral in G, then $G/L \in \mathfrak{F}$;*

(c) *If $\mathfrak{F}$ is a formation as well, then $G/Z^{\mathfrak{F}}(G) \in \mathfrak{F}$;*

(d) *Assume that the integrated H-function is also full and $\mathrm{Supp}(f) = \mathbb{P}$. If G/L is $\mathfrak{F}$-cohypercentral in G, then every chief factor above L is $\mathfrak{F}$-cocentral.*

Proof

(a) Since G/L is $\mathfrak{F}$-cohypercentral in G, there exists a G-chief series between L and G such that every p-chief factor in this series is $\mathfrak{F}$-cocentral in G. Hence $G_{f(p)\mathfrak{N}_p}$ covers every p-chief factor in this series, for all primes $p \in \pi = Supp(f)$. By the proof of "(ii) $\Rightarrow$ (i)" in Theorem 8.8, we see that $G_{f(p)\mathfrak{N}_p} \subseteq G_{\mathfrak{F}}$ for every $p \in \pi$. Therefore, $G_{\mathfrak{F}}$ covers every p-chief factor of G in this series and so it also covers G/L. Thus $G = LG_{\mathfrak{F}}$.

(b) Assume that $\mathfrak{F}$ is closed under homomorphic images and G/L is $\mathfrak{F}$-cohypercentral in G. Then by (a), $G = LG_{\mathfrak{F}}$ and so $G/L = LG_{\mathfrak{F}}/L \simeq G_{\mathfrak{F}}/L \cap G_{\mathfrak{F}}$. Since $G_{\mathfrak{F}} \in \mathfrak{F}$ and $\mathfrak{F}$ is closed under homomorphic images, $G_{\mathfrak{F}}/L \cap G_{\mathfrak{F}} \in \mathfrak{F}$ and consequently $G/L \in \mathfrak{F}$.

(c) Assume that $\mathfrak{F}$ is a formation. Then, since $Z^{\mathfrak{F}}(G) = \bigcap\{L \trianglelefteq G \mid G/L$ is $\mathfrak{F}$-cohypercentral in $G\}$, we obtain that $G/Z^{\mathfrak{F}}(G) \in \mathfrak{F}$ by (b).

(d) Supposed that G/L is $\mathfrak{F}$-cohypercentral in G and H/K is a p-chief factor of G between L and G. Then by (a), the $\mathfrak{F}$-radical $G_{\mathfrak{F}}$ of G covers G/L. Hence $G_{\mathfrak{F}}$ covers H/K. Since $H/K \in \mathfrak{N}_p$, a Sylow p-subgroup P of $G_{\mathfrak{F}}$ covers H/K. Because f is an integrated and full H-function, we have $f(p) \subseteq \mathfrak{F} \subseteq f(p)\mathfrak{N}_p\mathfrak{S}_{p'} = f(p)\mathfrak{S}_{p'}$ for every $p \in \mathbb{P}$. Hence $G_{\mathfrak{F}}/(G_{\mathfrak{F}})_{f(p)} \in \mathfrak{S}_{p'}$. But since $(G_{\mathfrak{F}})_{f(p)} = G_{\mathfrak{F}\cap f(p)} = G_{f(p)}$, the Sylow p-subgroup P of $G_{\mathfrak{F}}$ is also a Sylow p-subgroup of $G_{f(p)}$. Since P covers H/K, $G_{f(p)}$ covers H/K. Thus H/K is $\mathfrak{F}$-cocentral. This completes the proof.

The following example shows that $G/Z^{\mathfrak{F}}(G)$ is not $\mathfrak{F}$-cohypercentral, in general.

Example 8.12 Let $\mathfrak{F} = \mathfrak{N}$. Then $\mathfrak{N} = LR(f)$, where f is the H-function such that $f(p) = \mathfrak{N}_p$ for every prime $p \in \mathbb{P}$. Put $G = S_3 \times Z_2$, in which the symmetric group $S_3 = \langle s, t \rangle$, where $s^3 = 1$ and $t^2 = 1$. Let $Z_2 = \langle z \rangle$ and α be the map from S_3 to Z_2 such that $\alpha(s) = 1$ and $\alpha(t) = z$. Let $H = \{(k, \alpha(k)) \mid k \in S_3\}$. Then $|G : H| = 2$ and so $H \triangleleft G$.

Consider all possible chief series of G:

$$1 \triangleleft A_3 \times 1 \triangleleft S_3 \times 1 \triangleleft G,$$

$$1 \triangleleft A_3 \times 1 \triangleleft H \triangleleft G,$$

$$1 \triangleleft A_3 \times 1 \triangleleft A_3 \times Z_2 \triangleleft G,$$

$$1 \lhd 1 \times Z_2 \lhd A_3 \times Z_2 \lhd G.$$

It is easy to see that G/H and $G/S_3 \times 1$ are $\mathfrak{N}$-cohypercentral in G since they are covered by $G_{f(p)} = G_{\mathfrak{N}_2} = 1 \times Z_2$. Besides, the chief factor $G/A_3 \times Z_2$ is $\mathfrak{N}$-coeccentric since $G_{f(2)}$ avoids it. Analogously, $S_3 \times 1/A_3 \times 1$ and $H/A_3 \times 1$ are not $\mathfrak{N}$-cohypercentral in G. This shows that G/H and $G/S_3 \times 1$ are the only two $\mathfrak{N}$-cohypercentral factors of G. Therefore, $Z^{\mathfrak{N}}(G) = A_3 \times 1$. Obviously, $G/Z^{\mathfrak{N}}(G)$ is not $\mathfrak{N}$-cohypercentral in G.

Remark 8.13 If a class $\mathfrak{F}$ of finite groups is both a local formation and a local Fitting class, then the $\mathfrak{F}$-hypercenter of a group G in the theory of formations is different from the $\mathfrak{F}$-cohypercenter of G in the theory of Fitting classes, in general. For instance, let $\mathfrak{F} = \mathfrak{N}$ be the class of all finite nilpotent groups. Then $\mathfrak{N}$ is both a local formation and a local Fitting class. Let $G = S_3 \times Z_2$. Then, obviously, the $\mathfrak{N}$-hypercenter $Z_{\mathfrak{N}}(G)$ of G is $1 \times Z_2$ if $\mathfrak{N}$ is regarded as a formation. However, by the above example 8.12, we see that the $\mathfrak{N}$-cohypercenter $Z^{\mathfrak{N}}(G)$ of G is $A_3 \times 1$ if we regard $\mathfrak{N}$ as a Fitting class.

On the problem of cover-avoidance properties for $\mathfrak{F}$-injectors. For a local formation $\mathfrak{F}$, Carter and Hawkes [60] proved that every $\mathfrak{F}$-normalizer of G (see [135, Definition 2.6.2]) coves each $\mathfrak{F}$-central chief factor of G and avoids each $\mathfrak{F}$-eccentric chief factor of G. This leads to the following problem.

Problem 8.14. Let $\mathfrak{F} = LR(f)$ be a local Fitting class. Is it true that every $\mathfrak{F}$-injector of G covers each $\mathfrak{F}$-cocentral chief factor of G and avoids each $\mathfrak{F}$-coeccentric chief factor of G?

Let f be an H-function and $\pi = Supp(f)$. If $f(p) = f(q)$ for all primes $p, q \in \pi$, then f is said to be invariable.

In this section, we give the answer to Problem 8.14 in the case where $\mathfrak{F}$ has an invariable H-function (see the following Theorem 8.18(3)).

Definition 8.15 Let $\mathfrak{F} = LR(f)$ for some H-function f with $Supp(f) = \pi$ and $p \in \pi$. We say that a p-chief factor H/K of G is $f(p)$-covered in G if $H = K(V_{f(p)} \cap H)$, and $f(p)$-avoided if $K = K(V_{f(p)} \cap H)$ for some $\mathfrak{F}$-injector V of G.

Lemma 8.16 Let $\mathfrak{F} = LR(f)$ for some H-function f and $\pi = Supp(f)$. If $\mathfrak{X}$ is a nonempty Fitting class and $\mathfrak{X} \subseteq \bigcap_{p \in \pi} f(p)$, then $C_G(G_{\mathfrak{F}}/G_{\mathfrak{X}}) \subseteq G_{\mathfrak{F}}$ for any group G.

Proof The proof is analogous to the proof of [140, Theorem 3.2] and we omit the details.

Lemma 8.17 Let $\mathfrak{F} = LR(f)$ for some invariable H-function f and $Supp(f) = \pi$. If V is an $\mathfrak{F}$-injector of G, then $V_{f(p)} = G_{f(p)}$ for all $p \in \pi$.

Proof Since $G_{\mathfrak{F}} \lhd V$ and $f(p) \subseteq \mathfrak{F}$, by Lemma 8.3, $G_{f(p)} = (G_{\mathfrak{F}})_{f(p)} = G_{\mathfrak{F}} \cap V_{f(p)}$ for all $p \in \pi$. Hence $[V_{f(p)}, G_{\mathfrak{F}}] \leq G_{f(p)}$ and thereby $V_{f(p)} \leq C_G(G_{\mathfrak{F}}/G_{f(p)})$. Since the H-function f is invariable, $C_G(G_{\mathfrak{F}}/G_{f(p)}) \leq G_{\mathfrak{F}}$ by Lemma 8.16. Therefore, $V_{f(p)} = G_{f(p)}$, for every $p \in \pi$.

Theorem 8.18 (Guo, Vorobev [212]). *Let $\mathfrak{F} = LR(f)$ with $Supp(f) = \pi$ and f be a integrated H-function of $\mathfrak{F}$. Then the following statements hold:*

(1) If $p \in \pi$, then every p-chief factor of G is either $f(p)$-covered or $f(p)$-avoided in G.

(2) Assume that f is invariable and $p \in \pi$. Then
  i) A p-chief factor of G is $f(p)$-covered if and only if it is f-cocentral in G;
  ii) An $\mathfrak{F}$-injector of G covers every $f(p)$-covered chief factor of G.

(3) If f is a full and invariable H-function, then an $\mathfrak{F}$-injector of G covers each $\mathfrak{F}$-cocentral chief factor of G and avoids each $\mathfrak{F}$-coeccentric chief factor of G.

Proof

(1) Suppose that H/K is a p-chief factor of G and let V be an $\mathfrak{F}$-injector of G. Since $V \cap H \trianglelefteq V$, by Lemma 8.3, we have

$$(V \cap H)_{f(p)} = V_{f(p)} \cap (V \cap H) = V_{f(p)} \cap H.$$

But by [89, Theorem VIII.2.13], $V \cap H$ is an $\mathfrak{F}$-injector of H. Hence, by [89, Theorem VIII.2.9], every conjugate subgroup of $(V \cap H)_{f(p)}$ in G is a conjugate subgroup of $(V \cap H)_{f(p)}$ in H. Hence by Frattini argument, $G = HN_G((V \cap H)_{f(p)})$ and so $K(V \cap H)_{f(p)} \trianglelefteq G$. This implies that the p-chief factor H/K is either $f(p)$-covered or $f(p)$-avoided.

(2) The statement (i) can be directly obtained by Lemma 8.17 and the definition of f-cocentral chief factor. We now prove (ii). Assume that V is an $\mathfrak{F}$-injector of G and H/K is an $f(p)$-covered chief factor of G, that is, $H = K(V_{f(p)} \cap H)$. Since the H-function f is invariable, by Lemma 8.17, $V_{f(p)} = G_{f(p)}$ for all primes $p \in \pi$. But since $G_{f(p)} \subseteq G_{\mathfrak{F}} \subseteq F$ for every $\mathfrak{F}$-injector F of G, F covers H/K and so (ii) holds.

(3) Let H/K be an $\mathfrak{F}$-cocentral p-chief factor of G, where $p \in \pi$, and V an $\mathfrak{F}$-injector of G. Then $H \leq KG_{f(p)\mathfrak{N}_p}$. Since f is integrated, by the proof of $(ii) \implies (i)$ in Theorem 8.8, we see that $G_{f(p)\mathfrak{N}_p} \leq G_{\mathfrak{F}}$. Besides, by the definition of $\mathfrak{F}$-injector, $G_{\mathfrak{F}} \leq V$. Hence $H \leq KV$ and so V covers H/K.

Now assume that an $\mathfrak{F}$-injector V of G covers some p-chief factor H/K of G. Obviously, a Sylow p-subgroup P of V covers H/K. If $p \notin \pi$, then $\mathfrak{N}_p \nsubseteq \mathfrak{F}$ since $\mathfrak{F}$ is a local Fitting class and so the Sylow p-subgroup P of V is trivial. It follows that $H/K = 1$, which contradicts the choice of H/K. Hence, we can assume that $p \in \pi$. Since $V \in \mathfrak{F}$, we have $V \in f(p)\mathfrak{N}_p\mathfrak{S}_{p'}$. Since f is full, we have $f(p)\mathfrak{N}_p = f(p)$. This implies that $V/V_{f(p)} \in \mathfrak{S}_{p'}$ and hence every Sylow p-subgroup P of V is a Sylow p-subgroup of $V_{f(p)}$. Consequently, $P \leq V_{f(p)}$. By Lemma 8.17, $V_{f(p)} = G_{f(p)}$ for every $p \in \pi$ and obviously $G_{f(p)} \subseteq G_{f(p)\mathfrak{N}_p}$. Hence H/K is covered by $G_{f(p)\mathfrak{N}_p}$. This shows that H/K is $\mathfrak{F}$-cocentral in G. This completes the proof.

5.9   On ω-Local Fitting Classes and Lockett Conjecture

We call a Fitting class which is closed under homomorphic image a radical homomorph. A radical homomorph $\mathfrak{H}$ is said to be saturated if $G \in \mathfrak{H}$ whenever
$G/\Phi(G) \in \mathfrak{H}$. If a saturated radical homomorph is a formation, then we naturally
call it a saturated Fitting formation.

Let ω is a nonempty set of primes. A group G is called an ωd-group if its order
is divisible by at least one number in ω. Let $F^p(G) = G^{\mathfrak{N}_p \mathfrak{S}_{p'}}$, that is, $F^p(G)$ is the
$\mathfrak{N}_p \mathfrak{E}_{p'}$-residual of G. Let $\mathfrak{E}_{\omega d}$ be the class of the groups whose composition factor
is an ωd-group.

Let $\phi \neq \omega \subseteq \mathbb{P}$. Following Shmetkov and Skiba [367], the map

$$f : \omega \cup \{\omega'\} \longrightarrow \{\text{Fitting class}\}$$

is called an ω-*local Hartley function* (or simply, an ω-local H-function). Write
$LR_\omega(f) = \{G | G^{\omega d} \in f(\omega') \text{ and } F^p(G) \in f(p), \text{ for all } p \in \omega \cap \pi(G)\}$, where $G^{\omega d}$
is the $\mathfrak{E}_{\omega d}$-residual of G. Then we call a Fitting class $\mathfrak{F}$ an ω-*local Fitting class* if
there exists an ω-local H-function f such that $\mathfrak{F} = LR_\omega(f)$.

Note that if $\omega = \mathbb{P}$, then the ω-local Fitting class is just a local Fitting class. It is
easy to see that all the classes $\mathfrak{E}_\pi, \mathfrak{S}_\pi, \mathfrak{N}_\pi, (1)$ are local Fitting classes.

Support that $\mathfrak{X}, \mathfrak{Y}$ be two sets of groups. We denote by $\text{Fit}(\mathfrak{X})$ the Fitting class
generated by $\mathfrak{X}$ and let $\mathfrak{X} \vee \mathfrak{Y} = \text{Fit}(\mathfrak{X} \cup \mathfrak{Y})$, that is, the Fitting class generated by
$\mathfrak{X} \cup \mathfrak{Y}$.

Recall that the product $\mathfrak{M} \diamond \mathfrak{H}$ of two Fitting classes is the class of all groups
G with $G/G_{\mathfrak{M}} \in \mathfrak{H}$. When $\mathfrak{H}$ is factor group closed, then $\mathfrak{F} \diamond \mathfrak{H} = (G :
G$ has a normal subgroup $N \in \mathfrak{F}$ with $G/N \in \mathfrak{H})$ (see [89, p. 566]); in this case, in
order to write concise, we use $\mathfrak{F}\mathfrak{H}$ to denote the product $\mathfrak{F} \diamond \mathfrak{H}$ of two Fitting classes
$\mathfrak{F}$ and $\mathfrak{H}$ (see § 5.6).

For a nonempty Fitting class $\mathfrak{F}$, let $\mathfrak{F}^*$ is the smallest Fitting class containing $\mathfrak{F}$
such that the $\mathfrak{F}^*$-radical of the direct product $G \times H$ of two groups G and H is equal
to the direct product of the $\mathfrak{F}^*$-radical of G and the $\mathfrak{F}^*$-radical of H; and let $\mathfrak{F}_*$ is the
intersection of all Fitting classes $\mathfrak{X}$ such that $\mathfrak{X}^* = \mathfrak{F}^*$ (see [300], [89, Chapter X]).

A Fitting class $\mathfrak{F}$ is said to be *normal* if the $\mathfrak{F}$-radical $G_{\mathfrak{F}}$ is an $\mathfrak{F}$-maximal subgroup
of G for every group G (see [274] and [275]). We call a Fitting class $\mathfrak{F}$ a *Lockett
class* if $\mathfrak{F} = \mathfrak{F}^*$. A Fitting class of soluble groups is called a *soluble Fitting class*. In
[428], it was proved that every local Fitting class is a Lockett class.

It was proved by Lockett [300] that in the class $\mathfrak{S}$ of finite soluble groups, the
following inclusions hold for any Fitting class $\mathfrak{F}$:

$$\mathfrak{F}_* \subseteq \mathfrak{F} \subseteq \mathfrak{F}^* \text{ and } \mathfrak{F}_* \subseteq \mathfrak{F}^* \cap \mathfrak{X} \subseteq \mathfrak{F}^*,$$

where $\mathfrak{X}$ is some normal Fitting class.

In connection with the above inclusions, Lockett [300] proposed the following
problem which is now known as the Lockett conjecture.

Lockett Conjecture [300, p. 135]. In the class $\mathfrak{S}$ of all soluble groups, for every
Fitting class $\mathfrak{F}$, there is a normal Fitting class $\mathfrak{X}$ such that $\mathfrak{F} = \mathfrak{F}^* \cap \mathfrak{X}$.

It was been proved that Lockett Conjecture holds for any soluble normal Fitting class $\mathfrak{F}$ (see [89, X; 3.7], every soluble local hereditary Fitting class [50, 57], the classes $\mathfrak{X}\mathfrak{N}$, $\mathfrak{X}\mathfrak{S}_\pi\mathfrak{S}_{\pi'}$, where $\mathfrak{X}$ is an arbitrarily nonempty Fitting class [46] ans every soluble local Fitting class [428]. Later on, Gallego [110] extended the Lockett conjecture to the class $\mathfrak{E}$ of all finite groups and he proved that if $\mathfrak{F}$ is a local Fitting class and $\mathfrak{F} \subseteq \mathfrak{E}$, then $\mathfrak{F}_* = \mathfrak{F}^* \cap \mathfrak{E}_*$.

However, Berger and Cossey [51] constructed an example which shows that there exists a soluble nonlocal Fitting class (which is a Lockett class) for which the Lockett conjecture does not hold. Therefore, the following two problems naturally arise:

Problem 9.1. Does there exist a soluble non-normal Fitting class which is not a Lockett class but for which the Lockett conjecture holds?

Problem 9.2. For every ω-local local Fitting class $\mathfrak{F}$, does the equality $\mathfrak{F}_* = \mathfrak{F}^* \cap \mathfrak{E}_*$ hold?

The main purpose of this section is to give the answers to the above two problems.

Lemma 9.3 ([300] and [89, X]). *Let $\mathfrak{F}$ and $\mathfrak{H}$ be two Fitting classes. Then:*

a) *If $\mathfrak{F} \subseteq \mathfrak{H}$, then $\mathfrak{F}^* \subseteq \mathfrak{H}^*$ and $\mathfrak{F}_* \subseteq \mathfrak{H}_*$;*
b) $(\mathfrak{F}_*)_* = \mathfrak{F}_* = (\mathfrak{F}^*)_* \subseteq \mathfrak{F} \subseteq \mathfrak{F}^* = (\mathfrak{F}_*)^* = (\mathfrak{F}^*)^*$;
c) $\mathfrak{F}^* \subseteq \mathfrak{F}_*\mathfrak{A}$, *where $\mathfrak{A}$ is the class of all abelian groups;*
d) *If $\{\mathfrak{F}_i \mid i \in I\}$ is a set of Fitting classes, then $(\cap_{i \in I} \mathfrak{F}_i)^* = \cap_{i \in I}\mathfrak{F}_i^*$.*
e) *If $\mathfrak{H}$ is a saturated radical homomorph, then $(\mathfrak{F}\mathfrak{H})^* = \mathfrak{F}^*\mathfrak{H}$.*
f) *If $\mathfrak{F}$ is a homomorph (in particular, a formation), then $\mathfrak{F}$ is a Lockett class.*

For a class $\mathfrak{X}$ of groups, let $\mathrm{Char}(\mathfrak{X}) = \{p \mid p \in \mathbb{P} \text{ and } Z_p \in \mathfrak{X}\}$, where $\mathbb{P}$ is the set of all prime numbers. We call $\mathrm{Char}(\mathfrak{X})$ the characteristic of $\mathfrak{X}$.

Lemma 9.4 ([89, X]). *(i) $\mathrm{Char}(\mathfrak{F}^*) = \mathrm{Char}(\mathfrak{F})$, for every Fitting class $\mathfrak{F}$;*
(ii) If $\mathfrak{F}$ is a solvable Fitting class, then: a) $\pi(\mathfrak{F}) = \mathrm{Char}(\mathfrak{F})$; b) $p \in \mathrm{Char}(\mathfrak{F})$ if and only if $\mathfrak{N}_p \subseteq \mathfrak{F}$.

Lemma 9.5 *Let $\mathfrak{F}$, $\mathfrak{X}$ and $\mathfrak{Y}$ be Fitting classes. Then $\mathfrak{F}\diamond(\mathfrak{X}\cap\mathfrak{Y}) = (\mathfrak{F}\diamond\mathfrak{X})\cap(\mathfrak{F}\diamond\mathfrak{Y})$.*

Proof The proof is obvious and we omit the details.

Lemma 9.6 [367]. *Let $\mathfrak{F}$ be a Fitting class. Then the following statements are equivalent.*

(a) $\mathfrak{F}(F^p)\mathfrak{N}_p \subseteq \mathfrak{F}$, *for all $p \in \omega$, where*

$$\mathfrak{F}(F^p) = \begin{cases} \mathrm{Fit}(F^p(G) \mid G \in \mathfrak{F}), & \text{if } p \in \pi(\mathfrak{F}); \\ \emptyset, & \text{if } p \in \pi'(\mathfrak{F}). \end{cases}$$

(b) $\mathfrak{F} = LR_\omega(f)$, *where $f(\omega') = \mathfrak{F}$ and $f(p) = \mathfrak{F}(F^p)\mathfrak{N}_p$, for all $p \in \omega$;*
(c) $\mathfrak{F}$ *is an ω-local Fitting class.*

Lemma 9.7 *If $\mathfrak{F}$ and $\mathfrak{H}$ are Fitting classes, then $(\mathfrak{F}_* \cap \mathfrak{H}_*)_* = (\mathfrak{F} \cap \mathfrak{H})_*$.*

Proof By Lemma 9.3(b), $\mathfrak{F}_* \subseteq \mathfrak{F}$ and $\mathfrak{H}_* \subseteq \mathfrak{H}$. Hence, $\mathfrak{F}_* \cap \mathfrak{H}_* \subseteq \mathfrak{F} \cap \mathfrak{H}$. Now, by Lemma 9.3(a), $(\mathfrak{F}_* \cap \mathfrak{H}_*)_* \subseteq (\mathfrak{F} \cap \mathfrak{H})_*$. On the other hand, by Lemma 9.3, $(\mathfrak{F} \cap \mathfrak{H})_* \subseteq \mathfrak{F}_*$ and $(\mathfrak{F} \cap \mathfrak{H})_* \subseteq \mathfrak{H}_*$. Hence, by Lemma 9.3 again,

$$(\mathfrak{F} \cap \mathfrak{H})_* = ((\mathfrak{F} \cap \mathfrak{H})_*)_* \subseteq (\mathfrak{F}_* \cap \mathfrak{H}_*)_*.$$

Thus, $(\mathfrak{F}_* \cap \mathfrak{H}_*)_* = (\mathfrak{F} \cap \mathfrak{H})_*$.

Lemma 9.8 ([110, Theorem 4.8(c)]). *Let $\mathfrak{F}$ be a Fitting class. If there exists a Fitting class $\mathfrak{H}$ such that $\mathfrak{H}\mathfrak{N}_p \subseteq \mathfrak{F} \subseteq \mathfrak{H}\mathfrak{N}_p\mathfrak{E}_{p'}$, for all $p \in \mathrm{Char}(\mathfrak{F})$, then $\mathfrak{F}$ is a Lockett class.*

Lemma 9.9 ([110, Theorem 4.11]). *Let $\mathfrak{F}$ be the Fitting class such that $\mathfrak{F} = (\mathfrak{F}_*\mathfrak{E}_{p'} \cap \mathfrak{F}) \vee \mathfrak{H}\mathfrak{N}_p$, for some Fitting class $\mathfrak{H}$. Then $\mathfrak{F} \cap \mathfrak{E}_* \subseteq \mathfrak{F}_*\mathfrak{E}_{p'}$.*

Lemma 9.10 ([431]). *Let $\mathfrak{F}$ be an ω-local Lockett class. Then $\mathfrak{F}$ can be defined by a largest inner ω-local H-function F with every value $F(p)$ (for every $p \in \mathbb{P}$) is a Lockett class, moreover, $F(\omega') = \mathfrak{F}$ and $F(p)\mathfrak{N}_p = F(p) \subseteq \mathfrak{F}$ for all $p \in \omega$.*

The following theorem gives a positive answer to Problem 9.1.

Theorem 9.11 (Guo, Shum, Vorob'ev [184, Theorem A]). *Let $\mathfrak{F}$ and $\mathfrak{H}$ be soluble non-normal Fitting classes and $\mathfrak{H}$ is a saturated formation. Suppose that $\mathfrak{F}$ and $\mathfrak{F}^*\mathfrak{H}$ satisfy Lockett conjecture. If $\mathfrak{F} \cap \mathfrak{H} = (1)$, then*

(a) $\mathfrak{F}_*\mathfrak{H}$ satisfies Lockett conjecture;
(b) $\mathfrak{F}_*\mathfrak{H}$ is an non-Lockett class (In particular, $\mathfrak{F}_*\mathfrak{H}$ is a nonlocal Fitting class) when $\mathfrak{N}_p\mathfrak{N}_q \subseteq \mathfrak{F}^*$, for some distinct primes p, q.

Proof

(a) Since $\mathfrak{F}^*\mathfrak{H}$ satisfies Lockett conjecture, $(\mathfrak{F}^*\mathfrak{H})_* = (\mathfrak{F}^*\mathfrak{H})^* \cap \mathfrak{S}_*$. However, since $\mathfrak{H}$ is a saturated radical homomorph, by Lemma 9.3(e), $(\mathfrak{F}^*\mathfrak{H})^* = (\mathfrak{F}^*)^*\mathfrak{H}$. Moreover, by Lemma 9.3(b), $(\mathfrak{F}^*)^* = \mathfrak{F}^*$. Hence,

$$(\mathfrak{F}^*\mathfrak{H})_* = \mathfrak{F}^*\mathfrak{H} \cap \mathfrak{S}_*. \tag{5.19}$$

Now, by Lemma 9.3(b) and Lemma 9.7,

$$(\mathfrak{F}^*\mathfrak{H})_* = ((\mathfrak{F}^*\mathfrak{H})_*)_* = ((\mathfrak{F}^*\mathfrak{H}) \cap \mathfrak{S}_*)_* = ((\mathfrak{F}^*\mathfrak{H})_* \cap (\mathfrak{S}_*\mathfrak{H})_*)_* = (\mathfrak{F}^*\mathfrak{H} \cap \mathfrak{S}_*\mathfrak{H})_*.$$

Since $\mathfrak{H}$ is a Fitting formation, it is easy to see that $\mathfrak{F}^*\mathfrak{H} \cap \mathfrak{S}_*\mathfrak{H} = (\mathfrak{F}^* \cap \mathfrak{S}_*)\mathfrak{H}$ (see also [428, Lemma 4]). Thus,

$$(\mathfrak{F}^*\mathfrak{H})_* = ((\mathfrak{F}^* \cap \mathfrak{S}_*)\mathfrak{H})_*. \tag{5.20}$$

Since $\mathfrak{F}$ satisfies Lockett conjecture, by (5.19) and (5.20), we obtain $(\mathfrak{F}_*\mathfrak{H})_* = \mathfrak{F}^*\mathfrak{H} \cap \mathfrak{S}_*$. Now, by Lemma 9.3, $(\mathfrak{F}_*\mathfrak{H})^* = (\mathfrak{F}_*)^*\mathfrak{H} = \mathfrak{F}^*\mathfrak{H}$. Thus, $\mathfrak{F}_*\mathfrak{H}$ satisfies Lockett conjecture.

(b) Assume that $\mathfrak{F}_*\mathfrak{H}$ is an Lockett class, this is, $\mathfrak{F}_*\mathfrak{H} = (\mathfrak{F}_*\mathfrak{H})^*$. Then, by Lemma 9.3, $\mathfrak{F}_*\mathfrak{H} = \mathfrak{F}^*\mathfrak{H}$. However, since $\mathfrak{F}$ satisfies Lockett conjecture and $\mathfrak{H}$ is a formation, $\mathfrak{F}^*\mathfrak{H} \cap \mathfrak{S}_*\mathfrak{H} = (\mathfrak{F}^* \cap \mathfrak{S}_*)\mathfrak{H} = \mathfrak{F}^*\mathfrak{H}$, and consequently $\mathfrak{F}^* \subseteq \mathfrak{F}^*\mathfrak{H} \subseteq \mathfrak{S}_*\mathfrak{H}$. On the other hand, by Lemma 9.3(c) and (5.19), we know that if $G \in \mathfrak{F}^*$, then $G/G_{\mathfrak{S}_*} \in \mathfrak{A} \cap \mathfrak{S}_{\pi(\mathfrak{F}^*)}$. But, by Lemma 9.4, $\mathfrak{A} \cap \mathfrak{S}_{\pi(\mathfrak{F}^*)} = \mathfrak{A} \cap \mathfrak{S}_{\pi(\mathfrak{F})} \subseteq \mathfrak{N}_{\pi(\mathfrak{F})} \subseteq \mathfrak{F}$. Hence, $\mathfrak{F}^* \subseteq \mathfrak{S}_*\mathfrak{F}$ and by Lemma 9.5,

$$\mathfrak{F}^* \subseteq \mathfrak{S}_*(\mathfrak{F} \cap \mathfrak{H}) = \mathfrak{S}_*.$$

However, by our hypothesis, we have $\mathfrak{N}_p\mathfrak{N}_q \subseteq \mathfrak{F}^*$. This contradicts [89, X; 5.32]. Thus, $\mathfrak{F}_*\mathfrak{H}$ is a non-Lockett class. Since every local Fitting class is a Lockett class, $\mathfrak{F}_*\mathfrak{H}$ is a nonlocal Fitting class. Therefore the theorem is proved.

Now we generalize Lockett conjecture from the class $\mathfrak{S}$ of all soluble group to the class $\mathfrak{G}$ of all groups. In the class $\mathfrak{G}$, we say a Fitting class $\mathfrak{F}$ satisfies Lockett conjecture if $\mathfrak{F}_* = \mathfrak{F}^* \cap \mathfrak{G}_*$.

Theorem 9.12 (Guo, Shum, Vorob'ev [184, Theorem B]). *Let $\mathfrak{F}$ be a ω-local Fitting class. If* Char$(\mathfrak{F}) \subseteq \omega$, *then $\mathfrak{F}$ satisfies Lockett conjecture.*

Proof Since $\mathfrak{F}$ is ω-local, by Lemma 9.6, $\mathfrak{F}(F^p) \subseteq \mathfrak{F}$, for all $p \in \omega$. Consequently, $\mathfrak{F}(F^p) \subseteq \mathfrak{F}$, for all $p \in$ Char$(\mathfrak{F})$. It is easy to verify that the ω-local Fitting class $\mathfrak{F}$ can be defined as follows:

$$\mathfrak{F} = (\cap_{p\in\pi_2} \mathfrak{G}_{p'}) \cap (\cap_{p\in\pi_1} f(p)\mathfrak{N}_p\mathfrak{G}_{p'}) \cap f(\omega')\mathfrak{G}_{\omega d}, \qquad (5.21)$$

where f is an ω-local function of $\mathfrak{F}$, $\pi_1 = \omega \cap Supp(f)$ and $\pi_2 = \omega \setminus \pi_1$ (see also [367]). It follows that $\mathfrak{F} \subseteq f(p)\mathfrak{N}_p\mathfrak{G}_{p'}$, for every $p \in \pi_1$. Now, by Lemma 9.6(b), $\mathfrak{F} \subseteq \mathfrak{F}(F^p)\mathfrak{N}_p\mathfrak{G}_{p'}$, for all $p \in Supp(f) \cap \omega$. Since Char$(\mathfrak{F}) \subseteq \omega$, for every $p \in$ Char$(\mathfrak{F})$, the following inclusion holds:

$$\mathfrak{F}(F^p)\mathfrak{N}_p \subseteq \mathfrak{F} \subseteq \mathfrak{F}(F^p)\mathfrak{N}_p\mathfrak{G}_{p'}.$$

Thus, by Lemma 9.8, $\mathfrak{F}$ is a Lockett class. By Lemma 9.10, $\mathfrak{F}$ can be defined by a largest inner ω-local function F with $F(p)\mathfrak{N}_p = F(p) \subseteq \mathfrak{F}$, for all $p \in \omega$. This shows that for all $p \in \omega$,

$$F(p)\mathfrak{N}_p \subseteq \mathfrak{F} \subseteq F(p)\mathfrak{G}_{p'} = F(p)\mathfrak{N}_p\mathfrak{G}_{p'}. \qquad (5.22)$$

Now, we claim that $\mathfrak{F} \subseteq \mathfrak{F}_*\mathfrak{G}_{p'}\mathfrak{N}_p$. In fact, since $\mathfrak{G}_{p'}\mathfrak{N}_p$ is a saturated Fitting formation, $(\mathfrak{F}_*\mathfrak{G}_{p'}\mathfrak{N}_p)^* = (\mathfrak{F}_*)^*\mathfrak{G}_{p'}\mathfrak{N}_p$ by Lemma 9.3(e). However, by Lemma 9.3(b), $(\mathfrak{F}_*)^* = \mathfrak{F}^*$. Hence, by Lemma 9.8, $(\mathfrak{F}_*\mathfrak{G}_{p'}\mathfrak{N}_p)^* = \mathfrak{F}\mathfrak{G}_{p'}\mathfrak{N}_p$. On the other hand, $\mathfrak{F}_*\mathfrak{G}_{p'}\mathfrak{N}_p$ is a local Fitting class (see [428, Corollary 1]), hence it is also a Lockett class. Thus, $\mathfrak{F}_*\mathfrak{G}_{p'}\mathfrak{N}_p = \mathfrak{F}\mathfrak{G}_{p'}\mathfrak{N}_p$ and so $\mathfrak{F} \subseteq \mathfrak{F}_*\mathfrak{G}_{p'}\mathfrak{N}_p$. Hence our claim holds. This implies that for all $G \in \mathfrak{F}$, $G/G_{\mathfrak{F}_*\mathfrak{G}_{p'}} \in \mathfrak{N}_p$. Moreover, by (5.22), $G/G_{F(p)} \in \mathfrak{G}_{p'}$, for all $G \in \mathfrak{F}$. Obviously, if $G \in \mathfrak{F}$, then $G_{\mathfrak{F}_*\mathfrak{G}_{p'}} \in \mathfrak{F}$ and so $G_{\mathfrak{F}_*\mathfrak{G}_{p'}} = G_{\mathfrak{F}_*\mathfrak{G}_{p'}\cap\mathfrak{F}}$. Thus,

$$(G/G_{\mathfrak{F}_*\mathfrak{G}_{p'}})/(G_{\mathfrak{F}_*\mathfrak{G}_{p'}\cap\mathfrak{F}}G_{F(p)}/G_{\mathfrak{F}_*\mathfrak{G}_{p'}}) \simeq G/G_{\mathfrak{F}_*\mathfrak{G}_{p'}\cap\mathfrak{F}}G_{F(p)} \in \mathfrak{N}_p$$

and

$$(G/G_{F(p)})/(G_{\mathfrak{F}_*\mathfrak{G}_{p'}\cap\mathfrak{F}}G_{F(p)}/G_{F(p)}) \simeq G/G_{\mathfrak{F}_*\mathfrak{G}_{p'}\cap\mathfrak{F}}G_{F(p)} \in \mathfrak{G}_{p'}.$$

This induces that $G = G_{\mathfrak{F}_*\mathfrak{G}_{p'}\cap\mathfrak{F}}G_{F(p)}$. Obviously, $G_{\mathfrak{F}_*\mathfrak{G}_{p'}\cap\mathfrak{F}} \in (\mathfrak{F}_*\mathfrak{G}_{p'} \cap \mathfrak{F}) \vee F(p)$ and $G_{F(p)} \in (\mathfrak{F}_*\mathfrak{G}_{p'} \cap \mathfrak{F}) \vee F(p)$. Therefore, $G \in (\mathfrak{F}_*\mathfrak{G}_{p'} \cap \mathfrak{F}) \vee F(p)$ and so $\mathfrak{F} \subseteq$

$(\mathfrak{F}_*\mathfrak{G}_{p'}\cap\mathfrak{F})\vee F(p)$. The reverse inclusion is obvious. Thus, $\mathfrak{F}=(\mathfrak{F}_*\mathfrak{G}_{p'}\cap\mathfrak{F})\vee F(p)$.
It follows from Lemma 9.9 that

$$\mathfrak{F}\cap\mathfrak{G}_* \subseteq \mathfrak{F}_*\mathfrak{G}_{p'}. \tag{5.23}$$

We now prove that $\mathfrak{F}\cap\mathfrak{G}_* = \mathfrak{F}_*$. In fact, by Lemma 9.3, $\mathfrak{F}_* \subseteq \mathfrak{G}_*$ and $\mathfrak{F}_* \subseteq \mathfrak{F}$,
and consequently, $\mathfrak{F}_* \subseteq \mathfrak{F}\cap\mathfrak{G}_*$. Assume that $\mathfrak{F}\cap\mathfrak{G}_* \not\subseteq \mathfrak{F}_*$ and let G be a group
in $(\mathfrak{F}\cap\mathfrak{G}_*)\setminus\mathfrak{F}_*$ of minimal order. Then, G has a unique maximal normal subgroup
M such that $M = G_{\mathfrak{F}_*}$. Since $G\in\mathfrak{F}$, by Lemma 9.3(c), $G/M\in\mathfrak{A}$ and thereby,
G/M is a composition factor of order p. Hence, by [89, IX;1.7], $p\in Char(\mathfrak{F})$.
However, $G/M\in\mathfrak{G}_{p'}$ by (5.23), thus, $G/M\in\mathfrak{N}_p\cap\mathfrak{G}_{p'}=(1)$ and so $G=M$.
This contradiction shows that $\mathfrak{F}\cap\mathfrak{G}_* \subseteq \mathfrak{F}_*$. Thus $\mathfrak{F}_* = \mathfrak{F}\cap\mathfrak{G}_*$. Now, by Lemma
9.8 and (5.22), we have $\mathfrak{F} = \mathfrak{F}^*$. This induces that $\mathfrak{F}_* = \mathfrak{F}^*\cap\mathfrak{G}_*$ and so $\mathfrak{F}$ satisfies
Lockett conjecture. This completes the proof.

If we let $\omega = \mathbb{P}$, then, by Theorem 9.12, we immediately re-obtain the following
result.

Corollary 9.13 (Gallego [110]). *Every local Fitting class satisfies Lockett conjecture.*

If $\mathfrak{F} = \mathfrak{G}$ and $\omega = \mathbb{P}$, then by Theorem 9.12, we reobtain the following Berger's
result which positively answers the Laue's problem (see [274, Problem II]).

Corollary 9.14 ([50] and [89, X; 6.15]). $\mathfrak{G}_* = \mathfrak{G}^*\cap\mathfrak{G}_* = \mathfrak{G}\cap\mathfrak{G}_*$.

In Kourovka Notebook [317], Lausch proposed the following problem:

Lausch's Problem 9.15 ([317, Problem 8.30]). Let $\mathfrak{F}, \mathfrak{H}$ be a solvable Fitting class
such that $\mathfrak{F}\cap\mathfrak{G}_* = \mathfrak{F}_*$ and $\mathfrak{H}\cap\mathfrak{G}_* = \mathfrak{H}_*$. Will the equality $(\mathfrak{F}\cap\mathfrak{H})\cap\mathfrak{G}_* = (\mathfrak{F}\cap\mathfrak{H})_*$
hold?

For solvable local Fitting classes $\mathfrak{F}, \mathfrak{H}$, Vorob'ev in [428] has given an affirmative
answer to the Lausch's problem. Our following corollary gives an affirmative answer
to the Lausch's problem in the class $\mathfrak{G}$ of all groups (particularly, in $\mathfrak{S}$) when $\mathfrak{F}, \mathfrak{H}$
are ω-local Fitting classes and $Char(\mathfrak{F}\cap\mathfrak{H})\subseteq\omega$ (particularly, when $\mathfrak{F}, \mathfrak{H}$ are local
Fitting classes).

Corollary 9.16 *Let $\mathfrak{F}, \mathfrak{H}$ be two ω-local Fitting classes satisfying Lockett conjecture
and their characteristics be subsets of ω. Then $\mathfrak{F}\cap\mathfrak{H}$ satisfies Lockett conjecture and
$(\mathfrak{F}\cap\mathfrak{H})_* = (\mathfrak{F}\cap\mathfrak{H})\cap\mathfrak{G}_*$.*

Proof Obviously, $\mathfrak{F}\cap\mathfrak{H}$ is a ω-local Fitting class and $Char(\mathfrak{F}\cap\mathfrak{H})\subseteq\omega$. By Theorem
9.12, $\mathfrak{F}\cap\mathfrak{H}$ satisfies Lockett conjecture. By using the same arguments as in Theorem
9.12, we also see that $\mathfrak{F}\cap\mathfrak{H}$ is a Lockett class, that is, $(\mathfrak{F}\cap\mathfrak{H})^* = \mathfrak{F}\cap\mathfrak{H}$. Therefore,
$(\mathfrak{F}\cap\mathfrak{H})_* = (\mathfrak{F}\cap\mathfrak{H})\cap\mathfrak{G}_*$.

The following shows that there exists an ω-local Fitting class $\mathfrak{F}$ such that
$Char(\mathfrak{F})\subseteq\omega$ but $\mathfrak{F}$ is not a local Fitting class.

Example 9.17 Let $\mathfrak{F} = \mathfrak{H}\mathfrak{N}_p$, where $\mathfrak{H} = Fit A$ is a Fitting class generated by some
finite simple nonabelian group A. Then, $Char(\mathfrak{H}) = \emptyset\subset\pi(\mathfrak{H})$ (see [89, Exercise

IX, §1.4]). We claim that $\mathfrak{F}$ is ω-local, for $\omega = \{p\}$. Indeed, let

$$\mathfrak{F}(F^p) = \begin{cases} Fit(F^p(G) \mid G \in \mathfrak{F}), & \text{if } p \in \pi(\mathfrak{F}); \\ \emptyset, & \text{if } p \in \pi'(\mathfrak{F}). \end{cases}$$

Then $\mathfrak{F}(F^p) \subseteq \mathfrak{H}$ and hence $\mathfrak{F}(F^p)\mathfrak{N}_p \subseteq \mathfrak{H}\mathfrak{N}_p = \mathfrak{F}$. This leads to $\mathfrak{F}$ is ω-local and so our claim is established. Clearly, $\pi(\mathfrak{F}) = \pi(\mathfrak{H}) \cup \{p\}$ and $\text{Char}(\mathfrak{F}) = \{p\}$. Hence, $\text{Char}(\mathfrak{F}) \neq \pi(\mathfrak{F})$. However, by [110, 4.9(b)], $\mathfrak{F}$ is local if and only if $\text{Char}(\mathfrak{F}) = \pi(\mathfrak{F})$ and $\mathfrak{F}(F^p)\mathfrak{N}_p \subseteq \mathfrak{F} \subseteq \mathfrak{F}(F^p)\mathfrak{N}_p\mathfrak{G}_{p'}$. This shows that $\mathfrak{F}$ is not a local Fitting class.

The following example shows that Theorem 9.12 is not true without the condition "$\text{Char}(\mathfrak{F}) \subseteq \omega$."

Example 9.18 Let R be an extra-special (see [89, A; 20.3]) group of order 27 with exponent 3 and $M = PSL(2,3)$. Then, by [89, B; 9.16], M has a faithful irreducible R-module W over the field $GF(7)$. Let $Y = W \rtimes R$. Denote by A the automorphism group of R. Let $B = C_A(Z(R))$ and Q the quaternion subgroups of B and $X = Z(Q)Y$, respectively. Let

$$\mathfrak{M} = (G \mid O^{2'}(G/O_{\{2,3\}}(G)) \in S_n D_0(X)),$$

where $D_0(X)$ is the class of all finite direct products of the groups which is isomorphic to X. By a result in [51], $\mathfrak{K} = \mathfrak{M} \cap \mathfrak{G}_7\mathfrak{G}_3\mathfrak{G}_2$ is a Lockett class but for which Lockett conjecture does not hold, that is, $\mathfrak{K}^* = \mathfrak{K}$ and $\mathfrak{K}_* \neq \mathfrak{K}^* \cap \mathfrak{G}_*$. Let $\mathfrak{F} = \{\mathfrak{F}_i \mid i \in I\}$ be the family of all solvable Fitting classes satisfying the following conditions:

a) For every $i \in I$, the product $(\mathfrak{K})_*\mathfrak{F}_i$ is an ω_i-local Fitting class, for some $\emptyset \neq \omega_i \subseteq \mathbb{P}$;
b) $\mathfrak{F}_i \cap \mathfrak{F}_j = (1)$, for all $i, j \in I$ with $i \neq j$;
c) $\mathfrak{F}_i$ is a radical saturated homomorph, for all $i \in I$.

Obviously, $\mathfrak{F} \neq \emptyset$ (for example, we can take $\mathfrak{F}_i = \mathfrak{N}_{p_i}$, for all $p_i \in \mathbb{P}$). Assume that $\mathfrak{K}_i = (\mathfrak{K})_*\mathfrak{F}_i$ satisfies Lockett conjecture, for all $i \in I$. Then, for every $i \in I$, $((\mathfrak{K})_*\mathfrak{F}_i)_* = (\mathfrak{K}_*\mathfrak{F}_i)^* \cap \mathfrak{G}_*$. Since $\mathfrak{K}$ is a Lockett class, by Lemma 9.3, we see that $(\mathfrak{K}_*\mathfrak{F}_i)^* = (\mathfrak{K}_*)^*\mathfrak{F}_i = \mathfrak{K}^*\mathfrak{F}_i = \mathfrak{K}\mathfrak{F}_i$ and hence

$$\cap_{i \in I}(\mathfrak{K}_*\mathfrak{F}_i)_* = \cap_{i \in I}(\mathfrak{K}\mathfrak{F}_i \cap \mathfrak{G}_*) = (\cap_{i \in I}\mathfrak{K}\mathfrak{F}_i) \cap \mathfrak{G}_* = \mathfrak{K}(\cap_{i \in I}\mathfrak{F}_i) \cap \mathfrak{G}_* = \mathfrak{K} \cap \mathfrak{G}_*.$$

By Lemma 9.3, $(\mathfrak{K}_*\mathfrak{F}_i)_* \subseteq \mathfrak{K}_*\mathfrak{F}_i$, for all $i \in I$. Hence, $\cap_{i \in I}(\mathfrak{K}_*\mathfrak{F}_i)_* \subseteq \mathfrak{K}_*$. On the other hand, since $\mathfrak{K}_* \subseteq \mathfrak{K}_*\mathfrak{F}_i$ for every $i \in I$, $\mathfrak{K}_* = (\mathfrak{K}_*)_* \subseteq \cap_{i \in I}(\mathfrak{K}_*\mathfrak{F}_i)_*$ by Lemma 9.3. Thus $\cap_{i \in I}(\mathfrak{K}_*\mathfrak{F}_i)_* = \mathfrak{K}_*$. It follows that $\mathfrak{K}_* = \mathfrak{K} \cap \mathfrak{G}_*$, a contradiction. Thus, there exists $i_0 \in I$ such that $\mathfrak{K}_{i_0}$ is an ω_{i_0}-local Fitting class which does not satisfy Lockett conjecture. By the condition b), it is clear that $Char(\mathfrak{R}_{i_0}) \nsubseteq \omega_{i_0}$.

Example 9.18 illustrates that, in general case, the answer to Problem 9.2 is negative.

5.10   A Problem Concerning the $\mathfrak{F}$-Residuals of Groups

For the $\mathfrak{F}$-residuals of finite groups, it is well known that if $\mathfrak{F}$ is a hereditary formation, then $(A \times B)^{\mathfrak{F}} = A^{\mathfrak{F}} \times B^{\mathfrak{F}}$, for any groups A and B. It has also been proved by K. Doerk and T. Hawkes [88] that if a formation $\mathfrak{F}$ is soluble or saturated, then $(A \times B)^{\mathfrak{F}} = A^{\mathfrak{F}} \times B^{\mathfrak{F}}$ for any groups A and B. They also constructed an example to show that this result is generally not true for nonsoluble formations $\mathfrak{F}$. Thus, if a formation $\mathfrak{F}$ is hereditary, or soluble, or saturated, then the class $(G^{\mathfrak{F}}|G$ is a group) is closed under direct products.

In this connection, the following question naturally arises:

Shemetkov Problem 10.1. Let $\mathfrak{F}$ be a soluble, or saturated, or hereditary formation. Is it true that the class of $\mathfrak{F}$-residuals $(G^{\mathfrak{F}}|G$ is a group) is still closed under subdirect products?

This problem was proposed by L. A. Shemetkov in [363]. If this question of L. A. Shemetkov can be answered affirmatively, then we can further improve the theorem of Doerk and Hawkes in [88].

In this section, we prove that there exists a soluble, saturated and hereditary formation $\mathfrak{F}$ of groups such that the class $(G^{\mathfrak{F}}|G$ is a group) is not closed under subdirect products. Thus, a negative answer to the problem of L. A. Shemetkov is obtained. This means that the theorem of Doerk and Hawkes for $\mathfrak{F}$-residual of groups can not be further sharpened to subdirect products.

In order to prove the result, we need some well-known results on modules. Let F be a field, G a group and M a FG-module. Denote the radical of M by Rad(M), that is, Rad(M) is the intersection of all maximal submodules of M. By the socle soc(M) of M, we mean the sum of all minimal submodules of M. If q is a prime, then we use F_q to denote the field with q elements. An FG-module V is called a trivial FG-module if $C_G(V) = G$. We now cite the following useful results from the literature.

Lemma 10.2   (see [89, B, 3.14]). *Let H be a group and M an F_pH-module for some prime p. Then* Rad(M) $= \Phi(M \rtimes H) \cap M$.

Lemma 10.3   (see [89, B, 4.6]). *Every simple KG-module V is isomorphic with $P/$Rad(P) for some indecomposable projective KG-module P.*

Lemma 10.4   (see [89, B. 4.10]). *Every indecomposable projective FG-module V has an unique minimal submodule which is isomorphic to $V/$Rad(V).*

Lemma 10.5   (see [451] or [135, Theorem 5.3.7]). *Let G be a group, p be a prime. If V is a trivial simple F_pG-module and P_V is the projective envelope of V, then*

$$C_G(P_V) = O_{p'}(G).$$

Theorem 10.6   (Guo, Shum, Skiba [172]). *Let $\mathfrak{F} = \mathfrak{FF}$ be a soluble nonidentity formation. Then the class $(G^{\mathfrak{F}}|G$ is a group) is not closed under subdirect products.*

Proof Let Z_p be a group of prime order p in $\mathfrak{F}$ and B a simple group such that $B \notin \mathfrak{F}$. Then, by Chap. 1, Lemma 5.9, there exists a simple $F_p B$-module M_1 such that $C_B(M_1) = 1$. Let $A = M_1 \rtimes B$. Then M_1 is the only minimal normal subgroup of A and hence $O_{p'}(A) = 1$. If V is a trivial simple $F_p A$-module, then by Lemma 10.3, there exists an indecomposable projective $F_p A$-module P such that $P/\mathrm{Rad}(P)$ is isomorphic to V. Let $R = \mathrm{Rad}(P)$ and $G = P \rtimes A$. Then, by Lemma 10.2, $R = \Phi(G) \cap P$. Assume that $R = 1$. Then P is a trivial $F_p A$-module, that is, $C_A(P) = A$. However, by Lemma 10.5, we have $C_A(P) = O_{p'}(A) = 1$. This is a contradiction, and hence $R \neq 1$. Let $L = \mathrm{soc}(P)$. We now show that L is the only minimal normal subgroup of G. In fact, by Lemma 10.4, L is the unique minimal submodule of P. Because $C_A(P) = A$, we know that L is the only minimal normal subgroup of G contained in P. Assume that there exists a minimal normal subgroup D of G such that $D \nsubseteq P$. Then $D \subseteq C_G(P)$. Since P is an elementary abelian group, $P \subseteq C_G(P)$ and by Dedekind's modular law, we derive that $C_G(P) = C_G(P) \cap PA = P(C_G(P) \cap A) = PC_A(P) = PA = P$. This contradiction shows that L is the only minimal normal subgroup of G.

Now, let $\mathfrak{N}_p$ be a class of all p-groups. We claim that $\mathfrak{N}_p \mathfrak{F} = \mathfrak{F}$. To prove our claim, we first suppose that $\mathfrak{N}_p \mathfrak{F} \nsubseteq \mathfrak{F}$. Then, we can let E be a group of minimal order in $\mathfrak{N}_p \mathfrak{F} \backslash \mathfrak{F}$. Since $\mathfrak{F}$ is a formation, E has an unique minimal normal subgroup $X = E^{\mathfrak{F}}$. Because $E \in \mathfrak{N}_p \mathfrak{F}, X \in \mathfrak{N}_p$. Hence $X = P_1 \times P_2 \times \cdots \times P_t$ where $P_1 \simeq P_2 \simeq \cdots \simeq P_t \simeq Z_p \in \mathfrak{F}$. It follows that $X \in \mathfrak{F}$. But $\mathfrak{F} = \mathfrak{F}\mathfrak{F}$ and $X = E^{\mathfrak{F}} \in \mathfrak{F}$, we have $E \in \mathfrak{F}$. This contradiction shows that $\mathfrak{N}_p \mathfrak{F} \subseteq \mathfrak{F}$. Since $\mathfrak{F} \subseteq \mathfrak{N}_p \mathfrak{F}$ always, we establish our claim $\mathfrak{F} = \mathfrak{N}_p \mathfrak{F}$. Now, let M be a group and $T = (M^{\mathfrak{F}})^{\mathfrak{N}_p}$. In particular, $T \subseteq M^{\mathfrak{F}}$. Since T is a characteristic subgroup of $M^{\mathfrak{F}}$ and $M^{\mathfrak{F}} \trianglelefteq M$, we can easily derive that T is normal in M. However, by Chap. 1, Lemma 1.1, we have $M^{\mathfrak{F}}/T = (M/T)^{\mathfrak{F}} \in \mathfrak{N}_p$. This leads to $M/T \in \mathfrak{N}_p \mathfrak{F} = \mathfrak{F}$ and hence $T = M^{\mathfrak{F}}$, a contradiction. Thus, $(M^{\mathfrak{F}})^{\mathfrak{N}_p} = M^{\mathfrak{F}}$ for every group M, and so $M^{\mathfrak{F}} \subseteq T$. Therefore $T = M^{\mathfrak{F}}$.

Let $T = G_1 \times G_2$, where $G_1 \simeq G_2 \simeq G$, and let $\varphi : G_1 \rightarrow G_2$ be an isomorphism of the groups G_1 onto G_2. Then, by the theorem of Doerk–Hawkes in [88], we have $T^{\mathfrak{F}} = G_1^{\mathfrak{F}} \times G_2^{\mathfrak{F}}$. Let H be the diagonal of $G_1^{\mathfrak{F}} \times G_2^{\mathfrak{F}}$, i.e., $H = \{(a, a^{\varphi}) | a \in G_1^{\mathfrak{F}}\}$. Then, H is clearly a subdirect product of the groups $G_1^{\mathfrak{F}}$ and $G_2^{\mathfrak{F}}$. Since $A \simeq G/P$ and $A/M_1 \simeq B \notin \mathfrak{F}$, we have $G/P \notin \mathfrak{F}$ and $G^{\mathfrak{F}} \neq 1$. It follows that $L \subseteq G^{\mathfrak{F}}$. Since $L \simeq P/R$, by Lemma 10.4 and $P/R \simeq V$, we know that L is a trivial $F_p A$-module. Hence $L \subseteq Z(G)$. Let L_1 be a subgroup of G_1 such that $L_1 \subseteq Z(G_1)$ and $L_1 \simeq L$. Then, we can easily see that $L_1 \subseteq Z(G_1 \times G_2)$. Let $F = L_1 H$. Since $L_1 \cap H = 1$, we have $F = L_1 \times H$. It is clear that F is a subdirect product of the groups $G_1^{\mathfrak{F}}$ and $G_2^{\mathfrak{F}}$. Assume that there exists a group X such that $F = X^{\mathfrak{F}}$. Then, by using the above result, we have $F^{\mathfrak{N}_p} = F$. Since $F/H \simeq L \in \mathfrak{N}_p$, $F^{\mathfrak{N}_p} \subseteq H$ and hence $H = F$. This contradiction shows that $F \neq X^{\mathfrak{F}}$ for every group X. In other words, the class $(G^{\mathfrak{F}} | G$ is a group$)$ is not closed under subdirect products.

Theorem 10.7 (Guo, Shum, Skiba [172]). *Let $\mathfrak{F} = \mathfrak{F}\mathfrak{F}$ be a nonidentity soluble formation. If $\mathfrak{F}$ is not the class of all soluble groups, then the class $(G^{\mathfrak{F}} | G$ is a soluble group$)$ is not closed under subdirect products.*

Proof Assume $Z_p \in \mathfrak{F}$ for every prime p, where Z_p is a group of order p. Then $\mathfrak{N}_p\mathfrak{F} = \mathfrak{F}$ for all primes p (see the proof of Theorem 10.6). We now claim that $\mathfrak{F}$ is the class of all soluble groups. In fact, if it is not, then we can let G be a group of minimal order in $\mathfrak{S}\backslash\mathfrak{F}$, where $\mathfrak{S}$ is the class of all soluble groups. In this case, $G^{\mathfrak{F}}$ is the unique minimal normal subgroup of G and clearly, $G^{\mathfrak{F}} \in \mathfrak{N}_p$ for some prime p. It follows that $G \in \mathfrak{N}_p\mathfrak{F} = \mathfrak{F}$, a contradiction. This proves our claim. However, this contradicts our assumption that $\mathfrak{F}$ is not the class of all soluble groups. Therefore there exist groups Z_p and Z_q of prime orders p and q (where $p \neq q$), respectively, such that $Z_p \in \mathfrak{F}$ and $Z_q \notin \mathfrak{F}$. Now, the theorem can be proved by using the same arguments as Theorem 10.6.

In summing up Theorems 10.6 and 10.7, we obtain the following corollary.

Corollary 10.8 *There exists a saturated, soluble and hereditary formation $\mathfrak{F}$ such that the class $(G^{\mathfrak{F}}|G$ is a group) is not closed under subdirect products.*

Proof Let π be a set of primes and $\mathfrak{F}$ the class of all soluble π-groups. Then it is obvious that the class $\mathfrak{F}$ is a hereditary, soluble and saturated formation such that $\mathfrak{F} = \mathfrak{F}\mathfrak{F}$. Hence, by Theorem 10.6, the class $(G^{\mathfrak{F}}|G$ is a group) is not closed under subdirect products.

In view of Corollary 10.8, a negative answer to the problem of L. A. Shemetkov is given.

5.11 Additional Information and Some Problems

I. Results of Sect. 2 force us to the following questions.

Problem 11.1. Let $\mathfrak{M}\mathfrak{H} \subseteq \mathfrak{F}$ be the product of the formations $\mathfrak{M}$ and $\mathfrak{H}$, where $\mathfrak{F}$ is a one generated saturated formation. What we can say on the properties of $\mathfrak{M}$ and $\mathfrak{H}$? In particular, does it true that $\mathfrak{M}$ is nilpotent, metanilpotent or, at least, soluble if $\mathfrak{F}$ is nilpotent, metanilpotent, or soluble?

Problem 11.2. Let $\mathfrak{M} \diamond \mathfrak{H}$ be a nonidentity one-generated nilpotent Fitting class. Does it true then that $\mathfrak{H}$ is nilpotent?

Problem 11.3. Let $\mathfrak{M} \diamond \mathfrak{H}$ be a nonidentity one-generated local Fitting class. Does it true then that $\mathfrak{H}$ is metanilpotent?

II. In the paper [29], the theory of one-generated formations was developed for $\mathfrak{X}$-local formations (see [89, p. 374]).

III. The concept of multiply local formation was introduced and applied in the paper of Skiba [378].

IV. Following Anderson (see [89, VIII, 2.1]), a nonempty set $\mathcal{F}$ of subgroups of a group G is called a Fitting set, if the following conditions are satisfied: (i) if $T \; sn \; S \in \mathcal{F}$, then $T \in \mathcal{F}$; (ii) if $S, \; T \in \mathcal{F}$ and $S, \; T \trianglelefteq ST$, then $ST \in \mathcal{F}$; (iii) if $S \in \mathcal{F}$ and $x \in G$, then $S^x \in \mathcal{F}$.

Let $\mathcal{F}$ be a Fitting set of a group G and H a subgroup of G. Set

$$\mathcal{F}_H = \{S \leq H \mid S \in \mathcal{F}\}.$$

Then $\mathcal{F}_H$ is a Fitting set of H. Where, there is no danger of confusion, $\mathcal{F}_H$ is usually denoted by $\mathcal{F}$ (see [89, VIII, 2.3]).

It is easy to see that, for a given group, all results in Theorems 6.25–6.29 are also true when the condition "$\mathfrak{F}$ is a nonempty Fitting class" is replaced by "$\mathcal{F}$ is a Fitting set of a group G." Therefore, we have the following two theorems.

Theorem 11.4 (Guo, Li [152]). *Suppose that every chief factor of a group G is an $E_\pi^{\mathfrak{N}}$-group and $\mathcal{F}$ is a Fitting set of G with $\pi(\mathcal{F}) \subseteq \pi$. Then G has an $\mathcal{F}$-injector and any two $\mathcal{F}$-injectors are conjugate in G.*

Theorem 11.5 (Guo, Li [152]). *Suppose that every chief factor of a group G is an $E_\pi^{\mathfrak{N}}$-group and $\mathcal{F}$ is a Fitting set of G with $\pi(\mathcal{F}) \subseteq \pi$. Let V be an $\mathcal{F}$-injector of G. If $V \leq A \leq G$ and A possesses a Hall π-subgroup, then V is an $\mathcal{F}$-injector of A.*

V. An analog of Theorem 6.34 is also true for Schunck classes. Recall that a class of groups is said to be a Schunck class if $\mathfrak{F}$ is a homomorph and $\mathfrak{F}$ is primitively closed (see [135, Definition 2.3.1] or [89, p. 272]).

Theorem 11.6 (see [97] and [366, Theorems 11.22 and 11.23]). *Let $\mathfrak{H}$ be a nonempty Schunck class. Then the following statements hold:*
(1) If $G = HF(G)$ and $H \in \mathfrak{H}$, then H is contained in an $\mathfrak{H}$-covering subgroup of G;
(2) If K is a $\pi(\mathfrak{H})$-soluble normal subgroup of G such that $G/K \in \mathfrak{H}$, then G possesses exactly one conjugacy class of $\mathfrak{H}$-covering subgroups.

Using Theorem 11.6, we deduce the following result.

Theorem 11.7 (Guo, Al-Sharo, Shemetkov [144]). *Let $\mathfrak{H}$ be a nonempty Schunck class, $\pi = \pi(\mathfrak{H})$. Let K be a normal subgroup of G such that every G-chief factor of K is E_π^n-selected. Assume that G/K possesses exactly one conjugacy class of $\mathfrak{H}$-covering subgroups. Then G possesses exactly one conjugacy class of $\mathfrak{H}$-covering subgroups.*

The proof of Theorem 11.7 is similar to the proof of Theorem 6.34.

VI. In connection with Theorems 6.18, the following problems seam to be interesting:

Problem 11.8. Let $\mathfrak{F}$ be a Fitting class, $\pi = \pi(\mathfrak{F})$ and G a group. Suppose that $G/G_{\mathfrak{F}}$ is a π-separated group. Does it true that G has an $\mathfrak{F}$-injector and any two $\mathfrak{F}$-injectors of G are conjugate in G?

Problem 11.9. Let $\mathfrak{F}$ be a nonempty saturated formation and G a group. Assume that $G^{\mathfrak{F}}$ is $\pi(\mathfrak{F})$-separated. Does it true then that G possesses exactly one conjugacy class of $\mathfrak{F}$-covering subgroups?

VII. In [153], it was shown that the condition "local" in Theorem 7.8 is essential.

VIII. Theorem 7.8 allows us to describe the $\mathfrak{F}$-radical of a Hall π-subgroup of a π-soluble group.

Theorem 11.10 (Guo, Li [153, Theorem 4.1]). *Let $\mathfrak{F}$ be a local Fitting class defined an full integrated H-function F and Φ be a full integrated H-function of the local Fitting class $K_\pi(\mathfrak{F})$. Then, for every π-soluble group G and its Hall π-subgroup H, the following statements hold:*

(a) If $\sigma = Supp(F) = \{p : p \in \mathbb{P} \text{ and } \emptyset \neq F(p) \neq \mathfrak{F}\}$, then $G_{\mathfrak{F}} = G_F$;
(b) $H_{\mathfrak{F}} = G_\Phi \cap H$, where $G_\Phi = \Pi_{p \in \pi(G) \cap \sigma} G_{\Phi(p)}$.

Corollary 11.11 ([153]). *Let $\mathfrak{F} = LR(F)$ for a full integrated H-function F and $\mathfrak{F} \supseteq \mathfrak{N}$. Let H be a Hall π-subgroup of a group G. Then $H_{\mathfrak{F}} = \Pi_{p \in \pi} H_{F(p)}$.*

IX. In the early 60s, the factorization of classes of groups has been studied by Neumann [322] and Shmel'kin [372, 373], respectively. They proved the following interesting theorem for varieties of groups.

Theorem 11.12 (Neumann, Shmel'kin). *(1) If $\mathfrak{M}$ is a variety of groups which is distinct from the variety which consists of all groups, then there exist indecomposable varieties $\mathfrak{M}_1, \mathfrak{M}_2, \cdots, \mathfrak{M}_t$ of groups such that $\mathfrak{M} = \mathfrak{M}_1 \mathfrak{M}_2 \cdots \mathfrak{M}_t$.*
    (2) If $\mathfrak{M}_1 \mathfrak{M}_2 \cdots \mathfrak{M}_t = \mathfrak{F}_1 \mathfrak{F}_2 \cdots \mathfrak{F}_n$, where $\mathfrak{M}_1, \mathfrak{M}_2, \cdots \mathfrak{M}_t, \mathfrak{F}_1, \mathfrak{F}_2, \cdots, \mathfrak{F}_n$ are indecomposable varieties of groups which are all distinct from the variety which consists of all groups, then $t = n$ and $\mathfrak{M}_i = \mathfrak{F}_i$ for every $i \in \{1, 2, \cdots, t\}$.

In the light of with the above theorems of Neumann and Shmel'kin, the following problem concerning the factorization of formations was proposed by Shemetkov and Skiba [367] and also Guo [135].

Problem ([367, Problem 18, p. 22] and [135, Problem 16]). Let $\mathfrak{F}_1 \mathfrak{F}_2 \cdots \mathfrak{F}_t = \mathfrak{M}_1 \mathfrak{M}_2 \cdots \mathfrak{M}_n$, where $\mathfrak{F}_1, \mathfrak{F}_2, \cdots, \mathfrak{F}_t, \mathfrak{M}_1, \mathfrak{M}_2, \cdots \mathfrak{M}_n$ are s-indecomposable s-closed ω-saturated formations. Is it true that $n = t$ and $\mathfrak{F}_i = \mathfrak{M}_i$, for all $i = 1, 2, \cdots, t$?

Note that a formation $\mathfrak{F}$ is said to be s-indecomposable if $\mathfrak{F}$ can not be expressed by $\mathfrak{F} = \mathfrak{M}\mathfrak{H}$, where $\mathfrak{M}$ and $\mathfrak{H}$ are subgroup closed formations which are distinct from the class (1) of identity groups.

In [169], Guo and Shum have given the affirmative answer to above problem.

X. It is known also the following

Problem 11.13. Let $\mathfrak{R} = \mathfrak{M}\mathfrak{H}$ be the product of the formations $\mathfrak{M}$ and $\mathfrak{H}$ and this factorization of $\mathfrak{R}$ is noncancellable. Suppose that $\mathfrak{R}$ is a subformation of some one-generated saturated formation $\mathfrak{F}$. What we can say then about $\mathfrak{R}$? In particular, does it true then that $\mathfrak{M}$ is soluble?

This problem is still open, but in the paper [427] N. N. Vorob'ev proved the following result which gives a partial solution of this problem.

Theorem 11.14 *Let $\mathfrak{F}$ be a one-generated hereditary saturated formation. Suppose that $\mathfrak{M}\mathfrak{H} \subseteq \mathfrak{F}$, where $\mathfrak{M}$ and $\mathfrak{H}$ are nonidentity formations. Then the following hold:*

1) Every simple group in $\mathfrak{M}$ is abelian;
2) If $\mathfrak{H} \neq \mathfrak{M}\mathfrak{H}$, then $\mathfrak{M}$ is soluble.

XI. In connection with Sect. 5.9, N. T. Vorob'ev proposed the following open problems.

Problem 11.15. Let $\mathfrak{F}$ and $\mathfrak{H}$ be soluble Fitting classes. Is it true that $(\mathfrak{F} \cap \mathfrak{H})_* = \mathfrak{F}_* \cap \mathfrak{H}_*$?

Problem 11.16. It is true that $\mathfrak{X}_* = \mathfrak{X} \cap \mathfrak{S}_*$, for every soluble Fischer class $\mathfrak{X}$? (For the concept of Fischer class see [89, p. 601]).

XII. About the factorization of Fitting classes, there are the following open problems.

Problem 11.17 (N. T. Vorob'ev [316, Problem 11.25]). Do there exists a local Fitting class which is decomposable into a nontrivial product of Fitting classes and every factor in such a decomposition is nonlocal?

Problem 11.18 (N. T. Vorob'ev [316, Problem 14.31]). Is the lattice of Fitting subclasses of the Fitting class generated by a finite group finite?

Bibliography

1. Agrawal, R.K.: Finite groups whose subnormal subgroups permute with all Sylow subgroups, Proc. Am. Math. Soc. **47**, 77–83 (1975)
2. Agrawal, R.K.: Generalized center and hypercenter of a finite group. Proc. Am. Math. Soc. **54**, 13–21 (1976)
3. Al-Sharo K.A.: On nearly s-permutable subgroups of finite groups. Comm. Algebra **40**, 315–326 (2012)
4. Alsheik Ahmad, A.Y., Jaraden, J.J., Skiba, A.L.: On $\mathfrak{U}_c$-normal subgroups of finite groups. Algebra Coll. **14**(1), 25–36 (2007)
5. Arad, Z., Chilag, D.: A criterion for the existence of normal π-complements in finite groups. J. Algebra **87**, 472–482 (1984)
6. Arad, Z., Fisman, E.: On finite factorizable groups. J. Algebra **86**, 522–548 (1984)
7. Arad, Z., Ward Michael, B., New criteria for the solvability of finite groups. J. Algebra **77**, 234–246 (1982)
8. Arroyo-Jordá, M., Arroyo-Jordá, P., Martinez-Pastor, A., Pérez-Ramos, M.D.: On finite products of groups and supersolubility. J. Algebra **323**, 2922–2934 (2010)
9. Arroyo-Jordá, M., Arroyo-Jordá, P., Pérez-Ramos, M.D.: On Conditional Permutability and Saturated Formations. Proc. Edinburgh Math. Soc. **54**, 309–319 (2011)
10. Asaad, M.: On the solvability of finite groups. Arch. Math. **51**(4), 289–293 (1988)
11. Asaad, M.: Finite groups some of whose n-maximal subgroups are normal. Acta. Math. Hung. **54**, 9–27 (1989)
12. Asaad, M.: On maximal subgroups of Sylow subgroups of finite groups. Comm. Algebra **26**(11), 3647–3652 (1998)
13. Asaad, M.: Finite groups with certain subgroups of Sylow subgroups complemented. J. Algebra **323**(7), 1958–1965 (2010)
14. Asaad, M.: Semipermutable π-subgroups. Arch. Math. **102**, 1–6 (2104)
15. Asaad, M., Csörgö, P.: The influence of minimal subgroups on the structure of finite groups. Arch. Math. **72**, 401–404 (1999)
16. Asaad, M., Heliel, A.A.: On S-quasinormally embedded subgroups of finite groups. J. Pure Appl. Algebra **165**, 129–135 (2001)
17. Asaad, M., Heliel, A.A.: On permutable subgroups of finite groups. Arch. Math. **80**, 113–118 (2003)
18. Asaad, M., Ramadan, M.: Finite groups whose minimal subgroups are c-supplemented. Comm. Algebra **36**, 1034–1040 (2008)
19. Asaad, M., Ramadan, M., Shaalan, A.: Influence of π-quasinormality on maximal subgroups of Sylow subgroups of fitting subgroup of a finite group. Arch. Math. **56**, 521–527 (1991)
20. Asaad, M., Shaalan, A.: On the supersolvability of finite groups. Arch. Math. **53**, 318–226 (1989)

© Springer-Verlag Berlin Heidelberg 2015

W. Guo, *Structure Theory for Canonical Classes of Finite Groups*,
DOI 10.1007/978-3-662-45747-4

21. Asaadm, M. Ezzat Mohamed, M.: On c-normality of finite groups. J. Aust. Math. Soc. **78**, 297–304 (2005)
22. Baer, R.: Group elements of prime power index, Trans. Am. Math. Soc. **75**, 20–47 (1953)
23. Baer, R.: Classes of finite groups and their properties. Ill. J. Math. **1**, 115–187 (1957)
24. Ballester-Balinches, A.: Maximal subgroups and formations. J. Pure Appl. Algebra **61**, 223–232 (1989)
25. Ballester-Bolinches, A.: On saturated formations, theta-pairs, and completions in finite groups. Siberian Math. J. **37**, 207–212 (1996)
26. Ballester-Bolinches, A., Beidleman, J.C., Feldman, A.D., Ragland, M.F.: On generalized pronormal subgroups of finite groups. Glasgow Math. J. to appear
27. Ballester-Bolinches, A., Beidleman, J.C., Feldman, A.D., Ragland, M.F.: On generalized subnormal subgroups of finite groups. Math. Nachr. to appear
28. Ballester-Bolinches, A., Calvo, C., Romero, E.: A question from the Kourovka notebook on formation products. Bull. Aust. Math. Soc. **68**, 461–470 (2003)
29. Ballester-Bolinches, A., Calvo, C.: Factorizations of one-generated $\mathfrak{X}$-local formations. Sib. Math. J. **50**(3), 385–394 (2009)
30. Ballester-Bolinches, A., Calvo, C., Esteban-Romero, R.: On $\mathfrak{X}$-saturated formations of finite groups. Comm. Algebra **33**(4), 1053–1064 (2005)
31. Ballester-Bolinches, A., Calvo, C., Esteban-Romero, R.: Products of formations of finite groups. J. Algebra **299**, 602–615 (2006)
32. Ballester-Bolinches, A., Esteban-Romero, R., Asaad, M.: Products of Finite Groups. Walter de Gruyter, Berlin (2010)
33. Ballester-Bolinches, A., Ezquerro, L.M.: On the Deskins index complex of a maximal subgroup of a finite group. Proc. Amer. Math. Soc. **114**, 325–330 (1992)
34. Ballester-Bolinches, A., Ezquerro, L.M.: Classes of Finite Groups. Dordrecht, Springer, (2006)
35. Ballester-Bolinches, A., Ezquerro, L.M., Skiba, A.N.: On second maximal subgroups of Sylow subgroups of finite groups. J. Pure Appl. Algebra **215**(4), 705–714 (2011)
36. Ballester-Bolinches, A., Ezquerro, L.M., Skiba, A.N., Local embeddings of some families of subgroups of finite groups. Acta Math. Sinic. **25**, 869–882 (2009)
37. Ballester-Bolinches, A., Ezquerro, L.M., Skiba, A.L.: Subgroups of finite groups with a strong cover-avoidance property. Bull. Aust. Math. Soc. **79**, 499–506 (2009)
38. Ballester-Bolinches, A., Ezquerro, L.M., Skiba, A.L.: On subgroups of hypercentral type of finite groups. Isr. J. Math. 2013. **199**, 259–265 (2014)
39. Ballester-Bolinches, A., Guo, X.: On complemented subgroups of finite groups. Arch. Math. **72**, 161–166 (1999)
40. Ballester-Bolinches, A., Guo, X.: Some results on p-nilpotence and solubility of finite groups. J. Algebra **228**, 491–496 (2000)
41. Ballester-Bolinches, A., Pedraza-Aguilera, M.C.: Sufficient conditions for supersolvebility of finite groups. J. Pure Appl. Algebra **127**, 113–118 (1998)
42. Ballester-Bolinches, A., Perez-Ramos, M.D.: On $\mathfrak{F}$-subnormal subgroups and Frattini-like subgroups of a finite group. Glasgow Math. J. **36**, 241–247 (1994)
43. Ballester-Bolinches, A., Perez-Ramos, M.D.: Some questions of the Kourovka notebook concerning formation products. Comm. Algebra **26**(5), 1581–1587 (1998)
44. Ballester-Bolinches, A., Wang, Y.: Finite geoups with some C-normal minimal subgroups. J. Pure Appl. Algebra **153**, 121–127 (2000)
45. Ballester-Bolinches, A., Wang, Y., Guo, X.: c-supplemented subgroups of finite groups. Glasg. Math. J. **42**, 383–389 (2000)
46. Beidleman, J.C., Hauck, P.: Über Fittingklassen und die Lockett-Vermutung. Math. Z. **167**, 161–167 (1979)
47. Beidleman, J.C., Heineken, H.: A note on intersections of maximal $\mathfrak{F}$-subgroups. J. Algebra **333**, 120–127 (2010)

48. Beidleman, J.C., Smith, H.: A note on supersoluble maximal subgroups and theta-pairs. Publ. Mat. **37**, 91–94 (1993)

49. Belonogov, V.A.: Finite solvable groups with nilpotent 2-maximal subgroups. Math. Note. **3**(1), 15–21 (1968)

50. Berger, T.R.: Normal fitting pairs and Lockett's conjecture. Math. Z. **163**, 125–132 (1978)

51. Berger, T.R., Cossey, J.: An example in the theory of normal fitting classes. Math. Z. **154**, 287–294 (1977)

52. Bhattacharya, P., Mukherjee, N.P.: On the intersection of a class of maximal subgroups of a finite group II. J. Pure Appl. Algebra **42**, 117–124 (1986)

53. Bianchi, M., Mauri, A. G., Hauck, P.: On finite groups with nilpotent Sylow-normalizers. Arch. Math. **47**, 193–197 (1986)

54. Blessenohl, D.: Über formationen und Halluntergruppen endlicher auflösbarer Gruppen. Math. Z. **142**, 299–300 (1975)

55. Blessenohl, D., Gaschütz, W.: Über normale Schun- kund Fittingklassen. Math. Z. **118**, 1–8 (1970)

56. Brison, O. J.: Hall operators for fitting classes. Arch. Math (Basel) **33**, 1–9 (1979/1980)

57. Bryce, R. A., Cossey, J.: A problem in the theory of normal Fitting classes. Math. Z. **141**, 99–110 (1975)

58. Buckley, J.: Finite groups whose minimal subgroups are normal. Math. Z. **116**, 15–17 (1970)

59. Carocca, A.: p-supersolubility of factorized finite groups. Hokkaido Math. J. **21**, 395–403 (1992)

60. Carter, R., Hawkes, T.: The $\mathfrak{F}$-normalizers of a finite soluble group. J. Algebra **5**, 175–202 (1967)

61. Chen, G., Li, J.: The influence of X-semipermutability of subgroups on the structure of finite groups. Sci. China Ser. A Math. **52**, 261–271 (2009)

62. Chen, M.S.: A remark of finite groups. Tamkang J. Math. **8**, 105–109 (1977)

63. Chen, S., Guo, W.: S-C-permutably embedded subgroups. Int. J. Contemp. Math. Sci. **3**, 951–960 (2008)

64. Chen, X., Guo, W.: On $\hat{\theta}$-pairs for maximal subgroups of a finite group. J. Group Theory, **16**, 51–68 (2013)

65. Chen, X., Guo, W.: On weakly S-embedded subgroups and weakly τ-embedded subgroups. Sib. Math. J., **54**(5), 931–945 (2013)

66. Chen, X., Guo, W.: On the partial Π-property of subgroups of finite groups. J. Group Theory, **16**, 745–766 (2013)

67. Chen, X., Guo, W.: Finite groups in which SS-permutability is a transitive relation. Acta. Math. Hungar. **142**(2), 466–479 (2014)

68. Chen, X., Guo, W.: On the $\pi\mathfrak{F}$-norm and the $\mathfrak{H}$-$\mathfrak{F}$-norm of a finite group. J. Algebra **405**, 213–231 (2014)

69. Chen, X., Guo, W., Skiba, A.N.: Some conditions under which a finite group belongs to a Baer-local formation. Comm. Algebra **42**, 4188–4203 (2014)

70. Chen, X., Guo, W., Skiba, A.N.: On generalized $\mathfrak{U}$-hypercentral subgroups of a finite groups. Accepted in J. Algebra

71. Chen, X., Guo, W., Skiba, A.N.: On generalized embedded subgroups of a finite group. To appear Algebra and Logic

72. Chigura, N.: Number of Sylow subgroups and p-nilpotence of finite groups. J. Algebra **201**, 71–85 (1998)

73. Cobb, Ph.: Existence of Hall subgroups and embedding of π-subgroups into Hall subgroups. J. Algebra **127**, 229–243 (1989)

74. Cooney, M. P.: The nonsolvable Hall subgroups of the general linear groups. Math. Z. **114**(4), 245–270 (1970)

75. Cossey, J., Guo, X.: The existence of normal II-complements in finite groups. Comm. Algebra **23**, 4257–4260 (1995)

76. Čunihin, S.A.: On π-selected groups. Dokl. Akad. Nauk SSSR, **59**, 443–445 (1948)

77. Čunihin, S.A., On theorems of Sylow's type. Dokl. Akad. Nauk. SSSR. **66**, 65–68 (1949)
78. Čunihin, S. A.: On existence and conjugacy of subgroups of a finite group. Mat. Sbornik. **33**(1), 111–132 (1953)
79. Čunihin, S. A.: Subgroups of Finite Groups. Nauka i Tehnika, Minsk, 1964.
80. Čunikhin, S.A., Shemetkov, L.A.: Finite groups. J. Soviet Math. **1**(3), 291–332 (1973)
81. Curtis, C.W., Reiner, I.: Representation theorem of finite groups and associative algebras, vol. 11, Interscience, New York, 1962, (2nd ed, 1966; Pure and Appl. Math.)
82. D'Arcy, P.: Locally dedined fitting classes. J. Aust. Math. Soc. **20**(1), 25–32 (1975)
83. Deskins, W.E.: On maximal subgroups. Proc. Sympos. Pure Math. **1**, 100–104 (1959)
84. Deskins, W.E.: A condition for the solvability of a finite group. Ill. J. Math. **2**, 306–313 (1961)
85. Deskins, W.: On quasinormal subgroups of finite groups. Math. Z. **82**, 125–132 (1963)
86. Deskins, W.E.: A note on the index complex of a maximal subgroup. Arch. Math. **54**, 236–240 (1990)
87. Doerk, K.: Minimal nicht uberauflosbare, endlicher Gruppen. Math. Z. **91**, 198–205 (1966)
88. Doerk, K., Hawkes, T.: On the residual of a direct product. Arch. Math. (Basel) **30**, 458–468 (1978)
89. Doerk, K., Hawkes, T.: Finite soluble groups. Walter de Gruyter, Berlin (1992)
90. Du, N., Li, S.: On the strong theta completions for maximal subgroups. Chin. Ann. Math. Ser. A. **27**, 279–286 (2006) (in chinese)
91. Du, N., Li, S.: Influence of the strong theta completions on finite groups. Algebra Colloq. **17**, 59–64 (2010)
92. Du, N., Li, S.: On the $\mathfrak{F}$-abnormal maximal subgroups of finite groups. J. Pure Appl. Algebra **208**, 345–349 (2007)
93. Dutta, T.K., Sen, P.: Some characterisations of π-solvable groups using index-complex. Indian J. Pure Appl. Math. **33**, 555–564 (2002)
94. Dutta, T.K., Sen, P.: Some characterizations of solvable groups using θ-subgroup and some characteristic subgroups. Southeast Asian Bull. Math. **27**, 805–812 (2004)
95. Elovikov, A.B.: Factorizations of one-generated formations. Math. Notes. **73**(5), 684–697 (2003)
96. Elovikov, A.B.: Factorizations of one-generated partly step formations. **21**, 99–118 (2009)
97. Erickson, R.P.: Projectore of finite groups. Comm. Algebra **10**, 1919–1938 (1982)
98. Ezquerro, L.M.: A contribution to the theory of finite supersoluble groups. Rend. Sen. Mat. Univ. Padova. **89**, 161–170 (1993)
99. Ezquerro, L.M., Soler-Escrivà, X.: Some permutability properties related to $\mathcal{F}$-hypercentrally embedded subgroups of finite groups. J. Algebra **264**, 279–295 (2003)
100. Fan, Y., Guo, X., Shum, K.P.: Remarks on two generalizations of normality of subgroups. Chin. J. Contemp. Math. **27**(2), 1–8 (2006)
101. Fan, Y., Guo, X., Shum, K.P.: Remarks on two generalizations of normality of subgroups. Chinese Ann. Math. **27A**(2), 169–176 (2006)
102. Fedri, V., Serens, L.: Finite soluble groups with supersoluble Sylow normalizers. Arch. Math. **50**, 11–18 (1988)
103. Feng, X., Guo, W.: On $\mathfrak{F}_n$-normal subgroups of finite groups. Front. Math. China **5**, 653–664 (2010)
104. Feng, Y., Zhang, B.: Frattini subgroups relative to formation functions. J. Pure Appl. Algebra **64**, 145–148 (1990)
105. Fischer, B.: Klassen konjugierter Untergruppen in endlichen auflösbaren Gruppen. Habilitationschrift, Universität Frankfurt (1966)
106. Fischer, B., Gaschütz, W., Hartley, B.: Injektoren endlicher auflösbarer Gruppen. Math. Z. **102**, 337–339 (1967)
107. Foguel, N.: On seminormal subgroups, J. Algebra **165**, 633–635 (1994)
108. Friesen, D.: Products of normal supersoluble subgroups. Proc. Am. Math. Soc. **30**, 46–48 (1971)

109. Gagen, T.M.: Topics in finite groups, London Math. Soc. Lecture Note Series 16. Cambridge University Press, London (1976)
110. Gallego, M.P.: Fitting pairs from direct limites and the Lockett conjecture. Comm. Algebra **24**(6), 2011–2023 (1996)
111. Galt, A., Guo, W., Averkin, E., Revin, D.: On the local case in the aschbacher theorem for linear and unitary groups. Sib. Math. J. **55**, 239–245 (2014)
112. Gaschütz, W.: Über die ϕ-untergruppen endlicher Gruppen. Math. Z. **58**, 160–170 (1953)
113. Gaschütz, W.: Zur Theorie der endlichen auflösbaren Gruppen. Math. Z. **80**(4), 300–305 (1963)
114. Gaschütz, W.: Lectures of subgroups of Sylow type in finite soluble groups, (Notes on pure Mathematics, No. 11). Australian National University, Canberra (1979)
115. Gavrilyuk, A.L., Guo, W., Makhnev, A.A.: Automorphisms of Terwilliger graphs with $\mu = 2$. Algebra i Logika **47**(5), 330–339 (2008)
116. Gong, L., Guo, X.: On the intersection of the normalizers of the nilpotent residuals of all subgroups of a finite group. Algebra Colloq. **20**, 349–360 (2013)
117. Gorenstein, D.: Finite groups. Harper & Row Publishers, New York (1968)
118. Griess, R., Schmid, P.: The Frattini module. Arch. Math. **30**, 256–266 (1978)
119. Gross, F.: On the existence of Hall subgroups. J. Algebra **98**, 1–13 (1986)
120. Gross, F.: On a conjecture of Philip Hall, Proc. London Math. Soc. **52**(3), 464–494 (1986)
121. Gross, F.: Conjugacy of odd Hall subgroups. Bull. London Math. Soc. **19**, 311–319 (1987)
122. Gross, F.: Odd order Hall subgroups of the classical linear groups. Math. Z. **220**, 317–336 (1995)
123. Gross, F.: Hall subgroups of order not divisible by 3. Rocky Mt. J. Math. **23**, 569–591 (1993)
124. Grytczuk, A., Vorob'ev, N.T.: On Lockett's Conjecture for finite groups. Tsukuba J. Math. **18**(1), 63–67 (1994)
125. Guo, W.: On normalizers of Sylow subgroups. Dokl. Akad. Nauk Belarusi **37**, 22–24 (1993)
126. Guo, W.: Finite groups with specified properties of the normalizers of their Sylow subgroups. Chin. Ann. Math. Ser. A. **15**, 627–631 (1994)
127. Guo, W.: Finite groups with given normalizers of Sylow subgroups, Chin. Sci. Bull. **39**, 1951–1953 (1994)
128. Guo, W.: Finite groups with given indices of normalizers of Sylow subgroups. Sib. Math. J. **37**, 253–357 (1996)
129. Guo, W.: Finite groups with given normalizers of Sylow subgroups II. Acta Math. Sinica **39**, 509–513 (1996)
130. Guo, W.: Injectors of finite groups. Chin. Ann. Math. Ser. A. **18**, 145–148 (1997)
131. Guo, W.: The Theory of Group Classes. Science Press, Beijing (1997) (in Chinese)
132. Guo, W.: Local formations in which every subformation of type $\mathfrak{N}_p$ has a complements. Chin. Sci. Bull. **42**, 364–367 (1997)
133. Guo, W.: The groups generated by subnormal subgroups. Algebra Colloq. **5**(1), 41–48 (1998)
134. Guo, W.: On one question of the Kourovka Notebook. Comm. Algebra **28**, 4767–4782 (2000)
135. Guo, W.: The Theory of Classes of Groups. Science Press-Kluwer Academic Publishers, Beijing (2000)
136. Guo, W.: On one problem in the theory of step formations. Izv. VUZ Math. **9**, 33–37 (2001)
137. Guo, W.: On a problem of the theory of multiply local formations. Sib. Math. J. **45**, 1036–1040 (2004)
138. Guo, W.: Finite groups with nilpotent local subgroups. Chin. Ann. Math. Ser. **25A**, 217–224 (2004) (Translate to: Chin. J Contemp. Math. **25**(2), 193–200 (2004))
139. Guo, W.: Formations determined by Hall subgroups. J. App. Algebra Discret. Struct. **4**(3), 139–147 (2006)
140. Guo, W.: On $\mathfrak{F}$-radicals of finite π-soluble groups. Algebra Discret. Math. **3**, 48–53 (2006)
141. Guo, W.: On $\mathfrak{F}$-supplemented subgroups of finite groups. Manuscripta Math. **127**, 139–150 (2008)

142. Guo, W.: Finite groups with seminormal Sylow subgroups. Acta Math. Sinica English Series, **24**(10), 1751–1758 (2008)

143. Guo, W.: Some idears and results in group theory. In: Recent Developments in Algebra and Related Areas, pp. 111–118. Higher Education Press and International Press, Beijing-Boston (2009)

144. Guo, W., Al-Sharo, Kh.A., Shemetkov, L.A.: On F-covering subgroups of finite groups. Southeast Asian Bull. Math. **29**, 97–103 (2005)

145. Guo, W., Andreeva, D.P. Skiba, A.N.: Finite groups of Spencer height ≤ 3, to appear Algebra Colloquium

146. Guo, W., Chen, S.: Weakly c-permutable subgroups of finite groups. J. Algebra **324**, 2369–2381 (2010)

147. Guo, W., Feng, X., Huang, J.: New characterzations of some classes of finite groups. Bull. Malays. Sci. Soc. **34**(3), 575–589 (2011)

148. Guo, W., Hu, B., Monakhov, V.: On indices of subgroups of finite soluble groups. Comm. Algebra **33**, 855–863 (2005)

149. Guo, W., Huang, J., Skiba, A.N.: On G-covering subgroup systems for some saturated formations of finite groups. Comm. Algebra **41**, 2948–2956 (2013)

150. Guo, W., Legchekova, H.V., Skiba, A.N.: The structure of finite non-nilpotent groups in which every 2-maximal subgroup permutes with all 3-maximal subgroups. Comm. Algebra **37**, 2446–2456 (2009)

151. Guo, W., Legchekova, E.V., Skiba, A. N.: Finite groups in which every 3-maximal subgroup commutes with all maximal subgroups. Math. Notes **86**(3), 325–332 (2009)

152. Guo, W., Li, B.: On the injectors of finite groups. J. Group Theory **10**, 849–858 (2007)

153. Guo, W., Li, B.: On the Shemetkov Problem for Fitting Classes. Beitrage zue Algebra und Geometrie Comtributions to Algebra and Geometry **48**(1), 281–289 (2007)

154. Guo, W., Lu, Y., Niu, W.: S-embedded subgroups of finite groups. Algebra Logic **49**(4), 293–304 (2010)

155. Guo, W., Lutsenko, Y.V., Skiba, A.N.: On nonnilpotent groups with every two 3-maximal subgroups permutable. Sib. Math. J. **50**(6), 988–997 (2009)

156. Guo, W., Makhnev, A.A.: On distance-regular graphs without 4-claws. Doklady Math. **88**, 625–629 (2013). (Original Russian Text: Dokl. Akad. Nauk, **453**, 1–5 (2013))

157. Guo, W., Makhnev, A.A., Paduchikh, D.V.: Automorphisms of coverings of strongly regular graphs with parameters (81,20,1,6). Math. Notes **86**, 26–40 (2009)

158. Guo, W., Makhnev, A.A., Paduchikh, D.V.: On almost distance-transitive graphs. Doklady Math. **88**, 581–585 2013. (Original Russian Text: Dokl. Akad. Nauk, **452**, 599–906 (2013)).

159. Guo, W., Revin, D.O.: On the class of groups with pronormal Hall π-subgroups. Sib. Math. J. **55**, 415–427 (2014)

160. Guo, W., Revin, D.O., Vdovin, E.P.: Confirmation for Wielandt's conjecture to appear J. Algebra.

161. Guo, W., Safonov, V.G., Skiba, A.N.: On some construstions and results of the theory of partially soluble finite groups, Proceedings of the international conference on Algebra, 2010, World Scientific, 289–307.

162. Guo, W., Sel'kin, V.M., Shum, K.P.: Factorization theory of 1-generated ω-composition formations. Comm. Algebra **35**, 2347–2377 (2007)

163. Guo, W., Shemetkov, L.A.: Finite groups with the hypercentral condition. Dokl. Akad. Nauk Belurusi **36**, 485–486 (1992)

164. Guo, W., Shum, K.P.: On totally local formations of groups. Comm. Algebra **30**, 2117–2131 (2002)

165. Guo, W., Shum, K.P.: Formation operators on classes of algebra Comm. Algebra **30**, 3457–3472 (2002)

166. Guo, W., Shum, K.P.: Problems on product of formations. Manuscripta Math. **108**, 205–215 (2002)

167. Guo, W., Shum, K.P.: Frattini theory for classes of finite universal algebras of Malcev varieties. Sib. Math. J. **43** (2002), 1039–1046
168. Guo, W., Shum, K.P.: Lattices of Schunck classes of finite universal algebras. Algebra Colloquium **10**, 219–228 (2003)
169. Guo, W., Shum, K.P.: Products of ω-saturated formations. Italian J. Pure Appl. Math. **14**, 177–182 (2003)
170. Guo, W., Shum, K.P.: Uncacellative factorizations of composition formations. J. Algebra **267**, 654–672 (2003)
171. Guo, W., Shum, K.P.: A note on finite groups whose normalizers of Sylow 2,3-subgroups are prime power indices. J. Appl. Algebra Discrete Str. **3**, 1–9 (2005)
172. Guo, W., Shum, K.P., Skiba, A.N.: On $\mathfrak{F}$-residuals of finite groups. Bull Austral Math. Soc. **65**, 271–275 (2002)
173. Guo, W., Shum, K.P., Skiba, A.N.: G-covering subgroup systems for the classes of supersoluble and nilpotent groups. Isr. J. Math. **138**, 125–138 (2003)
174. Guo, W., Shum, K.P., Skiba, A.N.: G-covering systems of subgroups for classes of p-supersoluble and p-nilpotent of finite groups. Sib. Math. J. **45**(3), 433–442 (2004)
175. Guo, W., Shum, K.P., Skiba, A.N.: Conditionally permutable subgroups and supersolubility of finite groups. SEAMS Bull. Math. **29**, 493–510 (2005)
176. Guo, W., Shum, K.P., Skiba, A.N.: X-permutable maximal subgroups of Sylow subgroups of finite groups. Ukrain Math. J. **58**, 1299–1309 (2006)
177. Guo, W., Shum, K.P., Skiba, A.N.: Criterions of supersolubility for products of superaoluble groups. Publ. Math. Debrecen **68**(3–4), 433–449 (2006)
178. Guo, W., Shum, K.P., Skiba, A.N.: On primitive subgroups of finite groups. Indian J. Pure Appl. Math. **37**(6), 369–376 (2006)
179. Guo, W., Shum, K.P., Skiba, A.N.: X-quasinormal subgroups. Sib. Math. J. **48**, 593–605 (2007)
180. Guo, W., Shum, K.P., Skiba, A.N.: X-semipermutable subgroups of finite groups. J. Algebra **315**, 31–41 (2007)
181. Guo, W., Shum, K.P., Skiba, A.N.: Schur-Zassenhaus Theorem for X-permutable subgroups. Algebra Colloquium **15**, 185–192 (2008)
182. Guo, W., Shum, K.P., Skiba, A.N.: On solubility and supersolubility of some classes of finite groups. Sci. China Ser. A Math. **52**(2), 272–286 (2009)
183. Guo, W., Shum, K.P., Skiba, A.N.: Finite groups with some given systems of Xm-semipermutable subgroups. Math. Nachr. **283**, 1603–1612 (2010)
184. Guo, W., Shum, K.P., Vorob'ev, N.T.: Problems related to the Lockett conjecture on fitting classes of finite groups. Indagat. Math. New Ser. **19**, 391–399 (2008)
185. Guo, W., Shum, K.P., Xie, F.: Finite groups with some weakly S–supplemented subgroups. Glasg. Math. J. **53**, 211–222 (2011)
186. Guo, W., Skiba, A.N.: Factorizations of one-generated composition formations. Algebra Logic **40**(5), 545–560 (2001)
187. Guo, W., Skiba, A.N.: Two notes on the equations of the lattices of ω-local and ω-composition formations of finite groups. Izv VUZ. Math. **480**, 14–22 (2002)
188. Guo, W., Skiba, A.N.: On finite quasi-$\mathfrak{F}$-groups. Comm. Algebra **37**(2), 470–481 (2009)
189. Guo, W., Skiba, A.N.: On some classes of finite quasi-$\mathfrak{F}$-groups. J. Group Theory **12**, 407–417 (2009)
190. Guo, W., Skiba, A.N.: Finite groups with given s-embedded and n-embedded subgroups. J. Algebra **321**, 2843–2860 (2009)
191. Guo, W., Skiba, A.N.: On X-permutabe subgroups of finite groups. J. Xuzhou Normal Univ. (Natural Science Edition) **27**, 16–24 (2009)
192. Guo, W., Skiba, A.N.: On quasisupersoluble and p-quasisupersoluble finite groups. Algebra Colloquium **17**, 549–556 (2010)
193. Guo, W., Skiba, A.N.: Criteria of existence of Hall subgroups in non-soluble finite groups. Acta Mathematica Sinica, English Series **26**(2), 295–304 (2010)

194. Guo, W., Skiba, A.N.: On factorizations of finite groups with $\mathfrak{F}$-hypercentral intersections of the factors. J. Group Theory **14**, 695–708 (2011)
195. Guo, W., Skiba, A.N.: On G-covering subgroup systems of finite groups. Acta Mathematica Hungarica **133**(4), 376–386 (2011)
196. Guo, W., Skiba, A.N.: Finite groups with systems of Σ-embedded subgroups. Sci. China Math. **54**, 1909–1926 (2011)
197. Guo, W., Skiba, A.N.: New criterions of existence and conjugacy of Hall subgroups of finite groups. Proc. Am. Math. Soc. **139**(7), 2327–2336 (2011)
198. Guo, W., Skiba, A.N.: On the intersection of the $\mathfrak{F}$-maximal subgroups and the generalized $\mathfrak{F}$-hypercentre of a finite group. J. Algebra **366**, 112–125 (2012)
199. Guo, W., Skiba, A.N.: New characterizations of p-soluble and p-supersoluble finite groups. Studia Sci. Math. Hungarica **49**(3), 390–405 (2012)
200. Guo, W., Skiba, A.N.: On $\mathfrak{F}\phi^*$-hypercentral subgroups of finite groups. J. Algebra **372**, 275–292 (2012)
201. Guo, W., Skiba, A.N.: On the intersection of maximal supersoluble subgroups of a finite group. Truda Inst. Math. Dokl. Akad. Nauk. Belurusi. **21**(1), 48–51 (2013)
202. Guo, W., Skiba, A.N.: On Hall subgroups of a finite group. Cent. Eur. J. Math. **11**(7), 1177–1187 (2013)
203. Guo, W., Skiba, A.N.: Finite groups with generalized Ore supplement conditions for primary subgroups, to appear in J. Algebra (2015)
204. Guo, W., Skiba, A.N.: On boundary factors and traces of subgroups of finite groups, to appear in Communications in Mathematics and Statistics
205. Guo, W., Skiba, A.N.: On graduated properties of subgroups of finite groups, to appear
206. Guo, W., Skiba, A.N., Shum, K.P.: Lattices of subgroup and subsystem functors. Algebra Log. **45**, 403–414 (2006)
207. Guo, W., Skiba, A.N., Shum, K.P.: X-quasiormal subgroups. Sib. Math. J. **48**, 593–605 (2007)
208. Guo, W., Skiba, A.N., Yang, N.: SE-supplemented subgroups of finite groups. Rend. Sem. Mat. Padova. **129**, 245–263 (2013)
209. Guo, W., Skiba, A.N., Yang, N.: A generalized CAP-subgroup of a finite group, to appear Science China Math. (2015)
210. Guo, W., Tang, N., Li, B.: On $\mathfrak{F}$-z-supplemented subgroups of finite groups. Acta. Math. Scientia. **31B**(1), 22–28 (2011)
211. Guo, W., Vorob'ev, N.T.: On injectors of finite soluble groups. Comm. Algebra **36**, 3200–3208 (2008)
212. Guo W., Vorobe'v N.T.: On the theory of $\mathfrak{F}$-centrality of chief factors and $\mathfrak{F}$-hypercentre for fitting classes. J. Algebra **344**, 386–396 (2011)
213. Guo, W., Xie, F., Li, B.: Some open questions in the theory generalized permutable groups. Sci. China. Ser. A. Math. **52**10, 2132–2144 (2009)
214. Guo, W., Feng, X., Huang, J.: New characterizations of some classes of finite groups. Bull. Malays. Math. Sci. Soc. **34**(3), 575–589 (2011)
215. Guo, W., Xie, F., Lu, Y.: On g-s-supplemented subgroups of finite groups. Front. Math. China. **5**, 287–295 (2010)
216. Guo, W., Yu, X.: On $\mathfrak{F}_n$-normal subgroups of finite groups Siberian. Math. J. **52**(2), 197–206 (2011)
217. Guo, W., Zhu, L.: On formations with Shemetkov condition. Algebra colloq. **9**(1), 89–98 (2002)
218. Guo, X.: On Deskins's conjecture. J. Pure Appl. Algebra, **124**, 167–171 (1998)
219. Guo, X., Shum, K.P., Ballester-Bolinches, A.: On complemented minimal subgroups in finite groups. J. Group Theory **6**, 159–167 (2003)
220. Guo, X., Shum, K.P.: The influnce of minimal subgroup of focal subgroups on the structure of finite groups. J. Pure Appl. Algebra **169**, 43–50 (2002)
221. Guo, X., Shum, K.P.: Permutability of minimal subgroups and p-nilpotency of finite groups. Israel J. Math. **136**, 145–155 (2003)

222. Guo, X., Shum, K.P.: On p-nilpotency and minimal subgroups of finite groups. Sci. China (Series A) **46**, 176–186) (2003)
223. Guo, X., Shum, K.P.: p-nilpotencey of finite groups and minimal subgroups. J. Algebra **270** 459–470) (2003)
224. Guo, X., Shum, K.P.: Cover-avoidance properties and the structure of finite groups. J. Pure Appl. Algebra **181**, 297–308 (2003)
225. Guo, X, Shum, K.P.: Finite p-nilpotent groups with some subgroups c-supplemented. J. Aust. Math. Soc. **78**, 429–439) (2005)
226. Guo, X., Wang, J., Shum, K.P.: On semi-cover-avoiding maximal subgroups and solvability of finite groups. Comm. Algebra **34**, 3235–3244 (2006)
227. Guo, X., Guo, P., Shum, K.P.: On semi cover-avoiding subgroups of finite groups. J. Pure Appl. Algebra **209**, 151–158 (2007)
228. Hall, F.: A note on soluble groups. J. London Math. Soc. **3**, 98–105 (1928)
229. Hall, M.: The Theory of Groups. Macmillan, New York (1959)
230. Hall, M.: The Theory of Groups, Chelsea, New York (1976)
231. Hall, P.: On the Sylow systems of a soluble group. Proc. London math. Soc. **43**(2), 507–528 (1937)
232. Hall, P.: A characteristic property of soluble groups. J. London Math. Soc. **12**, 205–221 (1939)
233. Hall, P.: Theorems like Sylow's. Proc. London Math. Soc. **6**, 286–304 (1956)
234. Hao, L., Zhang, X., Yu, Q.: The influence of X-s-semipermutable subgroups on the structure of finite groups. Southeast Asian Bull. Math. **33**, 749–759 (2009)
235. Hao, L., Guo, W., Hu, B.: S-conditionally semipermutable subgroups of finite groups. Adv. Math. **39**, 623–628 (2010)
236. Hartley, B.: On Fischer's dualization of formation theory. Proc. London Math. Soc. **3**(2), 193–207 (1969)
237. Hartley, B.: A theorem of Sylow type for finite groups. Math. Z. **122**, 223–226 (1971)
238. Hartley, B.: Helmut Wielandt on the π-structure of finite groups. In: Wielandt, H., Huppert, B., Schneider, H. (eds.) Mathematische Werke/Mathematical Works, vol. 1: Group Theory, pp. 511–516 de Gruyter, Berlin (1994)
239. He, X., Li, Y. Wang, Y.: On weakly SS-permutable subgroups of a finite group. Publ. Math. Debrecen, **77**(1–2), 65–77 (2010)
240. Hu, B., Guo, W.: C-semipermutable subgroups of finite groups. Siberian Math. J. **48**, 180–188 (2007)
241. Huang, J., Guo, W.: s-conditionally permutable subgroups of finite groups. Chin. Ann. Math. **28A**, 17–26 (2007) (in Chinese)
242. Huang, J.: On $\mathfrak{F}_s$-quasinormal subgroups of finite groups. Comm. Allgebra **38**, 4063–4076 (2010)
243. Huo, L., Guo, W., Zhang, G.: Automorphisms of generalized orthogonal graphs of characteristic 2. Front. Math. China **9**(2), 303–319 (2014)
244. Huo, L., Guo, W., Ma, C.: Automorphisms of generalized orthogonal graphs of odd characteristic. Acta Math. Sinica. Chin. Ser. **57**, 1–18 (2014)
245. Huppert, B.: Normalteiler und maximale Untergruppen endlicher Gruppen. Math Z. **60**, 409–434 (1954)
246. Huppert, B.: Zur Sylowstruktur auflösbarer Gruppen. Arch. Math. **12**, 161–169 (1961)
247. Huppert, B.: Zur Sylowstruktur auflösbarerGruppen II. Arch. Math. **15**, 251–257 (1964)
248. Huppert, B.: Endliche Gruppen I. Springer-Verlag. Berlin (1967)
249. Huppert, B., Blackburn, N.: Finite Groups II. Springer-Verlag, Berlin (1982)
250. Huppert, B., Blackburn, N.: Finite Groups III. Springer-Verlag, Berlin (1982)
251. Isaacs, I.M.: Semipermutable π-subgroups. Arch. Math. **102**, 1–6 (2014)
252. Ito, N., Szép, J.: Uber die Quasinormalteiler von endlichen Gruppen. Acta Sci. Math. **23**, 168–170 (1962)
253. Janko, Z.: Endliche Gruppen mit lauter nilpotenten zweitmaximalen Untergruppen. Math. Z. **79**, 422–424 (1962)

254. Janko, Z.: Finite groups with invariant fourth maximal subgroups. Math. Z. **82**, 82–89 (1963)
255. Jaraden, J.J., Skiba, A.N.: On c-normal subgroups of finite groups. Comm. Algebra **35**, 3776–3788 (2007)
256. Johnson, D.L.: A note on supersoluble groups. Can. J. Math. **23**, 562–564 (1971)
257. Jahad Jumah: On factirizations of Bear-local Formations. Comm. Algebra **31**(10) (2003), 4697–4711.
258. Kamornikov, S.F.: On a class of lattice semigroup functors. Mat. Zametki **89**(3), 355–364 (2011)
259. Kamornikov, S.F., Selkin, M.V.: Subgroup functors and classes of finite groups, Belaruskaya Navuka, Minsk (2003)
260. Kazarin, L.S.: Sylow-type theorems for finite groups, structure properties of algebra systems, Kabardino-Balkarian Univ., Nalchik, pp. 42–52 (in Russia) (1981)
261. Kazarin, L.S.: On the product of finite groups. Sov. Math. Dokl. **27**, 354–357 (1983)
262. Kegel, O.H.: Produkte nilpotenter Gruppen. Arch. Math. **12**, 90–93 (1961)
263. Kegel, O.: Sylow-Gruppen and Subnormalteilerendlicher Gruppen. Math. Z. **78**, 205–221 (1962)
264. Kegel, O.H.: Zur Struktur mehrafach faktorisierbarer endlicher Gruppen. Math. Z. **87**, 409–434 (1965)
265. Kegel, O.H.: Untergruppenverbande endlicher Gruppen, die den subnormalteilerverband each enthalten. Arch. Math. **30**(3), 225–228 (1978) (Basel)
266. Knyagina, V.N., Monakhov, V.S.: On π'-properties of a finite group possessing a Hall π-subgroup. Sib. Math. J. **52**, 234–243 (2011)
267. Knyagina, V.N., Monakhov, V.S.: On permutability of n-maximal subgroups with Schmidt subgroups. Trudy Inst. Mat. i Mekh. UrO RAN **18**(3), 125–130 (2012)
268. Kondrat'ev, A.S.: A criterion for 2-nilpotency of finite groups. In: Subgroup Structure of Groups, Sverdlovsk, pp. 82-84 (1988) (in Russian)
269. Kondratev, A.S., Guo, V.: Finite groups in which the normalizers of the Sylow 3-subgroups are of odd or primary index. Sib. Math. J. **50**, 272–276 (2009)
270. Kosenok, N., Ryzhik, V.: On s-normal subgroups of finite groups. Probl. Algebra **17**, 209–210 (2001)
271. Kovaleva, V.A., Skiba, A.N.: Finite solvable groups with all n-maximal subgroups $\mathfrak{U}$-subnormal. Sib. Math. J. **54**(1), 65–73 (2013)
272. Kovaleva, V.A., Skiba, A.N.: Finite soluble groups with all n-maximal subgroups $\mathfrak{F}$-subnormal. J. Gr. Theory (2013). doi:10.1515/jgt-2013-0047.
273. Kurzweil, H., Stellmacher, B.: The Theory of Finite Groups: An Introduction. Springer-Verlag, New York (2004)
274. Laue, H.: Über nichtauflösbare normale Fittingklassen. J. Algebra **45**, 274–283 (1977)
275. Lausch, H.: On normal fitting classes. Math. Z. **137**, 131–136 (1974)
276. Lennox, J.C., Stonehewer, S.E.: Subnormal Subgroups of Groups. Clarendon Press, Oxford (1987)
277. Li, B.: On Π-property and Π-normality of subgroups of finite groups. J. Algebra **334**, 321–337 (2011)
278. Li, B., Guo, W.: On $\mathfrak{F}$-covering subgroups of finite groups. Proc. F. Scorina Gomel Univ. **27**(6), 11–15 (2004)
279. Li, B., Guo, W.: On some open problems related to X-permutability of subgroups. Comm. Algebra **39**, 757–771 (2011)
280. Li, B., Guo, W., Huang, J.: Finite groups in which Sylow normalizers have nilpotent Hall supplements. Sib. Math. J. **50**(2009), 667–673
281. Li, B., Skiba, A.N.: New characterizations of finite supersoluble groups. Sci. China Math. **51**, 827–841 (2008)
282. Li, B., Zhang, Z.: The influence of s-conditionally permutable subgroups on finite groups. Sci. China Ser. A Math. **52**, 301–310 (2009)

283. Li, D., Guo, X.: The influence of c-normality of subgroups on the structure of finite groups, II. Comm. Algebra **26**(1998), 1913–1922

284. Li, J., Chen, G., Chen, R.: On weakly S-embedded subgroups of finite groups. Sci. China Math. **54**(3), 449–456 (2011)

285. Li, S.: The Deskins index complex and the supersolvability of finite groups. J. Pure Appl. Algebra **144**, 297–302 (1999)

286. Li, S.: Finite non-nilpotent groups all of whose second maximal subgroups are TI-groups. Math. Proc. Royal Irish Acad. **100A**(1), 65–71 (2000)

287. Li, S., Liu, J.: A generalization of cover-avoiding properties in finite groups. Comm. Algebra **39**, 1455–1464 (2011)

288. Li, S., Shen, Z., Kong, X.: On SS-quasinormal subgroups of finite groups. Comm. Algebra **36**, 4436–4447 (2008)

289. Li, S., Shen, Z., Liu, J., Liu, X.: The influence of SS-quasinormality of some subgroups on the structure of finite groups. J. Algebra **319**, 4275–4287 (2008)

290. Li, S., Shen, Z.: On the intersection of the normalizers of derived subgroups of all subgroups of a finite group. J. Algebra **323**, 1349–1357 (2010)

291. Li, S., Zhao, Y.: On θ-pairs for maximal subgroups. J. Pure Appl. Algebra **147**, 133–142 (2000)

292. Li, S., Zhao, Y.: On s-completion of maximal subgroups of finite groups. Algebra Colloquium **11**, 411–420 (2004)

293. Li, Y., Qiao, S., Wang, Y.: A note on a result of Skiba. Sib. Math. J. **50**(3), 467–473 (2009)

294. Li, Y., Qiao, S., Su, N., Wang, Y.: On weakly s-semipermutable subgroups of finite groups. J. Algebra **371**, 250–261 (2012)

295. Li, Y., Qiao, S., Wang, Y.: On weakly S-permutably embedded subgroups of finite groups. Comm. Algebra **37**, 1086–1097 (2009)

296. Li, Y., Wang, Y.: On π-quasinormally embedded subgroups of finite groups. J. Algebra **281**, 109–123 (2004)

297. Liu, X., Ding, N.: On chief factors of finite groups. J. Pure Appl. Algebra **210**, 789–796 (2007)

298. Liu, X., Guo, W., Shum, K.P.: Products of finite supersoluble groups. Algebra Colloquium **16**(2), 333–340 (2009)

299. Liu, Y., Guo, W., Kovaleva, V.A., Skiba, A.N.: Criteria for p-solvability and p-supersolvability of finite groups. Math. Notes **94**(3), 117–130 (2013)

300. Lockett, P.: The fitting class $\mathfrak{F}^*$. Math Z. **137**, 131–136 (1974)

301. Lukyanenko, V.O., Skiba, A.N. : On τ-quasinormal and weakly τ-quasinormal subgroups of finite groups. Math. Sci. Res. J. **12**, 243–257 (2008)

302. Lukyanenko, V.O., Skiba, A.N.: On weakly τ-quasinormal subgroups of finite groups, Acta Math. Hungar. **125**(3), 237–248 (2009)

303. Lukyanenko, V.O. , Skiba, A.N. : On τ-quasinormal and weakly τ-quasinormal subgroups of finite groups. Math. Sci. Res. J., **12**, 243–257 (2008)

304. Lutsenko, Yu.V., Skiba, A.N.. Structure of finite groups with S-quasinormal third maximal subgroups. Ukrainian Math. J. **61**(12), 1915–1922 (2009)

305. Lutsenko, Yu. V., Skiba, A. N.. On the groups with permutable 3-maximal subgroups. Izv. F. Scorina Gomel State University, No.2, 112–116 (2008)

306. Lutsenko, Yu.V., Skiba A.N.: Finite groups with subnormal second and third maximal subgroups. Math. Notes. **91**(5), 680–688 (2012)

307. Maier, R., Schmid, P.: The embedding of quasinormal subgroups in finite groups, Math. Z. **131** 269–272 (1973)

308. Mal'cev, A.I.: Algebraic Systems, Nauka, Main Editorial Board for Physical and Mathematical Literature, Moscow (1970)

309. Malinowska, I.A.: Finite groups with sn-embedded or s-embedded subgroups. Acta Math. Hungar. **136**(1–2), 76–89 (2012)

310. Mann, A.: Finite groups whose n-maximal subgroups are subnormal. Trans Amer. Math. Soc. **132**, 395–409 (1968)

311. Miao, L.: Finite groups with some maximal subgroups of Sylow subgroups Q-suppplemented. Comm. Algebra, **35**, 103–113 (2007)

312. Miao, L., Chen, X., Guo, W.: Finite groups with c-normal subgroups. Southeast Asian Bulletin of Mathematics, **25**, 479–483 (2001)

313. Miao, L., Guo, W.: On the influence of the indices of normalizers of Sylow subgroups in the structure of finite p-soluble groups. Siberian Math. J., **43**, 120–125(2002)

314. Miao, L., Guo, W.: Finite groups with some primary subgroup $\mathfrak{F}$-s-supplemented. Comm. Algebra, **33**, 2789–2800 (2005)

315. Miao, L., Guo, W., Shum, K.P.: New criteria for p-nilpotency of finite groups. Comm. Algebra, **35**, 965–974 (2007)

316. Mazurov, V.D., Khukhro, E.I. (eds.): The Kourovka notebook, Unsolved problems in group theory, Institute of Mathematics, Siberian Branch of RAS, Novosibirsk, 1992 (or 1999).

317. Mazurov, V.D., Khukhro, E.I. (eds.): The Kourovka notebook, Unsolved problems in group theory, Institute of Mathematics, Siberian Branch of RAS, Novosibirsk (2002)

318. Mazurov, V.D., Khuhro, E.I.: The Kourovka notebook, Unsolved problems in group theory, Institute of Mathematics, Siberian Branch of RAS, Novosibirsk (2010)

319. Mazurov, V.B., Revin, D.O.: On the Hall D_π-property for finite group. Sib. Math. J. **38**, 106–113 (1997)

320. Monakhov, V.S.: Product of supersoluble and cyclic or primary groups, Finite Groups. Proc. Gomel Sem., Gomel, 1975–1977 (in Russian), "Nauka i Tekhnika", Minsk, 50–63 (1978)

321. Mukherjee, N.P., Bhattacharya, P.: On theta pairs for a maximal subgroup. Proc. Amer. Math. Soc. **109**, 589–596 (1990)

322. Neuman, X.: Varieties of groups, Ergeb. Math. Band 37. Springer-Verlag, Berlin (1967)

323. Niu, W., Guo, W., Liu, Y.: The influence of s-conditional permutability of subgroups on the structure of finite groups. J. Math Res. Exp. **31**(1), 173–179 (2011)

324. Ore, O.: Contributions in the theory of groups of finite order. Duke Math. **5**, 431–460 (1939)

325. Peng, F., Li, S., Li, K., Bai, Y.: On X-ss-permutable subgroups of finite groups. Southeast Asian Bull. Math. **35**, 285–295 (2011)

326. Petrillo, J.: CAP-subgroups in a direct product of finite groups. J. Algebr. **306**, 432–438 (2006)

327. Poljakov, L.J.: Finite groups with permutable subgroups in: Finite Groups, Nauka i Tekhnik, Minsk (1966)

328. Ramadan, M.: Influence of normality on maximal subgroups of Sylow subgroups of a finite group. Acta Math. Hungar. **59**, 107–110 (1992)

329. Ramadan, M., Mohamed, M.E. , Heliel, A.A.: On c-normality of certain subgroups of prime power order of finite groups. Arch. Math. **85**, 203–210 (2005)

330. Randolph, J.W.: An extension of a theorem of Janko on finite groups with nilpotent maximal subgroups. Can. J. Math. **23**, 550–552 (1971)

331. Reizik, V.N., Skiba, A.N.: On factorizations of one-generated p-saturated formations, Problems in Algebra, No.11, 104–116 (1997)

332. Ren, Y.: Notes on π-quasi-normal subgroups in finite groups. Proc. Amer. Math. Soc. **117**, 631–636 (1993)

333. Revin, D.O.: The D_π-property in a class of finite groups. Algebr and Logic **41**(3), 187–206 (2002)

334. Revin, D.O., Vdovin, E.P.: Hall subgroups of finite group, Ischia group theory 2004, Proceedings of a conference in honor of Marcel Herzog (Naples, Italy 2004), Contemp. Math. vol. 402, Amer. Math. Soc., Providence, RI; Bar-Ilan University, Ramat Gan, pp. 229–263 (2006)

335. Revin, D.O., Vdovin, E.P.: On the number of classes of conjugate Hall subgroups in finite simple groups. J. Algebr. **324**, 3614–3652 (2010)

336. Revin, D.O., Vdovin, E.P.: Existence criterion for Hall subgroups of finite groups. J. Gr. Theor. **14**, 93–101 (2011)
337. Robinson, D.J.S.: A Course in the Theory of Groups. Springer-Verlag, New York (1982)
338. Robinson, D.J.S.: The structure of finite groups in which permutability is a transitive relation. J. Austral Math. Soc. **70**, 143–159 (2001)
339. Rusakov, S.A.: Analogous of theorem of Sylow on existence and embedding subgroups. Siberian Math. J. **4**(5), 325–342 (1963)
340. Safonov, V. G.: ⑥-separability of the lattice of τ-closed totally saturated formations. Algebr L. **49**, 470–479 (2010)
341. Schmid, P.: Locale Formationen endlicher Gruppen. Math. Z. **137**, 31–48 (1974)
342. Schmid, P.: Subgroups permutable with all Sylow subgroups. J. Algebr. **82**, 285–293 (1998)
343. Schmidt, R.: Subgroup lattices of groups. Walter de Gruyter (1994)
344. Sel'kin, M.V.: On influence of maximal subgroups on the formation structure of finite groups. Finite groups, pp. 151–163. Nauka i Tehnika, Minsk (1975)
345. Sel'kin, M.V.: Maximal subgroups in the theory of classes of finite groups. Belaruskaya Navuka, Minsk (1997)
346. Sel'kin, V.M.: One-generated formations and their factorizations. Gomel, **24**, 1–23, preprint/GGU im. F. Skoriny (2002)
347. Sementovskii, V.G.: Injector of finite groups. In: Investigation of the normal and subgroups of structure of finite groups, pp. 166–170. Nauka i Tekhnika, Minsk (1984)
348. Shaalan, A.: The influence of π-quasinormality of some subgroups on the structure of a finite group. Acta Math. Hung. **56**(3–4), 287–293 (1990)
349. Shareshian, J., Woodroofe, R.: A new subgroup lattice characterization of finie solvable groups. J. Algebr. **351**, 448–458 (2012)
350. Shemetkov, L.A.: On Hall's theorem. Sov. Math. Dokl. **3**, 1624–1625 (1962)
351. Shemetkov, L.A.: On the existence and the embedding of subgroups of finite groups, Uspekhi mat. nauk, **17**(6), 200 (Russian) (1962)
352. Shemetkov, L.A.: A new D-theorem in the theory of finite groups. Sov. Math. Dokl. **6**, 89–93 (1965)
353. Shemetkov, L.A.: D-structure of finite groups. Mat. Sb. **67(109)**, 384–407 (1965)
354. Shemetkov, L.A.: Conjugacy and embedding of subgroups, Finite groups, Minsk, 1966, pp. 881–883.
355. Shemetkov, L.A.: Sylow properties of finite groups. Math. USSR-Sb. **5**(2), 261–274 (1968)
356. Shemetkov, L.A.: On Sylow properties of finite groups. Dokl. Akad. Nauk BSSR. **16**(10), 881–883 (1972)
357. Shemetkov, L.A.: Graduated formation of groups. Math. Sb. **94**(4), 628–648 (1974)
358. Shemetkov, L.A.: Two directions in the development of the theory of non-simple finite groups. Russian Math. Survey. **30**(2), 185–206 (1975)
359. Shemetkov, L.A.: Formations of finite groups. Nauka, Main Editorial Board for Physical and Mathematical Literature, Moscow (1978)
360. Shemetkov, L.A.: Composition formations and radicals of finite groups. Ukr. Math. Zh. **40**(3), 369–375 (1988)
361. Shemetkov, L.A.: Some ideas and results in the theory of formations of finite groups. Probl. Algebra 7, 3–38 (1992)
362. Shemetkov, L.A.: Frattini extentions of finite groups. Comm. Algebra **25**, 955–964 (1997)
363. Shemetkov, L.A.: Gaschütz products of group classes. Dokl. Akad. Nauka Belarus. **42**(3), 22–26 (1998)
364. Shemetkov, L.A.: On partially saturated formations and residuals of finite groyups. Comm. Algebra **29**(9), 4125–4137 (2001)
365. Shemetkov, L.A.: Generalizations of Sylow's theorem. Sibirsk. Mat. Zh. **44**, 1425–1431 (2003)
366. Shemetkov, L.A., Skiba, A.N.: Formations of Algebraic Systems, Nauka, Main Editorial Board for Physical and Mathematical Literature, Moscow (1989)

367. Shemetkov, L.A., Skiba, A.N.: Multiply ω-local formations and fitting classes of finite groups, Sib. Adv. Math. **10**(2), 114–147 (2000)

368. Shemetkov, L.A., Skiba, A.N.: On the $\mathfrak{X}\Phi$-hypercentre of finite groups. J. Algebr. **322**, 2106–2117 (2009)

369. Shemetkov, L.A., Vasil'ev, A.F.: Nonlocal formations of finite groups. Dokl. Akad. Nauk Belarusi **39**(4), 5–8 (1995)

370. Shen, Z., Li, S., Shi, W.: Finite groups all of whose second maximal subgroups are PSC-groups. J. Algebra Appl. doi:10.1142/S0219498809003291 (2009)

371. Shen, Z., Shi, W., Qian, G.: On the norm of the nilpotent residuals of all subgroups of a finite group. J. Algebr. **352**, 290–298 (2012)

372. Shmel'kin, A.L.: Wreath and varieties of groups. Izv. AN SSSR. Math. **29**, 149–170 (1965)

373. Shmel'kin, A.L.: Subgroups of varieties of groups. Dokl. Akad. Nauk. SSSR. **149**, 543–545 (1963)

374. Shmighirev, E.: On some questions of the theory of formations, finite groups. Nauka i tekhnika Minsk 213–225 (1975)

375. Sidorov, A.V.: On properties of the $\mathfrak{F}$-hypercentre of a finite group. Problems in Algebra **10**, 141–143 (1996)

376. Skiba, A.N.: A criterion for membership of a finite group in a local formation, Finite groups (Proc. Gomel Sem., Gomel University), pp. 166–169, Nauka i Tekhnika, Minsk, Belarus, 1978.

377. Skiba, A.N.: On products of formations. Algebra Logic **22**(5), 574–583 (1983)

378. Skiba, A.N.: Characteristics of soluble groups with given nilpotent length. Probl. Algebra (University Press) Minsk **3**, 21–31 (1987)

379. Skiba, A.N.: On non-trivial factorizations of one-generated local factorizations of finite groups. Proc. Int. Conf. Algebra Dedicat. Mem. A.I. Mal'cev, Novosibirsk, Aug. 21–26, (1989), PT. Providence (R.I.), pp. 363–374 (1992)

380. Skiba, A.N.: On nontrivial factorizations of a one-generated local formation of finite groups. Contemp. Math. 3, 363–374 (1992)

381. Skiba, A.N.: Algebra of Formations. Minsk, Belaruskaja Navuka (1997)

382. Skiba, A.N.: On factorizations of composition Formations, Math. Notes **65**(3), 326–330 (1999)

383. Skiba, A.N.: On weakly s-permutable subgroups of finite groups. J. Alegebra **315**, 192–209 (2007)

384. Skiba, A.N.: On two questions of L.A. Shemetkov concerning hypercyclically embedded subgroups of finite groups. J. Group Theory **13**, 841–850 (2010)

385. Skiba, A.N.: A characterrization of hypercyclically embedded subgroups of finite groups. J. Pure Appl. Algebra **215**, 257–261 (2011)

386. Skiba, A.N.: On the intersection of all maximal $\mathfrak{F}$-subgroups of a finite group. J. Algebra **343**, 173–182 (2011)

387. Skiba, A.N.: On the $\mathfrak{F}$-hypercentre and the intersection of all $\mathfrak{F}$-maximal subgroups of a finite group. J. Pure Appl. Algebra **216**(4), 789–799 (2012)

388. Skiba, A.N., Shemetkov, L.A.: On hereditarily indecomposable formations of groups. Dokl. Akad. Nauk BSSR, **33**(7), 581–582 (1989)

389. Skiba, A.N., Shemetkov, L.A.: Multiply $\mathfrak{L}$-composition formations of finite groups. Ukrainsk. Math. Zh. **52**(6), 783–797 (2000)

390. Skiba, A.N., Rizhik, V.N.: On factorizations of one-generated p-local formations. Probl. Algebra **11**, 104–106 (1997)

391. Skiba, A.N., Titov, O.V.: Finite groups with C-quasinormal subgroups. Sib. J. Math. **48**(3), 544–554 (2007)

392. Slepova, L.M.: On formations of $E^{\mathfrak{F}}$-groups. Dokl. Akad. Nauk BSSR. **21**, 285–289 (1977)

393. Spencer, A.E.: Maximal nonnormal chains in finite groups. Pacific J Math. **27**, 167–173 (1968)

394. Spitznagel, E.L., Jr.: Hall subgroups of centain families of finite groups. Math. Z. **97**, 259–290 (1967)
395. Srinivasan, S.: Two sufficient conditions for supersolvability of finite groups. Isreal J. Math. **35**(3), 210–214 (1980)
396. Stonehewer, S.E.: Permutable subgroups in infinite groups. Math. Z. **125**, 1–16 (1972)
397. Su, X.: Seminormal subgroups of finite groups. J. Math. (Wuhan) **8**(1), 7–9 (1988)
398. Su, N., Wang, Y., Li, Y.: A characterization of hypercyclically embedded subgroups using cover-avoidance property. J. Group Theory **16**, 263–274 (2013)
399. Su, N., Wang, Y.: On the intersection of the normalizers of the $\mathcal{F}$-residuals of subgroups of a finite group. Algebra Represent. Theory. doi:10.1007/s10468-013-9407-1
400. Suzuki, M.: The nonexistence of a centain type of simple groups of odd order. Proc. Am. Math. Soc. **8**(4), 686–695 (1957)
401. Suzuki, M.: Group theory, II. Springer-Verlag, New York (1986)
402. Thompson, J.G.: Nonsolvable finite groups all of whose local subgroups are solvable. Bull. Am. Math. Soc. **74**(3), 383–437 (1968)
403. Thompson, J.G.: An example of core-free quasinormal subgroup of p-group. Z. Math. **96**, 226–227 (1973)
404. Tyutyanov, V.N.: D_π-theorem for finite groups with composition factors. Vestsi Nats. Akad. Navuk Belarusi Ser. Fiz.-Mat. Navuk (1), 12–14 (2000) (in Russia)
405. Tyutyanov, V.N.: On theorems of the Sylow type for finite groups. Ukrainian Math. J. **52**(10), 1628–1633 (2000)
406. Vasil'ev, A.F.: Formations and their recognazing. Proc. Gomel State University **41**(2), 23–29 (2007)
407. Vasil'ev, V.A.: Finite groups with m-supplemented maximal subgroups of Sylow subgroups. Proc. F. Scorina Gomel State University **4**(67), 29–37 (2011)
408. Vasil'ev, V.A.: On p-nilpotency of one class of finite groups. Probl. Phys. Math. Tech. 3(16), 61–65 (2013)
409. Vasil'ev, A.F., Khalimonchik, I.N.: On subgroups lattices of subnormal type in finite groups. Tr. Inst. Mat. **21**(1), 25–34 (2013)
410. Vasilev, A.F., Kamornikov, S.F.: On the functor method for studying the lattices of subgroups of finite groups. Sib. Math. J. **42**, 30–40 (2001)
411. Vasiliev, A.F., Kamornikov, S.F.: The Kegel-Shemetkov problem on lattices of generalized subnormal subgrops of finite groups. Algebra Log. **41**, 228–236 (2002)
412. Vasil'ev, V.A., Skiba, A.N.: On one generalization of modular subgroups. Ukrainian Math. J. **63**(10), 1494–1505 (2012)
413. Vasil'ev, A.F., Vasil'eva, T.I.: Normalizer subgroup functors of finite groups. Vestnik BSU (1), 92–96 (2006)
414. Vasil'ev, A.F., Vasil'ev, T.I., Tyutyanov, V.N.: On the finite groups of supersoluble type. Sib. Math. J. **51**, 1004–1012 (2010)
415. Vasil'yev, A.F., Vasil'yev, T.I.: On finite groups with generally subnormal Sylow subgroups. Probl. Phys. Math. Tech. **9**(4), 86–91 (2011) (in Russian)
416. Vasil'ev, A.F., Vasil'ev, T.I., Tyutyanov, V.N.: On the products of P-subnormal subgroups of finite groups. Sib. Math. J. **53**, 47–54 (2012)
417. Vasil'ev, A.F., Vasil'ev, T.I., Tyutyanov, V.N.: On K-P-subnormal subgroups of finite groups. Math. Notes **95**, 31–40 (2014)
418. Vasil'ev, A.F., Vasil'ev, V.A., Vasil'ev, T.I.: On permuteral subgroups in finite groups. Sib. Math. J. **55**, 230–238 (2014)
419. Vdovin, E.P.: Carter subgroups of finite almost simple groups. Algebra Log. **46**, 90–119 (2007)
420. Vdovin, E.P., Revin, D.O.: Hall subgroups of odd order in finite groups. Algebra Log. **41**, 8–29 (2002)
421. Vdovin, E.P., Revin, D.O.: A conjugacy criterion for Hall subgroups of finite groups. Sib. Math. J. **51**, 402–409 (2010)

422. Vdovin, E.P., Revin, D.O., Theorems of Sylow type, Russian Math. Surveys **66**, 829–870 (2011)
423. Vdovin, E.P., Revin, D.O.: Pronormality of Hall subgroups in finite simple groups. Sib. Math. J. **53**, 419–430 (2012)
424. Vdovin, E.P., Revin, D.O.: On the pronormality of Hall subgroups. Sib. Math. J. **54**, 22–28 (2013)
425. Vdovin, E.P., Revin, D.O., Semetkov, L.A.: Formations of finite C_π-groups. St. Petersburg Math. J. **24**, 29–37 (2013)
426. Vishnevskaya, T.R.: On factorizations of one-generated p-local formations. Probl. Algebra **16**, 201–202 (2000)
427. Vorob'ev, N.N.: On factorizations of subformations of one-generated saturated finite varieties. Comm. Algebra **41**(3), 1087–1093 (2013)
428. Vorob'ev, N.T.: On radical classes of finite groups with the Lockett condition. Math. Zametki **43**, 161–168 (1988) (Translated in Math. Notes **43**, 91–94 (1988))
429. Vorob'ev, N.T.: On factorizations of non-local formations of finite groups. Probl. Algebra 6, 21–24 (1992)
430. Vorob'ev, N.T.: On the Hawkes conjecture for radical classes. Sib. Math. J. **37**(6), 1137–1142 (1996)
431. Vorob'ev, N.T.: On largest integated of Hartley's function, Proc. Gomel University. Probl. Algebra **1**, 8–13 (2000)
432. Vorob'ev, N.T.: Local products of fitting classes, Vesti AN BSSR. Ser. fiz-math. navuk **6**, 28–32 (1991)
433. Wang, L., Chen, G.: Some properties of finite groups with some (semi-p-)cover-avoiding subgroups. J. Pure Appl. Algebra **213**, 686–689 (2009)
434. Wang, P.: Some fufficient conditions of supersoluble groups. Acta Mathematica Sinica **33**, 480–485 (1990)
435. Wang, Y.: c-normality of groups and its properties. J. Algebra **180**, 954–965 (1996)
436. Wang, Y.: Finite groups with some subgroups of Sylow subgroups c-supplemented. J. Algebra **224**, 467–478 (2000)
437. Wang, Y., Guo, W.: Nearly s-normality of groups and its properties. Comm. Algebra **38**, 3821–3836 (2010)
438. Wang, Y., Wei, H.: $c^{\#}$-normality of groups and its properties. Algebr. Represent. Theory **16**, 193–204 (2013)
439. Wei, H., Wang, Y.: On c^*-normality and its properties. J. Group Theory **10**, 211–223 (2007)
440. Wei, H., Wang, Y.: On CAS-subgroups of finite groups. Israel J. Math. **159**, 175–188 (2007)
441. Wei, H., Wang, Y., Li, Y.: On c-supplemented maximal and minimal subgroups of Sylow subgroups of finite groups. Proc. Am. Math. Soc. **32**(8), 2197–2204 (2004)
442. Wei, X., Guo, X.: On SS-quasinormal subgroups and the structure of finite groups. Sci. China Math. **54**, 449–456 (2011)
443. Wei, X., Guo, X.: On finite groups with prime-power order S-quasinormally embedded subgroups. Monatsh. Math. **162**(3), 329–339 (2011)
444. Wielandt, H.: Zum Satz von Sylow. Math. Z. **60**, 407–409 (1954)
445. Wielandt, H.: Zum Satz von Sylow II. Math. Z. **71**, 461–462 (1959)
446. Wielandt, H.: Entwicklungslinien in der Strukturtheoric der endlichen Gruppen, pp. 268–278. Proc. Intern. Congress Math. (Edinburgh 1958). Cambridge University Press, New York (1960)
447. Wielandt, H.: Uber die Normalstruktur von mehrfach faktorisierbaren Gruppen. B. Austral Math. Soc. **1**, 143–146 (1960)
448. Wielandt, H.: Arithmetische Struktur und Normalstruktur endlicher Gruppen, Conv. Internaz. di Teoria dei Gruppi Finiti e Applicazioni (Firenze 1960), Edizioni Cremonese, Roma, pp. 56–65 (1960)
449. Wielandt, H.: Subnormal subgroups and permutation groups, Lectures given at the Ohio State University, Columbus, Ohio (1971)

450. Weinstein, M. (ed.): Between Nilpotent and Solvable. Polygonal Publishing House (1982)
451. Willems, W.: On the projectives of a group algebra Math. Z. **171**, 163–174 (1980)
452. Xu, M.: An Introduction to Finite Groups. Science Press, Beijing (1999)
453. Xu, M., Guo, W., Huang, J.: Weakly s-quasinormally embedded subgroups of finite groups. Chin. Ann. Math. Ser. **32A** , 299–306 (2011)
454. Xu, M., Zhang, Q.: On conjugate-permutable subgroups of a finite group. Algebra Colloq. **12**, 669–676 (2005)
455. Xu, Y., Li, X.: Weakly s-semipermutable subgroups of finite groups. Front. Math. China **6**(1), 161–175 (2011)
456. Yang, N., Guo, W.: On $\mathfrak{F}_n$-supplemented subgroups of finite groups. Asian-European J. Math. **1**, 627–629 (2008)
457. Yang, N., Guo, W.: On τ-primitive subgroups of finite groups. Siberian Math. J. **50**, 560–566 (2009)
458. Yang, N., Guo, W., Huang, J., Xu, M.: Finite groups with weakly S-Quasinormally embedded subgroups. J. Algebra Appl. **11**, (2012) doi:10.1142/S0219498811005786
459. Yang, N., Guo, W., Shemetkova, O.L.: Finite groups with S-supplemented p-subgroups. Siberian Math. J. **53**, 371–376 (2012)
460. Yang, N., Guo, W., Vorobev, N.T.: Factorizations of fitting classes. Front. Math. China **7**, 943–954 (2012)
461. Yi, X., Miao, L., Zhang, H., Guo, W.: Finite groups with some $\mathfrak{F}$-supplemented subgroups. J. Algebra Appl. **9**, 669–685 (2010)
462. Yi, X., Skiba, A.N.: On S-propermutable subgroups of finite groups. Bull. Malays. Math. Sci. Soc. to appear
463. Zagurskij, V.N., Vorob'ev, N.T.: Fitting classes with the given properties of Hall subgroups. Math. Zametki **78**(2), 234–240 (2005)
464. Zhang J.: Sylow numbers of finite groups. J. Algebra **176**, 111–123 (1995)
465. Zhang, Q.: s-semipermutability and abnormality in finite groups. Comm. Algebra **27**, 4514–4524 (1999)
466. Zhang, Q., Wang, L.: Finite nonabelian simple groups which contain a nontrivial s-semipermutable subgroups. Algebra Colloq. **12**, 301–307 (2005)
467. Zhang, Q., Wang, L.: The influence of s-semipermutable subgroups on the structure of a finite group (in Chinese). Acta Math. Sinica **48**, 81–88 (2005)
468. Zhang, Y. et al.: Translation and revision of *Between and nilpotent and soluble* (in Chinese). Publishing House of Wuhan University (1998)
469. Zhang, Y., Ballester-Bolinches, A., Guo, X.: On the Deskins index complex of a maximal subgroup of a finite group. J. Pure Appl. Algebra **136**, 211–216 (1999)
470. Zhao, T., Li, X., Xu, Y.: Weakly s-supplementaly embedded minimal subgroups of finite groups. Proc. Edinburg Math. Soc. **54**, 799–807 (2011)
471. Zhao, Y.: On the index complex of a maximal subgroup and the supersolvability of a finite group. Comm. Algebra **24**, 1785–1791 (1996)
472. Zhao, Y.: On Deskins's conjecture concerning the supersolvability of a finite group. J. Pure Appl. Algebra **124**, 325–328 (1998)
473. Zhao, Y.: On the Deskins completions, theta completions and theta pairs for maximal subgroups. Comm. Algebra **26**, 3141–3153 (1998)
474. Zhao, Y.: On the Deskins completions, theta completions and theta pairs for maximal subgroups II. Comm. Algebra **26**, 3155–3164 (1998)
475. Zhao, Y.: On the Deskins completions and theta completions for maximal subgroups. Comm. Algebra **28**, 375–385 (2000)
476. Zhu, L., Guo, W., Shum, K.P.: Weakly c-normal subgroups of finite groups and their properties. Comm. Algebra **30**, 5505–5512 (2002)

List of Symbols

$Aut_G(H/K)$ 6
$c(G)$ 39
$Soc(G)$ 17
$Com(\mathfrak{X})$ 4, 249
$Com(G)$ 4
$C(\mathfrak{X})$ 249
$O_\omega(G)$ 249
M_G 3
$F^*(G)$ 12
$\mathfrak{A}, \mathfrak{A}(n), \mathfrak{G}, \mathfrak{S}, \mathfrak{N}, \mathfrak{U}, \mathfrak{G}_\pi, \mathfrak{S}_\pi, \mathfrak{N}_\pi, \mathfrak{G}_p, \mathfrak{G}_{cp},$
$\quad \mathfrak{N}^r$ 1
$G^{\mathfrak{X}}$ 1
$G_{\mathfrak{X}}$ 1
$R(G)$ 4
$C^p(G)$ 4
$\mathfrak{X}(p)$ 2
$\Phi^2(G)$ 217
$\Phi^*(G)$ 7
$Z_\theta(G), Z_{\mathfrak{X}}(G), Z_{\mathfrak{X}\Phi}(G), Z_{\mathfrak{X}\Phi^*}(G), Z_{\pi\mathfrak{X}}(G)$
$\quad$ 8–9
$Z_n(G)$ 10
$Z^{\mathfrak{F}}(G)$ 320
$Z^*(G)$ 23
$\Delta^{\mathfrak{F}}(G)$ 19
$\mathfrak{N}^*$ 19
$\mathfrak{F}^*$ 20
$\mathfrak{F}_\pi^*$ 20
$Z_\infty(G), Z_{\mathfrak{N}}(G)$ 42
$Int_{\mathfrak{X}}(G)$ 42
$Int_{\mathfrak{F}}^*(G)$ 50
$\psi_0(N)$ 43
$l(G)$ 54
$l_p(G), l_\pi(G)$ 119
$\mathfrak{M}\mathfrak{H}$ 2, 296
$\mathfrak{M} \diamond \mathfrak{H}$ 2, 296
$genz^*(G)$ 57
$X(A), X_c(A)$ 68

$X_m(A)$ 69
A^B 119
H τ-p-emb G 139
H τ-emb G 139
$\tau(G), \tau_{pe}(G)$ 131, 140
τ_{pe}, τ_e 140
$H_{\tau G}, H_{peG}, H_{eG}, H_{psubeG}, H_{subeG}, H_{pseG},$
$\quad H_{seG}$ 140
H_{sG}, H^{sG} 170
$S_\tau \mathfrak{X}$ 285
$\pi(\mathfrak{F})$ 1
Σ_H 182
Σ_i 182
$\text{form}\mathfrak{X}$ 183
$h(G)$ 226
$h_{\mathfrak{U}}(G)$ 246
$LF(f)$ 2
$CF_\omega(f)$ 249
$LF_\omega(f)$ 249
$c_\omega\text{form}\mathfrak{X}$ 250
$c_\omega\text{form}G$ 250
$l_\omega\text{form}\mathfrak{X}$ 250
$l_\omega\text{form}G$ 250
$\mathfrak{H}(\omega'), \mathfrak{H}(p)$ 261
$\theta\text{form}\mathfrak{X}$ 273
$\theta\text{form}G$ 273
$\theta^l\text{form}\mathfrak{X}$ 274
$\tau\text{form}\mathfrak{X}$ 285
$\vee_\theta(\mathfrak{F}_i | i \in I)$ 273
$\cap(f_i | \in I)$ 274
θ^ℓ 274
l_n, l_n^τ, l_∞ 275
$\mathfrak{F}/_\theta(\mathfrak{H} \cap \mathfrak{F})$ 275
$\mathfrak{F}/_n^\tau(\mathfrak{F} \cap \mathfrak{N})$ 275
$L_\infty(\mathfrak{F})$ 277
$\mathfrak{F}_0$ 279
$\mathfrak{F}/_\infty\mathfrak{F}_0$ 275, 279

© Springer-Verlag Berlin Heidelberg 2015
W. Guo, *Structure Theory for Canonical Classes of Finite Groups*,
DOI 10.1007/978-3-662-45747-4

Index of Subjects

© Springer-Verlag Berlin Heidelberg 2015
W. Guo, *Structure Theory for Canonical Classes of Finite Groups*,
DOI 10.1007/978-3-662-45747-4

Printed in the United States
By Bookmasters